CAMBRIDGE LIBRARY COLLECTION

Books of enduring scholarly value

Travel and Exploration

The history of travel writing dates back to the Bible, Caesar, the Vikings and the Crusaders, and its many themes include war, trade, science and recreation. Explorers from Columbus to Cook charted lands not previously visited by Western travellers, and were followed by merchants, missionaries, and colonists, who wrote accounts of their experiences. The development of steam power in the nineteenth century provided opportunities for increasing numbers of 'ordinary' people to travel further, more economically, and more safely, and resulted in great enthusiasm for travel writing among the reading public. Works included in this series range from first-hand descriptions of previously unrecorded places, to literary accounts of the strange habits of foreigners, to examples of the burgeoning numbers of guidebooks produced to satisfy the needs of a new kind of traveller - the tourist.

Catalogue of the Special Loan Collection of Scientific Apparatus at the South Kensington Museum 1876

In 1876 the South Kensington Museum held a major international exhibition of scientific instruments and equipment, both historical and contemporary. Many of the items were retained and eventually formed the basis of important collections now held at the Science Museum, London. This is the 1877 third edition of the exhibition catalogue, which was expanded to include a 'large number of objects' received since the publication of the second edition, and which also included corrections in order to 'afford a complete record of the collection for future reference'. In two volumes and twenty sections comprising over 4,500 entries, the catalogue lists a huge variety of items, ranging from slide rules and telescopes to lighthouse parts and medical equipment. It gives detailed explanations of how they were used, and notes of their ownership and provenance, while the opening pages comprehensively record the contributing individuals and institutions in Britain, Europe and America.

Cambridge University Press has long been a pioneer in the reissuing of out-of-print titles from its own backlist, producing digital reprints of books that are still sought after by scholars and students but could not be reprinted economically using traditional technology. The Cambridge Library Collection extends this activity to a wider range of books which are still of importance to researchers and professionals, either for the source material they contain, or as landmarks in the history of their academic discipline.

Drawing from the world-renowned collections in the Cambridge University Library, and guided by the advice of experts in each subject area, Cambridge University Press is using state-of-the-art scanning machines in its own Printing House to capture the content of each book selected for inclusion. The files are processed to give a consistently clear, crisp image, and the books finished to the high quality standard for which the Press is recognised around the world. The latest print-on-demand technology ensures that the books will remain available indefinitely, and that orders for single or multiple copies can quickly be supplied.

The Cambridge Library Collection brings back to life books of enduring scholarly value (including out-of-copyright works originally issued by other publishers) across a wide range of disciplines in the humanities and social sciences and in science and technology.

Catalogue of the Special Loan Collection of Scientific Apparatus at the South Kensington Museum 1876

VOLUME 1

ANON

CAMBRIDGE
UNIVERSITY PRESS

CAMBRIDGE UNIVERSITY PRESS

Cambridge, New York, Melbourne, Madrid, Cape Town,
Singapore, São Paolo, Delhi, Tokyo, Mexico City

Published in the United States of America by Cambridge University Press, New York

www.cambridge.org
Information on this title: www.cambridge.org/9781108042413

This edition first published 1877
This digitally printed version 2012

ISBN 978-1-108-04241-3 Paperback

Science and Art Department
of the Committee of Council on Education.

CATALOGUE

OF THE

SPECIAL LOAN COLLECTION OF SCIENTIFIC APPARATUS

AT THE

SOUTH KENSINGTON MUSEUM.

MDCCCLXXVI.

THIRD EDITION.

LONDON:
PRINTED BY GEORGE E. EYRE AND WILLIAM SPOTTISWOODE,
PRINTERS TO THE QUEEN'S MOST EXCELLENT MAJESTY.
FOR HER MAJESTY'S STATIONERY OFFICE.

1877.

SCIENCE AND ART DEPARTMENT
OF THE COMMITTEE OF COUNCIL ON EDUCATION.
SOUTH KENSINGTON.

ESTABLISHED in connexion with the Board of Trade in March 1853 as a development of the Department of Practical Art, which in 1852 had been created for the re-organisation of Schools of Design. Placed under the direction of the Committee of Council on Education in 1856.

Lists of Presidents and Vice-Presidents.

Board of Trade.

1852. Rt. Hon. H. Labouchere, M.P., President.
„ Rt. Hon. J. W. Henley, M.P., President.
1853. Rt.Hon.Edward Cardwell, M.P., President.
1855. Rt. Hon. Lord Stanley of Alderley, Pres.

Committee of Council on Education.

1856. Rt. Hon. Earl Granville, K.G., Lord President.
„ Rt.Hon.W. E.Cowper,M.P.,Vice-President.
1858. Most Hon. Marquess of Salisbury, K.G.
„ Rt. Hon. Sir C. B. Adderley, K.C.M.G., M.P., Vice-President.
1859. Rt. Hon. Earl Granville, K.G.
„ Rt. Hon. Robert Lowe, M.P., Vice-Pres.

1864. Rt. Hon. H. A. Bruce, M.P., Vice-Pres.
1866. His Grace the Duke of Buckingham and Chandos.
„ Rt. Hon. H. T. Lowry Corry, M.P., Vice-President.
1867. His Grace the Duke of Marlborough, K.G.
„ Rt. Hon. Lord Robert Montagu, M.P., V.-P.
1868. Most Hon. the Marquess of Ripon, K.G.
„ Rt. Hon. W. E. Forster, M.P., Vice-Pres.
1873. Rt. Hon. Lord Aberdare.
„ Rt. Hon. W. E. Forster, M.P., Vice-Pres.
1874. His Grace the Duke of Richmond and Gordon, K.G., Lord President.
„ The Right Hon. the Viscount Sandon, M.P., Vice President.

OFFICE HOURS, TEN TO FOUR.

GENERAL ADMINISTRATION.

Secretary —Sir Francis R. Sandford, C.B.
Assistant Secretary.—Norman MacLeod.
Chief Clerk.—G. Francis Duncombe.
First-class Clerks.—A. J. R. Trendell; Alan S. Cole; F. R. Fowke; A. S. Bury.
Second-class Clerks.—J. B. Rundell; H. W. Williams; E. Belshaw; G. G. Millard; A. F. E. Torrens; O. J. Dullea.
Postal Clerk.—W. Burtt.
Clerk of Accounts.—Vacant.
Book-keeper.—T. A. Bowler.
Assistant Book-keeper.—E. Harris.

GENERAL STORES.

Storekeeper.—W. G. Groser. *Deputy.*—H. Lloyd.
Clerks.—J. Smith; F. Walters.

SCIENCE DIVISION.

Director.—Major Donnelly, R.E.
Occasional Inspectors.—F. J. Sidney, LL.D.; Capt. Harris, E.I.C. (*Navigation*).
Official Examiner.—G. C. T. Bartley.
Assistant Professional Examiner.—T. Healey.

Professional Examiners for Science.

Subjects.
I.—Practical, plane, and solid Geometry.—
II.—Machine Construction and Drawing.—W. C. Unwin, B.Sc.
III.—Building Construction.—Major Seddon, R.E.
IV.—Naval Architecture.—W. B. Baskcomb.
V.—Pure Mathematics.—C. W. Merrifield, F.R.S.; T. Savage, M.A.
VI.—Theoretical Mechanics.—Rev. John F. Twisden, M.A.
VII.—Applied Mechanics.—T.M.Goodeve,M.A.
VIII.—Acoustics, Light, and Heat.—J. Tyndall, LL.D., F.R.S.; F. Guthrie, F.R.S.
IX.—Magnetism and Electricity.—J. Tyndall, LL.D., F.R.S.; H. Debus, F.R.S.
X.—Inorganic Chemistry.—E. Frankland, D.C.L., Ph.D., F.R.S.

40075.

Subjects.
XI.—Organic Chemistry.—E. Frankland, D.C.L., Ph.D., F.R.S.
XII.—Geology.—A. C. Ramsay, LL.D., F.R.S.
XIII.—Mineralogy.—W. W. Smyth, M.A., F.R.S.
XIV.—Animal Physiology.—T. H. Huxley, LL.D., F.R.S.
XV.—Elementary Botany.—W. T. T. Dyer, M.A., B.Sc., F.L.S.
XVI. } —General Biology.—T. H. Huxley,
XVII. } LL.D., F.R.S.; W. T. T. Dyer, M.A., B.Sc., F.L.S.
XVIII.—Mining.—W. W. Smyth, M.A., F.R.S.
XIX.—Metallurgy.—J. Percy, M.D., F.R.S.
XX.—Navigation.—J. Woolley, LL.D.
XXI.—Nautical Astronomy.—J. Woolley, LL.D.
XXII.—Steam.—T. M. Goodeve, M.A.
XXIII. { Physical Geography.—D. T. Ansted, M.A., F.R.S.
{ Physiography.—
XXIV.—Principles of Agriculture.—H. Tanner, F.C.S.

ART DIVISION.

Director.—E. J. Poynter, R.A.
Assistant Director.—H. A. Bowler.
Occasional Inspectors.—S. A. Hart, R.A.; F. B. Barwell; W. B. Scott.
Official Examiner.—T. Chesman. B.A., LL.B.
Professional Examiners.—F. R. Pickersgill, R.A.; W. F. Yeames, A.R.A.; J. E. Boehm; Val. Prinsep; M. Morris; G. Atchison: J. Marshall,F.R.S.,F.R.C.S.; E. J. Poynter, R.A., and H. A. Bowler.
Assistant Professional Examiner.—J. A. D. Campbell.
Occasional Examiners.—G. M. Atkinson; G. R. Redgrave.
Inspectors of Local Schools of Science and Art.—R. G. Wylde; J. F. Iselin, M.A.; E. P. Bartlett; Captain W. de W. Abney, R.E., F.R.S.

Organising Master of Science and Art Classes.—J. C. Buckmaster. F.C.S.

a 2

NATIONAL ART TRAINING SCHOOL.

Principal.—E. J. Poynter, R.A.
Head Master.—J. Sparkes.
Registrar.—R. W. Herman.
Mechanical and Architectural Drawing.—H. B. Hagreen.
Geometry and Perspective.—E. S. Burchett.
Painting, Freehand Drawing of Ornament, &c., the Figure and Anatomy, and Ornamental Design.—J. Sparkes; C. P. Slocombe; T. Clack, and F. M. Miller.
Modelling.—F. M. Miller.
Lady Superintendent of Female Classes.—Miss Trulock.
Instructors.—Mrs. S. E. Casabianca: Miss Channon.
Occasional Lecturers.—E. Bellamy, F.R.C.S., Eng., *Anatomy*; Dr. Zerffi, *Historic Ornament*; R. W. Herman, *Principles of Ornamental Construction*; F. W. Moody, *The Figure.*
Teacher of Etching Class.—A. Legros.
Teacher of Wood Engraving Class.—C. Roberts.

SOUTH KENSINGTON MUSEUM.

Director.—P. Cunliffe Owen, C.B. (temp. absent).
Acting Director.—R. A. Thompson.
Assistant Directors.—Major E. R. Festing, R.E.; Col. Sir H. B. Sandford.
Director of New Buildings.—Major-Gen. Scott, C.B., F.R.S.
Decorative Artist.—R. Townroe.
Instructor in Decorative Art and Decorative Artist.—F. W. Moody.
Museum Keeper (Art Collections).—G. Wallis.
Museum Keeper (National Art Library).—R. H. Soden Smith, M.A., Trinity College, Dublin, F.S.A.
Museum Keeper (Educational Library and Collections).—A. C. King, F.S.A.
Assistant Museum Keepers.—W. Matchwick, F.L.S.; H. Sandham; R. Laskey; C. B. Worsnop; R. F. Sketchley, B.A., Exeter College, Oxford; H. E. Acton; J. W. Appell, Ph.D.; J. Barrett, B.A.; C. H. Derby, B.A.
Museum Clerks.—M. Webb; H. M. Cundall; L. Finding.
Technical and Special Assistants.—H. Vernon; A. Masson; W. E. Streatfeild; A. Reid; F. Coles; W. G. Johnson; G. H. Wallis; S. Cowper; O. Scott.
Superintendent for Examples and Publications.—J. Cundall.

BETHNAL GREEN BRANCH OF THE SOUTH KENSINGTON MUSEUM.

(Opened on June 24, 1872.)

GEOLOGICAL SURVEY.

Director-General.—A. C. Ramsay, LL.D., F.R.S.
Director for England and Wales.—H. W. Bristow, F.R.S.
Director for Ireland.—E. Hull, M.A., F.R.S.
Director for Scotland.—A. Geikie, F.R.S.
Naturalist.—T. H. Huxley, LL.D., F.R.S.
Palæontologist.—R. Etheridge, F.R.S.

ROYAL SCHOOL OF MINES AND MUSEUM OF PRACTICAL GEOLOGY.

Director of Museum of Practical Geology.—A. C. Ramsay, LL.D., F.R.S.
Keeper of Mining Records.—Robert Hunt, F.R.S.

Assistants.—Richard Meade; James B. Jordan.
Registrar, Curator, and Librarian.—T. Reeks.
Assistant Librarian.—T. Newton.
Assistant Curator.—A Pringle.

PROFESSORS.

Chemistry.—Edward Frankland, D.C.L., Ph.D., F.R.S.
Natural History.—T. H. Huxley, LL.D., F.R.S.
Physics.—F. Guthrie, B.A., Ph.D., F.R.S.
Applied Mechanics.—T. M. Goodeve, M.A.
Metallurgy.—J. Percy, M.D., F.R.S.
Geology.—J. W. Judd.
Mining and Mineralogy.—W. W. Smyth, M.A., F.R.S.
Mechanical Drawing.—Rev. J. H. Edgar, M.A.
Museum open every week-day but Friday, and on Saturdays and Mondays till 10 p.m., except from the 10th of August to the 10th of September.

EDINBURGH MUSEUM OF SCIENCE AND ART.

Director.—Prof. T. C. Archer, F.R.S.E.
Keeper of Natural History Collections.—Prof. R. H. Traquair, M.D.
Curator.—Alexander Galletly.
Assistant in Natural History Museum.—J. Gibson.
Assistant in Industrial Museum.—W. Clark.
Clerks.—C. N. B. Muston; T. Stock.

ROYAL COLLEGE OF SCIENCE, DUBLIN.

Dean of Faculty.—J. P. O'Reilly, C.E., M.R.I.A.
Secretary.—F. J. Sidney, LL.D.
Curator of Museum.—A. Gages.
Clerk.—G. C. Penny.

PROFESSORS.

Physics.—W. F. Barrett, F.C.S.
Chemistry.—R. Galloway, F.C.S.
Geology.—E. Hull, M.A., F.R.S.
Applied Mathematics.—H. Hennessy, F.R.S.
Botany.—W. R. McNab, M.D.
Zoology.—A. Leith Adams, M.B., F.R.S.
Descriptive Geometry and Drawing.—Thomas F. Pigot, C.E.
Mining and Mineralogy.—J. P. O'Reilly, C.E., M.R.I.A.
Demonstrator in Palæontology.—W. H. Baily, F.L.S.
Assistant Chemist.—W. Plunkett.
Assistant Physicist.—A. E. Porte.

ROYAL DUBLIN SOCIETY.

President.—His Grace the Duke of Leinster.
Secretaries.—G. J. Stone, A.M., F.R.S.; C. Kelly, J.P., Q.C.
Registrar and Assistant Secretary.—W. E. Steele, M.D.
Treasurer, &c.—H. C. White.
Director of Natural History Museum.—A. Carte, M.A., M.D.
Keeper of Minerals.—R. J. Moss, F.C.S.
Librarian.—W. Archer, F.R.S.
Temporary Assistant.—H. W. D. Dunlop, A.B., C.E.
Director of Botanic Gardens, Glasnevin.—D. Moore, Ph.D.

ZOOLOGICAL GARDENS, DUBLIN.

Secretary.—Rev. S. Haughton, M.D., D.C.L., F.T.C.D., F.R.S.

PHYSIOLOGICAL
LABORATORY
CAMBRIDGE

TO KENSINGTON GARDENS

NORTH WEST ENTRANCE

REFRESHMENT ROOM

LETTERS WITHIN CIRCLES REFER TO UPPER FLOOR

M
L
N
K
O
H
P
G
Q
F

Royal H
Ga

REFERENCE T

GROUND

A EDUCATIONAL CO

B.C. APPLIED MECHAN

D NAVAL ARCHITECTU

E LIGHT-HOUSE APP

F MAGNETISM AND EL

C ARITHMETIC AND GE

H.K. MEASUREMENT.

L. ASTRONOMY AND M

UPPER

M GEOGRAPHY, GEOLO

N BIOLOGY.

O CONFERENCE RO

P CHEMISTRY.

Q LIGHT, HEAT, SOUND, A

QUEEN'S GATE TERRACE.

PRINCE ALBERT'S ROAD.

ELVASTON PLACE

FROM THE GLOUCESTER ROAD
RAILWAY STATION

QUEEN'S GATE PLACE

E
D
C
B

100   50   10   100   200

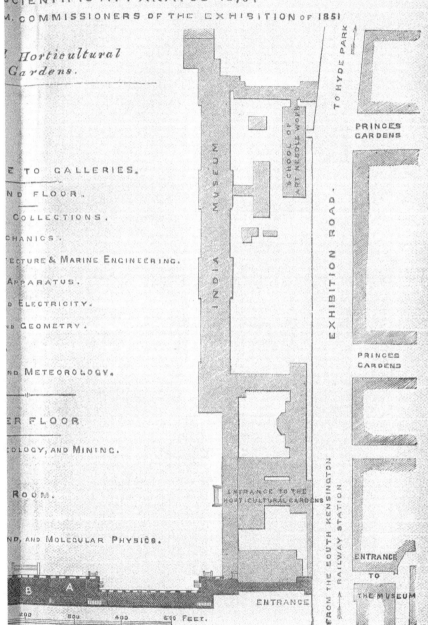

INGTON MUSEUM.

SCIENTIFIC APPARATUS 1876,

M. COMMISSIONERS OF THE EXHIBITION OF 1851

Horticultural
Gardens.

E TO GALLERIES.

ND FLOOR.

COLLECTIONS.

CHANICS.

ECTURE & MARINE ENGINEERING.

APPARATUS.

D ELECTRICITY.

ND GEOMETRY.

ND METEOROLOGY.

ER FLOOR

OLOGY, AND MINING.

ROOM.

ND, AND MOLECULAR PHYSICS.

B   A

200   300   400   500 FEET.

TO HYDE PARK

PRINCES GARDENS

EXHIBITION ROAD.

PRINCES GARDENS

INDIA MUSEUM

SCHOOL OF ART NEEDLE WORK

ENTRANCE TO THE
HORTICULTURAL GARDENS

FROM THE SOUTH KENSINGTON
RAILWAY STATION

ENTRANCE

ENTRANCE
TO
THE MUSEUM

CONTENTS.

PREFACE TO THE THIRD EDITION.

THE receipt of a large number of objects since the compilation of the Second Edition has rendered necessary the publication of a Third Edition, to afford a complete record of the collection for future reference. Every endeavour has been made to ensure its correctness. Slips from the former edition have been sent, wherever practicable, to the several contributors for their corrections, in order that all errors might, as far as possible, be eliminated.

Although a considerable amount of re-arrangement has been found necessary, it has been thought advisable to retain the numbers given to the objects in the former edition, as those numbers have already been quoted in the Handbook and other publications.

This edition has been revised and passed through the press by Mr. A. T. Atchison, to whom the thanks of my Lords are due for the great care and trouble he has bestowed upon it.

South Kensington Museum,
May 1877.

INTRODUCTION.

By Minute dated 22nd January 1875, the Lords of the Committee of Council on Education approved of a proposal to form a Loan Collection of Scientific Apparatus, which was to include not only apparatus for teaching and for investigation, but also such as possessed historic interest on account of the persons by whom, or the researches in which, it had been employed. Their Lordships then invited some of the leading men of science of the country— the Presidents of the learned Societies and others—to act on a Committee to consider the matter, and aid them with their advice. This Committee, to whose exertions the formation of the collection is so largely due, consisted of—

The Right Hon. the Lord Chancellor.

Professor F. A. Abel, F.R.S., President of the Chemical Society.

The Right Hon. Lord Aberdare, President of the Horticultural Society.

Capt. W. de W. Abney, R.E., F.R.S.

Professor H. W. Acland, M.D., F.R.S., President of the Medical Council of the United Kingdom.

Professor J. C. Adams, M.A., F.R.S.

Professor W. G. Adams, M.A., F.R.S.

Sir G. B. Airy, K.C.B., D.C.L., F.R.S., the Astronomer Royal.

Dr. G. J. Allman, F.R.S., President of the Linnæan Society.

Mr. J. Anderson, LL.D., C.E.

Mr. D. T. Ansted, M.A., F.R.S.

Professor E. Atkinson, Ph.D.

Professor R. Stawell Ball, LL.D., F.R.S.

Professor W. F. Barrett.

Rev. A. Barry, D.D.

Mr. W. B. Baskcomb.

Mr. H. Bauerman.

Mr. G. Bentham, F.R.S.

Mr. Hugh Birley, M.P.

Professor Bloxam.

Major Bolton.

Professor F. A. Bradley.

Mr. F. J. Bramwell, F.R.S.

Mr. T. Brassey, M.P.

Mr. H. W. Bristow, F.R.S.

Mr. C. Brooke, M.A., F.R.S.

Mr. G. Busk, F.R.S.

Major-General Cameron, C.B., F.R.S.

Dr. W. B. Carpenter, C.B., F.R.S.

Mr. C. O. F. Cator.

Mr. W. Chappell.

Mr. H. W. Chisholm, Warden of the Standards.

Lord Alfred Churchill, Chairman of Council of Society of Arts.

Mr. G. T. Clark.

Mr. Latimer Clark.

Professor R. Bellamy Clifton, M.A., F.R.S.

Sir Henry Cole, K.C.B.

Vice-Admiral Sir R. Collinson, K.C.B., Deputy Master of the Trinity House.

Dr. Debus, F.R.S.

Mr. Warren De La Rue, D.C.L., F.R.S.

Mr. G. Dixon, M.P.

Professor P. M. Duncan, M.B., F.R.S., President of the Geological Society.

Professor W. T. Thiselton Dyer, M.A., B.Sc.

Major-General F. Eardley-Wilmot, R.A., F.R.S.

The sections were—

1. Mechanics (including pure and applied mathematics).
2. Physics.
3. Chemistry (including metallurgy).
4. Geology, Mineralogy, and Geography.
5. Biology.

The Committees for the several sections are given at page xxv.

The question of classification, having been carefully considered at numerous meetings of these Sub-Committees, was brought before the General Committee on the 12th May, and the several schemes were referred to a special Sub-Committee, formed of three members from each sectional Sub-Committee. It was also decided to postpone the Exhibition, which it was originally intended to open in June 1875, to March 1876. The large number of objects sent from abroad, and the late period of their arrival, have necessitated a further postponement of the opening to May 1876.

The Sub-Committee appointed to revise and report on the classification of the Collection after three meetings, under the chairmanship of the President of the Royal Society, submitted a scheme of classification to the General Committee on June 22nd. After having been carefully considered, it was, with some slight alterations, approved, and is given at page xix. This programme was immediately issued, and the classification into sections is that adopted for the catalogue and exhibition, though the nature of the Galleries has necessitated some alteration in the order of the sections.

It had been the intention from the first to give the Loan Collection an international character, so as to afford men of science and those interested in education an opportunity of seeing what was being done by other countries than their own in the production of apparatus, both for research and for instruction—an opportunity which it was hoped would be of advantage also to the makers of instruments. As soon, therefore, as the programme had been definitely settled, steps were taken to interest foreign countries in the Exhibition ; and it was determined to obtain the co-operation of men of science on the Continent, who, while acting as members of the General Committee, should form special Sub-Committees charged with the due representation of the science of their respective countries.

It was necessary to take special precautions to prevent misunderstanding as to the character of the Collection. The mention of internationality at once suggested the idea of an International Exhibition similar in its character and arrangements to the numerous Industrial Exhibitions which have been held in various countries. A wrong impression of this kind would have entailed serious inconvenience.

In International Exhibitions a certain amount of space is allotted to each country. These spaces are then divided by the Commissioners of each country among its exhibitors, who display their objects—subject to certain general rules of classification—as they consider most advantageous, retaining the custody of their own property. The expenses of transport, arrangement, &c., are borne by the countries who exhibit. And the Exhibitions appeal naturally, more or less exclusively, to the industrial or trade-producing interests of those countries.

This was not the idea of the proposed Loan Collection at South Kensington. For that Collection it was desired to obtain not only apparatus and objects from manufacturers, but also objects of historic interest from museums and private cabinets, where they are treasured as sacred relics, as well as apparatus in present use in the laboratories of professors. The transport of all objects was undertaken by the English Government, and they were to be handed over absolutely to the custody of the Science and Art Department for exhibition ; the arrangement being not by countries but strictly according to the general classification.

So soon as the object and scope of the Collection were thoroughly understood, the Committee of Council on Education met with the most gratifying responses to their invitations, which were communicated officially through the Foreign Office. Her Majesty's Ministers at Paris, Berlin, St. Petersburgh, Vienna, Florence, Brussels, the Hague, Stockholm, Madrid, Berne, and Washington, have personally interested themselves in the matter. And the Foreign Governments have afforded every facility and encouragement in forwarding this strictly international undertaking.

The subjoined list of the foreign members of the General Committee speaks for itself by the eminence and European reputation of its members.

BELGIUM.

M. Stas, Membre de l'Académie Royale (President).

M. le Général Brialmont, Président de l'Académie Royale et Inspecteur Général du Génie.

M. Dewalque, Membre de l'Académie Royale, Professeur de Géologie et de Minéralogie à l'Université de Liége.

M. Maus, Membre de l'Académie, Inspecteur Général des Ponts et Chaussées.

M. Plateau, Membre de l'Académie Royale, F.R.S.

M. Schwann, Membre de l'Académie Royale, Professeur à l'Université de Liége.

M. Van Beneden, Membre de l'Académie et Professeur à l'Université de Louvain, F.R.S.

M. le Général Liagre, Secrétaire perpétuel de l'Académie Royale, et Commandant et Directeur des Études de l'École Militaire (Secretary).

FRANCE.

M. le Général Arthur Jules Morin, Membre de l'Académie des Sciences, Directeur du Conservatoire des Arts et Métiers (President).

M. Alexre. Edmond Becquerel, Membre de l'Académie des Sciences, Professeur au Conservatoire des Arts et Métiers, F.R.S.

M. Henri Marie Bouley, Membre de l'Académie des Sciences, Inspecteur Général des Écoles Vétérinaires.

M. Gabriel Auguste Daubrée, Membre de l'Académie des Sciences, Directeur de l'École des Mines.

M. Jean Louis Armand de Quatrefages de Bréau, Membre de l'Académie des Sciences, Professeur au Museum d'Histoire Naturelle.

M. Jean Baptiste Dumas, Secrétaire perpétuel de l'Académie des Sciences, F.R.S.

M. Hervé Auguste Etienne Albans Faye, Membre de l'Académie des Sciences, Président du Bureau des Longitudes.

M. Edmond Frémy, Membre de l'Académie des Sciences, Professeur au Museum d'Histoire Naturelle.

M. Jules Célestin Jamin, Membre de l'Académie des Sciences, Professeur à l'École Polytechnique.

M. Lenglet, Consul Général de France à Londres.

M. Urbain Jean Joseph Le Verrier, Membre de l'Académie des Sciences, Directeur de l'Observatoire, F.R.S.

M. Eugène Melchior Peligot, Membre de l'Académie des Sciences, Directeur des Essais à la Monnaie.

M. Henri Edouard Tresca, Membre de l'Académie des Sciences, Sous-Directeur du Conservatoire des Arts et Métiers (Secretary).

GERMANY.

I.—BERLIN COMMITTEE.

Dr. A. W. Hofmann, Professor of Chemistry, F.R.S. (President).
Dr. Beyrich, Professor of Geology.
Dr. du Bois-Reymond, Professor of Physiology.
Dr. Dove, Professor of Physics, F.R.S.

Dr. Förster, Director of the Observatory.

Dr. Hagen, President of the Board of Works.

T. G. Halske, Telegraphic Engineer.

Dr. Hauchecorne, Director of the School of Mines.
Dr. Helmholtz, Professor of Physics, F.R.S.
Dr. Kiepert, Professor of Geography.
Dr. G. Kirchhoff, Professor of Physics, F.R.S.
Dr. Kronecker, Professor of Mathematics.
Dr. C. D. Martius, Chemist.
Von Morozowicz, General.

Dr. Neumayer, Hydrographer of the Imperial Admiralty.
Dr. Reuleaux, Director of the Polytechnic Academy.
Dr. Schellbach, Professor of Mathematics.
Dr. Werner Siemens, Telegraphic Engineer.
Dr. Virchow, Professor of Pathology.
Dr. C. H. Vogel, Astronomer.
Dr. Websky, Professor of Mineralogy.

II.—COMMITTEE REPRESENTING OTHER CITIES AND TOWNS OF GERMANY.

Dr. Von Babo, Professor of Chemistry, Freiburg.
Dr. Beetz, Professor of Physics, Munich.
Dr. Buff, Professor of Physics, Giessen.
Dr. Clausius, Professor of Physics, Bonn, F.R.S.
His Excellency Dr. Von Dechen, Director of the Mining Department, Bonn.
Dr. Von Fehling, Professor of Chemistry, Stuttgart.
Dr. Von Feilitzsch, Professor of Physics, Greifswald.
Dr. Graebe, Professor of Chemistry, Königsberg.
Dr. Von Groddeck, Director of the School of Mines, Klausthal.
Dr. Heeren, Professor of Chemistry, Hanover.
Dr. Hittorf, Professor of Chemistry, Münster.
Dr. Karsten, Professor of Physics, Kiel.
Dr. Karsten, Professor of Physics, Rostock.
Dr. Knapp, Professor of Chemistry, Braunschweig.
Dr. Knoblauch, Professor of Physics, Halle.
Dr. Kölliker, Professor of Physiology, Würzburg, F.R.S.
Dr. Kundt, Professor of Physics, Strasburg.
Dr. Launhardt, Director of the Polytechnic School, Hanover.
Dr. Möhl, Cassel.

Dr. Poleck, Professor of Chemistry, Breslau.
Dr. Preyer, Professor of Physiology, Jena.
Dr. Von Quintus-Icilius, Professor of Physics, Hanover.
Dr. Reusch, Professor of Physics, Tübingen.
Dr. Romberg, Professor in the Nautical School, Bremen.
Dr. Rosenthal, Professor of Physiology, Erlangen.
Dr. Rümker, Director of the Observatory, Hamburg.
Dr. Serlo, Director of the Mining Department, Breslau.
Dr. C. Von Siemens, Professor in the Agricultural Academy, Hohenheim.
His Excellency Dr. Von Steinbeis, President, Stuttgart.
Dr. W. Weber, Professor of Physics, Göttingen, F.R.S.
Dr. Wiedemann, Professor of Physical Chemistry, Leipzig.
Dr. Winkler, Professor of Metallurgy, Freiberg.
Dr. Wöhler, Professor of Chemistry, Göttingen, F.R.S.
Dr. Wüllner, Professor of Physics, Aachen.
Dr. Zeuner, Director of the Polytechnic School, Dresden.
Dr. Zetzsche, Director of the Polytechnic School, Chemnitz.

ITALY.

Il Com. Blaserna, Professor of Physics and Rector of the Royal University of Rome.

Il Com. Cantoni, Professor of Physics at the Royal University of Pavia.

Il Cav. Respighi, Professor of Astronomy in the Royal University of Rome, and Director of the Observatory of the Campidoglio.

THE NETHERLANDS.

Professor Dr. P. L. Rijke, Conseiller d'État (President).

Professor Dr. H. G. de Sande Bakhuyzen.

Professor Dr. C. H. D. Buys Ballot.

Professor Dr. J. Bosscha.

Professor Dr. F. C. Donders, F.R.S., President of the Royal Academy of Science, Amsterdam.

Professor Dr. J. W. Gunning.

Professor Dr. R. A. Mees.

Professor Dr. V. S. M. Van der Willigen.

Dr. D. de Loos, Director of the Secondary Town-School of Leyden (Secretary).

NORWAY.

Professor Esmark.

Herr Mohn, Director of the Meteorological Institute of Norway.

Professor Waage.

RUSSIA.

M. Struve, Conseiller Privé, Directeur de l'Observatoire Central Nicolas (President).

M. Ovsiannikow, Membre de l'Académie.

M. Gadolin, Membre de l'Académie.

M. Gruber, Professeur de l'Académie de Médécine et de Chirurgerie.

M. Stubendorf, Colonel d'État-Major.

M. Wyschnegradsky, Professeur de l'Institut technologique.

M. Beilstein, Professeur de l'Institut technologique.

M. Barbot de Marny, Professeur de l'Institut des Mines.

M. Koulibine, Professeur de l'Institut des Mines.

SWITZERLAND.

Professor E. Wartmann (President).

Professor J. Amsler Laffon.

Professor D. Colladon-Ador.

Professor Dr. F. A. Forel.

Professor Dr. E. Hagenbach-Bischoff.

Professor Ad. Hirsch.

Professor Albert Mousson.

M. E. Sarasin-Diodati.

Professor L. Soret.

Colonel Gautier (Secretary).

AUSTRIA AND HUNGARY.

The Minister of Public Instruction has appointed Sectionschef Fidler to organise the contributions from these countries.

SPAIN.

No committee has been formed, but the Government has promised to contribute, and Señor Riano has been specially appointed to make the necessary arrangements.

UNITED STATES.

The Government has, through Mr. Fish, replied that it is in communication with the various departments and scientific institutions with the object of forwarding the Exhibition.

When men of this position in all branches of science have given their adhesion to the programme of such an Exhibition, its success might well be considered as secured. But these gentlemen did not rest satisfied with merely giving their names in recognition of its value: they have spared no time and labour in making the undertaking a real success. And the Lords of the Committee of Council on Education feel assured that, in offering them their thanks for their invaluable services, they convey not only their own sentiments but the grateful recognition of their labours by the country at large.

It will be readily understood from what has been said of the nature, scope, and method of the Exhibition that a large staff was required, in addition to the permanent staff of the Museum, to organise and arrange the collection in the limited time which could be afforded for that purpose. Special arrangements had, therefore, to be made; and their Lordships have great satisfaction in recording the names of those gentlemen who have rendered very valuable services—many of them as volunteers—greatly aiding the staff of the Museum in their laborious duties. These were Captain Abney, R.E.; Dr. Atkinson; Mr. Bartlett; Dr. Brunton; Dr. Biedermann; Professor Crum-Brown; Captain Fellowes, R.E.; Professor Carey Foster; Dr. Michael Foster; Herr Kirchner; Professor Goodeve; Dr. Guthrie; Commander T. A. Hull, R.N.; Mr. Iselin; Mr. Judd; Mr. Norman Lockyer; Dr. R. J. Mann; Mr. Clements Markham; Professor H. MacLeod; Professor Roscoe; Professor Shelley; Dr. Burdon Sanderson; Dr. Schuster; Dr. Voit; and Mr. R. Wylde.

To those men of science, who, in this matter and in the work of the General Committee and Sub-Committees, have given much valuable time, and have afforded them the benefit of their great knowledge and experience, the Lords of the Committee of Council on Education feel their best thanks are due, and they trust that the immediate success and future results of the Exhibition, which owes so much to them, will reward them for the labours which they have ungrudgingly devoted to it.

In order to make the Collection as useful and interesting as possible, a Handbook containing introductory notices to the several sections has been prepared. For writing these notices the Lords of the Committee of Council on Education have been fortunate in securing the services of gentlemen the mention of whose names will be a sufficient indication of the character of the work. These gentlemen are—

Capt. W. de W. Abney, R.E.
Professor W. Kingdon Clifford, M.A., F.R.S.
Capt. J. E. Davis.
Professor G. Carey Foster, B.A., F.R.S.
Professor Geikie, F.R.S.
Professor Goodeve, M.A.
Professor Guthrie, F.R.S.
Professor T. H. Huxley, LL.D., Secretary of the Royal Society;
Mr. J. Norman Lockyer, F.R.S.
Professor MacLeod.
Mr. Clements Markham, C.B., F.R.S.

Mr. N. Story Maskelyne, M.A., F.R.S.
Professor J. Clerk Maxwell, M.A., F.R.S.
Mr. R. H. Scott, M.A., F.R.S.
Professor H. J. S. Smith, M.A., F.R.S.
Mr. W. Warington Smyth, M.A., F.R.S.
Mr. H. C. Sorby, F.R.S.
Mr. W. Spottiswoode, M.A., LL.D., F.R.S.
Dr. W. H. Stone.
Professor P. G. Tait, M.A.

It had been originally proposed to exhibit the Collection of Scientific Apparatus in the South Kensington Museum; but various circumstances, which could not be foreseen, having rendered it necessary to abandon this intention, Her Majesty's Commissioners for the Exhibition of 1851, most liberally placed the galleries on the western side of the Horticultural Gardens at the disposal of the Science and Art Department. Though, unfortunately, these galleries are disconnected from the Kensington Museum, they are admirably adapted to the present purpose, and afford an accommodation which could not otherwise have been obtained.

By order,

F. R. SANDFORD,
Secretary, Committee of Council
on Education.

CLASSIFICATION OF THE COLLECTION.

Arithmetic.

Apparatus for teaching arithmetic. — Calculating machines.— Instruments for solving equations.—Slide rules.—Numbering and enumerating apparatus, &c.

Geometry.

Instruments used in geometrical drawing.—Methods of copying —Pantigraph, micrograph.—Peaucellier's cell and parallel motion —Machines for description of curves and specimens of the curves they describe, including geometric turning.—Instruments for giving graphic representations of phenomena.—Models to illustrate descriptive geometry.—Specimens to illustrate the process of making models according to a design.—Models to illustrate solid geometry, perspective, crystallography, &c. — Stereoscopic illustrations of solid geometry.

Measurement.

Of length.—Standard yard, metre, &c.—Comparator for standards of length (sight and touch). — Gauges, measuring wheels, steel tapes, &c.—Micrometers and verniers.—Cathetometers.

Of area.—Planimeters, &c.

Of volume.—Standard gallon, litre, &c. — Pipettes, burettes. —Meters for gas, water, &c.

Of angles. — Divided circles, theodolites, clinometers, goniometers, &c.

Of mass.—Standard pound, kilogramme, &c. — Vacuum and other balances.

Of density.—Specific gravity bottles, areometers, &c.

Of time.—Clocks and pendulums, chronometers, watches, and balance wheels.—Tuning forks for measuring small intervals of time.—Chronographs.

Of velocity.—Such as Morin's machine.—Strophometers, current meters, ships' logs, &c.

Of momentum.—Ballistic apparatus.

Of force.—Spring balances, pressure gauges, torsion balances, &c.

Of work.—Indicators, dynamometers, &c.

Kinematics, Statics, and Dynamics.

Elementary illustrations.—Position and displacement of a point, a rigid body, or a material system.—Composition and resolution of displacements.—Velocity and acceleration, their composition and

b 2

resolution.—Displacements of a connected system.—Principles of mechanism.—Rolling contact, sliding contact, belting, link connexions, shafting, universal joints, &c.—Transmission of work.—Relation between the displacement of two pieces of a machine and the forces which they transmit.—The mechanical powers.—Instruments for illustrating the laws of motion, such as pendulums, gyroscopes, dynamical tops.

Laws of fluid pressure; stability of floating bodies.

Discharge of fluids through orifices, and their motion in channels.

Hydraulic and pneumatic transmission of power.

Molecular Physics.

Instruments and apparatus employed in teaching, and in the investigations and observations connected with :—

Pressure on Matter.—Tension, Compression (piezometer) Torsion, Flexion ; Relation of volume to pressure ; Elasticity of liquids and gases.—Hardness (of solids and liquids), Toughness, Brittleness, Malleability, &c.

Communication of Pressure through Fluids.—Pressure of air, its consequences and applications.—Barometers, Air-pumps, Siphons, Suction-pumps, Spirators, &c. ; Pressure of water, its consequences and applications,—Levels, Side pressure, &c.

Density.—Methods of measuring densities of Gases, Vapours, Liquids, Solids.

Adhesion and Cohesion.—Condensation of gases in solids, Solution of gases in liquids, Mixing of gases with gases (Diffusion, Transpiration, &c.), Absorption of liquids by solids (Capillarity, &c), Absorption of liquids by gases (Evaporation, &c.), Mixing of liquids with liquids (Osmose, Diffusion Dialysis).—Evaporation of solids, Solution of solids, Mixture of solids with solids (Cementation, &c.).

Sound.

Instruments and apparatus employed in teaching, and in the investigations and observations connected with :—

Geometrical, Mechanical, and Optical methods of Illustrating the Laws of Wave Motion.—Progressive waves, Composition of Vibrations, Interference, Stationary waves.

Generation of Sound.—Fog-horn, &c.

Conduction of Sound.—Through solids, liquids, and gases, Stethoscopes.

Velocity of Sound.

Detection of Sound.—Sensitive flame, &c.

Reflexion and Refraction.—Ear trumpets, Acoustic lenses, &c.

Dispersion and Absorption.

Musical Sounds.—Pitch, Standards of Pitch, Standard Tuning Forks, &c.; Methods of measuring and comparing rates of vibration; Toothed wheels, Syrens, &c.; Vibration Microscopes, &c.; Methods of illustrating the nature of musical intervals; Manometric flames, Mirrored Tuning Forks, &c.

Musical Quality.—Illustrations of the different quality of the sounds of various instruments, Harmonics, and overtones, Resultant tones, Instruments for studying quality, Resonators, Phonautographs, &c.

Musical Instruments Illustrating the above.—Methods o exhibiting the mode of vibration of various instruments and the quality of the sounds yielded by them.

Light.

Instruments and apparatus employed in teaching, and in the investigations and observations connected with:—

Production.—Combustion, Electric discharge, &c.

Measurement of Intensity, Velocity.

Action of Matter on Light.—Reflection, Refraction, Dispersion, Achromatism, Direct vision prisms, Polarization, Absorption (colour), Fluorescence, &c.

Action of Light on Light.—Interference, Diffraction, Measurement of wave length (optical banks), &c.

Action of Light on Matter.—Photography, Radiometry, Phosphorescence, &c.

Technical Applications of Optical Principles.—Lighthouse—illumination, &c.

Heat.

Instruments and apparatus employed in teaching, and in the investigations and observations connected with :—

Sources of Heat.—Chemical, Electrical, Dynamical, Solar, Calorescence, &c.

Effects of Heat on Matter.—Changes of Temperature, Expansion and change of Elasticity, Liquefaction, Vaporization, &c.

Measurement of Temperature.—Thermometers, Pyrometers, &c.

Propagation of Heat.—Radiant Heat,—Radiometer, Reflexion, Refraction, Radiation, Absorption, Polarization; Conduction,—Solids, Liquids, Gases; Convection,—Ventilation, &c.

Effect of change of Molecular State on Temperature.—Freezing mixtures, Ice machines, &c.

Effect of change of Pressure and Volume.

Heat Quantity.—Unit of Heat, Calorimeters, Specific Heat, &c. Methods of determining Latent Heat.

Mechanical Equivalent of Heat.—Methods of determining. Illustrations of Thermodynamics.

Electrical Equivalent of Heat.—Methods of determining.
Analysis of Solar Radiation.

Magnetism.

Instruments and apparatus employed in teaching, and in the investigations and observations connected with :—
Natural Magnets.
Permanent Artificial Magnets.
Electro-Magnets.
Methods of Magnetization.—Effects of Magnetization. Conditions affecting intensity of Magnetization :—Temperature (chemical), Composition, Strains, &c.
Magnetic Induction of all Substances.—Diamagnetism.
Measurement of Intensity of Magnetization, Magnetic moment.
Terrestrial Magnetism.—Instruments for observation and automatic registration of the magnetic elements.

Electricity.

Instruments and apparatus employed in teaching, and in the investigations and observations connected with :—
Production and Maintenance of Difference of Potential.—Electrical machines acting by friction, induction (doublers, replenishers, &c., Holz's and Töppler's machines, &c.) ; Galvanic batteries ; Thermo-electric piles ; Magneto-electric machines. Other sources, such as Pyro-electricity, Pressure electricity, Cleavage, Capillarity, Osmose, &c.
Detection and Measurement of Difference of Potential.—Electroscopes, Electrometers, Standards of electro-motive force, Methods of comparison.
Accumulation of Electricity.—Insulators, Condensers, Accumulators, Effects due to accumulated electricity, Distribution on conductors, Polarization of dielectrics, &c.
Measurement of Electric Quantity.—Torsion balances, Standard accumulators, Methods of comparing electric capacities and dielectric coefficients.
Detection and Measurement of Electric Currents.—Galvanoscopes, Galvanometers, Voltameters, Electro-dynamometers, &c.
Resistance.—Standards of resistance, Methods of comparing resistances, Methods of establishing absolute standards (British Association unit appar.).
Effects of Electric Currents.—Production of light, heat, Electrolysis, Electro-diffusion. Action on magnets, soft iron (electro-magnets), Action of currents on currents.
Technical Application of Electricity.—Electric telegraph, &c.

Astronomy.

Star maps, catalogues, globes, orreries, &c.
Meridian instruments.
Arrangements for communicating true time.
Altazimuths, zenith-sectors, sextants, &c.
Equatoreal Telescopes { Reflectors.
{ Refractors.
Micrometers.
Driving clocks.
Special arrangements for—
    Celestial photography.
    Spectroscopic observations.
    Thermo-electric observations.
Siderostats.

Applied Mechanics.

As the Exhibition must be regarded as chiefly referring to education, research, and other scientific purposes, it must in this division consist principally of models, diagrams, mechanical drawings, and small machines, illustrative of the principles and progress of mechanical science, and of the application of mechanics to the arts.

Properties of Materials.
Structures at Rest and in Motion.
Prime Movers.
Reservoirs of Energy.
Regulators.
The Application of the Principles of Mechanics to Machinery as used in the Arts.
Shipping, Naval Architecture, and Marine Engineering.

Chemistry.

Scientific instruments, apparatus, and materials employed in the investigation and teaching of Chemical Science, and in the application of its principles to scientific purposes.

Diagrams and models.
Illustrations of analytical results.
Specimens of chemicals,—(a) organic, (b) mineral.
Apparatus and fittings for laboratory and lectures.
Apparatus for gravimetric and volumetric operations.
Apparatus for distillation and filtration.
Apparatus for operations by the dry or hot method, such as furnace, blowpipe, &c.
Refrigeratory apparatus.
Apparatus for spectrum analysis.

NOTE.—Operations of the following nature may be illustrated, viz. :

Organic analysis.
Mineral analysis.
Electrolysis.
Water analysis.
Gas analysis.
Spectrum analysis.
Methods of investigation connected with vegetation and respiration.

Meteorology.

Thermometers and barometers, of special construction.
Anemometers, rain gauges, hydrometers, &c.
Self-recording meteorological apparatus.
Illustrations of various systems of storm signals.
Weather maps.
Instruments illustrating the phenomena of atmospheric electricity.
Instrument stands.

Geography.

Instruments used in surveying.
Instruments used in Geodesy and Hydrography, including hypsometrical instruments, tide gauges, &c.
Projections, maps, charts, models, and globes.
Deep-sea sounding apparatus. Seismographical instruments.

Geology and Mining.

Instruments for field and underground surveying.
Typical collections of rock specimens, including vein stones.
Typical fossils arranged stratigraphically.
Maps in different stages, and finished maps.
Geological models, horizontal and vertical sections.
Diagrams and plates of fossils, and general geological diagrams used in lecture rooms.
Microscopic sections of rocks and minerals, and apparatus for cutting such sections.
Anemometers, water gauges, mining barometers, and thermometers.
Mining plans, sections, and models of workings.

Mineralogy, Crystallography, &c.

Goniometers.
Apparatus for studying and exhibiting the optical characters of crystals.

Sections for optical examination.

Blowpipe and other portable apparatus for determining mi · nerals.

Collections of crystals, models of crystals, plates of crystals, and apparatus for drawing them.

Educational collections of minerals, &c.

Diagrams and models for lecture rooms.

Biology.

1. Microscopes with accessory apparatus for biological research, &c.

2. Physiological apparatus for investigating—

    a. The growth and mechanical movements of living organisms and their parts.

    b. The chemical phenomena of living organisms.

    c. The electrical phenomena of living organisms.

    d. The functions of the nervous and other systems.

3. Apparatus for anatomical research.

4. Apparatus for collecting and preserving object of natural history.

5. Appliances for teaching biology.

A limited number of examples illustrating the performances of the apparatus will be admissible.

Sub-Committees of Sections.

SECTION I.—MECHANICS, INCLUDING PURE AND APPLIED MATHEMATICS AND MECHANICAL DRAWING.

Professor J. C. Adams, M.A., F.R.S.

Mr. J. Anderson, LL.D., C.E.

Professor R. Stawell Ball, LL.D., F.R.S.

Rev. A. Barry, D.D.

Mr. W. B. Baskcomb.

Mr. Hugh Birley, M.P.

Major Bolton.

Professor F. A. Bradley.

Mr. F. J. Bramwell, F.R.S.

Mr. T. Brassey, M.P.

Mr. H. W. Chisholm, Warden of the Standards.

Mr. G. T. Clark.

Mr. Latimer Clark.

Professor R. B. Clifton, M.A., F.R.S.

Sir Henry Cole, K.C.B.

Mr. G. Dixon, M.P.

Major-General F. Eardley-Wilmot, R.A., F.R.S.

Mr. D. Glasgow.

Professor T. M. Goodeve, M.A.

The Right Hon. Lord Hampton, G.C.B., F.R.S., President of the Institute of Naval Architects.

Mr. T. E. Harrison.

Sir J. Hawkshaw, F.R.S.

Mr. T. Hawksley, President of the Institute of Mechanical Engineers.

Mr. J. Hick, M.P.

Professor J. C. Maxwell, M.A., F.R.S.

Mr. C. W. Merrifield, F.R.S.

Mr. A. J. Mundella, M.P.

Dr. Pole, F.R.S.

The Right Hon. Lord Rayleigh, F.R.S.

Mr. J. Scott Russell, F.R.S.

Major Seddon, R.E.

Professor Shelley.

Mr. C. W. Siemens, D.C.L., F.R.S.

Professor H. J. S. Smith, M.A., F.R.S.

Mr. G. R. Stephenson, President of the Institute of Civil Engineers.
Professor P. G. Tait, M.A.
Mr. J. Torr, M.P.
Rev. J. F. Twisden, M.A.
Professor W. C. Unwin, B.Sc.

Sir C. Wheatstone, F.R.S. (since deceased).
Sir J. Whitworth, Bart., F.R.S.
Mr. Bennet Woodcroft, F.R.S.
Dr. J. Woolley, F.R.S.
Colonel H. Stuart Wortley.

Section II.—Physics.

Capt. W. de W. Abney, R.E.
Professor W. G. Adams, M.A., F.R.S.
Sir G. B. Airy, K.C.B., D.C.L., F.R.S., Astronomer Royal.
Professor E. Atkinson, Ph.D.
Professor R. Stawell Ball, LL.D., F.R.S.
Professor W. F. Barrett.
Mr. C. Brooke, M.A., F.R.S.
Mr. C. O. F. Cator.
Mr. W. Chappell.
Professor R. B. Clifton, M.A., F.R.S.
Vice-Admiral Sir R. Collinson, K.C.B., Deputy Master of the Trinity House.
Dr. Debus, F.R.S.
Mr. Warren De La Rue, D.C.L., F.R.S.
Mr. H. S. Eaton, President of the Meteorological Society.
Professor G. Carey Foster, B.A., F.R.S., President of the Physical Society.
Dr. J. H. Gladstone, F.R.S.
Professor Guthrie, Ph.D., F.R.S.
Mr. J. Baillie-Hamilton.

Mr. J. Hopkinson, B.A., D.Sc.
Mr. W. Huggins, D.C.L., F.R.S., President of the Royal Astronomical Society.
Lord Lindsay, M.P.
Mr. J. Norman Lockyer, F.R.S.
Reverend R. Main, M.A., F.R.S.
Dr. R. J. Mann.
Mr. C. W. Merrifield, F.R.S.
Dr. Pole, F.R.S.
The Right Hon. Lord Rayleigh, F.R.S.
Professor A. W. Reinold, M.A.
Earl of Rosse, D.C.L., F.R.S.
Mr. R. H. Scott, M.A., F.R.S.
Mr. W. Spottiswoode, M.A., LL.D., F.R.S.
Dr. W. H. Stone.
Lieut.-Colonel Strange, F.R.S. (since deceased).
Professor P. G. Tait, M.A.
Professor Tyndall, LL.D., F.R.S.
Mr. C. V. Walker, F.R.S., President of Society of Telegraphic Engineers.
Sir C. Wheatstone, F.R.S. (since deceased).
Dr. Woolley.

Section III.—Chemistry.

Professor F. A. Abel, F.R.S., Chemist to the War Department.
Professor Bloxam.
Sir Henry Cole, K.C.B.
Mr. Warren De La Rue, D.C.L., F.R.S.
Professor Frankland, Ph.D., D.C.L., F.R.S.
Dr. Gilbert, F.R.S.
Dr. J. H. Gladstone, F.R.S.

Professor Odling, M.A., F.R.S., President of the Chemical Society.
Dr. Percy, F.R.S.
Mr. J. A. Phillips.
The Right Hon. Lyon Playfair, Ph.D., C.B., M.P., F.R.S.
Professor Roscoe, Ph.D., F.R.S.
Dr. W. J. Russell, F.R.S.
Professor Williamson, Ph.D., F.R.S.

Section IV.—Physical Geography, Geology, and Mineralogy.

Mr. D. T. Ansted, M.A., F.R.S.
Mr. H. Bauerman.
Professor F. A. Bradley.
Mr. H. W. Bristow, F.R.S.
Major-General Cameron, C.B., F.R.S.
Mr. C. O. F. Cator.

Vice-Admiral Sir R. Collinson, K.C.B., Deputy Master of the Trinity House.
Professor P. M. Duncan, M.B., F.R.S., President of the Geological Society.

Mr. H. S. Eaton, President of the Meteorological Society.

Sir P. De M. G. Egerton, Bart., M.P., F.R.S.

Mr. R. Etheridge, F.R.S.

Captain Evans, R.N., C.B., F.R.S., Hydrographer of the Navy.

Mr. J. Evans, F.R.S.

Mr. D. Forbes, F.R.S.

Professor Hughes.

Lieut.-General Sir H. James, R.E., F.R.S.

Dr. R. J. Mann.

Mr. N. Story-Maskelyne, M.A., F.R.S.

Mr. C. W. Merrifield, F.R.S.

Professor Miller, M.A., LL.D., F.R.S.

Professor Morris.

Professor Prestwich, F.R.S.

Professor A. C. Ramsay, LL.D., F.R.S.

Major-General Sir H. C. Rawlinson, K.C.B., F.R.S., President of the Royal Geographical Society.

Mr. R. H. Scott, M.A., F.R.S.

Mr. W. Warington Smyth, M.A., F.R.S.

Mr. H. C. Sorby, F.R.S., President of the Royal Microscopical Society.

Major-General R. Strachey, C.S.I., F.R.S.

SECTION V.—BIOLOGY.

The Right Hon. Lord Aberdare, President of the Royal Horticultural Society.

Professor H. W. Acland, M.D., F.R.S., President of the Medical Council of the United Kingdom.

Dr. G. J. Allman, F.R.S., President of the Linnæan Society.

Mr. G. Bentham, F.R.S.

Mr. C. Brooke, M.A., F.R.S.

Mr. G. Busk, F.R.S.

Dr. W. B. Carpenter, C.B., F.R.S.

Professor W. T. Thiselton Dyer, M.A., B.Sc.

Professor Flower, F.R.S.

Professor Michael Foster, M.D., F.R.S.

Colonel Lane Fox, F.R.S., President of the Anthropological Institute.

Mr. A. H. Garrod, M.A., F.R.S.

Mr. A. C. L. G. Günther, M.A., M.D., F.R.S.

The Hon. Alan Herbert.

Dr. J. D. Hooker, C.B., President of the Royal Society.

Professor T. H. Huxley, LL.D., F.R.S.

Professor E. Ray Lankester, M.A., F.R.S.

Mr. W. K. Parker, F.R.S.

Mr. G. W. Royston-Pigott, M.A., M.D., F.R.S.

Professor W. Rutherford, M.D., F.R.S.

Professor J. S. Burdon Sanderson, M.D., F.R.S.

Mr. H. C. Sorby, F.R.S., President of the Royal Microscopical Society.

Mr. F. H. Wenham.

LIST OF CONTRIBUTORS, WITH ADDRESSES.

** *The numbers refer to the Pages of the Catalogue.*

UNITED KINGDOM.

ABEL, F. A., F.R.S., *Royal Arsenal, Woolwich*; 377–8.

ABERCROMBY, HON. R., 21, *Chapel Street, Mayfair, London*; 684.

ABNEY, CAPT. W. DE W., R.E., *Chatham*; 234, 424, 707.

ABRAHAM, C., *University of Edinburgh*; 595.

ACLAND, DR., F.R.S. *on behalf of the Radcliffe Trustees, Oxford*; 851, 861.

ADAMS, PROF. A. L., M.A., F.R.S., *Royal College of Science, Dublin*; 994.

ADAMS, J. C., *Cambridge University*; 400.

ADIE, P., 15, *Pall Mall, London*; 55, 444, 677, 751, 756, 872.

ADMIRALTY (Hydrographic Department), *Whitehall, London*; 95, 287–9, 731–8, 774–6, 795–8.

AERONAUTICAL SOCIETY OF GREAT BRITAIN (*F. W. Brearey, Hon. Sec.*), *Maidenstone Hill, Greenwich*; 108.

AGRICULTURAL SOCIETY OF ENGLAND, ROYAL, 12, *Hanover Square, London*; 474.

AHRBECKER, H. C., 117, *Stamford Street, London*; 74, 531.

ALLEN, H., *Homewood, South Orange, New Jersey, U.S.A.*; 429.

ALLPORT, S. B., 50, *Whittall Street, Birmingham*; 59.

AMHURST-TYSSEN, W. A., *Diddlington Hall, Norfolk*; 471.

ANATOMICAL DEPARTMENT, *Oxford University Museum*; 983–4.

ANDREWS, DR., F.R.S., *Queen's College, Belfast*; 162, 168, 169, 268, 325, 570, 575, 671.

ANDREWS, W., *Indo-European Telegraph Company, Old Broad Street*; 368.

APPS, A., 433, *Strand, London*; 216, 311, 760.

ARCHBUTT, J. AND W. E., 201, *Westminster Bridge Road, London*; 76.

ASTON AND MANDER, 25, *Old Compton Street, Soho, London*; 1, 2, 20.

ASTRONOMER ROYAL, THE (Sir G. B. Airy, K.C.B., F.R.S.), *Greenwich, London*; 413, 421–2.

ASTRONOMICAL SOCIETY, ROYAL, *Burlington House, Piccadilly, London*; 413.

AUSTEN, MAJOR G., 17, *Bessborough Gardens, S.W.*; 803.

AUSTIN AND HUNTER, *Sunderland*; 513, 528.

AUTOTYPE COMPANY, 36, *Rathbone Place, London*; 243–4, 433.

AVELING AND PORTER, 72, *Cannon Street, London; and Rochester*; 459.

BABBAGE, MAJOR-GENERAL, *Dainton House, Bromley, Kent*; 5, 6, 532–3.

BADEN-POWELL, MRS., 1, *Hyde Park Gate South, London*; 230.

BAGOT, A. C., *Pembroke College, Cambridge*; 10.

BAILEY, WALTER, 176, *Haverstock Hill, London*; 217.

BAILEY, W. H., & Co., *Albion Works, Salford, Manchester*; 98, 267, 445, 456, 548.

BAKER, W. CLINTON, *Bayfordbury, Herts.*; 420.

BALDWIN, A. C., 37, *Chester Square, S.W.*; 391, 419.

EDWARDS, B. J., & CO., 6, *Lincoln Terrace, Kilburn, London*; 232.

ELDER, J., & CO., *Govan, near Glasgow*; 498.

ELKINGTON & CO., 22, *Regent Street, London*; 379.

ELLIOTT BROTHERS, 449, *Strand, London*; 2, 3, 51, 57, 74, 75, 77, 95-6, 99, 103, 109, 147, 152, 181-3, 229, 250, 259, 261, 279, 281, 290, 298, 300, 305, 308, 316 -8, 327, 333-4, 339-45, 370, 375, 419, 465, 475-6, 681, 689, 693, 699, 708, 722, 747, 751, 780, 870.

ENOCK, F., 30, *Russell Road, Holloway*; 924, 978.

ERARD, MESSRS., 18, *Great Marlborough Street, London*; 192.

ESSEX AND CHELMSFORD MUSEUM, *Chelmsford*; 903-4.

ETLER, C., *1st Battalion Grenadier Guards, Chelsea Barracks, S.W.*; 302.

EVANS, L., *Hemel Hempsted*; 8, 17.

FAIJA, H., 4, *Great Queen Street, Westminster*; 103.

FARADAY, MRS., *Barnsbury Villa, 320, Liverpool Road, London*; 382-3.

FARWIG, J. F., 36, *Queen Street, Cheapside, London*; 252.

FAULKNER, J., 13, *Great Ducie Street, Strangeways, Manchester*; 281-2, 319.

FELLOWES, I. H., 46, *South Quay, Great Yarmouth*; 515.

FLANNERY, J. F., 6, *Broadway Chambers, Westminster*; 103, 524-5.

FLETCHER, A. E., 21, *Overton Street, Liverpool*; 97-8, 657.

FLETCHER, T., F.C.S., *Museum Street, Warrington*; 252-3.

FORSTER, P., *Sunderland*; 515.

FOSTER, ROBERT, *Sunderland*; 74.

FOWLER, J., C.E., 2, *Queen Square Place, Westminster*; 548.

FRANCIS, G., C.E., *Chester*; 745.

FRANKLAND, PROF., Ph.D., D.C.L., F.R.S., *Royal School of Mines, Jermyn Street, London*; 580-6, 623-4, 672.

FRODSHAM, C., & CO., 84, *Strand, London*; 118.

FROUDE, W., F.R.S., *Chelston Cross, Torquay*; 487-9.

FULLER, G., *Belfast*; 381.

FULTON, J., *University of Edinburgh*; 592-3.

GAIRDNER, 45, *South Bridge, Edinburgh*; 923.

GALLETLY, A., *Museum of Science and Art, Edinburgh*; 159, 160.

GALLOWAY, PROF. R., *Royal College of Science, Dublin*; 658.

GALTON, F., F.R.S., 42, *Rutland Gate, London*; 13, 14, 21, 178, 767, 769, 802.

GARDNER, J. STARKIE, *Park House, St. John's Wood Park, London*; 88, 437-8, 549, 552, 866-7.

GARDNER, J., AND SONS, 453, *Strand, London*; 250.

GARNER, R., F.R.C.S., *Stoke-upon-Trent*; 257, 904.

GARNHAM & CO., *Sash Court, Wilson Street, Finsbury, London*; 376.

GASKELL, DEACON, & CO., *Widnes, Lancashire*; 644-5.

GEIKIE, PROF., DIRECTOR, *Geological Survey of Scotland, Edinburgh*; 827.

GEOGRAPHICAL SOCIETY OF LONDON, ROYAL, 1, *Savile Row, Burlington Gardens, London*; 798, 801.

GEOLOGICAL MUSEUM, *Jermyn Street*; 668.

GEOLOGICAL SECTION, UNIVERSITY MUSEUM, *Oxford*; 844.

GEOLOGICAL SOCIETY OF LONDON, *Burlington House, Piccadilly, London*; 815-21.

GEOLOGICAL SURVEY OF ENGLAND AND WALES, *Jermyn Street, London*; 821-3.

GEOLOGICAL SURVEY OF SCOTLAND (Prof. Geikie, F.R.S., Director), *Edinburgh*; 827.

GEOLOGICAL SURVEY OF THE UNITED KINGDOM (A. C. Ramsay, LL.D., F.R.S., Director-General), *Jermyn Street, London*; 823-8.

GEORGE, CAPT. C., *Royal Geographical Society*, 1, *Savile Row, London*; 754-8, 769.

GLADSTONE, J. H., Ph.D., F.R.S., 17, *Pembridge Square, London*; 579-80.

GLASGOW MECHANICS' INSTITUTION; 451, 460, 497.

GOODMAN, G. H., 55, *Penrose Street, Newington, S.E.*; 472.

HOGGAN, G. & F. E.. M. D., 13, *Grenville Place, Portman Square*; 977.

HOLMAN, D. S., *Philadelphia, U.S.A.*; 928.

HOLMES, J. M., *Town Hall Chambers, Birmingham*; 549.

HOLMES, N. J., 8, *Gt. Winchester Street Buildings, E.C.*; 534-5.

HOLT, H. P., C.E., *Royal Insurance Buildings, Leeds*; 75, 453, 456, 469, 474, 498.

HOOKER, J. D., M.D., P.R.S., *Royal Gardens, Kew, Surrey*; 112, 904.

HOPKINS & WILLIAMS, 16, *Cross Street, Hatton Garden, E.C.*; 620.

HORNER, C., *Fern Villa, Mortlake, Surrey*; 231.

HOROLOGICAL INSTITUTE, BRITISH, 35, *Northampton Square, London*; 123-5.

HOW, J., & Co., 5, *St. Bride Street, Fleet Street*, late 2, *Foster Lane, London*; 166, 214, 280, 290, 305, 313, 339, 380-1, 629, 839, 884, 907, 916, 979.

HOWE, W., *Clay Cross, Chesterfield*; 146.

HULL, E., DIRECTOR, (*Geological Society of London*); 815-21.

HULL, PROF. E., M.A., F.R.S., *Royal College of Science, Dublin*; 813.

HUNT, R., *Craven Hotel, Strand*; 550.

HUNTER & ENGLISH, *Bow*; 525.

HUSBANDS, H., *Bristol*; 743, 753.

HUTCHINSON, J., & Co., *Windes, Lancashire*; 645-7.

HUXLEY, PROF., F.R.S., *Royal School of Mines, Jermyn Street, London*; 993-4.

HYDRAULIC ENGINEERING Co., *Chester*; 455.

INDIA, SECRETARY OF STATE FOR, *India Office, Whitehall, London*; 729-30, 787-9.

IRELAND, GENERAL VALUATION OF, *Dublin*; 803.

IRELAND, ROYAL COLLEGE OF SCIENCE FOR, *Stephen's Green, Dublin.* (*See* ROYAL COLLEGE OF SCIENCE.)

JACKSON, REV. J. C., 67, *Amherst Road, Hackney, London*; 116, 391.

JACKSON, M., 65, *Barbican, London*; 1009-10.

JEBB, G. R., *Chester*; 553.

JELLETT, REV. J. H., B.D., *Trinity College, Dublin*; 217.

JOHNSON, MATTHEY, & Co., *Hatton Garden, London*; 53, 88, 641, 669.

JOHNSTON, W. & A. K., 18, *Paternoster Row, E.C.*; 1034-5.

JORDAN, J. B., *Museum of Practical Geology, Jermyn Street, London*; 681, 815, 851.

JOULE, J. P., D.C.L., F.R.S., *Broughton, Manchester*; 163, 281, 287, 290.

KEMPE, A. B., B.A., 7, *Crown Office Row, Inner Temple, London*; 145-6.

KESSELMEYER, CH. A., *Manchester*; 439.

KEW COMMITTEE OF THE ROYAL SOCIETY, *Kew Observatory, Surrey*; 212, 261, 291, 293-6, 299, 332-3, 398, 421, 424, 694, 697, 701, 711, 738-9, 768.

KEW MUSEUM, *Royal Gardens, Kew, Surrey*; 982.

KING'S COLLEGE, LONDON, COUNCIL OF, *Somerset House, Strand, London*; 13, 147, 150, 162-3, 166, 202, 204, 223, 237, 258, 283, 301-2, 311, 321, 324, 342, 368-9, 382, 426, 450, 464-5, 479, 529; 547, 553, 702, 1046.

KINGSTON, G. T., *Meteorological Office of the Dominion of Canada; Toronto, Canada*; 695-6.

KNOBEL, E. B., F.R.A.S., F.G.S. 20, *Avenue Road, Regent's Park, London*; 406, 438.

KULLBERG, V., 135, *Liverpool Road, N.*; 127.

LADD, W., & Co., 11 *and* 12, *Beak Street, Regent Street, London*; 223, 224, 313, 909.

LAIDLER, W. H., 9, *Edward Street; Bow Common, London*; 60.

LAING, J., *Deptford Yard, Sunderland*; 517.

LAIRD BROTHERS, *Birkenhead Iron Works, Birkenhead*; 500-5, 531.

LASLETT, T. N., 97, *Maryon Road, Charlton, S.E.*; 748.

LATHBURY, R., JUN., *Park House, Chiswick, Middlesex*; 499.

AUSTRO-HUNGARIAN EMPIRE.

ZMURKO, DR. L., *Prof. of Mathematics, University, and Polytechnic Institute, Lemberg*; 18, 20, 21.

ZULKOWSKY, PROF. C., *Imp. and Royal Technical High School, Brünn*; 639.

BELGIUM.

ARENS, ANTOINE MARIANUS, *Rue de Bruxelles, Namur*; 1012.

BRUYLANTS, G., *Laboratory of General Chemistry, University of Louvain*; 592.

COUGNET, J., *Ixelles, Brussels*; 165.

GÉRARD, A. J., 5, *Place St. Lambert, Liège*; 100, 107–8, 128.

GOCHET, PROF. A. M., *Normal School, Carlsbourg, Luxembourg*; 813, 1026.

HENRY, L., *Professor of Chemistry, University of Louvain*; 592.

LE BOULENGÉ, MAJOR, 23, *Thier de la Fontaine, Liège*; 56, 100.

LEURS, LIEUT.-GENERAL, 9, *Rue de la Longue-haie, Brussels*; 100.

MALAISE, PROF. C., MEMBER OF THE ROYAL ACADEMY OF BELGIUM, *State Agricultural Institute, Gemblous, Province of Namur*; 836.

MARTINOT, A., *Nismes, Mariembourg*; 1011.

NAVEZ, LIEUT. COL., *Schaerbeeth, Brussels*.

PIRON, FRÈRE M., *Director of the Normal School, Carlsbourg*; 13, 39.

RENARD, A., 11, *Rue des Récollets, Louvain*; 845.

SACRÉ, E., 30, *Rue Cantersteen, Brussels*; 83.

SCHWANN, PROF. T., 11, *Quai de l'Université, Liège*; 954–5, 1002.

SIMONAN AND TOOVEY, *Chaussé de Lille, Tournai*; 236.

VAN RYSSELBERGHE, F., *Paymaster, Royal Navy, Antwerp, late Professor at the School of Navigation, Ostend*; 74, 709.

VAN SCHERPENZEEL THINI, J., *Director of Mines, 34, Rue Nysten, Liège*; 878, 880.

FRANCE.

ALVERGNIAT FRÈRES, 10, *Rue de la Sorbonne, Paris*; 250, 322, 641.

AUZOUX, DR., 96, *Rue de Vaugirard, Paris*; 982–3.

BAUDIN, 276, *Rue St. Jacques, Paris*; 171, 723.

BECQUEREL, M. E., 47, *Rue Cuvier, Paris*; 237, 425.

BERTHELOT, M., 97, *Boulevard St. Michel, Paris*; 1035.

BIARD, M., 10, *Rue Mont Theibord, Paris*; 813.

BONIS, MADAME, 18, *Rue Montmartre, Paris*; 317.

BONTEMPS, *Telegraph Inspector, Paris*; 377.

BOURDON, C., 74, *Rue du Faubourg du Temple, Paris*; 128, 146, 166, 456, 470.

BOURGOGNE, E., 34, *Rue Cardinal Lemoine, Paris*; 977.

BRÉGUET, 81, *Boulevard Mont Parnasse, Paris*; 173, 280–1, 306–7, 316, 328, 426, 669, 711, 781, 878.

CACHELEUX, 6, *Rue des Vieilles Haudriettes, Paris*; 549.

CARRÉ, E., 24, *Rue d'Assas, Paris*; 166, 299, 328.

CHAMEROY, 162, *Faubourg St. Martin, Paris*; 83.

CLEUET, 196, *Rue d'Allemagne, Paris*; 469.

COLLEGE OF FRANCE, PARIS; 147, 182, 222, 255, 258, 268, 312, 327.

COLLIN & CO., 6, *Rue de l'Ecole de Médecine, Paris*; 945.

COLLOT, BROTHERS, *Boulevard de Montrouge, Paris*; 79, 83, 87.

CONSERVATOIRE DES ARTS ET MÉTIERS, *Paris*; 3, 12, 21, 128, 162–9, 182, 196, 234, 255, 261, 268, 274, 295, 305, 307, 324, 338, 381, 414, 420, 427, 430, 460, 487, 638, 683, 696, 743, 887.

CRÉTÉS, M., 66, *Rue de Rennes, Paris*; 204, 227, 930, 933.

DAUBRÉE, M., *Membre de l'Institut, Director of the School of Mines, Paris*; 847.

D'Abbordre, A., 120, *Rue du Bac, Paris*; 746–7.

Delagrave, M., 58, *Rue des Écoles, Paris*; 813.

Delesse, M., *Chief Engineer of Mines, Paris*; 811–2, 844.

Deleuil, 42, *Rue des Fourneaux, Paris*; 85, 87.

Department of Lighthouses, *France*; 538–45.

Dépôt of Marine Charts and Plans, *Paris*; 813.

Deprez, M., 16, *Rue Cassine, Paris*; 370, 467.

Desains, Membre de l'Institut, 78, *Rue d'Arras, Paris*; 274.

Deschiens, 123, *Boulevard St. Michel, Paris*; 372.

Digeon, 13, *Rue de Marseille, Paris*; 153, 446, 476.

Dolfus, E., 9, *Rue St. Fiacre, Paris*; 470.

Duboscq, J., 21, *Rue de l'Odéon, Paris*; 183, 190, 205–6, 215, 218, 240, 250, 328, 337, 420.

Dumoulin-Froment, 85, *Rue Notre Dame des Champs, Paris*; 21, 60, 75, 101, 374, 781.

Dumas, J., 69, *Rue St. Dominique, St. Germain, Paris*; 568.

Eastern Railway of France, *Paris*; 111.

École de Pharmacie, *Rue de l'Arbalète, Paris*; 223.

École Polytechnique, *Paris*; 63, 169, 182, 191, 203, 255, 268, 281, 343, 383.

Enfer Fils, 10, *Rue de Rambouillet, Paris*; 669.

Erhard, 12, *Rue Duguay Trouin, Paris*; 236, 381.

Evrard, 30, *Rue des Blancs Manteaux, Paris*; 413.

Faculty of Sciences, *Paris*; 168, 275.

Feil, 56, *Rue Lebrun, Paris*; 215.

Fizeau, M., Membre de l'Institut, 3, *Rue de la Vieille Estrapade, Paris*; 234, 237.

Fontoure, H., 92, *Rue St. Georges, Paris*; 314.

Fortin Hermann, MM., 122, *Boulevard Mont Parnasse, Paris*; 56, 66, 458.

French Commission for Observing the Transit of Venus in 1874; 413, 423.

Gavard, A., 70, *Quai des Orfèvres, Paris*; 16, 21.

Gillot, Madame, 179, *Rue du Faubourg, St. Martin, Paris*; 234.

Girard, A., *Prof., Conservatoire des Arts et Métiers, 292, Rue St. Martin, Paris*; 918.

Golaz, 24, *Rue des Fosses, St. Jacques, Paris*; 169, 268–72, 620–3, 625–6.

Gondolo & Co. (misprinted Tondola), 9, *Boulevard du Palais, Paris*; 129.

Goupil & Co., 9, *Rue Chaptal, Paris*; 234.

Gros, 94, *Rue de Montreuil, Paris*; 467.

Guéroult, G., 2, *Rue de Vienne, Paris*; 192..

Guyot D'Arlincourt, 102, *Rue Neuve des Mathurins, Paris*; 373–4.

Hanicque De St. Senoch, 19, *Rue Demours, Paris*; 810.

Hayaux du Tilly, 15, *Rue de Lisbonne, Paris*; 803, 813.

Hirn, G. A., 3, *Logelbach, Colmar*; 110.

Honzeau, M., *Rouen*; 663.

Isoard Fils, 78, *Rue St. Maur, Paris*; 59.

Jamin, 24, *Rue Soufflot, Paris*; 280.

Jannettaz, 9, *Rue Linne, Paris*; 275–6.

Janssen, M., Membre de l'Institut, 33, *Rue Labat, Paris*; 423.

Kastner, F., 43, *Rue de Clichy, Paris*; 179.

Laboratoire du Collége de France, *Place Cambrai, Paris*; 590.

Lancelot, 11, *Rue des Poitevins, Paris*; 196.

Larrey, Baron, 91, *Rue de Lille, Paris*; 927.

Launay, Prof., *Lycée, Caen*; 802.

Laurent, Leon, 21, *Rue de l'Odéon, Paris*; 214, 218, 221, 223, 226, 248, 255, 277, 391, 412, 914–5.

Laussedat, Col., Prof., *Conservatoire des Arts et Métiers, Paris*; 768.

Lemaire Douchy, 64, *Rue Taitbout, Paris*; 879.

GERMANY.

KUMMER, PROF. DR. E. E., *Berlin*; 153.

KUNDT, PROF. DR., *Strasburg*; 211, 291.

LANDOIS, PROF. DR., *Münster*; 990.

LANDOLT, PROF. DR. (Polytechnic School), *Aix-la-Chapelle*; 224, 620, 1041-2.

LANDSBERG & WOLPERS, *Hanover*; 8, 59, 368, 1010.

LAQUEUR, PROF., *Strasburg*; 934-5.

LABAULX, PROF. von, *Breslau*; 604, 772-3, 849, 869.

LASPEYRES, PROF. (Polytechnic School), *Aix-la-Chapelle*; 96.

LEITZ, E., *Wetzlar*; 910-11.

LENTZ, E. A., 36 & 37, *Spandauer Strasse, Berlin*; 638, 1036.

LEPSIUS, PROF. DR., *Royal Library, Berlin*; 158, 159.

LEYBOLD, E. (Successors to), *Cologne*; 1047-54.

LEYSER, G. MORITZ, *Leipsic University*; 926.

LIEBERMAN, PROF. C., *Berlin*; 604.

LINGKE, A., & Co. (M. Hildebrand and E. Schramm), *Freiberg, Saxony*; 396-7, 745, 753, 873, 875, 884.

LIST, DR. K., *Hagen*; 173.

LISTING, PROF. DR., *Göttingen*; 219.

LOCHMANN, P., *Zeitz*; 456, 459.

LÖCKERMANN, DR., *Hamburg*; 430.

LOHDE, L., 33, *Haide Strasse, Berlin*; 37.

LÖHMANN, R., 3, *Brückenstrasse, Berlin*; 84.

LOHSE, DR., C. (*Astronomer of the Royal Astro-Physical Observatory*), *Potsdam*; 425-6.

LOMMEL, PROF. DR., *Erlangen*; 247-8.

LUCAE, PROF. A., *Berlin*; 183.

MAGNUS, DR., *Breslau*; 934.

MAJER, F., *Strasburg*; 923.

MARBURG, MATH. AND PHYSICAL INSTITUTE (Prof. Melde); 53.

MATTHIESSEN, PROF. DR., *Rostock*; 714.

MEIDINGER, PROF. DR. H., *Carlsruhe*, 278, 367.

MEISSNER, A. (Müller and Reinecke), *Berlin*; 60, 746, 749, 751.

MELDE, PROF. DR., *Mathematical and Physical Institute, Marburg*; 53.

MERCK, E., *Darmstadt*; 605.

MEYER, L., *Berlin*; 263.

MEYER, PROF. L., *Carlsruhe*; 599.

MEYER, DR. O. E. (The University), *Breslau*; 176.

MEYER (*Town School*), *Halle*; 54.

MICHAELIS, PROF. A., *Carlsruhe*; 600.

MINISTERIAL COMMISSION FOR THE SCIENTIFIC EXPLORATION OF THE GERMAN SEAS, *Kiel*; 771.

MITSCHERLICH, PROF. A., *Münden, Hanover*; 164, 213, 262, 570, 629, 641, 886, 889, 894, 1036.

MITTELSTRASS BROTHERS, *Magdeburg*; 319.

MÖBIUS, PROF. DR., *Kiel*; 925.

MÖHL, DR. H. (*Royal High School of Industry*), *Cassel*; 839, 848, 883.

MÖLLER, L., *Giessen*; 888.

MÜLLER, DR. F., *Osnabrück*; 704.

MUNICH, UNIVERSITY OF; 54, 84, 86, 106, 263, 275, 629.

NARTEN, DR. W. (Royal High School of Industry), *Cassel*; 15, 748, 754, 875.

NAUTICAL OBSERVATORY-"DEUTSCHE SEEWARTE" (Dr. Neumayer, Director), *Hamburg*; 259, 293, 681, 712, 721-2, 802.

NÖRDLINGER, PROF. DR., *Hohenheim, Wurtemberg*; 89, 445-6, 1001.

OPPEL, PROF., J. J., *Frankfort-on-Maine*; 36, 179, 188, 245-6, 427, 429, 431.

OPPEL, DR. K. *Frankfurt*; 429.

OPPENHEIM, PROF. A., *Berlin*; 605.

ORTH, PROF. DR., *Berlin*; 835.

OSTERLAND, C., *Freiberg, Saxony*; 638, 667, 873, 875.

OTT & CORADI, *Kempten, Bavaria*; 15, 77, 750, 753.

PANSCH, DR., *Kiel*; 989.

PERNET, DR. (Assistant, Cabinet of Physics), *Breslau*; 262-3.

PETTENKOFER, PROF. DR. MAX, *Munich*; 950.

PFAFF, PROF. DR., *Erlangen*; 255, 887, 897.

PHARMACEUTICAL INSTITUTE OF THE UNIVERSITY, *Breslau*; 599.

PHYSICAL INSTITUTE, *Freiburg*; 255.

PHYSIOLOGICAL INSTITUTE, *Prague*; 954.

PIEL, H., *Bonn*; 897.

e

WIECKE, DR., *Cassel*; 38, 39.
WIENER, PROF. DR. C., *Carlsruhe*; 37.
WINKEL, R., *Göttingen*; 912.
WINKLER, PROF., *Freiberg*; 610.
WINNECKE, PROF. DR., *Strasburg*; 404, 432, 740.
WINTER, E., *Hamburg, Eimsbüttel*; 297.
WOHLER, PROF. DR., *Göttingen*; 601.
WOHLERS (Successor to Campbell), *Hamburg*; 930, 936.

WOLFF & SONS, *Heilbronn and Vienna*; 639, 641, 672, 1036.
WÜLLNER, PROF. DR. (Polytechnic School), *Aix-la-Chapelle*; 331, 460.
ZEISS, C., *Jena*; 211, 215, 230, 909-10.
ZIEGLER, DR. A., *Freiburg, Baden*; 898, 984.
ZIMMER BROTHERS, *Stuttgart*; 89, 744, 759.
ZORN, W., *Berlin*, 17, *Schöneberger Str.*; 174.

HOLLAND.

ASSEN SECONDARY GOVERNMENT SCHOOL; 118, 258, 278.
BAKHUYZEN, H. G. VAN DE SANDE, *Director, Observatory, Leyden*; 392-4, 398-9, 428-9, 438.
BECKERS SONS, *West Zeedyk, Rotterdam*; 82.
BLEEKRODE, DR. L., *The Hague*; 299, 1041-2.
BOOGAARD, PROF.DR. J. A., *Director of the Museum of Anatomy, Academy of Leyden*; 900.
BOSSCHA, PROF. J., *Royal Polytechnic School, Delft*; 255, 342, 345, 579, 885, 888, 921.
BRONDGEEST, DR., *Physiological Laboratory and Ophthalmological School, Utrecht*; 958.
BUYS-BALLOT, PROF., *Utrecht*; 53, 127, 184, 264, 391, 413, 700, 766, 904, 934.
DE LOOS, DR. D., *Director of the Secondary Town School, Leyden*; 259, 671.
DONDERS, PROF., *Physiological Laboratory and Ophthalmological School, Utrecht*; 178, 318, 957-60, 962-70.
ENGELMANN, PROF., *Physiological Laboratory and Ophthalmological School, Utrecht*; 167, 317, 957, 959-60.
FERHAAR, A. T., *Utrecht*; 977-8.
GRONEMAN, DR. F. G., *Director of the Secondary Government School, Groningen*; 149.

GUNNING, DR. J. W., *Professor of Chemistry "Athenæum Illustre," Amsterdam, and Scientific Adviser to the Treasury Department, Holland, Amsterdam*; 156-7, 259.
HARTING, PROF. DR. P., *University of Utrecht*; 938.
HOOGEWERFF, S., PH. D., *Rotterdam*; 176, 255.
HUIZINGA, PROF., *Director of the Physiological Laboratory, University of Groningen*; 940, 976.
MEES, PROF. R. A., *Director of the Physical Laboratory, University of Groningen*; 162, 189, 190, 335, 475.
MULDER, DR. M. E., *Groningen*; 920.
OTTMANS, H., 141, *Amstel Hooge Sluis, Amsterdam*; 901.
OUDEMANS, PROF. A. C., *Royal Polytechnic School, Delft*; 591.
ROYAL POLYTECHNIC SCHOOL (Prof. J. Bosscha), *Delft*; 255, 342, 345, 579, 885, 888, 921.
RIJKE, PROF. DR. P. L., *Director of the Cabinet of Physics, University of Leyden*; 81, 131, 178, 246, 255, 321, 410, 413, 427, 462, 900.
SCHOOL, SECONDARY GOVERNMENT, *Assen*; 118, 258, 278.
SCIENTIFIC SOCIETY OF ZEELAND (G. N. de Stoppelaar, Sec.), *Middelburgh*; 900.
SNELLEN, DR., *Physiological Laboratory and Ophthalmological School, Utrecht*; 748, 961-2, 964-5, 967.
SNŸDERS, J. A., *Lecturer, Royal Polytechnic School, Delft*; 176, 936.

ITALY.

NORWAY.

RUSSIA.

SPAIN.

SWITZERLAND.

Mousson, Prof. A., *Zurich*; 167, 226.

Pictet (Raoul) & Co., *Geneva*; 274.

Ramboz and Schuchardt, *Geneva*; 717.

Recordon, Prof. E., 53, *Terrassière, Geneva*; 428, 430, 474, 813.

Renevier, Prof. E., *Lausanne, Switzerland*; 829.

Sarasin, G., *Tour de Balessart, Geneva*; 18, 442, 766.

Schmid, A., *Engineer, Zurich*; 79.

Soret, J. L., *Geneva*; 219, 229, 274, 570.

Soret, Perrot, and Sarasin (De la Rive Collection), *Geneva*; 172, 224, 383-6.

Stapff, Dr. F. M., *Geological and Mining Engineer, St. Gothard Railway*; 2.

Studer, Prof. B., *Commission of Switzerland, Geological Survey, Berne*; 829.

Wartmann, E., *Professor of Natural Philosophy, University of Geneva*; 273, 327-8, 370-1.

Wolf, Prof. R., *Director of the Observatory, Zurich*; 722.

CATALOGUE.

SECTION 1.—ARITHMETIC.

WEST GALLERY, GROUND FLOOR, ROOM G.

I.—SLIDE RULES.

1. Slide Rule, of boxwood, arranged by Mr. Dixon, Lowmoor Ironworks.   *Aston & Mander.*

In addition to the lines of the ordinary slide rule this instrument contains :
    Lines of common and hyperbolic logs and numbers.
    Lines of sines, cosines, and numbers.
    Lines of cubes and roots, direct.
    A copy of Dixon's "Slide Rule Practice" is issued with each rule.

2. Slide Rule, of ivory, showing the actual and racing tonnage of yachts.   *Aston & Mander.*

The length and breadth of beam being "set together," as directed in the instructions, the racing tonnage of yachts of any size is shown as marked.

3. Slide Rule, of boxwood, adapted to brickwork measurement in all its branches, cubing stone, &c.   *Aston & Mander.*

In this adaptation of the rule to brickwork measurements, all the results are obtained by one setting, viz., "length to height"; while, immediately opposite, any thickness will be found; the superficial area in square feet; the contents in rods of reduced work 1½ bricks, in cubic feet, in cubic yards, and the number of bricks required.

4. Slide Rule, of boxwood, adapted to timber measurement in all its branches, giving the superficial or cubic contents of round and unequal sided timber, St. Petersburg standard, price, &c.   *Aston & Mander.*

5. Slide Rule, of boxwood, with reversible slides, movable inverted lines, &c. Arranged by Chas. Hoare. *Aston & Mander.*

Uses explained in Hoare's "A. B. C. of Slide Rule Practice."

6. Slide Rule, of ivory, with reversible slides, movable inverted lines, &c. Arranged by Chas. Hoare.   *Aston & Mander.*

Uses explained in Hoare's "A. B. C. of Slide Rule Practice."

7. Slide Rule, of ivory, adapted for use in iron and steel plate and sheet rolling mills. Designed by Chas. Hoare.

Aston & Mander.

This rule will show directly the precise net and waste weight of iron and steel plates, and sheets, of any size, shape, and thickness. It may be applied to all ordinary metals, and to find areas, cubic contents, liquid capacity, &c.

8. Slide Rule, of boxwood, adapted for sheet iron and steel manufacturers. The dimensions, thicknesses, and weights are given both in English and metrical standards. Designed by Chas. Hoare. *Aston & Mander.*

The length of the sheet or plate (on the slide) being first set to the width, then immediately below any thickness (on the top lines) will be found (on the slide) the actual weight of the sheet either in pounds (avoirdupois) or kilogrammes, metrical or English measures being used without previous conversion.

9. Slide Rule, of ivory, adapted for use in iron and steel-bar rolling mills, showing instantly the precise net and waste weights for bars of any length, size, and form. Designed by Chas. Hoare.

Aston & Mander.

24. Scales, of boxwood, to show cubes, squares, and roots, areas, diameters, circumferences, and decimal equivalents. Designed by Chas. Hoare. *Aston & Mander.*

The bevel edged set square is used to read the divisions, and dispenses with the need of voluminous printed tables.

9a. Three Slide Rules. *Elliott Brothers.*

10. Estimator. A slide rule, by which the volume of prismoidal bodies (embankments, ditches, cuttings, &c., occurring in the construction of railroads, canals, fortifications, &c.,) is calculated mechanically.

Dr. F. M. Stapff, Geological and Mining Engineer at the St. Gotthard Railway.

This instrument, invented by the exhibitor, is patented in Sweden and the United States of America.

11a. Timber Rule, for finding the content of timber of any form, regular or irregular. The rule has eight gauge points or divisors for reducing dimensions in inches to contents in square feet. *Dring and Fage.*

11b. " Verie " or Excise Officer's Rule.

Dring and Fage.

Verie is probably a corruption of " Vero," a revenue officer who made an alteration in the method of laying down some of the lines on the rule; previously to which they were called Everard's rules.

The lines on the rule are the A, B, C, D, MD, (or malt depth) 6x or variety lines, viz., 1st, 2nd, 3rd, 4th, Dr. Hutton's and Dr. Young's, and two ullage lines (segment standing and segment lying).

The A, B, C, and D lines are commonly called Gunter's lines (from Gunter, the celebrated mathematician, who was the first to apply a logarithmic line to the instrument for the solution of arithmetical problems) of which the A, B, and C, are merely repetitions of each, and laid down to single radius, and the D to double radius.

The MD line is similar to the A, B, and C, but is a broken line of two radii, with the figures and divisions in an inverted order (reading from right to left), commencing at 2218 · 192 in the right-hand radius, and ending at the same point in the left-hand radius, 2218 · 192 being the number of inches in a bushel. By the method in which this line is arranged and used in conjunction with the A, B, and C lines the contents in bushels of rectangular and similar figures may be found at one operation.

The X or variety lines or lines of special gauge points (invented by Mr. Woolgar) for finding the mean diameter of a cask whatever its form; these lines commence at 18 · 789, the circular gauge point, and are extended according to each variety to which they may be applied.

The ullage lines are rules for finding the contents of a cask by comparison with a standard cask holding 100 gallons, a form nearest those frequently occurring in practice.

It cannot be ascertained by whom these lines were invented.

The fixed gauge points on the rule are those for the imperial gallon and bushel, both square and round.

These rules are principally used by excise officers and maltsters. So admirable is the arrangement, that nearly every problem to which the principle of the slide rule is applicable can be solved on one of these rules.

11c. Slide Rule, invented by Mr. Coulson, of Redan, used for setting out railway curves, finding the weights of materials from their specific gravities, breaking strains, &c.

Dring and Fage.

The applications of this rule are so varied that the author's description of them exceeds 400 octavo pages of closely printed matter.

12. Slide Rule, by M. Mabire.

Conservatoire des Arts et Métiers, Paris.

12a. Cylindrical Reckoning Rule. (The property of the Conservatoire des Arts et Métiers.)

M. Mannheim, Professor at the Polytechnic School, Paris.

13. Calculating Rules, 1 of 50 cm., 1 of 36 cm., 1 of 26 cm., as arranged by M. Mannheim.

M. Tavernier Gravet, Paris.

13a. Small Cylindrical Calculating Machine. Arranged by M. Mannheim. *Conservatoire des Arts et Métiers, Paris.*

14. Pocket Calculator, arranged by Major-General A. De Lisle, R.E., for the use of engineers. *Elliott Brothers.*

This slide rule is useful for finding the weight of various materials, with the help of the small tables on the back, for checking bills of quantities, and for all approximate calculations required in engineering practice. The slides are:—

On Face.

On Stock—A. The ordinary logarithmic line.
      I. The same inverted.
On Slides—Upper I. Inverted line.
     „ D. Line of squares.
    Lower B. Ordinary logarithmic lines.
      Tan ⎫
      Sin ⎭ Trigonometric lines.

Special Marks—
  M. Modulus of logs. to find prop. parts of logs.
  A. Reciprocal of M. to find hyperbolic logs.
  S″ To find length of arcs, &c.
  R′ Radius for minutes.
  R″ Radius for seconds.

On Back.

On Stock—D. Line of squares.
On Slide—E. Line of cubes read with D on line of $\frac{3}{2}$ powers read with A.
    F. Line of $\frac{5}{2}$ powers read with D, or of $\frac{4}{}$ powers read with A.
     Tables and useful Numbers.

Line E with A gives variation in depth of water running over weir due to alteration of length of weir. Neville's Hydraulics, page 22, 3rd edition.

A	143 = d	
E	220 = l	60 = l′
A		6 = d

Line F with D assists in finding the dimensions of a pipe or channel, with a given hydraulic inclination to discharge a given quantity from the calculated discharge of a pipe or channel of known dimensions and the same inclination. Thus, if a pipe 4″ diameter discharge 15 cubic feet per minute, what diameter will discharge 33 cubic feet? Neville's Hydraulics, page 245.

D		
F	15 = D	33 = D′
D	4 = d	5.48 = d′

The two slides on the face working together solve the following equations :—

$$x = \frac{ab}{cde} \; ; \; x = \frac{abc}{de} \; ; \; x = \frac{abcd}{e}$$

15. Slide Rule, of boxwood, with double slide.

              Renaud-Tachet, Paris.

16. Routledge's Original Engineers' Slide Rule and manuscript book of instructions for using it.

    H. M. Commissioners of Patents.

17. Kentish's Compound Slide Rule.

Dring and Fage.

This is a new and ingenious arrangement of Gunter's lines, by means of which problems in trigonometry and navigation can be solved, in addition to those ordinarily done on the slide rule.

17a. Dr. Roget's Slide Rule of Involution.

W. H. Prosser.

This rule exhibits at one view all the powers and roots of any given number. It is a measure of the powers of numbers, in the same way as Gunter's scale is a measure of their ratios. Described in Phil. Trans. 1815, Part 1.

17b. Slide Rules (3), with double sliders, being suggested improvements on the ordinary slide rule, giving greater clearness in reading off, and avoiding complication in the lines.

W. H. Prosser.

17c. Glass Slide Rule, invented by Léon Lalanne.

W. H. Prosser.

This rule is made of two slips of card, upon which the scales are printed. The slider, also made of card, has scales, constants, and gauge points printed on both sides, and moves between the two slips. The whole is enclosed between two pieces of glass.

17d. Slide Rule, with only one slider, adapted for the pocket-book. Arranged by J. W. Woollgar. *W. H. Prosser.*

18. Salleron's Slide Rule for reduction of volumes of gases to standard temperature and pressure.

The Council of the Yorkshire College of Science.

19. Salleron's Slide Rule for reducing barometric heights to standard temperature.

The Council of the Yorkshire College of Science.

II.—CALCULATING MACHINES.

20. Calculating Machine, adapted to trigonometrical computations, invented by Sir Samuel Morland (1625–1695), and constructed by Henry Sutton and Samuel Knibbs of London, in 1664. Formerly belonging to Mr. C. Babbage, F.R.S.

Major-General Babbage.

On the lid of this machine is the following inscription :
" Machina Cyclologica Trigonometrice Quâ Tribus datis, reliqua omnia in Triangulis Planis Quæsita faciliter atque unico intuitu expediuntur—a Samuele Morlando inventa—Anno Salutis MDCLXIII.

21. Calculating Machine, designed by Viscount Mahon, afterwards third Earl Stanhope (1753–1816), and constructed by James Bullock in 1775. Formerly belonging to Mr. C. Babbage, F.R.S.   *Major-General Babbage.*

22. Calculating Machine, designed by Viscount Mahon, afterwards third Earl Stanhope (1753–1816), and constructed by James Bullock in 1777. Formerly belonging to Mr. C. Babbage, F.R.S.   *Major-General Babbage.*

23. Babbage's Calculating Machine; or Difference Engine.   *H.M. Board of Works.*

This machine was invented by the late Mr. Charles Babbage, F.R.S., who was born on the 26th December 1791, and died on the 18th October 1871.

Its construction was commenced in 1823 by authority, and at the cost of the Government, and was carried on for several years under Mr. Babbage's gratuitous supervision. The work was suspended in 1833, and after many delays, Mr. Babbage was informed in November 1842 that the Government regretted the necessity of abandoning the machine, alleging the expense of its completion as the ground for their decision.

At the time of its suspension about 17,000*l.* had been expended by Government upon its construction, and a large part of the machinery had been made. The small portion now exhibited was put together in 1833, prior to the suspension of the work, in order to show the action of the machinery.

The whole engine, when completed, was intended to have had 20 places of figures and 6 orders of differences.

This machine was expressly designed for the purpose of calculating and printing tables, and not to perform single arithmetical sums.

If a single article is wanted, it is not, generally speaking, worth while to construct a machine to make it ; but, when large numbers are required, their production comes within the true province of machinery, and in this sense the Difference Engine is emphatically a machine for manufacturing tables.

The mode in which the Difference Engine calculates tables is, by the continual repetition of the simultaneous addition of several columns of figures to other columns, in the manner more particularly described below, and printing the result.

In the small portion put together, and now exhibited, the figure opposite the index on the lowest wheel visible, in all cases, represents units ; the figure on the next wheel above, tens; that on the one above it, hundreds; the next thousands, and so on.

The right hand column of wheels shows the result of the calculation or the tabular number; for instance, series of squares, cubes, or logarithms, &c. appear upon it, according to the nature of the calculation the machine is making.

The next or central column represents the First Difference, and the left hand column the Second Difference. At the bottom of the central column is a figure wheel, covered, which can be used as a third difference, so as to enable this portion of the machine to calculate tables of which the Third Difference does not exceed 9. This will be better understood if this last wheel is supposed to represent the lowest wheel of a fourth column of figures standing beyond the left hand side of the machine, as it would be if it formed part of the complete machine.

This arrangement is effected by a movable platform, with axles, and gearing wheels upon them, which are used for adding from the third difference wheel at the bottom of the central column to the second difference which is shown on the left hand column. The effects capable of being produced by this mechanism, when the gearing is altered, and the loose wheels belonging to it are put into gear with certain figure wheels, is explained in Babbage's Ninth Bridgewater Treatise, together with the new views which it opened up to him upon the subject of natural laws.

The three upper wheels of the left hand column are separated from the rest of the machine, and are employed in counting the natural numbers. In other words, they register the number of calculations made by the machine, and give the natural numbers corresponding with the respective terms of the table.

Four half turns of the handle, two backwards and two forwards, are required for each calculation, and the words "calculation complete" come round upon a wheel at the top of the central column to show when this is done. This wheel also shows, by the word "adjust," in what position of the handle the figure wheels may be freely moved by hand, in order to introduce different numbers or a different table.

24b. Cabinet, containing tablets for making mathematical calculations. *Archæological Museum, Madrid.*

Rose wood cabinet, inlaid with ivory. In three divisions are thirty small drawers, containing ivory plates with numbers and divisions for making mathematical calculations. In the inside are the arms of the monastery of the Escorial. Milanese work of the 16th century.

25. Panometer, or Calculating Machine.
Edward Grohmann, Vienna.

By this extremely simple apparatus, various arithmetical computations can be performed with great readiness.

25a. Calculating Machine for Multiplication.
P. Nicholas Dadiane, St. Petersburg.

26. Calculating Machine, for performing complex arithmetical operations ; invented by M. Thomas of Colmar.
Professor Hennessy, F.R.S.

26a. Calculating Machine for Adding, Subtracting, Multiplying, and Dividing.
Theodore Esersky, St. Petersburg.

26b. Small Calculating Machine, encased in a pocketbook. *Theodore Esersky, St. Petersburg.*

26c. Ten Copies of Multiplication and Division Tables.
Theodore Esersky, St. Petersburg.

27. Wertheimber's Calculating Machine, applicable to wheel work. Patent, No. 9616—1843.
The Committee, Royal Museum, Peel Park, Salford.

28b. Original Calculating Machine, known as "Napier's Bones." *Lord Napier and Ettrick.*

One of the earliest attempts to construct a calculating machine, made by John Napier, the inventor of logarithms. His method of calculating by rods was published in a volume (which is exhibited with the machine) at Edinburgh in 1617. The apparatus was commonly known as "Napier's Bones."

28. " Napier's Bones " or Rods. Made about 1700.
Dring and Fage.
Invented by Baron Napier, the originator of logarithms, used for performing division and multiplication.

28a. Set of " Napier's Bones." 16th century.
Lewis Evans.

29. Calculating Disc, size about 18 centimeters, with double-divided circle ; constructed on the system of Prof. Sonne.
Landsberg and Wolpers, Hanover.

30. Calculating Disc, with index of logarithms.
Landsberg and Wolpers, Hanover.

31. Calculating Disc, pocket apparatus.
Landsberg and Wolpers, Hanover.

32. Calculating Circle, 0·08 meter in diameter, with single scale of brass. *Rudolf Weber, Aschaffenburg.*

33. Calculating Circle, 0·15 meter in diameter, with single square and cubic scale. *Rudolf Weber, Aschaffenburg.*

The circles are on account of their continuous scale more convenient and more accurate than straight slide rules. They are, therefore, peculiarly adapted as pocket instruments for practical purposes, and can be relied on to be as accurate as logarithms to four places.

34. Cubing Circle, 0·08 meter in diameter, for ascertaining the cubical contents of trees in forests.
Rudolf Weber, Aschaffenburg.

The cubing circle is to be noted as giving the index numbers for obtaining the cubical contents of *standing* (not felled) timber ; these have been obtained from practical experiments carried out by the Government Department of Forests in Bavaria on more than 40,000 trunks of different kinds of trees. The circle may be relied on for great accuracy in forest valuation.

35. Calculating Instrument, invented by Sir S. Morland.
Bennet Woodcroft, F.R.S.

35a. Calculating Planisphere.
Royal College of Science for Ireland, Dublin.

35b. McFarlane's Calculating Planisphere.
The Committee, Royal Museum, Peel Park, Salford.

36. Calculating Machine, designed by the Vicar Hahn of Echterdingen, in 1770-1776 ; constructed by his son, Court Mechanician in Stuttgart, in 1809 ; fourth specimen.
Her Highness the Duchess of Urach.

The machine which is exhibited is on exactly the same principle as that of the one now in general use which was invented by Thomas, the only difference being that in Thomas's machine the numbers are placed in straight lines, and in that of Hahn in a circle. It must have served as a model for the machine of Thomas. The machine is to the present day in perfect order, and works calculations up to numbers of 12 digits.

37. Logarithmic Calculation Apparatus, with *one* folding scale, equivalent to five meters in length.

Prof. Gustav Hermann, Aix-la-Chapelle.

Calculation by means of this instrument is effected with the use of only one scale. The two revolving arms are used like a compass. When the logarithm of a quotient $\frac{a}{b}$ has been fixed between the arms, the plate must be turned until the one arm is brought to a factor c, the product $\frac{a}{b}c$ will then be read on the other arm. This arrangement admits of the scale being made as long as may be necessary by breaking it into lengths, without rendering the instrument inconvenient. In the exhibited instrument ten circles are used, by which means the scale attains a length of five meters, and is accurate up to $\frac{1}{10000}$. In using the instrument care must be taken to mark the number of each scale circle, which can be fixed by small sliding buttons. The number of the circle on which the result is to be read is found by the same rules as the characteristic of a logarithm, on the supposition that the ten scale-circles form a graphic table of logarithms of all the natural numbers, the base of the system being $\sqrt[10]{10}$.

38. Arithmetical Disc, a very simple calculating machine, with accompanying description. *Prof. Prestel, Emden.*

39. Calculating Machine, of the last century.

The Royal Gewerbe Academy, Berlin (Director, Prof. Reuleaux).

This calculating machine formed part of the legacy of Hofrath Beireis, the well-known physicist and chemist in the 18th century, and is very similar to the calculating machine, No. **36.** The following description which accompanies this calculating machine is especially interesting, as it was probably drawn up and written by the maker.

On the Use of the Calculating Machine.

There are on the small number-discs [Zahlen-Scheiblein] in the smallest [circles], as well as on the large, in the great circles, the numbers 1, 10, 100, 1,000, and so on ; these indicate that the numerals on the first disc, where the (1) is marked, are units, on the second disc, where the numeral (10) is marked, the tenfold of the numbers, and on the third, the hundredfold, and so on ; and this applies also to those arcs on the largest circles. And if on its first six discs respectively the following numerals were placed, 357,862, they would indicate multiples of the numbers 1, 10, 100, which are on the respective discs. On the sixth disc, where the numeral (3) is placed, is to be found the number 100,000 ; thus indicating three times a hundred thousand. On the fifth disc the numeral 5 represents 50,000, similarly the numeral 7, 7,000, and so on ; this would, in the first place, be "*ciphering.*"

Now follows, in the second place, *addition.* In calculating use is made of the same numbers that are placed in the openings on the number disc, the black numbers on the large discs are used for addition and multiplication, and at the commencement noughts must be placed in all the apertures. As an example, the following numbers will now be added together.

$$\begin{array}{r} 352 \\ 643 \\ 574 \\ \hline 1,569 \\ \hline \end{array}$$

These numbers must be placed on those arcs which are found consecutively in the outer circle. The handle must then be turned, when the numerals 352 will have come in the place of the noughts, on the three first large discs under the opening. Then the second and third row are placed on the discs and by turning the handle added to the preceding.

In the third place comes subtraction, and it must first of all be understood that the red numbers are to be used on the larger discs; for the rest the method is very easy.

The following is an example:—

        The numbers 5,786, as the larger numbers are placed in red figures

5,786   in the openings of the large discs; (6) in the units, (8) in the tens,

3,524   and so on. The smaller number, which is to be subtracted, must

——   be placed on the arcs, in the same order as was shown for addition.

2,262   And in the same manner likewise must the handle be turned round

══   *once*, and then all is done, and the remainder appears in the

        openings in which the larger amount was previously to be seen.

For multiplication (as for addition) the black numbers must be used on the large discs, and before anything else, plain noughts must be placed in the openings. And because in this method of reckoning small discs are also

        used, noughts must at first be placed on them. Let the numbers

365   365, as in the case of addition, be placed on the arcs. One of the

24   hands, working from the centre, points to the units on the small

——   discs, and shows that multiplication will be effected with the units

1,460   as long as the hand is unaltered. Now if the handle be turned

7,300   round once, the numbers 365 will appear on the larger discs, instead

——   of the noughts which were there previously. But on the smaller

8,760   discs where the hand points, the number (1), and this proves,—that

══   the numbers 365 have been carried once under the aperture of the

        large disc. The handle is again turned, and there will be seen on

the large disc the numbers 730 and on the smaller one the number (2). But it must in the units' place be multiplied by (4), and so the handle has to be turned four times round, and there will then appear on the large disc the number 1,460, and on the smaller the number (4), and therefore, in this problem, enough has been done with the units. Since 365 had to be multiplied by 24, the operation is repeated twice in the tenfold numbers, and the desired result obtained; this is effected in the following way :—

The hand acting from the centre must be placed on the second disc, which corresponds to tens. Near to the spot where the handle has its resting-point, there is at the circumference of the machine a steel catch pressed down in a notch; this must be lifted up, and so turned round at the circumference of the machine that the hand working from the centre comes to point on the *tens* disc, and as in this case the catch will be pressed into the newly found notch, the handle being turned round twice, the product 8,760 will be found in the openings of the large discs.

For division the red figures must be used, as in the case of subtraction. As an example take the number 8,760, which was obtained by the first multiplication, and divide it by 365. The number 8,760 must be placed on the large discs, in red figures, under the openings. The divisor 365 must be placed on the arcs; the first digit of the divisor (3) must be turned round under the first digit of the dividend, viz., under the number (8); on the small discs noughts must everywhere be placed. The catch must be left in the notch. Now 3 can be taken from 8, the handle is turned once round, and the numbers 5,110 appear, which is the remainder after 3,650 has been taken from 8,760. 3 can again be taken from 5; the handle is again turned round once more, and the remainder 1,460 appears. Since 3 cannot be taken from 1, the catch is lifted, and 3, the first digit of the divisor, brought under the 4 in the dividend. The handle is then turned 4 times, and the remainders, 1,095, 730, 365, and 0, appear in succession, being those due to the successive subtractions of 365 from 1,460. The value of the quotient is seen to be as follows, 3,650 or 365 × 10 is subtracted twice, giving 20 as the first part of the quotient, then 365 is subtracted 4 times, giving 4 as the second part, with no remainder; thus the quotient sought is 24.

40. Tide Calculating Machine.

Sir William Thomson, F.R.S.

The object is to predict the tides for any port for which the tidal constituents have been found by the harmonic analysis from tide gauge observations: not merely to predict the times and heights of high water, but the depth of water at any and every instant, showing it by a continuous curve, for a year, or for any number of years in advance.

This object requires the summation of the simple harmonic functions representing the several tidal constituents to be taken into account, which is performed by the machine in the following manner:—For each tidal constituent to be taken into account the machine has a shaft with an overhanging crank, which carries a pulley pivoted on a parallel axis adjustable to a greater or less distance from the shaft's axis, according to the greater or less range of the particular tidal constituent for the different ports for which the machine is to be used. The several shafts, with their axes all parallel, are geared together so that their periods are to a sufficient degree of approximation proportional to the periods of the tidal constituents. The crank on each shaft can be turned round on the shaft and clamped in any position: thus it is set to the proper position for the epoch of the particular tide which it is to produce. The axes of the several shafts are horizontal, and their vertical planes are at successive distances one from another, each equal to the diameter of one of the pulleys (the diameters of these being equal). The shafts are in two rows, an upper and a lower, and the grooves of the pulleys are all in one plane perpendicular to their axes. Suppose, now, the axes of the pulleys to be set each at zero distance from the axis of its shaft, and let a fine wire or chain, with one end hanging down and carrying a weight, pass alternately over and under the pulleys in order, and vertically upwards or downwards (according as the number of pulleys is even or odd) from the last pulley to a fixed point. The weight is to be properly guided for vertical motion by a geometrical slide. Turn the machine now, and the wire will remain undisturbed, with all its free parts vertical and the hanging weight unmoved. But now set the axis of any one of the pulleys to a distance $\frac{1}{2}$ T from its shaft's axis, and turn the machine. If the distance of this pulley from the two on each side of it in the other row is a considerable multiple of $\frac{1}{2}$ T, the hanging weight will now (if the machine is turned uniformly) move up and down with a simple harmonic motion of amplitude (or semi-range) equal to T in the period of its shaft. If, next, a second pulley is displaced to a distance T′, a third to a distance T″, and so on, the hanging weight will now perform a complex harmonic motion equal to the sum of the several harmonic motions, *each* in its proper period, which would be produced separately by the displacements T, T′, T″. Thus, if the machine was made on a large scale, with T, T′... equal respectively to the actual semi-ranges of the several constituent tides, and if it is turned round slowly (by clockwork, for example), so that each shaft goes once round in the actual period of the tide which it represents, the hanging weight would rise and fall exactly with the water level as affected by the whole tidal action. This, of course, could be of no use, and is only suggested by way of illustration. The actual machine is made of such magnitude that it can be set to give a motion to the hanging weight equal to the actual motion of the water level reduced to any convenient scale: and provided the whole range does not exceed about 30 centimètres, the geometrical error due to the deviation from perfect parallelism in the successive free parts of the wire is not so great as to be practically objectionable. The proper order for the shafts is the order of magnitude of the constituent tides which they produce, the greatest next the hanging weight, and the least next the fixed end of the wire: this so that the greatest constituent may have only one pulley to move, the second in magnitude only two pulleys, and so on. In the actual machine there are 10 shafts, which, taken in order from the hanging weight, give respectively the following tidal constituents:

1. The mean lunar semi-diurnal.
2. The mean solar semi-diurnal.
3. The larger elliptic semi-diurnal.
4. The luni-solar diurnal declinational.
5. The lunar diurnal declinational.
6. The luni-solar semi-diurnal declinational.
7. The smaller elliptic semi-diurnal.
8. The solar diurnal declinational.
9. The lunar quarter-diurnal, or first shallow-water overtide of mean lunar semi-diurnal.
10. The luni-solar quarter diurnal, shallow-water tide.

The hanging weight consists of an ink bottle with a glass tubular pen which marks the tide level in a continuous curve on a long band of paper moved horizontally across the line of motion of the pen, by a vertical cylinder geared to the revolving shafts of the machine. One of the five sliding points of the geometrical slide is the point of the pen sliding on the paper stretched on the cylinder, and the couple formed by the normal pressure on this point, and on another of the five, which is about 4 centimètres above its level and $1\frac{1}{2}$ centimètres from the paper, balances the couple due to gravity of the ink bottle and the vertical component of the pull of the bearing wire, which is in a line about a millimètre or two farther from the paper than that in which the centre of gravity moves. Thus is ensured, notwithstanding small inequalities of the paper, a pressure of the pen on the paper very approximately constant, and as small as is desired.

Hour marks are made on the curve by a small horizontal movement of the ink bottle's lateral guides, made once an hour; a somewhat greater movement, giving a deeper notch, to mark the noon of every day.

The machine may be turned so rapidly as to run off a year's tides for any port in about four hours.

It is intended that each crank shall carry an adjustable counterpoise, to be adjusted so that when the crank is not vertical the pulls of the approximately vertical portions of wire acting on it through the pulley which it carries shall, as exactly as may be, balance on the axis of the shaft, and that the motion of the shaft shall be resisted by a slight weight hanging on a thread wrapped once round it and attached at its other end to a fixed point. This part of the design, planned to secure against " lost time " or " back-lash " in the gearings of the shafts, and to preserve uniformity of pressures between teeth and teeth, teeth and screws, and ends of axles and " end-plates," was not carried out, but can easily be applied to the machine now exhibited.

The general plan of the screw gearing for the motions of the different shafts is due to Mr. Légé, the maker of the machine. The construction has been superintended throughout by Mr. Roberts, and to him is due the whole arithmetical design of the gearing to give with sufficient approximation the proper periods to the several shafts.

40a. Sir William Thomson's Instrument for Harmonic Analysis of Tidal Observations, and for other uses. It includes a disc-ball and cylinder integrator of Professor James Thomson. For explanations, reference may be made to proceedings of the Royal Society for February 3, 1876.

Sir William Thomson.

41. Pascal's Calculating Machine (1642).

Conservatoire des Arts et Métiers, Paris.

42. Petroff's Arithmetical Apparatus.

M. Petroff, Kalouga.

43. Arithmometer, with measuring apparatus, and the full size skeleton of the square metre and cubic metre, folding up by means of a hinge. *Frère Memoire Piron.*

43a. Counting Machine.

P. N. Dadiane, St. Petersburg.

43b. Calculating Machine.

P. N. Dadiane, St. Petersburg.

45. Model of Gas Meter Counting Machine.

Council of King's College, London.

46. Cavendish's Original Counting Machine.

Council of King's College, London.

III.—MISCELLANEOUS.

48. Apparatus for the Statistical Treatment of large numbers of Seeds, &c., to sort them rapidly into classes differing by regular gradations of magnitude, with the view of testing how far the relative numbers in the several classes accord with the results of the Law of Error or Dispersion.

Francis Galton, F.R.S.

It consists of a square box, having parallel bars fixed across its top at equal distances apart. An equal number of similarly arranged bars are connected by means of rods running along their ends, like the bars of a gridiron, thus forming a framework that is laid on the top of the box. Hence there are two systems of parallel bars in the same plane, one of which is fixed and the other movable. When the frame is pulled forwards as far as it can go, each of its bars presses along its whole length against one of the fixed bars, and when it is pushed gently back the framework bars separate simultaneously and equally from the fixed bars, and any objects that may have been laid between their edges, and are small enough, will drop through. The bars are bevelled along their opposite faces, in order to receive these objects. The framework is moved by a screw turned by a ratchet wheel, which is itself moved by the to-and-fro action of a handle between stops, one of which is adjustable at pleasure. Hence, every time the handle is worked, the space between the bars is widened by a definite space, and all the seeds, &c., whose diameter is greater than the original and less than the final space, will drop through. A tray, divided into compartments, slides beneath the box ; it is pushed forward through the space of one compartment before giving a fresh movement to the handle, and thus the seeds become sorted into the different compartments. (This instrument was used to illustrate a lecture before the Royal Institution on Friday evening, February 27, 1874.)

49. Apparatus affording Physical Illustration of the action of the Law of Error or of Dispersion.

Francis Galton, F.R.S.

Shot are caused to run through a narrow opening among pins fixed in the face of an inclined plane, like teeth in a harrow, so that each time a shot passes between any two pins it is compelled to roll against another pin in the row immediately below, to one side or other of which it must pass, and, as the arrangement is strictly symmetrical, there is an equal chance of either event. The effect of subjecting each shot to this succession of alternative courses is, to disperse the stream of shot during its downward course under conditions identical with those supposed by the hypothesis on which the binomial law of error is founded. Consequently, when the shot have reached the bottom of the tray, where long narrow compartments are arranged to receive them, the general outline of the mass of shot there collected is always found to assimilate to the well-known bell-shaped curve, by which the law of error or of dispersion is mathematically expressed. (This arrangement was devised, by the exhibitor, to illustrate a lecture before the Royal Institution on Friday evening, February 27, 1874.) When using the machine, tilt it backwards and all the shot will be returned to the receptacle at the top ; then set it in its proper position, and the shot will run to the opening whence they distribute themselves. It is now necessary to press to and fro a button at the top of the frame, which sets a small rake in action, which prevents the shot from getting jammed at the mouth of the opening.

50. Practical Approximation to the value of the circumference in terms of the diameter, by means of a right angled triangle having one acute angle $=27° 35' 49.636''$.

Edward Bing, Riga.

For the purpose of effecting this object, as well as for answering kindred questions, use is made of a triangle, specimens of which are here exhibited, and of which one angle is a right angle and another is defined by an equation.

SECTION 2.—GEOMETRY.

West Gallery, Ground Floor, Room G.

I.—INSTRUMENTS USED IN GEOMETRICAL DRAWING.

52. Pantograph, by Breithaupt and Son.
Royal High School of Industry, Cassel. (*W. Narten.*)

This pantograph was made for the geodetic collection of the school, in the year 1866, by Messrs. Breithaupt and Son, Cassel. It is used for enlarging and reducing maps and plans. The peculiarity of its construction is the movement of all the arms between pairs of points, by which means friction is as far as possible avoided. The employment of tubes instead of the usual rectangular bars is also to be recommended, by which means bending, which creates errors in the use of the instrument, is avoided; besides which, the weight of the whole is considerably decreased, thus also lessening friction in the movement of the points.

The peculiar construction of this pantograph was invented and carried out by Messrs. Breithaupt and Son, and the instrument possesses great accuracy and facility of use.

53. Pantograph. *Renaud Tachet, Paris.*

54. Pantograph, with free hanging arms of new construction.
Ott and Coradi Kempten, Baviera.

By means of this instrument figures on a reduced or an enlarged scale can be transferred either to paper, stone, or metal.

These pantographs differ in their construction from other similar instruments by not resting on friction-rollers, but are freely suspended by means of metal wires from cast-iron curved standards; thus only a small portion of the weight of the instrument rests on the table. The advantages of this construction are these : easy and secure management of the instrument; any ordinary table may be used of a size sufficient to afford room for the stands, the original, and copy; the accuracy of the graphic representation is greater at less cost.

The guidance of the instrument is so easy and so accurate that with a little practice every outline can be reproduced. Drawings, likewise, can be transferred to substances measuring a certain height, such as lithographic stones, it being only necessary to place frame and original correspondingly higher. In producing enlarged copies the guiding peg and the drawing pencil must be exchanged, and the releasing cord fastened accordingly. The guidance, in making enlarged copies, is also performed with the handle of the tracing pencil with the same accuracy as when making reduced copies.

54a. Horizontal Pantograph, traversing a surface of 36 inches in length by 20 inches in width. Reduction from $\frac{1}{2}$ to $\frac{1}{12}$.
L. Oertling.

54b. Pantograph, large model, with double scale and reverse action, belonging to the Indian service. Four of these large instruments are now in use. *M. Adrian Gavard, Paris.*

54c. Frame, containing the **Drawings** of **Pantographs** and **Pantopolygraphs** made by the exhibitor.

M. Adrian Gavard, Paris.

54d. Pantograph by Adrian Gavard, Paris.

S. J. Hawkins.

Size 56 centimeters.

A method of raising and lowering the tracer A by a rack and pinion movement, for use when the reduction of a drawing has to be made on tracing paper, enables the tracer to pass over any irregularity of surface, and thus prevents injury to the paper.

There is also a method of ascertaining when the tracer A, the pencil or central point B, and the fulcrum C, are in a straight line, for use with maps or drawings of irregular scales, and for which there are not any corresponding divisions on the Pantograph. The points A, *b*, C, D, *e* are removed when the instrument is adjusted, and the two small screws *f, f* screwed into the head of the tracer and fulcrum.

The instrument is adjusted for the reduction of Ordnance maps 25·34 inches to the mile to 5 chains to the inch or 16 inches to the mile.

54e. Pantograph made at Madrid by Rostriaga, and which has been used at the mines of Almadin since the end of the last century. *Royal School of Mines, Madrid.*

Compass of Proportion, called also military compass, **invented by Galileo** in 1596.

Royal Institute of " Studii Superiori," Florence.

On one side are engraved the arithmetical lines, the geometrical lines, the stereometrical lines, and the metallic lines ; on the opposite side are the polygrafical lines. the tetragonical lines, and the joined lines for the squaring of figures comprised by right angles and curves together. By means of it 40 important operations can be carried on, and it has in addition a quadrant with a squadron of bombardiers, and transversal lines to take the inclination of the scarp of any wall.

Galileo presented compasses of a similar kind, in the year 1598, to the Prince of Holstein, Sagredo, and Bentivoglio, who was afterwards made Cardinal, to Ottone Brahe, to the Count of Luxemburg, and many other gentlemen, both Italians and foreigners, who had gone to Padua to follow the lectures of so eminent a philosopher. He also presented one made of silver to the Archduke Ferdinand of Austria, and to the Landgrave of Hesse.

55. Large Collection of Mathematical Instruments for **Geometrical** and **Fortification Drawing,** as well as for **Artillery purposes.** The property of His Highness the Prince Pless, Fürstenstein. *The Breslau Committee.*

This ancient collection, dating from the commencement of the last century, is remarkable for the excellent workmanship and good state of preservation of the instruments.

It contains 19 compasses and 11 accessory parts, 28 rules and scales, two of the same with two keys for fortification drawing, eight triangles and set squares, 10 protractors, two pantographs, and 52 other instruments. In all, 134 pieces.

56. Case of Mathematical Instruments.
Renaud Tachet, Paris.

57. Proportional Compasses. *Renaud Tachet, Paris.*

58. T-squares, Set Squares, and Curves.
Renaud Tachet, Paris.

59. Diagonal Scale.
Geneva Association for the Construction of Scientific Instruments.

60. Scales made of **Mica,** for use in geometrical drawing.
Max. Raphael, Breslau.

275. Meter-measures, constructed of mica.
Max. Raphael, Breslau.

These measures have the advantage of being transparent, and may serve for copying geometrical drawings. Owing to their remaining unaffected by the ordinary changes of temperature they may also be used as *standard* measures.

61. Perspective Apparatus invented by James Watt.
Bennet Woodcroft, F.R.S.

62. Set of Mathematical Instruments, with all the modern improvements ; as used by professional draughtsmen, &c. ; illustrated by diagrams of work performed. *Wm. Ford Stanley.*

62a. Magazine Case of Drawing Instruments.
Henry Porter.

64. Case of Mathematical Instruments, probably Dutch, made at the beginning of the 18th century. *Lewis Evans.*

65. Two Magazine Cases of Drawing Instruments.
Mark Eames.

65a. Large Magazine Case of Instruments.
G. W. Strawson.

65b. Magazine Cases of Instruments (2).
G. W. Strawson.

65c. Morocco Case of Instruments. *G. W. Strawson.*

65d. Napier's Compasses (Electrum). *G. W. Strawson.*

66. Beam Compass, T-squares, Set Squares, and Curves. *Bock and Handrick, Dresden.*

67. Models of Mathematical Instruments. The ortho-compass and the addition compass. *Prof. L. Zmurko, Lemberg.*

The first of these instruments is constructed so that the points of the compass are always parallel to each other, and perpendicular to the surface of the paper. The second is a compass which can be used also as a protractor, as it contains an apparatus which indicates the amount of opening between the arms.

69a. Photographs of Mathematical Instruments.
Otto Fennel, Cassel.

69d. Révoil Tele-iconograph, altered for perspective draw-ings enlarged to 20 times on a horizontal plane-table.
M. Georges Sarasin, Geneva.

The instrument consists of a telescope, adapted to a Wollaston *camera lucida*, and fixed on a stand arranged so as to make it a mathematical or scientific instrument; while on a separate stand is placed a plane table for drawing. In order to facilitate the exact grouping of the partial perspectives in accordance with a general cylindrical perspective, and capable of being developed, and in order to permit of drawing while the telescope is inclined at great angles, the following additions have been made to the Révoil model:—1st. A tightening ring with an adjusting screw, which fixes the prism to any point on the thread of the screw by which it is fixed to the eye piece. 2d. A web of six threads crossed at right angles in the focus of the object glass for the purpose of setting the partial images in a straight line and in a direction in accordance with the horizontal or vertical. 3d. A spirit level on the telescope stand to ensure the verticality of the axis of rotation. 4th. A socket and rack joint, permitting the height of the prism above the drawing to be determined whatever be the angle of the telescope and, consequently, the scale of the drawing. 5th. A graduated scale with vernier, giving a reading to five minutes on the horizontal limb. 6th. A method of attaching the instrument to its stand, so as to be at the same time firm and easy to work.

69e. Patent Dotting Pen.
E. O. Richter and Co., Chemnitz, Saxony.

The arrangement is as follows:—
A small toothed wheel adapted to the kind of dotted lines to be drawn is attached to a plate, which, rolling on the paper, lifts a lever, which is again dropped by means of a spring. Attached to this arm is a drawing-pen, easily adjustable, by means of a hinge, to the correct position suited to the wheel. The wheel itself is held in position by a small, somewhat elastic plate, which can be displaced a little for fitting the different reserve wheels on which the kind of lines to be drawn depends.

69f. Patent Compasses with stationary centre point.
E. O. Richter and Co., Chemnitz, Saxony.

These compasses differ from others, chiefly by the centre-point being stationary on the paper, so that the tracing-pen, resting on the paper by its own gravity, moves about the centre-point as axis, whilst the movable pen can be displaced, without removing the compasses from the paper. By these advantages speedy and neat tracing will be achieved, even when the smallest circles are to be drawn.

69g. Patent Compasses, with drawing pen and leaden tube.
E. O. Richter and Co., Chemnitz, Saxony.

69h. Patent Diamond Compasses, for lithography.
E. O. Richter and Co., Chemnitz, Saxony.

69i. Patent Diamond Compasses with drawing pen, for lithography. *E. O. Richter and Co., Chemnitz, Saxony.*

69j. Patent Hatching Ruler (for shading by cross-lines in drawing and engraving).
E. O. Richter and Co., Chemnitz, Saxony.

A ruler is attached to a cylinder which, rolling on a plate, draws the former after it. By means of an endless screw, working in a wheel, the cylinder is moved forward. A cogwheel is attached to the endless screw, in which a bar catches, and by the depression of which the mechanism is put in motion. The depression can be regulated by a screw, by which means the various distances of the lines are obtained. One finger is sufficient for working the ruler by simply pressing on the screw, and holding it until the line is drawn.

69k. Six Setter Diamonds, for lithography.

69l. Five Machine Diamonds, for lithography.

69m. Six Scratching Diamonds, for lithography.
E. O. Richter and Co., Chemnitz, Saxony.

69n. Perspectograph. *Lieut.-General M. W. Smith.*

This instrument is employed to determine the perspective position of a point on the surface of a picture corresponding to any point in nature, the actual position of which is ascertained by means of ground plans, elevations, sections, actual measurement, or otherwise, or else assumed in the composition of a picture. A horizontal line is drawn across the picture and a point assumed upon it to represent the point or centre of view, the perspective point (the two ordinates, of which one parallel and the other perpendicular to the horizontal line are determined by the instrument) is then laid down from the point of view by scale and offset. The manipulation of the instrument is very simple and easily acquired; and as the perspective representation of any number of points, constituting lines, planes, solids, &c., can be rapidly transferred to the picture, the most complicated problems of perspective, whether rectangular or oblique, can be performed without lines of geometrical construction, or the transference of the subject from a plan previously prepared to the picture. A detailed description of the working of these instruments, as well as of the system upon which their construction depends, will be found in " Engineering," 1876.

69o. Pointfinder. *Lieut.-General M. W. Smith.*

This instrument is used in sketching from nature and is employed to determine a point on the surface of the picture corresponding to any point in nature to which the sights of the instrument may be directed, as follows :—
Having drawn a line horizontally across the paper to represent the horizontal line, and assumed a point upon it as the point of view, the instrument is set to zero both on the horizontal, and side vertical graduated arcs; and levelled by means of the small plumb bob, the sights being at the same time upon any

point in nature which may be chosen as the point of view. The sights of the instrument are then directed to any point in nature which it may be desirable to introduce into the picture. This may be done by the lateral and vertical movements of the horizontal disc and plate to which the sights are attached; the divisions and subdivisions indicated by the zero mark on the horizontal disc, and index on the side arc, are then read off and transferred to the picture by means of the graduated scale and offset, the zero of the scale being placed to the point of view and the bevelled edge to the horizontal line, the offset is then moved along the edge of the scale till the bevelled edge corresponds with the division or subdivision read off from the circular disc, when a slight dot with the point of a pencil at the division or subdivision on the offset corresponding to the reading on the side vertical arc will indicate the point which is the perspective representation of the point in nature. The scale can be increased, if desired, by multiplying the number of divisions or subdivisions read off on the graduated arcs by 2, 3, 4, or any other common multiplier. The offset also may be dispensed with by placing the edge of the scale first to the horizontal line, for the first reading making a slight mark on the line with a pencil at the number indicated, and then perpendicular to it for the second reading, this may be the most convenient mode of laying down the points when the sketching book is held in the hand. The stand of the instrument folds up, and the apparatus when packed is very light and portable, consisting of camp stool, drawing book, instrument and stand.

443a. Plate Glass Sector, designed for the purpose of plotting angles on plans or charts where it is necessary to see the work under the sector, and the divisions being on the side next the paper no variation in pricking off can take place.

Thos. F. Chappé, M. Inst. C.E.

Boxwood Beam Compasses.

Bock and Handrick, Dresden.

244. Scales, of boxwood, showing the equivalents of English and Foreign measures of length.   *Aston & Mander.*

The bevel edged set square slide is used to show the divisions coinciding; and the equivalent values of English and Foreign measures of length may thus be readily obtained.

245. Plotting Scales. Ivory. Two specimens, to show fine and accurate dividing.   *Aston & Mander.*

No. 1 shows two chains to the inch, represented by 200 divisions to the inch.

No. 2 shows one chain to the inch, represented by 100 divisions to the inch.

II.—INSTRUMENTS FOR TRACING SPECIAL CURVES.

70. Conograph. An instrument by which the various conic sections may be drawn.

    a. Ellipso-Parabolograph.
    b. Hyperbolograph.

Dr. Lawrence Zmurko, Lemberg.

This instrument consists of two movements independent of, and perpendicular to, each other; the first of these is set in action by turning a disc, the second by means of a spring. These movements are so contrived that the extent of the second motion shall be such a function of that of the first as to cause a conic section to be described.

71. Cycloidograph. An instrument for tracing cycloids.
Dr. Lawrence Zmurko, Lemberg.

71a. Instrument for tracing with accuracy ellipses and spirals up to 25 centimetres. *M. Adrian Gavard, Paris.*

72. Instrument for drawing **Conic Sections.**
Edward Uhlenhuth, Anclam, Pommerania.

This instrument, which was invented by the exhibitor, shows in the first place the formation of the parabola. By altering the arrangement, the construction of the ellipse and hyperbola easily follows.

73. Elliptic Compass. *Renaud-Tachet, Paris.*

74. Colonel Peaucellier's Compound Compass.
Conservatoire des Arts et Métiers, Paris.

75. Compound Geometric Chuck, producing the kinematic retrogressive parabola, by continuous motion; either on a moving plane by a fixed point, or on a fixed plane by a moving point. *Henry Perigal.*

76. Machine for **Compounding two Simple Harmonic Curves.** Invented and constructed by the exhibitor.
A. E. Donkin.

A strip of paper is wound round the cylinder; the little glass pen moving backwards and forwards on it draws one curve, a similar motion of the cylinder the other. Since both move at once the curves are combined, and the result rendered visible to the eye by the revolution of the cylinder.

A. Eccentric for giving simple harmonic motion to pen.
B.   "   "   "   cylinder C.
D. } wheels for determining relative numbers of vibrations of pen and
E. } cylinder.
F. Wheel for transmitting slow motion to pinion G which turns the cylinder.
H. Idle wheel.
I. Change wheels to supply different ratios of vibration of pen and cylinder.

76c. Bow and Scale for Exhibiting Elliptic Functions. *A. G. Greenhill.*

76e. Spherical Rules and Squares for Spherical Drawing. *Dumoulin Froment, Paris.*

76a. Rough Model of the Trace-Computer, designed by the Exhibitor, for the use of the Meteorological Office.
Francis Galton, F.R.S.

Given two ordinates having the same abscissa, the instrument, of which this is a working model, pricks out a third ordinate that shall be some desired function of the other two. The original instrument was contrived for the use of the Meteorological Office, where it is employed to derive the trace for humidity from the traces of the dry and wet bulb thermometers. It consists of a horizontal slab, whose upper surface has been shaped, as hereafter described, in accordance with the numerical tables that have been calculated from the desired function, the height of its surface at each point being the tabular value corresponding to the two entries severally represented by the distance of that point from the front and from one side of the slab. The plate that carries the two traces is placed horizontally on a frame that travels in front of the slab. Two slides move at right angles to this plate, and have microscopes attached to them, that traverse the paper along ordinates having the same abscissa. One of these slides is rigidly connected with a frame on which the slab is able to move from front to back; when this slide is pushed, the frame and the slab together are pushed with it. When the other slide is pushed, it also gives a sidelong movement to the slab on the frame by means of a toothed wheel acting on a rack. Thus the particular point of the slab that corresponds to the values of the two ordinates is brought vertically below a descending rod, and this is caused to drop gently on the surface of the slab by touching a treadle. The vertical space through which the rod descends is consequently the function required. The rod carries a horizontal pricker, with which it makes a dot on a plate held vertically in the same stage that carries the plate on which the two traces are drawn. The slabs can readily be fashioned by instrument-makers, who possess the necessary apparatus, according to any required tables. They are drilled to the requisite depth at various points, and are afterwards smoothed down. In the machine in use at the Meteorological Office, there are many additional appliances not shown in this rough model.

76b. Rule, with joint, which serves to curve an elastic plate into an arc of the circle, of any radius.

Professor Tchebicheff, University of St. Petersburg.

3099. Intersecting Compasses (Arcograph).

Geodetic Institute of the Royal Polytechnic School at Munich, Prof. Dr. von Bauernfeind.

The Arcograph serves to describe upon a given chord a circle the arc of which contains a given angle. The exhibitor, by this invention, has supplied the wants of the practical geometrician in solving, graphically, and without construction Pothenot's problem and all the problems which are described in geometry as "Rückwärtseinschneiden". See Bauernfeind's "Elemente der Vermessungskunde," 5th edition, Vol. II., pp. 167–173.

III.—MODELS OF FIGURES IN SPACE.

COLLECTION OF MODELS OF RULED SURFACES, CONSTRUCTED BY M. FABRE DE LAGRANGE, IN 1872, FOR THE SOUTH KENSINGTON MUSEUM.

This collection illustrates the principal types of the class of surfaces which can be traced out in space by the motion of a straight line.

These surfaces, on account of the facility with which they can be constructed and represented, and of the ease with which their intersections can be determined, are of more consequence than any others in the geometry of the Industrial Arts. It is only in small work, which can be put into the lathe, that the class of surfaces of revolution approaches them in respect of general utility. The most important surfaces of all, the plane, the right cylinder, the right cone, and the common screw, belong to both classes.

The representation of the surfaces by means of silk threads is of course only approximate; an approximation of the same character as the representation of a curve by a dotted or chain line, Fig. 1, or by a series of right lines touching the actual curve, Fig. 2.

The models are constructed with especial reference to the possibility of changing their shape, by moving some of the supports of the strings, by altering the lengths or positions of certain parts, or by converting upright forms into oblique. This possibility of *deformation*, as the process is technically called, greatly enhances the value of the models, by allowing them to represent a much greater variety of surfaces than if they were fixed. They are, however, too delicate to be much pulled about, and, unless they are very cautiously handled, the strings are apt to become entangled or break. They should never be used except by a person who understands them, and they should not be shifted without some good reason.

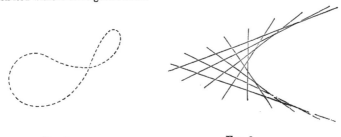

Fig. 1.   Fig. 2.

Fig. 1 is an example of the first, and Fig. 2 of the second. In both cases, the curve, although not actually drawn, is indicated with sufficient approximation for most practical purposes. Models Nos. 10 and 30 also afford illustrations of the principle exhibited in Fig. 2.

Geometrical drawings of most of the surfaces represented by these models are contained in BRADLEY's *Practical Geometry* (2 vols., oblong folio, published by Chapman and Hall). Many of them will also be found in the French treatises on practical and descriptive geometry, such as LEROY, ADHÉMAR, LEFEBURE DE FOURCY, DE LA GOURNERIE, and in their treatises on *Stereotomy* and *Stone-cutting* (*coupe des pierres*). Many of them are also given in SONNET's *Dictionnaire des Mathématiques Appliquées*. A catalogue of this collection of models, with an appendix containing an account of the application of analysis to their investigation and classification, was prepared for the South Kensington Museum in 1872, by Mr. C. W. Merrifield, F.R.S. The following descriptions are extracted from this catalogue :—

77. Hyperbolic Paraboloid generated by a single system of right lines.

Two bars, each pierced with holes, equally spaced. One bar is fixed, the other swings round an axis, which, moreover, can be inclined at different angles to the fixed bar.

When the bars are parallel the strings indicate a plane. When they are clined to one another, but still in the same plane, the strings still indicate a plane ; but when the bars are not in the same plane, the surface is the hyperbolic paraboloid.

The surface is sometimes called the *twisted plane*. But it must not be supposed that it can be made by bending a plane. On the contrary, when the surface *is* twisted, no two of the strings lie in the same plane, and, therefore, no part of the surface is plane. It can neither be flattened nor made from a plane, without stretching or contraction.

The hyperbolic paraboloid is the natural surface proper for a ploughshare.

78. Hyperbolic Paraboloid.

Two bars, pierced with holes at equal distances, the holes being connected by two different systems of strings. The surface, as well as the arrangement, is very nearly the same as in No. 77, only that there are two paraboloids instead of one. As the movable bar swings round, one paraboloid opens out while the other closes up. If the bars are swung so as to be in the same plane, one system of strings describes a plane by parallel lines, and the other by lines radiating from a point. If one bar is now turned so as to be end for end, we still get a plane, the set of parallel lines now passing through a point, while the set which previously passed through a point has now become parallel.

The pair of paraboloids intersect in three right lines. There is also a fourth intersection on the " line at infinity."

79. Hyperbolic Paraboloid.

Two bars equally spaced ; each turns on an arm perpendicular to itself, and one arm swings on a pillar. These arms can be ranged in one plane, and also turned end for end.

80. Hyperbolic Paraboloid generated by two systems of right lines.

A skew quadrilateral with four equal sides, each pierced with the same number of holes, equally spaced. The model exhibits the double generation of the surface. The plane containing two of the sides turns about hinges connecting it with the plane of the other two sides. By closing or opening this hinge the paraboloid opens out or closes. When completely open, it forms a plane divided into *diamonds*. When completely closed it again forms a plane, but the division is no longer uniform. The strings then become tangents to a plane parabola.

81. Hyperbolic Paraboloid.

A skew quadrilateral turning upon four hinges with parallel axes or pins.

The difference between this and the last is not in the kind of surface or mode of generation, but in the manner of *deforming* the surface. In No. 80 the lengths of the strings alter; while in this model they remain unaltered. Moreover, although the surface flattens in two ways, yet in both ways the strings become tangents to a plane parabola instead of parallel.

This model is well adapted for showing the leading sections of the solid All sections parallel to the pins of the hinges are plane parabolas, which degenerate into right lines when taken also parallel to the brass bars. Any other sections, whether perpendicular to the hinges or inclined to them, give hyperbolas, which degenerate into a pair of right lines when the plane of section is a tangent to the surface.

It may be worth while to remark that there is nothing absurd in the tangent plane to a surface cutting that surface, as a student unaccustomed to those subjects might at first think. On the contrary, when a surface is bent one way in one direction, and the other way in the opposite direction, the tangent plane *must* cut it. In this case, the plane passing through any two intersecting strings is a tangent plane, and evidently cuts the surface along each string.

If we imagine two planes parallel to the hinge pins, and each bisecting a pair of opposite bars, we obtain the *asymptotic planes* of the paraboloid, each of which is the assemblage of the asymptotic lines of the hyperbolas parallel to the principal hyperbolic section. Their being asymptotic has reference to these hyperbolas, and not to the parabolic character of the surface.

82. Hyperbolic Paraboloid.

A skew quadrilateral, with its opposite sides equal in length, and pierced with holes at equal distances.

Nearly similar to No. 81, but differently mounted, and with the sides of different lengths, the alternate sides only being equal. It is virtually a slightly different aspect of the same surface as No. 81.

83. Hyperbolic Paraboloid.

A skew quadrilateral, with all its sides equal, and pierced holes at equal distances.

As far as the curved surface is concerned, the same as No. 81. But the hinges are altered in direction, and the model shows plans and elevations of the right line generators of the surface. The rings also show parabolic sections of the surface.

In consequence of the alteration in the direction of the hinges, the spacing of the inclined bars, although equidistant, is at a different pitch from that of the horizontal bars.

84. Hyperbolic Paraboloid.

A skew quadrilateral, with all its sides equal, and pierced with holes at equal distances. It shows the plans and elevations of the right line generators. The rings show the parabolas of the *principal sections*.

No. 83 represents one quarter of what is here shown. The upper corners of Nos. 83 and 84 correspond; but the lower corner of the former corresponds with the middle ring of the latter.

85. Hyperbolic Paraboloid.

A skew quadrilateral, with all its sides unequal. The surface is the same as Nos. 83 and 84, but the proportions and the portion of the surface chosen for representation are different. The quadrilateral base being irregular, the strings alter in length as the surface is deformed by closing the hinges.

86. Hyperbolic Paraboloid.

Skew quadrilateral, pivoting on a single hinge. Intended to show the construction of the parabola connecting two roads which meet obliquely. This construction is used by engineers in laying out roads.

87. Hyperboloid of one Sheet.

Two rings or circles, in parallel planes, are pierced with equally spaced holes. In a certain position the threads give, 1st, a cylinder; and 2ndly, a cone.

The upper ring turns round a pin at its centre. In turning it, the cylinder closes in and the cone opens out, each altering into a hyperboloid of one sheet. We can go on turning the ring until these coincide in one hyperboloid, of which we thus get both systems of generating lines.

If the rings are set on a slope the hyperboloid is elliptic. If the rings are horizontal the hyperboloid is one of revolution.

Sloping one ring, so as not to be parallel with the other, gives rise to some curious ruled surfaces, but these are not in general hyperboloids.

88. Hyperboloid of one Sheet.

Two rings of different radius, in parallel planes, are divided into the same number of equal parts. The smaller and upper ring turns round a pin at its centre. In a particular position of the rings, the threads give two cones. Turning the ring transforms each of the cones into a hyperboloid, and when the two hyperboloids coincide, we get the two systems of right line generators.

The same stand also has a model of a hyperboloid with only one set of strings. By turning the upper ring either way it deforms into a cone; in the one case with its vertex between the rings, and in the other with its vertex a a considerable height above the rings.

Both these can have their upper rings moved along the top bar so as to incline the surfaces. We still get cones and hyperboloids, but it is only when the rings are horizontal, and centre to centre, that we get surfaces of revolution.

89. Hyperboloid of one Sheet, with its asymptotic cone.

90. Hyperboloid of one Sheet, with its asymptotic cone.

The tangent plane to the cone is also drawn. It meets the hyperboloid in two parallel right lines.

One of these right lines is the line of contact of a hyperbolic paraboloid with the hyperboloid, and the tangent plane is one of the director planes of the paraboloid, both systems of generating lines of which are exhibited.

91. Hyperboloid of one Sheet.

A slight variation from No. 90. The paraboloid only shows one system of right line generators, and the tangent plane is made by parallel instead of radiating lines.

92. Hyperboloid of one Sheet, and its tangent paraboloid.

This shows the transformation of a cylinder and its tangent plane into a hyperboloid and its tangent paraboloid.

93. Conoid, with its director plane. The director curve is a plane curve.

By shifting the position of the brasses the conoids deform into different conoids or other allied surfaces.

94. Conoid, with a director cone. The director curve is of double curvature.

By shifting the position of the brasses the conoids deform into different conoids or other allied surfaces.

95. Conoid, showing both sheets of the surface.

By shifting the position of the brasses the conoids deform into different conoids or other allied surfaces.

96. Conoids. Model showing the transformation of a cylinder into a conoid and back again. Also model showing the transformation of a cone into a conoid and back again. It is to

be noticed that the head lines of the two conoids, that is to say, the right line in which the two sheets of each conoid meet, are perpendicular to one another.

The transformation is effected by making the upper semicircle turn through two right angles.

97. Conoids.
Intersection of two equal conoids having a common director plane. The horizontal intersection is a plane ellipse.

98. Conoid, in contact with a hyperbolic paraboloid.

99. Conoids.
Two equal circles in parallel planes, divided equidistantly, are connected by threads, so as to form four surfaces.

A cylinder.   A conoid.
A cone.   A second conoid.

The director planes, as well as the head lines, of these conoids are at right angles to one another.

100. Conoids.
Two equal circles in parallel planes are connected by threads so as to form four surfaces.

A cylinder.
A cone.
A conoid.
A second conoid, with its director plane and line at right angles to those of the former.

Same arrangement as No. 99, except that the lower ring is replaced by a plane of section a little higher up. The section gives,—

For the cone, a circle smaller than the upper ring.
For the cylinder, a circle of the same size as the upper ring.
For the conoids, two ellipses turned crosswise.

101. Model
exhibiting the simultaneous transformation of a conoid into a cylinder, a cylinder into a conoid, the paraboloid touching the conoid into the tangent plane of a cylinder, and the tangent plane of a cylinder into the tangent paraboloid of a conoid, and reciprocally.

The changes may be arranged as follows :—

From.	Into.
Conoid.	Cylinder.
Tangent paraboloid.	Tangent plane.
Cylinder.	Conoid.
Tangent plane.	Tangent paraboloid.

These changes are all effected simultaneously by one movement, which can be reversed.

102. Model
exhibiting the transformation, first, of a conoid into a cylinder ; second, of the tangent paraboloid of the conoid into the tangent plane of the cylinder.

103. French Skew Arch (*biais passé*).

The inner drum, of yellow thread, represents this surface. It is a skew surface, with a right line director ; and its faces, the planes of the two semicircles, are usually parallel, although the model permits them to be placed obliquely to one another. The horizontal line joining the centres of the two large semicircles is the right line director.

The construction for any one of the generating lines is as follows :—Draw a plane through the right line director at any selected obliquity. It will, of course, give the radii of the outside circles, and the line joining the points at which it cuts the inside semicircles will be a generator of the surface. This line will evidently pass through the director line, because it is in the same plane with it.

In stone or brickwork, the sides of the voussoirs, will be given by the auxiliary plane in question. When the openings are parallel the voussoir joints are therefore plane, and the simplicity thus gained is the chief reason for adopting this form of skew arch. It is usual to take the right line direct or perpendicular to the openings, and symmetrical to them, that is to say, passing through the middle point of the parallelogram of the springing plane.

When the openings are not parallel the voussoir joints shown by the model are deformed into hyberbolic paraboloids. This deformation is, however, very slight, and in practical work would be avoided altogether by adhering to the principle of drawing a plane through the director line.

The opening of the voussoirs is usually determined by dividing the outer semicircle into equal parts.

This form of arch is inconvenient when the obliquity and the length of the barrel are excessive, for the generators are not generating lines of the cylinder containing the opening semicircles, but chords of it, and, therefore, at the middle, falling considerably inside it. The arch, therefore, droops in the middle, and this would be ugly and inconvenient if the proportions were excessive.

104. Staircase Vault for a square wall (*vis St. Gilles carrée*).

105. Staircase Vault.

Model for exhibiting some properties of this ruled surface, by showing how it is obtained from the deformation of a cylinder (*douelle de la vis St. Gilles carrée*).

106. Cylinder with Helix and developable Helixoid.

The helix is simply a screw thread. The developable helixoid, shown by the purple threads, is the surface swept out by the right line tangents of the helix. If we consider that each gore can be turned a very little bit about the thread which separates it from the next gore, we see that the surface can be flattened out or developed into a plane, without any crumpling. This happens because every two consecutive generating lines meet one another on the helix. That is why its surface is called developable. Its section by a horizontal plane is the involute of the circle.

The model allows the pitch of the helix to be shortened by lowering the upper plate, and the cylinder can also be inclined. When oblique, however. the curve which replaces the helix is not such a screw thread as can be turned in the lathe.

107. Skew Helixoid.

This surface is described by a right line, which always passes through the axis of a cylinder, and makes a constant angle with that axis. It also passes through a helix or screw thread traced on the cylinder. The model only shows the surface, not the cylinder. The section by a horizontal plane is the spiral of Archimedes. It is the surface of what is known as the screw with a triangular thread.

This is not the commonest form of the skew helixoid ; that is best seen on the underside of a screw staircase, or on the driving face of a common screw propeller. In these, two generating lines are at right angles to the axis.

The surface may also be considered as generated by a line which makes a constant angle with a given fixed line, and moves up that line, and at the same time turns round it, at uniform rates.

108. Skew Surface with its tangent paraboloid, capable of transformation into another skew surface while the paraboloid deforms into a plane.

This is (for a certain position of the lower semicircle) a skew surface with a director plane, the plane being vertical. The director curves are: one of them a circle divided equidistantly, the other a semicircle divided so as to keep the strings parallel to the director plane.

109. Intersection of Two Cones having double contact with one another, that is to say, having a pair of tangent planes in common.

The consequence of their having double contact is that their curve of intersection breaks up into two plane ellipses.

The vertices of the cones slide along a rule which turns on a universal joint. See also model No. 114.

110. Common Groin. Intersection of two cylinders having a pair of common tangents. The model may be set square or oblique.

111. Intersection of Two Cylinders, one piercing the other so as to give two separate loops of intersection.

112. Intersection of Two Cylinders, having a common tangent, so as to give a curve having a double point at the point of contact.

113. Intersection of Two Cylinders, neither completely piercing the other, so as to give only one loop of intersection.

114. Intersection of Two Cones, having double contact, along a pair of plane ellipses.

115. Groin. Oblique intersection of two splayed vaults of the same spring.

116. Pair of Intersecting Planes, which, by pulling the brass ball so as to give simultaneous rotation to the two upper rods, deform into paraboloids first, and then into planes described by radiating strings.

117. Intersecting Cylinder and Plane. By pulling the brass ball the head brasses rotate together, and the cylinder deforms into, first, a hyperboloid, and then a cone, while the plane deforms into, first, a paraboloid, and then again into a plane with radiating lines.

118. Pair of Intersecting Cylinders on circular bases. By pulling the brass ball the head brasses rotate together, and the cylinders deform, first, into hyperboloids, and then into cones.

119. Pair of Intersecting Cylinders on irregular bases. By pulling the brass ball the head brasses rotate together, and the cylinders deform, becoming at last cones.

120. Groin.

Model showing the deformation of a common groin, both obliquely, and by splaying the vaults. The model shows not only the intersection, but the plans of the intersection and of the generating lines.

121. Helix or Screw-thread.

Model showing the transformation of the right line generators of a right cylinder into screw threads of various pitch or obliquity.

The pitch of a screw is the distance between two successive turns, measured in a direction parallel to the axis. When this distance is small, the screw is said to have a fine pitch ; when great, a coarse or high pitch.

COLLECTION OF MODELS CONTRIBUTED BY THE LONDON
MATHEMATICAL SOCIETY.

123. Plücker's Models (14) of certain quartic surfaces, representing the equatorial form of complex surfaces.

London Mathematical Society.

At the meeting of the British Association at Nottingham, in 1866, Prof. Plücker read a paper on "Complexes of the Second Order." On this occasion he showed a series of models constructed by Epkens, of Bonn, of which the above are copies made for Dr. Hirst, and presented to the London Mathematical Society.

The following is Prof. Cayley's description of the models, extracted from Nos. 37 and 38 of the Mathematical Society's Proceedings, vol. iii., pp. 281–285, supplemented by a description of models A, B, C, D, E, F, drawn up by Prof. Henrici.

The Society possesses a series of 14 wooden models of surfaces, constructed under the direction of the late Prof. Plücker, in illustration of the theory developed in his posthumous work " Neue Geometrie des Raumes gegründet " auf die Betrachtung der geraden Linie als Raum-elemente," Leipzig, 1869.

These, all of them, represent, I believe, equatorial surfaces, viz., eight represent cases of the 78 forms of equatorial surfaces, " deren Breiten-Curven " eine feste Axenrichtung besitzen," vol. ii. pp. 352–363 ; the remaining models, A, B, C, D, E, F, I have not completely identified. I propose to go into the theory only so far as is required for the explanation of the models.

In a " complex," or triply infinite system of lines, there is, in any plane whatever, a singly infinite system of lines enveloping a curve ; and if we attend only to the curves the planes of which pass through a given fixed line, the locus of these curves is a " complex surface." Similarly, there is through any point whatever a single infinite series of lines generating a cone ; and if we attend only to the cones which have their vertices in the given fixed line, then the envelope of these cones is the same complex surface. In the case considered of a complex of the second degree, the curves and cones are, each of them, of the second order ; the fixed line is a double line on the surface, so that (attending to the first mode of generation) the complete section by any plane through the fixed line is made up of this line twice, and of a conic. The surface is thus of the order 4 ; it is also of the class 4 ; the surface has, in fact, the nodal line, and also 8 nodes (conical points), and we have thus a reduction $= 32$ in the class of the surface.

In the particular case where the nodal line is at infinity, the complex surface becomes an equatorial surface; viz. (attending to the first mode of generation), we have here a series of parallel planes each containing a conic, and the locus of these conics is the equatorial surface.

It is convenient to remark that, taking a, b, h, to be homogeneous functions of (x, w) of the order 2 ; f, g, of the order 1; and c of the order 0 (a constant); then the equation of a complex surface is—

$$\begin{vmatrix} y & z & 1 \\ y & a & h & g \\ z & h & b & f \\ 1 & g & f & c \end{vmatrix} = 0 ;$$

and that, writing $w = 1$, or considering a, h, b ; f, g ; c, as functions of x of the orders 2, 1, 0 respectively, we have an equatorial surface.

A particular form of equatorial surface is thus, $bcy^2 + caz^2 + ab = 0$, or taking $c = 1$, this is $by^2 + az^2 + ab = 0$, where a, b, are quadric functions of x.

The surface is still, in general, of the fourth order; it may, however, degenerate into a cubic surface, or even into a quadric surface; the last case is, however, excluded from the enumeration. The section by any plane parallel to that of yz is a conic; the section by the plane $y = 0$ is made up of the pair of lines $a = 0$, and of the conic $z^2 + b = 0$; that by the plane $z = 0$ is made up of the pair of lines $b = 0$, and of the conic $y^2 + a = 0$; the last-mentioned planes may be called the principal planes, and the conics contained in them principal conics. The surface is thus the locus of a variable conic, the plane of which is parallel to that of yz, and which has for its vertices the intersections of its plane with the two principal conics respectively. And we have thus the particular equatorial surfaces considered by Plücker, vol. ii. pp. 346–363 (as already mentioned), under the form

$$\frac{y^2}{Ex^2 + 2Ux + C} + \frac{z^2}{Fx^2 - 2Rx + B} + 1 = 0$$

and of which he enumerates 78 kinds, viz.: these are—

1 to 17. Principal conics, each proper.
18 to 29. One of them a line-pair.
30 to 32. Each a line-pair.
33 to 39. Principal conics, each proper, but having a common point.
40 to 43. One of them a line-pair, its centre on the other principal conic.
44 to 61. One principal conic, a parabola.

Model 2.   The form of the equation is here,—

$$\frac{y^2}{l^2[(x-a)^2+\beta^2]} - \frac{z^2}{l'^2[(x-a')^2+\beta'^2]} = 1$$

viz., the principal conics are one of them a hyperbola, the other imaginary ;
hence the generating conic has always two, and only two, real vertices, viz.,
it is always a hyperbola.   There are no real lines.

Model 3.   The form of the equation is—

$$\frac{y^2}{l^2[(x-a)^2+\beta^2]} + \frac{z^2}{l'^2[(x-a')^2+\beta'^2]} = 1;$$

viz., the principal conics are each of them a hyperbola ; the generating conic
has four real vertices, viz., it is always an ellipse.   There are no real lines.

Model 4.   The form of the equation is—

$$\frac{y^2}{l^2(x-\gamma)(x-\delta)} + \frac{z^2}{l'^2[(x-a')^2+\beta'^2]} + 1 = 0.$$

The principal conics are one of them an ellipse, the other imaginary ; for
values of x between γ and δ, the variable conic has two real vertices, or it is
a hyperbola ; for any other values it is imaginary, so that the surface lies
wholly between the planes $x=\gamma$, $x=\delta$.   The surface contains the real lines
$y=0$, $x=\gamma$, and $y=0$, $x=\delta$.

Model 9.   The form of the equation is—

$$\frac{y^2}{l^2(x-\gamma)(x-\delta)} + \frac{z^2}{l'^2(x-\gamma')(x-\delta')} + 1 = 0$$

where, say the values γ, δ, lie between the values γ', δ', the principal conics
are each of them an ellipse, the vertices (on the axis or line $y=0$, $z=0$) of
the one ellipse lying between those of the other ellipse.   The variable conic
for values of x between γ and δ has four real vertices, or it is an ellipse ; for
values beyond these limits, but within the limits γ', δ'—say, from γ to γ' and
from δ to δ'—there are two real vertices, or the conic is a hyperbola ; and
for values beyond the limits γ', δ', the variable conic is imaginary.
There are four real lines $(y=0, x=\gamma)$, $(y=0, x=\delta)$, $(z=0, x=\gamma')$, $(z=0,$
$x=\delta')$.   The surface consists of a central pillow-like portion, joined on by
two conical points to an upper portion, and by two conical points to an under
portion, the whole being included between the planes $x=\gamma'$, $x=\delta'$.

Model 13.   The form of the equation is—

$$\frac{y^2}{l^2(x-\gamma)(x-\delta)} - \frac{z^2}{l'^2(x-\gamma')(x-\delta')} + 1 = 0;$$

the values γ', δ', lying between γ, δ ; the principal conics are one of them a
hyperbola, the other an ellipse, the vertices (on the axis or line $y=0$, $z=0$)
of the hyperbola lying between those of the ellipse.   The variable conic, for
values of x between γ', δ', has two real vertices, or it is a hyperbola ; for
the values, say, from γ' to γ, and δ' to δ, there are four real vertices, or
the conic is an ellipse ; for values beyond the limits γ, δ, there are two
real vertices, and the conic is a hyperbola.   There are the four real lines
$(y=0, x=\gamma)$, $(y=0, x=\delta)$, and $(z=0, x=\gamma')$, $(z=0, x=\delta')$.   The surface
consists of eight portions joined to each other by eight conical points, but the
form can scarcely be explained by a description.

Model 32.   The form of the equation is—

$$\frac{y^2}{l^2(x-\gamma)^2} + \frac{z^2}{l'^2(x-\gamma')^2} = 1;$$

viz., the principal conics are each of them a line-pair, the variable conic is always an ellipse.

There are the two real nodal lines ($y=0$, $x=\gamma$) and ($z=0$, $x=\gamma'$), each of these being in the neighbourhood of the axis crunodal, and beyond certain limits acnodal; the surface is a scroll, being, in fact, the well-known surface which is the boundary of a small circular pencil of rays obliquely reflected, and consequently passing through two focal lines.

Model 34. The equation is—

$$\frac{y^2}{l^2(x-\gamma)(x-\delta)} + \frac{z^2}{l'^2(x-\gamma')(x-\delta')} + 1 = 0$$

where $x=\delta$ is not intermediate between the values $x=\gamma$ and $x=\gamma'$; say the order is δ, γ, γ'. The surface is thus a *cubic* surface; the principal conics are ellipses, having on the axis a common vertex, at the point $x=\delta$, and the remaining two vertices on the same side of the last-mentioned one. The variable conic for values between δ and γ has four real vertices, or it is an ellipse; for values between γ and γ' two real vertices, or it is a hyperbola; and for values beyond the limits δ, γ', it is imaginary. There are on the surface the two real lines ($y=0$, $x=\gamma$) and ($z=0$, $x=\gamma'$). The surface consists of a finite portion joined on by two conical points to the remaining portion.

Model 40. The form of equation is—

$$\frac{y^2}{l^2(x-\gamma)(x-\delta)} + \frac{z^2}{l'^2(x-\delta)^2} + 1 = 0$$

The surface is thus a *cubic* surface; the principal conics are, one of them an ellipse, the other a pair of imaginary lines intersecting on the ellipse; for values of x between γ and δ, the variable conic has thus two real vertices, and it is a hyperbola; for values beyond these limits it is imaginary, and the whole surface is thus included between the planes $x=\gamma$ and $x=\delta$. There are the two real lines ($y=0$, $x=\gamma$) and ($z=0$, $x=\delta$).

Taking $l^2=l'^2=1$, the surface is—

$$\frac{y^2}{(x-\gamma)(x-\delta)} + \frac{z^2}{(x-\delta)^2} + 1 = 0$$

which is a *particular* case of the parabolic cyclide.

The equatorial surfaces, not included in the preceding 78 cases, Plücker distinguishes (vol. ii. p. 363) as "gedrehte" or "tordirte," say, as twisted equatorial surfaces, the equation of such a surface is—

$$by^2 + 2hyz + az^2 + ab - h^2 = 0$$

where $b = Fx^2 - 2Rx + B$
$a = Ex^2 + 2Ux + C$
$h = Kx^2 - Ox - G$ (or in particular $= -Ox - G$).

Model A. is such a surface, being a twisted form of Model 9.
Model B. belongs to the case $a=0$; viz., the form of the equation is—

$$by^2 + 2hyz - h^2 = 0.$$

The variable conic is a hyperbola, the direction of one of the asymptotes being constant (vol. ii. p. 368).

There are, moreover, (p. 372) equatorial surfaces in which the variable conic is always a parabola, and where there are on the surface four real or imaginary singular lines.

In Model C the singular lines are all four real, but two of them coincide with the nodal line at infinity. Consequently, the variable parabola has its axis in a fixed direction. Its vertex moves along a hyperbola which has one asymptote in that fixed direction. The other two singular lines are on opposite sides of this asymptote and parallel to it. When the plane of the variable parabola passes through one of these lines, the parameter vanishes and changes sign. When it passes through the above-mentioned asymptote, the parabola reduces to the line at infinity and the plane becomes asymptotic to the surface. The latter consists of four parts, two on opposite sides of the asymptotic plane between this and one of the singular lines respectively, the other two extending from the singular lines to infinity.

The remaining three models, D, E, F, represent twisted surfaces. Of the four singular lines two are in each case imaginary. The remaining two are real on the first, coincident on the second, and imaginary on the third. Model D consists, therefore, of three, Model E of two, and Model F of one part.

The models are copies from some constructed by Epkens of Bonn. They were presented to the London Mathematical Society by Dr. Hirst, F.R.S. They have been remounted under the direction of Prof. Henrici, by M. Nolet, a student of University College, London.

Some account of complexes and complex surfaces will be found in Dr. Salmon's Geometry of Three Dimensions (3rd edition, pp. 405, 493, 566, 570).

123a. Rough Model of Steiner's Surface.

Prof. Cayley.

Steiner's surface is the quartic surface represented by the equation $\sqrt{x} + \sqrt{y} + \sqrt{z} + \sqrt{w} = 0$; where the co-ordinates x, y, z, w, of a point are proportional to arbitrary multiples of the perpendicular distances from four given planes; in the model, x, y, z, w are proportional to the perpendicular distances from the faces of a regular tetrahedron, the co-ordinates being positive for a point inside the tetrahedron.

The surface may be regarded as inscribed in the tetrahedron, touching each face along the circle inscribed in the face. The general form is that of the tetrahedron with its summits rounded off, and with the portions within the inscribed circles scooped away down to the centre of the tetrahedron, in such wise that the surface intersects itself along the lines drawn from the centre to the mid-points of the sides (or, what is the same thing, the lines joining the mid-points of opposite sides). The lines in question produced both ways to infinity are nodal lines of the surface, but as regards the portions outside the tetrahedron, they are acnodal lines, without any real sheet through them; and these portions of the lines are not represented in the model.

The sections by a plane parallel to a face of the tetrahedron are trinodal quartics, which (as the position of the plane is varied) pass successively through the forms :

1. Four acnodes.
2. Trigonoid, with three acnodes.
3. Tricuspidal.
4. Trifoliate, with three crunodes, cis-centric.
5. Do.   with triple point at centre.
6. Do.   with three crunodes, trans-centric.
7. Twice-repeated circle.

The three nodes are in each case the intersections of the plane by the nodal lines, and the twice-repeated circle is the circle inscribed in the face of the tetrahedron.

123b. Model of a Cubic Surface.

Prof. O. Henrici, F.R.S.

The equation to this surface is $xyz = h^3 (x + y + z - 1)^3$. There are 3 biplanar nodes as shown on the model. The 27 straight lines on the surface are all real, but coincide 9 to each with the 3 black lines drawn on the model.

123c. Sylvester's Amphigenous Surface, a surface of the ninth order. *Prof. O. Henrici, F.R.S.*

This surface is connected with the reality of the roots of equations of the ninth degree.

123d. Model representing the Right Lines on a Surface of the Third Class, having a tangent-plane touching along a conic (the singularity dualistically corresponding to a double-point of the second order).

Elling B. Holst, Stipendiary of the University of Christiania.

The model is composed of twenty-one wires, six of which, painted light red, lie in the same plane and touch the conic in points painted dark red. Through the fifteen points of intersection of these six lines the others, painted white, pass, again intersecting three and three, and are the lines in which the surface cuts itself. All points on these lines have therefore two tangent-planes ; where the latter are imaginary the lines are black. The black is in part laid on schematically, especially where the black part contains the point at infinity. The parabolic curve consists of the conic aforesaid and two species of cuspidal curves, viz. : —

1. One curve passing the dark red points and having cusps in those six limiting-points between black and white which are nearest to the conic, the curve therefore having a zig-zag course.

2. Four closed branches having cusps in the other twelve limiting-points.

All these parabolic curves together separate ten distinct ellipsoidally curved parts from the surface everywhere else hyperboloidally curved.

124. Models. A series illustrative of Plücker's Researches in Geometry of Three Dimensions. See explanation No. 123.

Prof. Hennessy, Dublin.

126. Model of the ruled cubic surface called the **Cylindroid.**

Dr. Robert S. Ball, LL.D., F.R.S.

This surface was discussed by Plücker in connexion with the theory of the linear-complex. The kinematical and physical significance of the surface will be found in the "Theory of Screws." The equation of the surface is $z(x^2 + y^2) - 2mxy = 0$.

125. Diagrams (48) showing the Fundamental Principles of the exhibitor's " Organic Geometry of Form."

Prof. Franz Tilser, Prague.

The above work demonstrates the necessity for a reform in geometry, and furnishes the necessary basis for establishing a new system adapted to satisfy the requirements of an exact science. To the above are added 7 " Paragram " Tablets, representing in natural organic connexion a synopsis of the principal elements to be observed in every graphical representation.

127. Models (6) illustrating the relative bases of Descriptive Geometry and the Organic Geometry of Form.

Prof. Franz Tilser, Prague.

128. Drawings. A collection, executed by the Students of the Bohemian Polytechnic Institute, illustrative of the instruction received in the subject of Organic Geometry of Form.

Prof. Franz Tilser, Prague.

129. Two specimens of Wire Stereometrical Models, with letters on cork.

Prof. J. Joseph Oppel, Frankfort-on-Maine.

130. Two specimens of Wire Trigonometrical Models, with letters.   *Prof. J. Joseph Oppel, Frankfort-on-Maine.*

131. Two specimens of Wooden Stereometrical Models, with letters.

Prof. J. Joseph Oppel, Frankfort-on-Maine.

The auxiliary lines, diagonals, &c. are distinguished by wires of different colours or thicknesses. They are in many cases movable, so that the perfect figure can be constructed before the eyes of the pupil.

Auxiliary planes are also distinguished by their colour. The angular points are provided with metal pins, to which letters on cork discs can be attached, so as to be turned upright towards the observer.

These models have proved highly serviceable for instruction during the past 20 years.

132. Large Model of an Ellipsoid, of white cardboard, on a turned stand.   *Prof. Dr. A. Brill, Munich.*

133. Cardboard Models of Surfaces of the second order, on frames. Made up of circular sections. The sections are attached to each other.   *Prof. Dr. A. Brill, Munich.*

This collection of models consists of :—

1. **Ellipsoid** having 20 circular sections.
2. **Ellipsoid** having 30 circular sections.
3. **Hyperboloid** of one sheet.
4. **Hyperboloid** of two sheets.
5. **Elliptic Paraboloid.**
6. **Cone** in two sheets.
7. **Hyperbolic Paraboloid.**

141. Series of Cardboard Models of Surfaces, of the second order, in a cardboard box. The sections are not attached to each other.   *Prof. Dr. A. Brill, Munich.*

These models, Nos. 132, 133, 141, are distinguished from those in common use by their mobility, by means of which each one represents not only a single ellipsoid or hyperboloid, but a series of surfaces of one or the other kind. For when the angle of inclination of the circular sections is altered, in a direction easily recognised by pressing or drawing out the model, there will be obtained a simple but infinite system, the individual forms of which can be converted from a flat figure through gradually-changing solid bodies to just such another figure with a different relation of axes, without, however, losing its properties.

The equations representing these systems of surfaces are in rectangular co-ordinates :—

For central surfaces :

$$\frac{x^2}{a^2} + \frac{y^2}{\cos^2\psi}\left(\frac{1}{a} - \frac{1}{k}\right) + \frac{z^2}{k\sin^2\psi} = 1, \text{ or } = 0 \text{ (cone)}.$$

For the elliptic paraboloid :

$$\frac{x^2}{a^2} + \frac{y^2}{a^2\cos^2\psi} - \frac{2z}{k\sin\psi} = 0$$

For the hyperbolic paraboloid :

$$\frac{x^2}{a^2\cos^2\psi} - \frac{y^2}{a^2\sin^2\psi} - \frac{2z}{k} = 0$$

Where 2ψ is the inclination of the circular sections, and a and k are real constants. From the first equation it appears that among the series of ellipsoids there will always be a sphere.

142. Model of a Surface of the third order, made in plaster of Paris, with 27 real right lines.

Prof. Dr. Christian Wiener, Carlsruhe.

The construction of the model is described on a placard fixed to the model.

143. Model of the same surface of the third order, in discs of card-board, with 27 real right lines.

Prof. Dr. Christian Wiener, Carlsruhe.

144. Poinsot's Star Polyhedra. *Dr. M. Doll, Carlsruhe.*

These models show the star dodecahedron with 20 points, the star dodecahedron with 12 points, the icosahedron, and dodecahedron.

148. Curvilinear central surface of the Ellipsoid, in four separate pieces. Proportions of the axes of the ellipsoid, 3 : 4 : 5. *Ludwig Lohde, Berlin.*

149. Dupin's Cyclide, according to the calculation of Professor Kummer, at Berlin. Model 0·094 m. diameter.

(*See* Monatsbericht der Akademie der Wissenschaften zu Berlin, 1863, pp. 328 and 336.) *Ludwig Lohde, Berlin.*

150. Kummer's Cyclide. *Ludwig Lohde, Berlin.*

151. Minimum-surface in a recurring number of tetrahedral surfaces.

(Submitted to the Berlin Academy of Sciences by Professor Kummer, on the 6th April 1865.) *Ludwig Lohde, Berlin.*

152. Maximum of Attraction of the Earth's Surface.

Ludwig Lohde, Berlin.

153. Geometric Body, executed in plaster of Paris, called **" Podoid "**; a transcendental curved surface, which is determined by the variable parallel co-ordinates μ, ϕ, and κ, whose equation represents the elliptic function

$$\mu = \int_0^\phi \frac{d\varphi}{\sqrt{(1 - \kappa^2\sin^2\varphi)}} = F(\phi, \kappa).$$

The construction in plaster of Paris embraces the limits $\kappa = +1$ to $\kappa = -1$ and $\varphi = 0$ to $\phi = \pi$.

Prof. Dr. Edward Heis, Münster, Westphalia.

154. The same **Podoid,** executed on a smaller scale, embracing the limits $\kappa = +1$ to $\kappa = -1$, $\varphi = 0$ to $\varphi = 2\pi$.

Prof. Dr. Edward Heis, Münster, Westphalia.

155. Right double circular Cone, of white wood.

Prof. Borchardt, Berlin.

On the one sheet of the double cone are shown, by sections, the circle, the ellipse, and the hyperbola; on the other, the circle, the parabola, and the corresponding hyperbola. The model takes to pieces at the sections.

156. Elliptic Cone, of white and brown wood.

Prof. Borchardt, Berlin.

On the oblique cone are shown the two circular sections, and the elliptic, hyperbolic, and parabolic sections. At the sections of the ellipse and parabola the model takes to pieces; the other sections are shown by the lines defined by the dark and light wood.

157. Ruled Surface of the fourth degree.

Prof. Borchardt, Berlin.

This model represents a surface of the fourth order determined by the equation—

$$\frac{3x^2}{(z-a)^2} + \frac{3y^2}{(z+a)^2} = 1.$$

The surface has two double right lines, between which lies a finite sheet of the surface as shown on the model, whilst beyond each double right line there extends a second and third infinite sheet of the surface. Every horizontal section of the surface is an ellipse. Of these are shown the circular section corresponding to $z = 0$, and the two ellipses corresponding to $z = \pm\frac{a}{2}$.

The model can be taken to pieces at each of these sections.

158. Rectangular Parallelopiped, intersected by a skew surface. *Prof. Borchardt, Berlin.*

159. Right Circular Cylinder, with spiral surface intersecting it. *Prof. Borchardt, Berlin.*

These five models, Nos. 155–159, were executed by the late *Ferd. Engel,* known from the drawings, which he has furnished to *Prof. Schellbach's* " Darstellende Optik."

160. String Model, representing a hyperboloid of one sheet. On it are shown the principal ellipse, the asymptotic cone, and a tangential surface, in threads of different colours.

Dr. Wiecke, Cassel.

This model represents by means of strings (kept tight by springs) of different colours the hyperboloid of one sheet and its principal auxiliary surfaces. The two sides of the surface are shown by the green and red strings respectively; the principal ellipse is given by the points at which the strings pass through the network stretched on the frame; the asymptotic cone is shown by yellow, and a tangent plane by white strings.

161. Model in plaster of Paris, representing the eighth part of the former (No. 160) with a developable normal surface, lines of curvature, and edge of regression. *Dr. Wiecke, Cassel.*

This plaster model represents the eighth part of the surface of an hyperboloid of one sheet; it is constructed on the principal ellipse, and shows the principal axes. It is also attempted to demonstrate on this hyperboloid the lines of curvature of the first and second kind, first investigated by Monge On this account the hyperboloid is bounded on the side opposite to the principal ellipse by a normal surface of which the directrix is one of the lines of curvature of the first kind. The normals are drawn in this normal surface, and produced to meet in the edge of regression, which with two of the normals will then become the boundaries of the normal surface.

161a. Collection of 45 geometrical solids in cut crystal, for purposes of demonstration. *Madame Wentzel.*

162. Intuitive Method of Projection, by movable planes. Cardboard models (19), practically illustrating problems of space. *Frère Memoire Piron.*

162a. Open Frames containing **Photographs** for teaching by projection. *J. and A. Molteni, Paris.*

162b. Projection Apparatus, polyorama for superposed images. *J. and A. Molteni, Paris.*

IV.—REPRESENTATION OF FIGURES IN SPACE BY MEANS OF DRAWINGS ON A PLANE.

163. Diagrams and Models, illustrative of **Descriptive Geometry,** executed by the Frères de la Doctrine Chrétienne, of Paris. *Prof. Pigot, Dublin.*

164. Drawings, executed in the college by the students, showing the nature of the courses of **Descriptive Geometry** and engineering. *Prof. Pigot, Dublin.*

165. Specimens of a series of **simple folding models** for illustrating the various propositions in **Descriptive Geometry.** *Prof. Osborne Reynolds.*

These are specimens of a series of models designed for illustrating the various propositions in descriptive geometry. They are especially designed for lecturing purposes, for which their simple construction, and the capability which they possess of folding into small compass, well adapts them.

These models contain a complete drawing for each proposition. The horizontal and vertical planes are hinged together, so that they can be folded flat, and the lines, planes, and surfaces are represented by coloured strings, which assume their positions when the planes are at right angles.

1. Illustrating the relation between the projections, traces, directions, and lengths of straight lines.
2. Illustrating the relation between the traces of planes, their inclinations, and intersections.
3. Illustrating the relation between the projections of a line and the traces of a plane; also the normal to a plane.
4. Illustrating the relation between the projections of lines and the angles between them.
5. Illustrating the relation between the traces of a cone and the traces of its tangent plane, and hence the method of drawing a plane having a given inclination.
6. Model, capable of opening out flat, so as to show how the horizontal and vertical planes may be represented on a drawing.

166. Stereoscopic Figures, for demonstration and use in the study of stereometry and spherical trigonometry. Edited by Julius Schlotke. *L. Friederichsen and Co., Hamburg.*

166a. Stereograms of the Lines of Curvature of Surfaces. Drawn by the exhibitor. *Prof. J. Clerk Maxwell.*

Lines of curvature of the cyclide of Dupin (4).
Curves on a sphere (4) :—
   α Two systems of orthogonal circles.
   β Concyclic spherical ellipses. (This represents a spherical harmonic of the second degree.)
   γ Confocal spherical ellipses.
   δ The projections of a spherical ellipse on the three principal planes.
Quadric surfaces (4) :—
   Elliptic paraboloid, hyperbolic paraboloid, ellipsoid, and the surface of centres of ellipsoid.
Fresnel's wave surface (3):—
   The lines are in the direction of the vibrations of the two polarized rays.
Steiner's surface (2).
Twisted cubics (2) :—
   Curve $\begin{cases} x = a \sin 2\theta \text{ (vertical).} \\ y = b \sin 3\theta \text{ (horizontal).} \\ z = c \cos 5\theta \text{ (perpendicular to paper).} \end{cases}$
   Curve $\begin{cases} x = a \sin 2\theta. \\ y = b \sin 3\theta. \\ z = c \sin 7\theta. \end{cases}$
Icosahedron in octahedron.
22 in all.

166b. Real Image Stereoscope for showing the above.
                                              Prof. Clerk Maxwell.

   The observer places his eyes about two feet from the large lens, and sees the united real images of the figures at or near the surface of the lens.

167. The principal Problems of Descriptive Geometry, represented by stereoscopic figures, by Julius Schlotke.

L. Friederichsen and Co., Hamburg.

168. Stereoscopic representation of a number of the most important crystals, their combinations, &c., by Julius Schlotke.

L. Friederichsen and Co., Hamburg.

The stereoscopic figures of Schlotke are as yet the only ones of their kind in use for illustrating instruction in descriptive geometry and crystallography, in polytechnic and other higher educational institutions, where they are much appreciated.

More particularly the division of crystallography is recommended, as it renders unnecessary the usual expensive models, and, better than those models, demonstrates the combinations and growth of crystals.

A stereoscopic apparatus is placed near the objects.

SECTION 3.—MEASUREMENT.

I.—SPECIAL COLLECTIONS.

COLLECTION OF STANDARD MEASURING APPARATUS CONTRIBUTED
BY THE STANDARDS DEPARTMENT OF THE BOARD OF TRADE.

A.—*Comparing Apparatus, &c. for Standard Weights and
Measures.*

169. Comparing Apparatus for End-Standards of length.
Used by Mr. Sheepshanks in the work of the Commission for
Restoration of the Imperial Standards, 1844–1850. Constructed by
Troughton and Simms.

The standard, and compared end-bars are placed successively on the **V**
supports, with one defining end in contact with the left hand stud and the
other defining end with the suspended gravity-piece interposed between it
and the screw on the right hand. The micrometer screw is to be gently
pressed forward until it just holds up the gravity piece in position, thus
ensuring constant pressure for each observation. The readings of the micrometer being taken, the difference of the two readings shows the difference in
length of the two end bars to less than 0·0001 inch, which is the value of
one division of the micrometer.

To obtain results with scientific precision the temperature of the measuring
axis of each bar during the comparison should be known, as well as its rate of
expansion. The temperature and length of the bar connecting the stud and
gravity piece and of the metal of the apparatus should also be constant.

170. Two Lever Frames, with rollers for supporting standard bars. Such lever frames are used for supporting all the Imperial Standard yards made by the Commission for restoring the
Standards. Constructed by Troughton and Simms.

Each bar is supported on the eight rollers of the two lever-frames, which
are placed symmetrically under the bar, so that the upward pressure of each
of the eight different rollers is necessarily equal, and the length between the
defining points of the bar is not altered by its flexure. Equal intervals of
supports $= \dfrac{\text{length of bar}}{\sqrt{n^2-1}}$, where n is the number of supports.

171. Double Micrometer Microscope for comparing the
smaller subdivisions of standards of length. Constructed by
Troughton and Simms.

It has a movable eye-piece with a double lens, sliding upon a horizontal
plate, and two micrometers ; and has two object-glasses, each with a double

lens, sliding on a horizontal plate parallel to the other plate. The measuring field is about two centimetres in extent, or a little less than 1 inch. Value of one division of each micrometer = 0·00003097 inch, or 0·0007866 millimetres.

172. Apparatus for determining the Expansion of Standard Bars. Constructed by Troughton and Simms.

The trough containing the steel bar with projecting points, distant 1 yard and 1 metre respectively, is filled with melting ice to secure constant length at the temperature of 0° C. The standard bar is placed in the lower trough, with two standard thermometers, and is raised gently against the points. Their impression shows the constant length on the bar at its noted temperature in ordinary air. Next fill the lower trough with melting ice, and take impressions to show the constant length on the bar when at 0° C. Then fill the lower trough with water, and raise to boiling point, or other less high temperature, by the heat from the gas jets underneath, and take impressions to show the constant length on the bar when at 100° C., &c. From the difference of these lengths accurately measured under micrometer microscopes, the rate of expansion of the bar is deduced.

173. Large Callipers for measuring diameter and depth of cylindrical or other measures. Constructed by Troughton and Simms.

These are made on the same principle as the instruments used for measuring shot and the bore of guns at Woolwich. They measure diameters up to 24 inches and within 0·001 inch by the aid of a vernier.

174. Model of Sub-divided Yard with comparing apparatus, for the use of local inspectors of weights and measures. Constructed by Troughton and Simms.

The tested yard measure is placed with its zero defining line immediately under that of the standard. By running the eye-piece along the upper guide bar, each defining line is accurately compared and differences determined to less than 0·01 inch by means of the small supplementary sub-divided inch measure placed also under the eye-piece. This apparatus is described and illustrated in Appendix III., 7th Annual Report of Warden of the Standards, 1873.

175. Spherometer for measuring spherical curves, with true gun-metal plane. Used for measuring the flexure of the middle of the glass disc placed upon the Imperial Standard bushel. Constructed by Troughton and Simms.

When the horizontal plane is made to rest with its three triangular flattened points upon the true plane, the central screw with its micrometer head is accurately adjusted in the same plane, and its reading noted. By substituting for the plane the surface to be tested, its convexity or concavity is determined from the difference of the reading of the micrometer, either − or + . Value of one revolution of the screw = 0·01 inch, and of one division of the micrometer = 0·0001 inch. By interposing a bright beam of light between the point of the screw and the surface tested, and by estimation of 0·1 division, accurate measurements have been made to 0·00001 inch. This instrument is described in Appendix X., 6th Annual Report of Warden of the Standards, 1872.

176. Cathetometer for vertical measurements. Constructed by Troughton and Simms.

For example, for accurately reading a barometer or manometer: place the cathetometer at a convenient distance, and adjust the cross wire of the upper telescope to the level of the mercury in the glass tube, and that of the lower telescope to the level of the mercury in the reservoir. The difference of the two readings on the graduated scale of 42 inches gives the length of the column of mercury to 0·001 inch, by aid of the vernier.

177. Stereometer for ascertaining the density of bodies by determining their volume. Constructed by Troughton and Simms.

This instrument was invented by M. Say (Annales de Chimie, t. xxiii. p.1,1797), for determining the specific gravity of gunpowder, and was used with some improvements by Professor W. H. Miller (See Phil. Trans. 1856, part iii. p. 800.) for determining the density of the platinum *Kilogramme des Archives*, during his work of restoring the imperial standard pound. The solid body tested is placed in the receiver communicating with the upper end of a vertical glass tube, the lower end of which communicates with that of a second glass tube having its upper end open to the air. The body should nearly fill the receiver, which is screwed up air-tight in its place. Mercury is poured into the second tube, and can be discharged by a stopcock at its lower end. Differences in the relative height of the mercury in the two tubes are noted by means of the cathetometer, as indicating the volume of compressed air under the two conditions, when the body is in the receiver and when it is removed. The volume of the body is deduced from the volume of the mercury contained in the tube between the different heights noted.

178. Balance of new construction oscillating with steel springs. This has been recently constructed by Oertling from a design of Mr. Artingstall.

Its principle is, that, instead of the beam and pans being suspended on knife-edges, thin elastic steel springs are used, and adjustments of knife edges from time to time are thus avoided. It is similar in construction to Steinheil's silk ribbon balance. Its advantages as a balance for weighings, where extreme scientific accuracy is not required, consist in its simplicity and durability; but it appears to be wanting in the sensibility and stability requisite for a balance of precision.

179. Model Kit of Apparatus for Local Inspectors of Weights and Measures. Constructed by Oertling.

This portable collection of all the necessary apparatus for comparing imperial weights and measures has been taken from the *Nécessaire des Vérificateurs*, employed in France for verifying metric weights and measures, with a view to its adoption in this country. It includes a Septimal Balance by means of which a weight of 56 lbs. is compared against 8 lbs., the sum of the standard weights contained in the kit.

180. Experimental Gasholder for determining the internal temperature. Constructed by Messrs. Wright & Co.

By raising and lowering the bulb of the thermometers, the tubes of which are made to slide through the top of the gasholder, the temperature of the gas or air at various heights inside the bell can be read off through the glass side, and the mean temperature determined.

181. King's Pressure Gauge, showing mechanical pressure of gas or air. Constructed by Mr. Sugg.

Standards Department, Board of Trade.

The amount of the mechanical pressure of gas or air contained in a gas-holder is shown by this pressure gauge, when it is put in communication with the gasholder by an air-tight tube. The surface of the water in the cistern of the pressure gauge is depressed by the force of the gas or air, and alters the level of a metal cup floating on it. A cord is attached to the float, and passes over a pulley, the spindle of which, aided by friction rollers, carries a pointer moving on a graduated dial, and thus indicates the amount of pressure in hundredths of an inch.

Specimens of Standard Weights and Measures.

182. Copy of Standard Weight, 112 lbs., of Queen Elizabeth.

One of two similar bronze weights deposited in one of the old Treasuries of the Exchequer, and fully described in App. IV. to the 7th Annual Report of the Warden of the Standards.

183. Gilt Steel Yard, line measure, of the same form as the imperial standard yard. Constructed by Troughton and Simms.

Well-holes are cut down to the mid-depth of the bar, where the defining

lines are cut upon gold studs, thus the measure being taken

at the middle portion of the central transverse line, intercepted between the two longitudinal lines. These lines, including the two transverse guide lines, one on each side of the defining line, are 0·01 inch apart.

184. Steel Yard End Measure, showing the form of end-standard yard adopted by the Standards Commission. Con-structed by Troughton and Simms.

The form of the defining ends is that of a spherical surface, whose centre is the centre of the division-line at the middle of the bar's length. The material of the defining end is a highly polished plug of agate, shrunk into a slightly conical hole at the end of the steel bar.

185. Steel Foot End-Measure. Constructed by Troughton and Simms.

186. Two Steel 6-inch End-Measures, one finished and one unfinished. Constructed by Troughton and Simms.

187. 10-Foot Measuring Rod, of pine wood, bound with brass. Constructed by Troughton and Simms.

188. 3-Foot Measuring Rod, of pine wood, bound with brass. Constructed by Troughton and Simms.

189. 1 lb. Avoirdupois Weight, of gun-metal, electro-plated with nickel.

Constructed as an experiment of coating brass or bronze with unoxidisable metal. Oxidation, however, is found to occur at points on the surface of the bronze under the nickel coating.

190. 1 Kilogram Weight, of gun-metal, nickel-plated.

191. Set of Glass Avoirdupois Weights, from 7 lbs. to 1 oz., made experimentally of green bottle glass, not subject to hydroscopic influences. The larger weights adjusted with lead shot.

192. Set of Metric Weights, from 1,000 grammes to 1 gramme. Constructed by Salleron, Paris, of opaque glass, adjusted with mercury to the density of brass weights, and hermetically sealed.

193. Specimen of an Enamelled Iron Weight of 56 lbs., made to resist oxidation, by De Grave, Short, and Co.

194. Specimen of a Patent Brass-cased Iron Weight of 14 lbs.

195. Section of a Patent Brass-cased Iron Weight of 14 lbs., showing mode of construction.

196. Copy of Standard Cubic Foot nickel-plated, with filling apparatus. Constructed by G. Glover & Co.

This is a copy of the standard cubic foot bottle, the primary unit from which the gas-measuring standards were derived. It was verified by weighing its contents of distilled water = 62·321 pounds avoirdupois, according to Sir G. Shuckburgh's determination of the weight of a cubic inch of water. It is used as a direct transferrer of a cubic foot of gas or air, which is driven out from it by raising the cistern and thus introducing water from underneath up to the defining line of a cubic foot. By this arrangement, the nearly undisturbed surface of the water is carried upwards and gradually through the entire height of the bottle, without risk of forming air bubbles.

197. Copy of five cubic feet Gas-measuring Standard, made of anti-corrosive metal, by G. Glover and Co., with scale of capacity graduated in feet and minute fractional parts.

The bell is equipoised when at various depths of its immersion in the water of the cistern by a balance, a portion of which hangs from a cord working in a groove in the circumference of a cycloidal wheel, and attached to the axis of the wheel from which the bell is suspended.

198. Copy of a Standard Test Dry Gas Meter, with testing table. Constructed by G. Glover & Co.

Such test gas meters are authorised to be used for testing stationary meters, where the larger gas measuring standards cannot conveniently be used. The accompanying testing table shows it fitted with thermometers and pressure gauges, and with stand pipes for outlet and inlet communications.

199. Model of a Petroleum Testing Apparatus, for ascertaining the temperature at which its inflammable vapour ignites. Designed by T. W. Keates, Esq., and proposed as a standard for use in accordance with authoritative uniform regulations. Constructed by How & Co.

200. Brass Scale of 41 inches, divided into tenths, and of a metre divided into millimetres ; both scales at 62° Fahr. Constructed by Dollond, and now the property of Mr. Petrie.

This possesses some scientific interest, having been compared several times with Shuckburgh's scale by Capt. Kater. He found in 1830, and again in 1831, that 36 inches of the scale = 35·99893 inches of the imperial standard yard, afterwards destroyed in 1834 ; and assuming the scale of inches to be perfectly correct, that the metre = 39·37045 inches.

Compared at the Standards Office in February 1876, with the bronze official yard, which has a standard metre at 0° C. marked on the same bar. From the mean of six comparisons, 36 inches of the scale = 35·99961 inches of the new imperial standard yard, and the metre = 0·999684 metre, being 0·316 millimetre less than the *Mètre des Archives* at the normal temperature of 0° C.

223b. Fraudulent Balance, seized from a butcher's shop by an inspector of weights and measures.

Standards Department, Board of Trade.

An illustration of the principle of the balance. One of the suspending hooks of an ordinary equal-armed balance is bent outwards thus lengthening that arm of the beam, and enabling the butcher to make about 14 oz. of meat counterbalance a 1 lb. weight on the other end of the beam.

Set of Old Standard Measures lent by the Mayor and Corporation of the City of Winchester.

201. Very old Steelyard Weight, date unknown. Found at Hyde Abbey, Winchester.

202. Set of Standard Troy Weights, from 256 oz. to 1 oz., of Queen Elizabeth. Dated 1588, being the year in which she granted a charter to the city.

203. Set of Standard Weights (avoirdupois), 56 lbs., 7 lbs., 8 lbs., 2 lbs., 1 lb., of Queen Elizabeth, dated 1588, being the year in which she granted a charter to the city. From the Muniment Room, Winchester.

204. Standard Weights (56 lbs., 28 lbs., 14 lbs., and 7 lbs.). Supposed to be of the date of Edward III. From the Muniment Room, Winchester.

205. Standard Yard Measure. Henry VII. From the Muniment Room, Winchester.

206. Standard Quart and Pint of William III. Dated 1700. From the Muniment Room, Winchester.

207. Standard Gallon, Quart, and Pint of Queen Elizabeth, dated 1601. From the Muniment Room, Winchester.

208. Standard Winchester Bushel, given to the Corporation by Henry VII. in the year 1487.

209. Standard Winchester Gallon, given to the Corporation by Henry VII., in the year 1487.

SET OF MEASURING INSTRUMENTS CONTRIBUTED BY SIR J. WHITWORTH, BART., D.C.L., F.R.S.

217. True Planes. The true plane is the foundation and source of all truth in mechanism.

The patent hexagonal surface plate is constructed so as to be supported and suspended from three points, and remains true in either position. The original true planes first exhibited by Sir Joseph Whitworth at the meeting of the British Association at Glasgow, in 1840, were rectangular, and were ribbed, so as to allow of their being supported on three points ; but when large rectangular surface plates were suspended from the two handles a perceptible alteration took place, and they were no longer as true as when supported on the three points.

212. Whitworth's Workshop Measuring Machine, for making difference gauges from correct cylindrical· standards of size.

One division of the micrometer wheel represents $\frac{1}{10000}$ of an inch, one quarter of a division, viz., $\frac{1}{40000}$ of an inch can be distinctly felt and gauged.

No proper size of bearing can be made for an axle to work in without having a difference gauge of such size as experience has proved to be best.

214. External and Internal Standard Cylinder Gauge, 1 inch in diameter.

The standard gauges are usually made from $\frac{1}{10}$th to 2 inches diameter, but they are also made for larger diameters.

They are a necessary adjunct to the workshop measuring machine when making difference gauges.

210. Box of Standard Lengths, of end measure, 1 inch to 12 inches.

Either these standard lengths of end measure or the cylindrical standard gauges are used for adjusting the workshop measuring machine ; for large dimensions these are preferable.

213. Box of Cylindrical External and Internal Difference Gauges, differing by $\frac{1}{5000}$th of an inch in diameter.

By means of these a workman can feel his way step by step and so make the bore of any number of barrels or tubes exactly the same diameter. They illustrate the importance of small differences in size ; while one fits, another $\frac{1}{5000}$ of an inch less in diameter appears not to fit at all.

A tight fit is not a proper fit, there must be a certain difference in diameter between an axle and the bearing in which it has to work ; what the difference should be depends on a variety of circumstances which experience alone can determine.

213a. Ten Standard Flat Surface Gauges, from $\frac{1}{10}$th of an inch to $\frac{1}{100}$th of an inch varying $\frac{1}{1000}$th of an inch.

They are standards for the thickness of sheet metal and the diameter of small wire; they serve for the correction of the wire gauge.

Cylindrical standards are not made less than $\frac{1}{10}$th of an inch in diameter.

216. Standard Screw Gauge of Whitworth thread.

This gauge is to show the form of the Whitworth standard thread. The two cylindrical parts give the exact diameter of the top and the bottom of the screw threads in universal use; and the angle of the thread is 55°, and is rounded off $\frac{1}{6}$th of its depth.

211. Millionth Measuring Machine.

The screw of this machine has 20 threads to the inch, the screw wheel 200 teeth, and the micrometer wheel is divided into 250, therefore each division represents the one-millionth of an inch.

The end of the fast headstock and the end of the movable headstock are true planes parallel to each other; the ends of the piece to be measured must also be parallel true planes; the feeling piece is a piece of steel about $\frac{2}{10}$ths thick, its sides being parallel true planes, and it is introduced between the standard to be measured and the true plane at the end of the fast headstock; when the proper adjustment has been made the movement of the micrometer wheel one division, viz., one millionth of an inch, will cause the feeling piece to be suspended—friction overcoming gravity.

The power of measurement and the true plane are the two great elements in practical mechanics.

An idea may be formed of the millionth of an inch, from the fact that if a sheet of foreign letter paper were divided into 4,000 thicknesses, each thickness would represent the millionth of an inch.

APPARATUS USED BY DR. JOULE, F.R.S., FOR ASCERTAINING THE MECHANICAL EQUIVALENT OF HEAT.

218. Revolving Electro-magnet, used in 1843 for ascertaining the **Mechanical equivalent of Heat.**

Part of the apparatus used in 1843 for the determination of the mechanical equivalent of heat: viz., a revolving piece, holding a glass tube filled with water, and containing an electro-magnet. This worked between the poles of a powerful magnet; and the heat evolved by the rotating electro-magnet was measured by the rise of temperature of the water. In this manner the quantity of heat lost by the circuit was ascertained when the machine worked as an engine; and, on the other hand, the quantity of heat produced when work was done on the machine was also measured. 833 ft. lbs. was the mechanical equivalent of a degree Fahr. in 1 lb. of water, as determined by these first experiments.

219. Calorimeter, containing a **revolving agitator.** This was employed in the experiments on the heat evolved by the friction of water, made in 1849. The equivalent arrived at was 772 ft. lb.

40075.   D

220. Cast-iron Vessel, containing **Friction Disk,** to revolve under mercury. Used in 1849 to determine the mechanical equivalent of heat by the friction of cast-iron against cast-iron. The equivalent arrived at was 775 ft. lb.

221. Electro-magnet consisting of a broad plate of half-inch iron, having a bundle of copper wires coiled round it. Employed in the first determination of the mechanical equivalent of heat.

222. Apparatus for determining the temperature of water at its maximum density.

Used in the experiments on atomic volume and specific gravity by Playfair and Joule (Memoirs of the Chemical Society, vol. iii., 1846). It consists of two tall vessels, connected together by a stop-cock at the bottom, and a trough at the top. A minute difference of the temperature of the water in one of the vessels from that of the maximum density, determines a flow through the trough to the vessel still nearer the temperature of maximum density. The temperature of water at maximum density was thus shown to be 39·1.

223. Paddle Apparatus, by means of which Dr. Joule. determined the dynamical equivalent of heat. Described in Philosophical Transactions for 1850, page 65.   *Sir William Thomson.*

II.—MEASUREMENT OF LENGTH.

A. STANDARD SCALES.

223a. Model of an Ancient Egyptian Standard Cubit, dated in the reign of King Horus, 9th Pharaoh of the 18th dynasty (1657 B.C.).   *Mrs. Chisholm.*

The ancient standard measure, of which this is a copy, was found in the ruins of Memphis, and is now in the Royal Museum at Turin. It is a Royal cubit of seven palms or 28 digits. The total length of this end standard measure is 523·5 millimetres or 20·6 inches, and agrees very nearly with that of several other ancient Pharaonic cubits still existing, as well as with the length of the Royal cubit as deduced by Sir Isaac Newton from the dimensions of the Great Pyramid, the mean length being 525 millimetres. The original natural cubit, or cubit of a man, of 6 palms is also marked upon this measure, being equal to 463 millimetres or 18·24 inches, and also the ancient Egyptian foot of 16 digits, or ⅔ of the natural cubit, and equal to 12·16 inches, or 1·013 English foot. The great span of 14 digits and the small span of 12 digits are also marked.

227. Standard Scale, in porcelain, showing the relations of modern British and ancient Great Pyramid inches.

Prof. Piazzi Smyth.

This scale was prepared to order by M. Casella, of London. It exhibits side by side 25 modern British inches and the same number of ancient Great Pyramid inches, similarly subdivided.

The 0 divisions of both sets of inches coincide at the left hand exactly, but from thence the gradual growth of the difference of 0·001 of an inch per inch in favour of the Great Pyramid scale may be traced, until at the 25th inch the difference amounts to 0·025 of the British inch. At that point, however, it is to be noted that 25 Great Pyramid inches are just one 10 millionth of the earth's semi-axis of rotation, or the nearest earth commensurable and most scientific unit ever yet proposed as a standard of length.

228. Standard Five-Inch Scale, in smoky agate, for microscope sight. *Prof. Piazzi Smyth.*

The material, which came from Brazil, and was worked up and divided to order by M. Jules Salleron, Paris, was chosen as being a natural product of almost infinite age, and therefore settled condition, harder than steel and utterly unoxidisable. The particular standard length adopted is shown for microscopic sight. It depends not on one pair only, but on 20 available pairs of lines, each five inches apart, drawn with a fine diamond point, and is intended to typify both one fifth of the sacred cubit of Israel and one 50 millionth of the earth's semi-axis of rotation.

229. Standard Five-Inch Scale, in white chalcedony, for microscope sight. *Prof. Piazzi Smyth.*

This material, which came out of some ancient Roman palace, is chosen for the same reasons as that last described. The scale is divided on the same system.

230. Standard Five-Inch Scale, in red porphyry, for microscope sight. *Prof. Piazzi Smyth.*

This material came from an Imperial Roman palace to which it had been taken under the Cæsars from some far more ancient Egyptian temple. It was originally quarried by the Egyptians of 3,500 years ago in the rich porphyry district between Thebes and the Red Sea, and it has been adopted for the same reasons as the preceding examples, and the scale is divided on the same system.

232. Standard of Length, derived from the earth's polar axis, which is unique and common to all terrestrial meridians. *Prof. Hennessy.*

This standard, proposed by Professor Hennessy, is a bronze bar, which, at 15° of temperature centigrade, is equal to the fifty millionth part of the earth's axis.

233. Steel Chain, of fifty links, whose total length is the millionth part of the earth's axis, or very nearly 500·5 English inches. It is nearly equal to the half chain of two perches in Irish plantation measure. *Prof. Hennessy.*

235. Standard Yard Measure, German Silver, with one chamfer divided to inches and 10ths, for temperature 60° Fahr. *Elliott Brothers.*

239. Steel Tape Measure, 66 ft. For testing tapes, divided to feet and inches on one side and links on the other. *Elliott Brothers.*

D 2

247. Measure of Length, according to natural principles.
Hans Baumgartner, Basle.

250. Half Metre, maple, with points.
Geneva Association for the Construction of Scientific Instruments.

251. Brass Metre. (Grand Duchy of Baden Model.)
Geneva Association for the Construction of Scientific Instruments.

252. Steel 2-Metre Standard, with points. (German Model.)
Geneva Association for the Construction of Scientific Instruments.

253. Brass Standard Metre. (Swiss Model.)
Geneva Association for the Construction of Scientific Instruments.

The Geneva Association for the Construction of Scientific Instruments possesses in its laboratories a machine for the division of straight lines, to the construction of which it has endeavoured to apply all the improvements of modern science. These efforts have been crowned with success, and the increasing reputation of this machine, which may be considered as the most complete at present existing, has obtained for the Geneva Association orders for metrical standards from several European Governments.

The machine is worked automatically, that is, all the process of dividing is done mechanically. Thus, apart from the inaccuracy consequent on the temperature of the operator, it avoids the errors proceeding from inattention or fatigue on his part. Mechanical action has, moreover, the advantage of being more regular, seeing that the motive power is always equal.

An ingenious contrivance enables the correction of errors due to a change of outward temperature during the process of division, to be effected ; thus, at any temperature a correct graduation corresponding to 0° is obtained. By the same means, a division of any length may be made, although the pitch of the thread of the screw of the machine, and the length of the division required, may be incommensurable.

The pitch of the screw has been thoroughly examined and corrected, so as to guarantee accuracy to the $\frac{1}{100}$ of a millimetre.

This machine for dividing straight lines has been used to effect the normal division of the large machine for dividing circles which stands by its side on the same bed of concrete. This application has been the means of exactly ascertaining the coefficient of dilatation of the machine for dividing straight lines. The maximum of error found in the division of the normal circle was less than a second. It is impossible to expect greater accuracy when it is remembered that the arc of a second on the circumference of the divided circle represents about $\frac{1}{400}$ of the millimetre.

259. Standard Metre, with rack motion to be used as a machine for dividing other metres.
Geneva Association for the Construction of Scientific Instruments.

This instrument may be used both as a comparing apparatus, and as a machine for dividing fractions of the metre, for the use of comptrollers of weights and measures. A small pointer or cutter is traversed, by means of a rack, along the meter, while a very simple lock action enables the millimetric displacements of the indicator to be registered without reading the divisions.

259b. Iridio-platinum Standard Metre, in course of manufacture. *Johnson, Matthey, and Co.*

259c. Section of Metre when finished, showing the form determined upon by the International Commission.

Johnson, Matthey, and Co.

260. 6-foot Measuring Rod, for uneven ground, for engineering and scientific purposes. Designed by the exhibitor, and made by Messrs. T. Cooke & Sons. *Edward Crossley.*

The apparatus consists of a wooden rod 6 ft. in length, with metal terminations containing spherical cups fitting on to spherical heads upon tripod stands. Three tripod stands are required. Each terminates in a flat ring upon which the base of the short pillar carrying the spherical head is horizontally adjustable, and to which it can be clamped. The rod is supported by two tripod stands, while the third is set forward to receive the rod in its next position. The inclination of the rod is read off to half-a-minute in each position by means of a level and arc attached to the centre of the rod. The true horizontal distance is then obtained by applying a tabulated correction for each inclination of the rod.

This instrument will give an accuracy of 1 in 10,000 on any sort of ground, even with a gradient of one in four.

266. Ivory Pocket Measures. *T. Hawksley.*

Containing the Fahrenheit and centigrade temperature scales ; English inches divided to $\frac{1}{50}$ in., and centimetres divided into millimetres ; designed for the purpose of introducing into everyday use the decimal system of measurement.

269. Foot-scale-plate. A rectangular brass-plate containing twenty different foot-scales, made in 1769 by Adam Steitz in Amsterdam. It is a copy from the original deposited in the Town Hall of Amsterdam. *Prof. Buys-Ballot, Utrecht.*

269a. Meter Diagram. *A. and F. Stanley, New York.*

289. Meter Scale with double divisions, for 0° and 20° C., by Breithaupt and Son, in Cassel.

Mathematical and Physical Institute, Marburg (Prof. Dr. Melde).

295. 2-Meter Standard Measure in steel.

F. W. Breithaupt and Son, Cassel.

This normal 2-meter standard is an end measure, as well as a line measure, with divisions throughout in centimeters, and on both the end decimeters in millimeters. It was graduated on the longitudinal dividing machine, constructed by George Breithaupt in the year 1850, for the temperature of 0° Celsius, as far as 0·01 mm. precision.

Such normal double meters have been made in large numbers on steel, as well as simple normal meters on brass, for the Imperial Commission of Normal Weights and Measures.

With this longitudinal dividing machine a length of one meter may be graduated *uninterruptedly* even to the smallest subdivisions.

296. Standard Scales in Rock Crystal, viz.:

  a. Scale of 10 cm.

  b. Scale of 15 cm.

  c. Scale of 20 cm.

The scales are cut parallel to the axis of the crystal, and are divided into millimeters ; the first and the last millimeters are divided into ten parts. The graduation has been carried out by Mr. Brauer in St. Petersburg.

Hermann Stern, Oberstein, Principality of Birkenfeld.

299. Meter in Steel, line measure divided into millimeters.

L. Steger, Kiel.

301. Apparatus for comparing Standard Measures of Length, by Stollenreuther.

University of Munich (Prof. von Jolly).

302. Meter in the form of a ruler with subdivisions.

M. Meyer, Teacher of Mathematics at the Gymnasium, Halle.

In order that the subdivisions of the meter should be clearly understood by the scholars, one side of the square has no divisions, that side is of the exact length of a meter ; the second side is divided into decimeters only, the third into centimeters, and the fourth into millimeters.

303. Schönemann's Measuring Wedge, reading to ·01 mm.

Gewerbe Schule at Halle (Director, Kohlmann).

304. Schönemann's Measuring Wedge, reading to ·01 mm.

Kleemann, Mechanical Engineer, Halle.

306. Meter-scale in Brass.

Prof. Baron von Feilitzsch, Greifswald.

This scale was constructed at the workshops of Messrs. F. W. Breithaupt and Son in Cassel (Province Hesse), and is remarkable for its great accuracy.

It is divided, on silver-plated brass, into centimeters, and at both ends into millimeters.

309. Original Meter Scale, iron. One of the forty specimens which were delivered to the members of the Meter-Com-

mission in the year vii. of the French Republic; formerly in the possession of Tralles. *Prof. Dr. Dove, Berlin.*

Original Meter Scale by Tralles (of iron).—One of the 40 standards which were delivered to the Commissioners.

Iron Meter à touts.—This meter, which was presented to Hapler by Tralles, was one of the three which the latter had made at the same time, by Lenoir, with the 15 which were distributed among the members of the Commission.

After the completion of the new measurement of degrees performed by Delambre and Mechain, the real length of the meter was determined by the Commission, consisting of, Swinden, Tralles, Laplace, Legendre, Cizcar, Mechain, and Delambre, in their report of the 6th Flórèal year 7, to be 443·295936, and the distance of the Pole, by assuming an oblateness of $\frac{1}{334}$, from the equator having been calculated to be 5130 ys toises, it was legally accepted as "*mètre orai et definitif*" at 443·296.

"Cette unité nommée mètre qui est le dixmillionème partie du quart du méridian revient selon les anciennes mesures à 3 pieds 11·296 lignes, en employant la toise du Peron à 13 degrés du thermomètre à mercure divisé en 80 parties."

312. Standard Meter on brass in mahogany case.
Ed. Sprenger, Berlin.

313. Standard Meter on Steel. *Ed. Sprenger, Berlin.*

314. Standard Meter on Wood. *Ed. Sprenger, Berlin.*

315. Standard Tape Measure, 20 meters.
Ed. Sprenger, Berlin.

315a. Standard Stirling Ell, believed to be a copy of the standard Scottish ell adjusted at Edinburgh, 26th of February 1755. *The Burgh of Stirling.*

315b. Standard Meter. *Bock and Handrick, Dresden.*

Two Standard Meters (boxwood).
Bock and Handrick, Dresden.

2-Meter Standard Measure.
Bock and Handrick, Dresden.

B. Telemeters.

226. Telemeter. For measuring the distance of inaccessible objects. *Patrick Adie.*

This instrument, the first of its name, was patented by Mr. Adie in 1863. It consists of two powerful telescopes at the ends of a fixed base; the united rays, by total reflection, give simultaneous observation in the eye-pieces.

234. Telemeter, for determining distant inaccessible points by one observation. Manufactured by Adie & Son, Pall Mall.
Prof. Pigot.

242. Nolan's Range Finder.
1. Two-angle measures.   Right and left.
2. Two Y supports.
3. Two tripods.
4. Two tripod buckets.
5. Two leather boxes with straps to contain items 1 and 2.
6. A 50 yards measuring tape.
7. A metal calculating roller.
8. Two magnifying glasses.
9. A leather case with strap to contain items 6, 7, 8.
10. A leather numnah which fits under the saddle of item 1, and on which the two boxes, item 5, are strapped.

War Office.

243. Two Instruments for Measuring Distances. Constructed by Dr. Meyerstein.

Prof. W. Klinkerfues, Göttingen.

262. Telemeter with prism by Col. Goulier for the rapid measurement of distance.   *M. Tavernier Gravet, Paris.*

263. Pocket Telemeter. By M. Gautier.

M. Tavernier Gravet, Paris.

264. Telemeters.   *Fortin Hermann Bros., Paris.*

285. Collection of War Telemeters. These instruments, which are based on the speed of transmission of sound, are intended for measuring distances in the field.

Le Boulengé, Liége.

307. Instrument for Measuring Distances, according to the systems of Kleinschmidt and Breithaupt.

Royal Museum, Cassel (Dr. Pinder, Director).

The instrument for measuring distances was constructed entirely of brass by J. C. Breithaupt during the second half of the 18th century. It consists of a rail of 0·978 m. in length, serving as measuring-base, on both ends of which is attached a movable telescope for sighting the object the distance of which is to be determined. From the known length of the base, and the indicated angles which the adjusted telescopes form with the base line, the distance sought is ascertained by trigonometrical calculation.

C. Gauges and Callipers.

11. Set of Gauging Instruments.   *Dring and Fage.*

Head rod. For ascertaining the head diameter of a cask, and working out the contents.

Bung rod and slide. For finding the bung diameter and diagonal of a cask. The rod is divided into inches and tenths, with a line of imperial area and diagonal line ; this last gives the approximate content without calculation, and is computed on the assumption that most casks are similar to one another in form, and therefore vary as the cubes of their like dimensions.

Long callipers used for finding the internal length of a cask from head to head.

Cross calliper. Used for finding the external diameter of a cask.

Stave gauge. For finding the thickness of the stave in a cask.

236. Sliding Calliper Gauge, with tangent screw and vernier for reading $\frac{1}{1000}$th of an inch inside and outside measurement. *Elliott Brothers.*

237. Decimal Gauge, German silver, with screw and ratchet motion for measuring to $\frac{1}{1000}$th of an inch. *Elliott Brothers.*

246. Aerial Spider Line Micrometer of great delicacy, measuring an object to the 100,000th of an inch.

Dr. Royston-Pigott, F.R.S.

The image of sets of spider lines of a recording micrometer placed beneath the stage, is formed by a half inch objective, five inches from the spider lines. This image is in fact a miniature diminished exactly seven times. The micrometer reads to the 1–20,000th of an inch, $\frac{1}{20000}''$; consequently the image is measured seven times more minutely. This would be the 1–140,000th, or $\frac{1}{140000}$th of an inch (English). On the whole, therefore, the instrument may be said to measure to the 1–100,000th, *i.e.*, $\frac{1}{100000}$th of an inch.

These aerial spider lines are made to move about the object to be measured at the will of the observer, and come into the focus of the microscope by regulating the plane of the aerial spider lines.

246a. Wollaston's Single Lens Micrometer.
Wollaston Collection, Cavendish Laboratory, Cambridge.

(Phil. Trans., 1813, p. 119.)

256. Callipers, for clock and watch making.
Geneva Association for the Construction of Scientific Instruments.

Much used in clock and watch making for measuring thicknesses. This instrument gauges to $\frac{1}{60}$th of a line, or $\frac{4}{100}$th of a millimetre.

The divisions traced on the steel arc are not equal, but are calculated to measure equal increments of the interval between the two nibs. They increase therefore with the chords.

257. Curious Steel Callipers for very accurate measurement, by Paull of Geneva, 1777. *Royal Society.*

267. Apparatus for **measuring** the exterior **diameter** of the **Gun Barrel** and the interior diameter of the rings to be shrunk on the same ; constructed by G. Brauer.

Arsenal of St. Petersburg.

The apparatus consists of a bar, with two adjustable arms, which is suspended across the cannon ; one arm being brought into contact on one side with the surface of the cannon, the other arm with its contact lever is brought into contact with the surface of the cannon on the other side, in such a manner that this contact lever at its upward and downward motion, by means of the vertical screw at the greatest diameter, indicates zero. In order to exactly determine the diameter, the divided movable scale is adjusted, after taking off the apparatus from the barrel of the cannon, between the two arms, so that the contact lever again indicates zero. The same scale serves also for measuring the interior diameter of the rings. The apparatus has been constructed and made by G. Brauer at St. Petersburg.

267a. Photograph of an Apparatus for **measuring** the eccentricity of the chamber and the curve of the **bore** of **cannons.**

G. Brauer, St. Petersburg.

The apparatus consists of two parts :—

1. A body, which is pressed into the mouth of the cannon by means of an endless screw. In this screw a telescope is fixed which can be turned about its own axis, and is provided with a filar micrometer and a position circle.

2. A piece, which can be slided along the barrel, being turnable about the axis of the bore, and in whose centre is a glass plate with engraved cross. This cross is viewed through the telescope before mentioned, and the determination of the position of the cross on the filar micrometer indicates the elements for determining the curve of the bore and the eccentricity of the chamber.

The apparatus was constructed by the exhibitor for the Russian Marine Artillery Department.

267b. Instrument for **Measuring** the **Bore of Cannons** (Etoile Mobile).   *G. Brauer, St. Petersburg.*

This instrument consists of a ring, in which two parallel rods slide longitudinally side by side, and one of which carries the scale, the other its vernier. The sliding rods are pressed asunder by means of springs, so that their exterior steel ends touch the side of the bore to be measured. As this contact must not take place during adjustment, the sliding rods are brought together by a bolt. A screw perpendicularly under the sliding rods is intended for adjusting them according to the greatest diameter, which has to be done afresh at each measurement. If the chamber of a breech-loading gun is to be measured with the apparatus a set of two rings has to be added, of which the one in front carries a telescope for viewing the scale. The two exterior rings are joined to each other by four bars, and these bars have a movement in the centre ring, which during the operation of measuring is pressed into the back part of the cannon bore by means of four screws, and the whole apparatus is then moved as required.

267c. Apparatus constructed for **measuring** the exterior **diameter** of small **cylinders** with an accuracy of 0·001 inch.

G. Brauer, St. Petersburg.

This apparatus, which was employed in the experiments on the elasticity of gun-metals, steel, cast-iron, &c., in Russia, is provided with an immovable pillar and a contact-lever, which can be adjusted by means of a screw-movement. The cylinder to be measured is placed between the two, and the screw turned until the lever points to zero, and then the reading is effected by the vernier of the longitudinal scale.

267a. Apparatus for Measuring the Breeches of Large Guns.   *M. Gadolin, St. Petersburg.*

267d. Apparatus for **Measuring** the **Length** of the **Impressions** made by the **Rodman Scale.**

Technological Institute at St. Petersburg.

The copper plate, on which is the impression to be measured, is placed on the slide of the apparatus, and then one end of the impression after the other is brought under the cross thread of the microscope by means of the screw of the slide, when the reading can be made on the head of the screw. By means of a micrometer eyepiece also smaller dimensions can be measured.

The apparatus belongs to the Technological Institute at St. Petersburg.

267e. Micrometer for Guns and Tubular Objects.

Samuel B. Allport.

This is a tube, provided with three spring arms, radially disposed round its end, between which a cone is inserted. The cone is connected with a screw at the other end of the tube, whereby it is projected or withdrawn, and so contracts or expands the spring arms.

The angle of the cone bears such a relation to the pitch of the screw that the expansion or contraction of the arms when inserted in a tube will indicate the variations in its bore in thousandths of an inch by suitable divisions on the head of the screw.

267e. Apparatus for measuring the eccentricity of the pivots of a polar or transit axis, with an approximation of ·00001 inch.

A. Hilger.

271. Calliper, with **Dial,** of the English inch measure, divided into eighths. *M. Isvard fils.*

272. Calliper, with **Dial** of two centimetres, divided into tenths of a millimetre. *M. Isvard fils.*

283. Cylindrical Gauges differing in diameter by one ten thousandth of an inch. Other gauges and specimens of surfaces.

Royal School of Mines.

284. Universal Calliper, with slide and reverse action.

Geneva Association for the Construction of Scientific Instruments.

Instrument of measurement, for ascertaining equally the thickness, the inner diameter, and length of tubes.

286. Apparatus for measuring accurately the **Diameter** of **Wires,** for testing whether pivots and other turned objects are perfectly circular in form, and for the determination of the error when they are not truly circular.

Landsberg and Wolpers, Hanover.

287. Apparatus for measuring the Thickness of thin metal plates, sheets of paper, &c.

Landsberg and Wolpers, Hanover.

288. Calliper-Compasses for larger measurements.

Landsberg and Wolpers, Hanover.

288a. Photographs, showing two kinds of machines for measuring with great precision the alterations in shape produced in metals by tension and compression. *Dumoulin Froment, Paris.*

291. Calliper Apparatus, for accurately determining diameters and lengths up to 150 mm.
A. Meissner (H. Müller and F. Reinecke), Berlin.

$\frac{1}{25}$ millimeter can be obtained by direct reading by means of a microscope, and the $\frac{1}{250}$ part of a millimeter by estimation.

293. Collection of Timber Callipers for the use of foresters.   *C. Staudinger and Co. (F. W. von Gehren), Giessen.*

A collection of tree-callipers ("Baumkluppen"), mostly in use for the purpose of comparison with those of Staudinger's construction, by many authorities recognised as the best.   A list of the names is added to the collection.

298. Calliper Compasses, with plane contact lever.
Physical Institute of the University of Kiel (Prof. Dr. G. Karsten).

This apparatus, which is in the possession of the Physical Science Institute of the University of Kiel, was constructed in 1832 by Repsold, and made use of by Schumacher for comparing the platinum kilogramme of the archives with the Danish.

(*See* Schum. Astronom. Jahrbuch, 1836, p. 243.)

A description by G. Karsten of the instrument will be found in "Vom Maasse und vom Messen," vol. I. of the "Encyclopædie der Physik," p. 506 and following.

308. Apparatus for **measuring** the **Thickness** of **Thin Plates.**   *R. Fuess, Berlin.*

308a. Improved Patent Measuring Gauge, with patent releasing arm.   *Wm. Henry Laidler.*

This gauge is constructed to enter a hole drilled in a plate; the arm will clear any rough edge or burr, and when the measurement is taken the arm can be released, and the instrument withdrawn, without altering or interfering with the indication. Each division on the vernier shows $\frac{1}{1000}$ part of an inch.

308b. Improved Ivory Calliper Gauge, with Engineer's Slide combined.   *Wm. Henry Laidler.*

D. Cathetometers.

241. Differential Cathetometer, an apparatus designed for measuring variations in the length of solid bodies, particularly of rods and wires.   *Dr. Heinrich Streintz, University of Gratz.*

The principle on which this apparatus is based is, reading by reflection from two mirrors. Two levers, having small mirrors *ss* attached to them perpendicular to their axis, are turned by the flat ends of the bar to be measured as indicated in the drawing. If a telescope and a scale are placed, at some distance, in such a position that the image of the scale reflected by the mirror is visible through the telescope, each variation in the position of the point at the end of the lever will be magnified to a degree indicated by the quotient, the numerator of which represents the double distance of the mirror from the scale, and the denominator that of the point from the axis of rotation.

As the latter distance can be diminished to one centimeter in the apparatus, and as, moreover, the telescope with the scale can be placed at any distance within which distinct images will be seen, say five metres, a shifting of the reflected image by one millimeter will be equal to displacement of the end-surface of the bar to be measured by 0·001 millimeter. As, however, the tenths of the millimeter even can be pretty accurately determined, the reading will be correct as far as the 10,000th part.

There is no doubt that the correctness of the reading with this apparatus can be carried still further, if mirrors of superior quality and powerful telescopes are employed.

In measuring the variations in the length of a wire a flat surface must be given to that part to which the wire is to be suspended, as well as to the part on which the weights are placed, and to which the levers are to be applied.

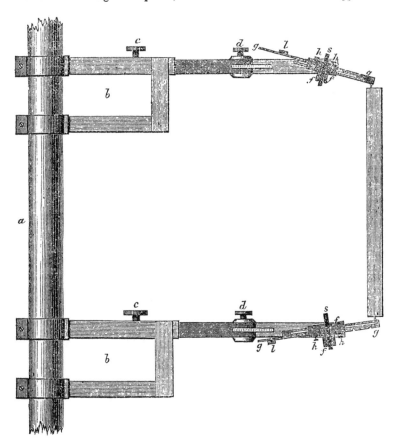

The arrangement of the apparatus is as follows :—

To a solid brass pedestal, which rests on three adjusting screws, a strong glass tube a $1\frac{1}{4}$ meter long is cemented, to which two brass bars bb are clamped. Each of these brass bars consists of two parts, of which one moves or slides in the other in such a manner that it can be either lengthened or shortened. The set screw c serves for fixing the chosen length. By means of the joint d a horizontal rotatory or veering movement of the fore-part of the bar can be effected. Close to its free end there is on each side a steel point— the two forming together an axis—which is held by a bow ff carrying two cups, in such a manner that the bow can be easily but surely turned round this axis.

The lever gg which must be firmly connected with the bow, has longitudinally a slit, or slide, with two sliding pegs placed in a level position with the axis and fastened to the bow, along which the lever can be moved, so that at whatever distance from the axis the extreme end of the lever may be fixed it must always turn with the bow around the axis.

In order not to be obliged to take the measurement of the length of the lever afresh at each experiment, it is provided at its upper surface with conical-shaped cavities in which the screws hh catch. These conical-shaped cavities would be, properly speaking, visible only in a drawing of a vertical section, but not in a front view, as represented in the sketch, but they have been marked in the drawing for the purpose of rendering the description more intelligible.

The measurement of the length of the lever at the different cavities is accomplished by means of a spherometer. The lower bar b is arranged in a manner quite similar to the upper bar.

In order to meet the requirement that the levers should but lightly press against or touch the end surfaces in the manner indicated in the drawing, small balancing blocks ll can be attached at any point to the levers.

For the purpose of reading, two telescopes with vertical scales are required, which must be placed in juxtaposition, that is to say, by the side of each other. Presuming the staff to move upwards and downwards without varying its length, the difference in the reading in the upper and in the lower mirror will naturally be of the same value in every position of the staff.

A glass tube has been chosen to serve as a column a, because glass possesses a very small coefficient of expansion. Moreover, in using the instrument, the tube must be filled with water and two thermometers placed in it, by means of which any change in the temperature that may take place during the process of measuring can be accurately determined.

A similar, although less perfect, apparatus has been employed by the exhibitor in two experiments already, namely, "as regards the variations in " the elasticity and the length of a wire under the influence of a galvanic " current." *See* Transactions of the Academy of Sciences at Vienna, Vol. LXVII., Part II., April 1873 ; and "respecting the moderation of " the torsion oscillations of wires." *See* Transaction of the Academy of Sciences at Vienna, Vol. LXIX., Part II., March 1874. Extracts of both treatises have also been published in Pogg. Ann.

The apparatus can be employed in measuring the coefficients of expansion, coefficients of elasticity, the after effects of elasticity, the expansion produced by magnetism, &c., and will secure in every case an accuracy not hitherto attained, not only by reason of the correctness of the readings, but also on account of the correction of temperature rendered possible through the employment of the glass column.

The measurement is likewise very easy of accomplishment, since a manipulation such as is the case with ordinary cathetometers is not required, as the variations taking place in the wire can be perceived through the telescope directly magnified and projected on the scale.

In most cases it will only be necessary in making such experiments to know exactly the absolute length of the body to be measured in equal per cents, as well as the elongation, for measuring which a good scale, or a very simple cathetometer, is all that will be required.

241a. Original Cathetometer by Dulong.
Polytechnic School, Paris.

241b. Cathetometer, with two Levelling Micrometer Telescopes.
Physical Science Cabinet of the Imperial Academy of Sciences, St. Petersburg.

241c. Drawing of a small Cathetometer, used by Prof. Mendeleeff in his investigations on the tension of gases.
Prof. Mendeleeff.

In order to eliminate a source of many errors the eyepiece is fixed in the telescope, and the whole cathetometer has to be put at the required distance from the object to be observed.

The telescope is provided with a micrometer screw

258. Great Cathetometer, for reading differential levels more than a metre apart.
Geneva Association for the Construction of Scientific Instruments.

This instrument is composed of a tripod supporting a central rod, which bears on its upper part the brass column, or prismatic piece, along which the telescope moves in a right line. The dimensions of the column are great, so as to avoid all flexure. The division on the silver plate is in millimeters, and the vernier of the slide gives readings to the 50th of a millimeter. This instrument has two levels ; the one placed between the rings that support the telescope, the other placed perpendicularly to the first upon the table situated at the base of the column.

The universities of Berlin, Rome, Dorpat, Neuchatel, &c. have instruments of this pattern.

273. Cathetometer, by Casella. The telescope moves on a girder-shaped brass bar, to which the scale is attached, and is furnished with a micrometer eyepiece, by means of which readings can be taken without moving the telescope. The instrument is supported by a massive iron frame-work.
Prof. A. W. Rücker, Leeds.

292. Cathetometer. *C. Bamberg, Berlin.*

The principal division of the instrument is executed on silver to centimeters. The division into single millimeters has been made on a scale connected with the principal slide, whose divided surface is on a level with that of the centimeter graduation. The millimeter-scale moves with the principal slide, which carries the means for reading and adjustment (microscope and telescope). The reading of the meter-division is effected (as far as 0·001 mm.) by means of the eyepiece-micrometer of the microscope. The micrometrical displacement of the principal slide, which is balanced in all positions round the longitudinal axis of the scale column, takes place by a peculiar contrivance, which avoids all one-sided pressure. The slide with the clamp is balanced by a counter-weight suspended from the ceiling or a trestle, so that its ascending and descending motion is effected with great facility.

294. Photograph of a Cathetometer, constructed by Staudinger and Co.

C. Staudinger and Co. (F. W. von Gehren), Giessen.

The peculiarities of the construction may be learned from the photographs. The instrument has an available graduated length of one meter; the column with the counter-weight turns completely around the long vertical axis, and is provided with adjustments, reversing telescope, and water-level.

305. Cathetometer.

Prof. Baron von Feilitzsch, Greifswald.

The cathetometer consists of a central axis, and a prism turning round the same. For placing the central axis in a vertical position a cylindrical water-level, indicating to 10 seconds, is employed. A scale on silver one meter in length, and divided throughout into millimeters, is inlaid into the prism. Sliding along this is a telescope, likewise fitted with a cylindrical water-level, the supporter of which is provided with a vernier indicating $\frac{1}{20}$ mm.

There is also a water-level, for regulating the direction of the prism.

310. Cathetometer, so arranged as to be used for horizontal measurement.   *Prof. Dr. Dove, Berlin.*

311. Cathetometer, by Breithaupt and Son, Cassel, with riding level.   *Polytechnic School, Cassel (Dr. E. Gerland).*

The following improvements, contributing partly to more minute readings with the apparatus, partly affording means of correction of the several parts, have been added to the well-known constructions.

The firmly placed central axis, around which the long frame and prism turns, can be placed vertically by a special cylindrical water-level, indicating to 10 seconds, and which is fastened to the frame independently of other parts, in order that the vertical position of the axis required in very fine measurements may be readily ensured; the more so, as all other observations are based on the correct adjustment of this water-level. The vertical position of the axis is effected in the same manner as with an ordinary levelling instrument, and any deviations of the water-level are corrected half on the adjusting screw of the same, and half by the regulating screws of the tripod.

The prism, the inlaid silver scale of which, 1 meter in length, is throughout divided into millimeters, and fitted with a vernier for $\frac{1}{20}$ mm., can be placed in a horizontal position and parallel to the face of the scale. By means of adjusting screws, and a reversible riding level, the telescope can be placed in the required position.

If this is done, the bubble of the telescope water-level will remain unchangeably in the centre during the rotation of the whole instrument on its central axis, as well as during the upward and downward motion of the slider.

A very severe proof consists in sighting a distant object with the telescope, which is then reversed in its sockets, and the apparatus turned round 180°, at which the object should be intersected again by the eyepiece cross.

The immovable cross in the ocular is cut on glass, in order to prevent hygroscopical and other interruptions. For the purpose of obtaining the rectangular position of the telescope, the supporter may also be placed with one end between points, while an elevation screw is fixed to the other. The essential point for effecting the before-mentioned correction by employing the attaching or adjusting water-level consists simply in adjusting the water-level axis exactly to the leaning face by means of the correction arrangement marked *a* in the drawing. The proof is effected by reversing the angle vertically, the water-level thus turning between its points. If after the proper attachment the bubble deviates from the centre, half of this deviation must be corrected by the regulating screws of the tripod, and the other by the correction arrangement *a*. It is, however, to be mentioned that, *previous* to the above proof, the parallel position of the water-level axis towards its points of attachment is to be examined, which can effected by reversing between its two points, and thereby a deviation of the bubble, if there be any, will be removed half by the adjusting screw *b*, and the other half by the arrangement *a*. Finally, there remains the examination and correction of the water-level sideways to be made, which is done in the usual manner by the screw *c*. This attaching or adjusting water-level may also be recommended for other purposes, for instance, in mounting of machines, &c.

Regarding the peculiar construction of the aforesaid adjusting water-level, the suspension between two points, in general, it may be remarked that the same has been derived from the compensation-level constructed by F. W. Breithaupt and Son some years ago (vide Dingler's Polytechn. Journal, vol. CLIV. p. 401). In what manner this principle has been adopted in other mechanical workshops, and represented partly as an invention of their own, has been proved by an article in Carl's Repertorium, vol. IX., p. 127, by the addition of an arrangement or simplification totally at variance with the construction.

E. Dividing Engines.

248. Instrument for dividing Mathematical Scales or Rules.

H. M. Commissioners of Patents

This instrument is to be used for dividing scales according to the French, Swiss, or English measures of length, and is provided with a vernier for obtaining the smaller divisions of the scale. It can also be adapted to the production of diagonal scales.

248a. Dividing Engine, made by the late Mr. Bryan Donkin, F.R.A.S., in the year 1828. *Bryan Donkin.*

The principle involved in the construction of the machine is the employment of a compensating arrangement, by which great accuracy is obtained, notwithstanding the inequalities of the screw used in the machine for advancing the cutting tool. The machine consists, first, of a table moving upon wheels on a railway. To the under side of the table is attached a clasp nut in two parts, moved by the main screw, which is below the table, and exactly parallel with the line of motion. To effect the compensation the table consists of an upper and lower plate, the upper one being capable of a small motion independent of the lower plate. The lower plate carries the fulcrum of a bent lever, whose arms are at right angles and as 50 to 1. This lever moves in a vertical plane, so that the longer arm lies by gravity alone on the undulating edge of the compensation bar. The upper plate is pressed endways against the shorter arm of the bent lever by means of a spring keeping them always in contact. By a kind of parallel motion the two plates are attached so as to allow of the very small motion required in the upper plate independently of the lower. The compensating bar, which is of the length of the screw, has 50 narrow slips of metal placed upon it, each having an adjusting screw by which the ends of the pieces may be placed in a continuous line, or above or below the line, as required by the mode of adjustment. This bar is carried by a pivot at one end, and the other end is raised or depressed by a screw, which adjusts the compensating bar to the total length moved through by the guide screw.

In the case of dividing a scale, the swing frame carrying the cutter or diamond point is attached to the framing of the machine, the scale to be divided being placed upon the upper plate.

In the case of cutting a screw, the tool holder is fixed upon the upper plate and the screw to be cut is placed between centres parallel to the motion of the table and to the guide screw, having motion imparted to it by a train of wheels connecting it with the screw of the machine. The compensating bar being adjusted for total length, and the small pieces of metal upon the same being adjusted to the intermediate errors of the guide screw, it will be seen that by the passage of the longer arm of the lever over the edge of the compensating bar, a slight motion will be imparted to the upper plate independent of the lower, so that, in other words, if by the error of the screw the lower table is moved through too great a space, the upper table is made to move (by the action of the lever) through a space equal thereto in the contrary direction, and *vice versâ*.

Note.—A description somewhat more in detail and of the manner of adjustment will be found in "Holtzapffel's Turners' and Mechanics' Manufactures," 2nd vol., p. 651 *et seq.*

Many scales were divided and many screws cut by this machine, of which some were given to various scientific friends, and Sir J. Whitworth, amongst others, had a scale and a screw about the year 1843 which have served him as standards.

265. Machine for **dividing right lines,** by Nicholas Fortin.
MM. Fortin Hermann Bros., Paris.

This machine is the one constructed by the celebrated inventor in 1787, and used in the works connected with the adoption of the metrical system.

The pitch of the screw is exactly one millimetre. (Fortin's machine for dividing circles, as well as this machine, was presented to the Conservatoire des Arts et Métiers by MM. Fortin Hermann Bros., in 1876.)

297. Micrometer Dividing Machine.
Voigt and Hochgesang (Gust. Voigt), Göttingen.

The pitch of the screw is $\frac{1}{4}$ mm., its head is divided into 200 parts; each part, therefore, corresponds to $\frac{1}{800}$ mm., reading by the vernier to $\frac{1}{10}$ of this value. By a spring fixed in the nut "loss of time" is completely removed.

The tracing appliance is constructed in the simplest manner possible. The tracing point—a diamond—is lifted by a mechanical contrivance, and let down again.

The slide allows of drawing a line of 30 millimeters in length.

The slide which carries the tracing point moves without greasing between six finely polished carnelian plates; by this arrangement any errors, which might be caused by clotted grease, will be rendered absolutely impossible.

F. Tide Registers.

327b. Patent Indicator, for tanks or reservoirs.

John Nicholas.

This gauge is similar to that last described, but the atmosphere giving comparatively a constant pressure the stand pipe can be dispensed with. The brass tube referred to in the previous description may be seen in the tank attached. It is not necessary to pierce or employ a tank when attaching one of these gauges, and the small pipes can be laid in the walls in a similar manner to gas tubes. In some cases one tube is sufficient, the water column being balanced by mercury in a metal tube at the back of the gauge. This gauge is suitable for tanks upon the roofs of mansions or hotels, where engines are used for pumping.

255. Registering Water-mark, of new construction, which records the curve of the water-level and its mean height.

Lieutenant-General Baeyer, President of the Geodetic Institute at Berlin.

Invented by F. H. Reitz, civil engineer, of Hamburg. The apparatus was made in the factory of Pape and Dennert. The clockwork is by Knoblich.

278a. Magneto-Electric Water Level Indicator.

Siemens and Halske, Berlin.

A float which rises or falls with the level of the water in the reservoir or tank communicates motion by a metallic chain to a magneto inductor, which, generating electric currents, works at any distance an indicator connected by a cable or insulated wire.

1695. Apparatus for making contact to show the height of water with float, rod-chain, counterpoise, and water tube.

C. & E. Fein, Stuttgart.

This is self-acting, and registers at any distance the water-level in a reservoir, &c.

It consists of five parts:—

(1.) The float with chain and counter weight which when acted on by the rise or fall of the water impart their motion to the contact arrangement

(2.) The contact arrangement which communicates the motion of the float to the recording instrument by opening or closing the circuit.

(3.) The recording instrument; this shows the level of the water at all times, the pointer being acted on by the motion to and fro of two electro-magnets.

(4.) The conducting wire.

(5.) The battery.

279. Three Gauges, in enamel cast iron, for registering the height of a river or lake. *De Dietrich and Co., Niederbronn.*

The first of these on the Niederbronn pattern is in black and white and graduated to centimetres, the second on the Nancy pattern is graduated in black and white for every two centimetres, and the third on the Paris pattern is in blue and white graduated to five centimetres.

These water-mark plates are fixed by means of iron clamps to piers, vertical embankments, &c., and serve for the observation of the level of the water in rivers, canals, lakes, and reservoirs.

Placed at proper distances from one another in the chief water-courses and its tributaries, they enable the rise of the water to be observed, and consequently timely warning to be given, by telegraph or otherwise, to the inhabitants of the districts concerned.

280. Recording Tide Gauge, with self-acting indications of the mean height of the water (system of F. H. Reitz, Hamburg); executed by Dennert and Pape, in Altona.

Royal Prussian Geodetic Institute, Berlin.

The tide-measuring system exhibited by the Royal Prussian Geodetic Institution of the European measurement of a degree, at the instance of its president, General Baeyer, and constructed by Dennert and Pape of Altona, with clock by T. Knoblich, of Hamburg, has a graphic apparatus for registering the tide-curve and an arrangement by which the mean water-level is indicated automatically. The registration of the water-level is effected by means of diamond points upon a cylinder placed horizontally for the accurate division of the arc.

The mean water-level is indicated by means of two agate rollers with divisions, which slide upon a horizontal glass disc turned by the clock of the tide-measurer, moved to and fro by the rising and falling water, and by the rotation of the glass disc, and may be read off at any desired intervals of time.

The calculation otherwise necessary of the mean water-level (the true level of the sea) from the indications of the registering apparatus is saved by the above-mentioned mechanical arrangement, and effected automatically with very great accuracy by the tide-measurer.

The determination of the form and dimensions of the earth undertaken by the European Committee for re-measuring degrees of longitude and latitude, also contains the determination of the mean sea level at points on different coasts, and its comparison by means of accurate levellings. During the last few years the Committee has endeavoured, in consequence of these examinations, to study the different apparatus and to promote a more exact observation of the tide corresponding to the exact levellings recently taken.

These circumstances were the cause of a commission from his Excellency General Baeyer, President of the Central Office of the European Committee for higher geodetic purposes, and the Royal Prussian Geodetic Institution, to F. R. Reitz, instructing him to prepare a tide apparatus according to his system. The instrument now exhibited was made by Dennert and Pape, of Altona, and the clockwork by Theodor Knoblich, of Hamburgh. It is intended by the Imperial German Admiralty to place this new instrument, after the close of the Exhibition, on the Isle of Sylt, Schleswig.

The commission given by his Excellency General Baeyer referred to the construction of a new instrument, combining a registering apparatus and mechanical means of determining the mean level of the sea. The apparatus here described is therefore a combination of both objects.

A buoy A, moved up and down by the tide in a vertical shaft, turns a disc C, about a horizontal axis, by means of a copper wire B. During the ebb the wire B descends and turns the disc C; during the flood the same is effected in the opposite direction by a weight D turning the disc E, connected with the disc C upon the same axis. In order to reduce the movement of the buoy A in a certain proportion, a pinion F, on the axis of the disc C and E, moves a rack G, in a horizontal direction. On one end of G a diamond-point H is fixed, on the other end two rollers J, J, with a horizontal axis.

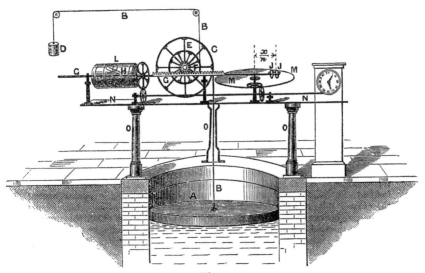

Fig. 1.

The clock of the instrument turns the cylinder round its axis in 24 hours by means of a combination of wheels, and the glass disc M in six hours.

The tide-curves are engraved upon the cylinder L with the diamond-point H. The rollers J, J move on the glass disc M by the combined action of the buoy A, the weight D, and the clockwork.

All the different parts of the apparatus are fixed upon the same plate N, N of cast iron, planed at the necessary points to insure their invariable position. The plate N rests on three columns of cast iron, placed upon the coping of the shaft.

The cylinder is covered with blackened chalk paper, whereon the diamond engraves the tide-curves as white fine lines on a black ground.

The paper on the cylinder L is divided in half hours, and from meter to meter of height. For this purpose a proper self-acting apparatus is constructed, by which both divisions are made with great exactness. Diamond-points are used for this purpose also. It is necessary, for the sake of distinctness, to renew the covering paper of the cylinder once a month. To avoid waste of time a second exactly similar cylinder is prepared and carefully divided, to replace the former cylinder.

For the observation of the constants of the apparatus, it is necessary to note complete revolutions of the disc C and glass disc M. For this purpose two indices are applied.

The circumference of the disc C, measured on the axis of the copper wire B, is exactly two meters in length.

The apparatus for the determinations of the mean height of the sea is explained by the following remarks:—

It is the question how to define the mean height of the water for a certain period.

This height would be the arithmetical mean between high and low water, supposing a regular form of the tide-curves. In fact, the real observed tide-curves differ very much from this theoretically-defined regular form (curve of sines). To show this the tide-curves of Cuxhaven, Southampton, and Ipswich will suffice as examples. The irregularities in these curves are evident and easily to be seen in figure 2, where the curve of sines is drawn for the purpose.

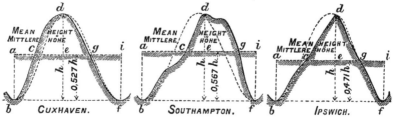

Fig. 2.

If h be the difference between high and low water, the mean height is for Cuxhaven $0 \cdot 527\,h$, for Southampton $0 \cdot 567\,h$, and for Ipswich $0 \cdot 471\,h$ instead of $0 \cdot 5\,h$, the amount of the mean height, supposing a regular form of the curve. The mean height of the sea is one of the most important results of the tide observations. It is the only datum by which to define the invariability, or the measure of the variation, of height of the continent and islands.

The line of mean height $acgi$ must be in such a position that the areas abc and gfi (figure 2) together equal the area $cdge$. This is to be found by a tedious calculation of areas of the tide-curves drawn on the cylinder L.

The apparatus here described gives the requisite data for an easy determination of the mean height by means of two rollers (JJ) (one of them con-

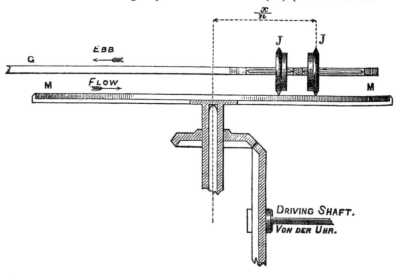

trolling the other) moving in accordance with the level of the water on a disc of glass (M). This disc is turned round its axis by the clockwork in 6 hours. The axis of the rollers J, J is parallel to the direction of their motion on the disc M. The number of the revolutions of the rollers is noted in certain periods by means of the divided rim (the rim is divided in 100 parts, tenths being estimated) and a numerical apparatus showing revolutions up to 100.

The height of the water may be taken from the point when the roller J stands in the centre of the disc M.

If the height of the water taken from this point be x, and the diminution of the movement of the buoy by the pinion F be $\frac{1}{n}$

$$\int \frac{x}{n} \delta\phi = \frac{1}{n} \int x\delta\phi \quad \ldots \quad (1.)$$

is the expression for the movement of a point in the circumference of the roller J in a period during which the disc M revolves through an arc ϕ.

The expression $\int x\delta\phi$ is the area of a figure with the ordinates x and the base ϕ (x representing the height of the water and ϕ the time). The mean value of x or the height of a rectangle of equal area with this figure and the base ϕ is found by dividing by ϕ. This mean value of x is the mean height to be found, equal m suppose. Then—

$$m = n\,\frac{\frac{1}{n}\int x\delta\phi}{\phi} \quad \ldots \quad (2.)$$

The movement of a point of the circumference of the roller J is equal to $\frac{1}{n}\int x\delta\phi$. This is equal to the product of the circumference of the roller J called p and the difference of readings on the margin of the roller J at the beginning and end of the period. Hence, if the readings are called a_1 and a_2,

$$m = \frac{n\,.\,p(a_2 - a_1)}{\phi} \quad \ldots \quad (3.)$$

If z is the number of seconds corresponding to the arc ϕ, and b the constant arc through which the disc M revolves in a second:

$$m = \frac{n\,.\,p}{b}\,.\,\frac{(a_2 - a_1)}{z} \; ; \quad \ldots \quad (4.)$$

Finally, if $\frac{np}{b} = c$:

$$m = c\,\frac{a_2 - a_1}{z} \quad \ldots \quad (I.)$$

the mean height of the sea.

The single constant c is easily to be determined by experiment and also with great exactness without the knowledge of the dimensions of the apparatus, as follows : In a certain position of the roller J a number of revolutions is made by the disc M, representing a number of seconds z_1 (a revolution is made in 21,600 seconds). At the beginning and end of these revolutions the readings (a_1 and a_2) of the roller J are noted. After this a certain length (l) of the copper wire upon the disc C is unrolled (measured by means of the circumference of the disc C, equal 2 meters, and the index, or directly). In the new position of the roller J a number of revolutions of the disc M again is made, representing a certain number of seconds (z_2), and also the corresponding two readings (a_3 and a_4) of the position of the roller at the beginning and end of these revolutions are now observed. All the requisite data for the determination of c are now obtained. If the two values of m corresponding to the two positions of the roller are m_1 and m_2 and their difference is equal l, from equation (I.),

$$m_1 = c \cdot \frac{a_2 - a_1}{z_1} \;\; ; \;\;\; m_2 = c \cdot \frac{a_4 - a_3}{z_2},$$

therefore :

$$m_2 - m_1 = l = c \left(\frac{a_4 - a_3}{z_2} - \frac{a_2 - a_1}{z_1} \right) ; \text{ finally,}$$

$$c = \frac{l}{\dfrac{a_4 - a_3}{z_2} - \dfrac{a_2 - a_1}{z_1}} \quad \ldots \quad \text{(II.)}$$

When the constant c for both rollers is calculated, the constant difference in the results for m is fixed by a number of revolutions of the disc M in the same position of the rollers. The difference of m found in this way, of course, is constant.

The equations (I.) for the apparatus exhibited, calculated as described, are : For the roller on the left side :

$$m = 8656 \cdot 632 \frac{a_2 - a_1}{z} \text{ meters.}$$

For the roller on the right side :

$$m = 8655 \cdot 983 \frac{a_2 - a_1}{z} \text{ meters.}$$

The roller on the right side gives a constant difference for the value m given with the roller on the left side equal $1 \cdot 3252$ meter. Respecting the correction of the apparatus, it is only necessary to make the axis of the rollers parallel with the direction of the movement on the disc M. In this parallel position this movement, of course, has no influence on the revolution of the rollers. The parallel position of the axis of the rollers may be tried by experiment, by moving them on the disc M, turning the disc C. During this experiment the disc M must remain at rest. No motion of the rollers will then be observed. It seems requisite, at first sight, that the roller J move through the centre of the disc M. But this is not necessary, an approximation only being needed for practical purposes. By a sidewards position no difference in the revolution of the rollers is effected. This is proved in the following way :—

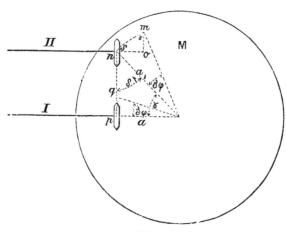

Fig. 4.

As to the two positions I. and II. of the rollers, it needs be shown that a motion $d\phi$ of the disc M causes the same effect.

In the position I. the motion of the roller is :

$$pq = a\delta\phi.$$

In the position II. the motion of the roller is not equal nm.

But :

$$mo = nm \cos j$$

consequently :

$$\cos j = \frac{a}{a_1} \text{ and } nm = a_1\delta\phi_1$$

therefore :

$$mo = a, \, d\phi\frac{a}{a_1} = a\delta\phi = pq.$$

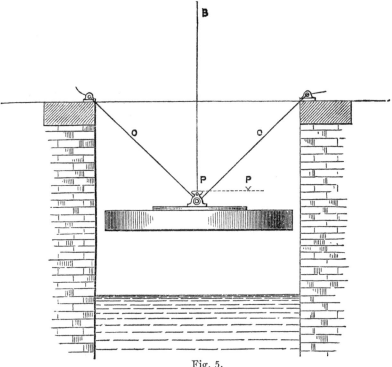

Fig. 5.

The motion of the rollers, therefore, in the two positions is equal.

To connect the distance of the points from which the mean height is measured by the instrument to a point given by levelling, the height of a point P of the buoy over the water is carefully measured.

This may be very exactly done by putting the buoy into a vessel filled with water. The buoy afterwards is lifted a little by the weight D (Figure 1). This is easy to calculate from the area of the buoy, the weight D, and the radius of the discs C and E. The buoy then is fixed at a certain height by a wire O. In this position the height of P. below a point given by levelling is carefully measured. The absolute height of the water represented by the chosen position of the buoy is now known. After this the mean height corresponding to the same position (viz., the constant height of this position) is calculated according to the result by a number of 10 or more revolutions of the disc M. If the measured height corresponding to the position of the buoy be equal h, and if the height given by the instrument be equal m, $h-m$ is the absolute height (according to the levelling) from which the mean height given by the instrument is taken.

The diamond-point can then be fixed in a position to get a corresponding diagram on the cylinder L to the absolute height given by levelling.

F. H. Reitz,

Hamburgh, May 1876.   Civil Engineer in Hamburgh.

281. Self-recording Tide-gauge, improved.

H. C. Ahrbecker.

In this instrument the whole of the paper can always be seen, and requires renewal only once a month. The clock goes for 32 days.

This instrument was designed to obviate the disadvantages which occur in working the ordinary pattern.

A float which rises and falls with the tide in an iron tube is attached to one end of a chain that passes over a wheel, at the other end a counterpoise hangs. When the float rises or falls it communicates its motion (by means of the before-mentioned wheel) to a sliding pencil that moves across a strip of paper drawn under it by the clock.

In the model exhibited the distance between two lines across the paper, $=2$ hours, $=1$ ft. in the length.

This instrument can be made to work without any attention whatever for 12 months if necessary.

282. " Maréegraphe," or tide-gauge.   *Van Rysselberghe.*

G. Miscellaneous Length-measuring Instruments.

238. Measuring Wheel for determining distance by registering the number of revolutions ; the upper index pointing out every single and the lower every 100 revolutions.

Elliott Brothers.

238a. Odometer or Way Measurer in gilt metal case elaborately chased, an early example, probably made in the second half of the 16th century.   *Alexander Nesbitt.*

In Beckmann's History of Inventions there is a description of two instruments resembling this, which belonged to the Emperor Rudolph II. (1576–1612).

415b. Fare and Distance Indicator for Street Cabs.

Robert Foster, Sunderland.

This is an instrument for measuring the distance travelled by a cab or other vehicle to which it may be attached. A driving band taking its motion from the road wheel actuates counting wheels, and so pointers are made to indicate on dials the distance passed over and the fare. There is also an appliance for registering on a slip of paper all the fares taken during the day. The pointers can be brought back to zero by the driver, but they cannot be moved forward except by the motion of the vehicle.

240. Improved Measuring Wheel or Mile Meter.
Elliott Brothers.

240a. Micrometrical Divisions in English and Metric Measure. *Dumoulin Froment, Paris.*

270. Holt's Diagrammeter. This instrument is specially made for measuring the ordinates of indicator-diagrams 5″ long, and is used much after the manner of a parallel rule, the registering nut on the screw being first placed at zero ; when it is required to register a measurement the break key is depressed, and when all the measurements have been taken the distance the nut has travelled gives the mean ordinate. *Henry P. Holt.*

274. Spherometer (by Salleron), to read to ˙001 mm.
The Council of the Yorkshire College of Science, Leeds.

276. The Wealemefna, E. R. Morris's patent.—A pendant for the watch-chain.
The Morris Patents Engineering Works, Birmingham.

To measure, it is merely necessary to advance the Wealemefna over the object, when the large hand will register the inches and fractions of an inch, and the small one the feet. The instrument registers to 25 feet.

276a. Schlagenheit's Measurer for Curved and Straight Lines. *S. J. Hawkins.*

This instrument has a small wheel A, the periphery measuring one inch and divided into five equal parts, indicated by fine points, marking when in use each length by a slight indentation upon the map or plan. At each revolution of the wheel a small spring B is struck, indicating that one inch has been traversed. Attached to the instrument is a small scale C, ⅕th of an inch in length, divided into 10 parts for measuring distances less than one division of the wheel.

276b. Opisometer. Instrument ordinarily used for the above purpose. *S. J. Hawkins.*

277. Measuring Instrument, E. R. Morris's patent. (Silver medal awarded at Manchester, 1875.) For the use of architects, surveyors, builders, contractors, timber merchants, &c., &c., and for general measuring purposes, in place of the rule or tape. It measures to 100 ft., and weighs under 3 oz.
The Morris Patents Engineering Works, Birmingham.

To use the instrument it is merely necessary to advance it along the object to be measured, when the large hand will register the inches and fractions of an inch on the outer dial, the smaller hand on the inner dial, the feet and the smallest hand on the recessed dial, the tens of feet travelled over. The instrument registers to 100 feet. Price, electro-silver, in leather case, 16s. 6d.

278. Chartometer, E. R. Morris's patent. (Silver medal awarded at Manchester, 1875.)

The Morris Patents Engineering Works, Birmingham.

The only instrument that measures and registers distances on maps, plans, scaled drawings, &c., and that is adapted for various scales. By guiding the small steel wheel along any route on a map, the hand registers the actual distance in miles, yards, &c., according to the dial in use and the scale of the map, which should correspond. To deal with a map of a difficult scale, the glass front is opened by pressing a spring ; the dial removed, and another corresponding to the fresh scale slipped into its place. A set of dials adapted to the scales of all the Ordnance maps, and the usual scales of travelling maps, &c., &c., is contained in a recess of the leather case, beneath the instrument.

277a. Pedometer, of the latest and most approved form.

J. and W. E. Archbutt, Westminster.

This instrument has pendulum action, and is worn suspended in the waist-coat pocket ; it is provided with a regulator whereby it can be set to accurately record distances walked.

277b. Improved form of Pedometer, by Dollond, in which the direct chain action is substituted for the lever ; made in the early part of the nineteenth century.

J. and W. E. Archbutt, Westminster.

277c. Pedometer or instrument for accurately registering distances walked. This instrument was invented and made by Spencer and Perkins in the latter part of the eighteenth century. *J. and W. E. Archbutt, Westminster.*

290. Scale for Measuring Curves, Eschenauer's patent.

Hermann Schäfer, Darmstadt.

The curve scale is intended for engineers, steam boiler makers, surveyors, architects, and others, for copying maps, plans, &c.

It will be of great advantage in the projection of railway lines, the curve scale requiring only to be adjusted to the situation in order to ascertain how the line can be most favourably traced, and expensive cuttings avoided. In regard to such surveys, as well as in the control or examination of railway lines already traced and sketched (for which purposes either the curves, cut of certain radii, or compasses, are used at present), the employment of the curve scale will save the trouble of trial, since the correct one can be immediately determined and read by means of this instrument.

Boiler manufacturers, also, and almost all engaged in technical pursuits, will find the curve scale very useful for determining the radius of an arc of a circle, of which three points are given, as, for instance, in curved steam boiler bottoms.

In fact, in all cases where part of an arc of a circle, or three points of the same, are given, the radius can be read direct, and without loss of time, in a manner hitherto unknown.

If it be desired to take the radius of a given curve by means of the scale, the middle bar of the same is placed on the curve line, and the scale is then moved so far upwards or downwards until the curve line meets in three commensurably described points of the scale. The number indicated gives the radius of the curve in centimeters, if the curve is drawn in its natural size.

If, however, the drawing of which the radius of the curve is to be determined is, as is usually the case, on a reduced scale, the radius indicated must be multiplied with the proportional number of the reduced scale.

For example, if the drawing should be to the scale of $\frac{1}{2500}$ of the natural size, and the curve radius on the curve scale is indicated with 52·5 cm., the actual radius of the curve will be 52·5 × 2,500 = 131,250 cm., or 1,312·5 meters; or, should the drawing be to the scale of $\frac{1}{500}$, and the curve scale indicates 43 cm., the radius of the curve will be 43 × 500 = 21,500 cm., or 215 meters.

The curve scale can likewise be used as a reduction scale of every other measure which is to be calculated in meter measure, as the radius in meters can always be read directly, no matter in what scale the drawing is made.

This is a great saving of labour, which is very much facilitated if, as often is the case, old maps and drawings are to be made use of.

In using the curve scale it will sometimes happen that the curve to be ascertained does not exactly meet the line drawn on the scale, but will fall between two lines. In this case the smaller division can, as the radii are marked progressively by 0·5 cm., be easily estimated by eye after a little practice.

For example, the curve of the radius of 1,110 meters, at a proportionate scale of $\frac{1}{2500}$, lies between 88·5 and 89·0 of the curve scale, and amounts to nearly 88·9.

As, however, in most cases, round numbers, without fractions, are chosen for the radii, the radius can always be determined with the greatest accuracy.

1093a. Ellipsometer.

Before the eyepiece of the glass, a double refracting prism is made to turn until a wire, moving perpendicularly to the principal section of the prism, passes through the two intersecting points of the two reflections of the ellipse. An index shows at the moment the position of the prism.

Graphometer of Botti.

The Royal Institute of "Studii Superiori," Florence.

III. MEASUREMENT OF AREA.

316. Amsler's Planimeter, for calculating with perfect accuracy the areas of plans, maps, or other plane surfaces, in square inches and metrical measure. *Elliott Brothers.*

317. Polarplanimeter. *Ott and Coradi, Kempten, Baviera.*

By means of the polarplanimeter the superficial contents of any kind of figures drawn on paper, no matter what their outline may be, can be ascertained by mere tracing more exactly and quickly than by any other method.

The inventors of this instrument are respectively J. Amsler, Schaffhausen, and Ch. Starke, Vienna. Ott and Coradi's construction is a combination of both, embracing the excellences of each. It differs from Amsler's instrument by the pole (axis) of the instrument not being formed by an inserted point of a needle, but by a steel ball embedded in a metal cylinder, thus giving it a firmer position; and, moreover, by the axis of the roller being lodged in a horizontal frame, and the dividing circle of the roller as well as the indicating wheel being free at the top, thereby affording much easier and more accurate reading than Amsler's instrument. This arrangement has the advantage that for simple calculation the zero point of the roller can be placed exactly on the zero point of the vernier, when the tracing pencil is at the commencement of the figure. The weight can be separated from the instrument, by withdrawing the bolt, and placed in the case by itself. The runner carrying the axis of the polar arm can be moved along the whole length of the quadrangular bar, by which means at every longitudinal scale desired a round number can be obtained for the value of the vernier unit (for example, scale 1·500 vernier unit, 2 square meters, or scale 1·1440 vernier unit, 5 square fathoms). The tracing bar is divided into ½ mm., and the runner sliding on the same carries on one side a vernier, on the other an index. For adjustment with the index, the most usual or specially desired longitudinal scales are marked with lines on the bar; by means of the vernier and the divisions on the bar, proportions of measure not previously given can be easily inserted and noted down; in the same manner, in the case of plans which have been drawn on shrunk paper, the area can be retained in its actual size by a corresponding movement of the runner, and the position of the vernier noted down for a certain amount of shrinking.

318. Planimeter, divided on a glass plate.

F. W. Breithaupt and Son, Cassel.

The planimeter consists of a network marked on a glass plate for a certain scale of the meter measure.

319. Wetli's Planimeter.

Physiological Institute of the University of Halle (Prof. Bernstein, Director).

The planimeter is fitted together by placing the six-toothed movement into the centre of the divided disc, whilst the central point of the small glass disc moves at the other end in the screw of the ring encircling the divided disc. Next, the slide with the large glass disc is placed on the three-railed track in such a manner that the horizontal glass disc comes underneath the smaller vertical one; the latter is then, by means of the screw which is fixed on the ring, regulated in such a manner that it is easily carried along with the horizontal disc by friction.

The pointer moving on the same axis with the divided disc indicates the superficial contents of the figure in square millimeters. The small toothed wheel records every 1,000 square millimeters of the surface.

————

IV. MEASUREMENT OF VOLUME.

STANDARD MEASURES.

319a. Casts of a Collection of Roman Measures to hold liquids.   *Archæological Museum, Madrid.*

The originals, of alabaster, are preserved at the Archæological Museum of Madrid. They were discovered at the end of the last century in the Torre del Mar, Province of Malaga, Spain.

322a. Series of Standard Measures of Capacity, in copper, with glass discs, from the centilitre to the double decalitre. (11 measures.)

Messrs. Collot Brothers, Boulevard de Montrouge, Paris.

324. The Standard Pint, popularly known as " The Stirling Jug." *The Smith Institute, Stirling.*

This measure was entrusted to the town by Act of (the Scottish) Parliament, in the year 1437. Sometime previous to 1745 it had been borrowed by a coppersmith for the purpose of making others, and as he joined the insurgents in " 45 " it was lost sight of. On his not returning, his effects were sold, with the exception of a few that were thrown into a garret as rubbish ; among these, in 1752, the Stirling Jug was found, after some years of patient and unwearied search (by Rev. A. Bryce, of Kirknewton). It is made of brass, and is in the shape of a hollow truncated cone, weighing 14 lbs. 10 oz. 1 dr. 18 grs. Scottish troy. Diam. of mouth 4.17 English in., of the bottom 5·25 in., and depth 6 in. On the front, near the mouth, is a shield in relief, bearing a lion *rampant*, the Scottish national arms, near the bottom is another, bearing an ape *passant gardant*, supposed to be the arms of the foreign maker.

324a. Russian Standard Measures of Capacity (Vedro, $\frac{1}{2}$ V., $\frac{1}{4}$ V., $\frac{1}{10}$ V., $\frac{1}{100}$ V.). *Siemens and Halske, Berlin.*

These measures, made of bronze, have a conical shape, newly adopted in Russia, for standard and trade measures of capacity. In these measures the inner diameter, A B, of the bottom is equal to an inner side, A C, and double diameter, C D, of the orifice. By very simple contrivance, such trade measures might be verified, approximately, by (linear) measurement of A B, A C, and C D.

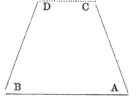

325. Set of Standard Measures for Alcohol, conical shaped, in order to diminish the possibility of evaporation of the liquid. *Siemens and Halske, Berlin.*

322. Measures of Capacity, according to natural principles. *Hans Baumgartner, Basle.*

WATER METERS.

321. Schmid's New Water Meter. *A. Schmid, Zürich.*

This meter consists of two of Schmid's patent hydraulic motors, coupled at right angles, and enclosed in a water-tight casing. They are set in motion by the force of the fluid they have to measure. At each revolution a volume equal to the contents of four cylinders must pass. The pressure required to keep tight the oscillating surfaces of the cylinders is furnished by the difference of pressure at inlet and outlet, which is thus self-regulating. The meter is also kept in motion by the difference of pressure. The frictional resistance is the same with all pressures of the fluid under measurement, and, according to the size of the meter, is represented by a water head of 3 to 16 ft. The different parts of the meter are constructed of materials not liable to chemical influence.

The chief advantages of this meter are :—

1. The velocity of the engine is exactly in proportion to the quantity flowing through the meter.

2. According to the most careful experiments, the error, if any, does not exceed 1 per cent.

321a. Siemens' and Adamson's Patent Water Meter.
Guest and Chrimes, Rotherham.

This meter has a great resemblance to the motive-power machine known as Barker's Mill. The water passes down through a funnel into the measuring drum, and in passing outward through the curvilinear channels of the same causes it to revolve, delivering a certain quantity of water at each revolution of the drum, and this is indicated by worm wheel and gearing, in gallons, feet, or any other units required, on a dial plate properly divided and prepared for the purpose.

The meter is exhibited in section, so that the internal arrangements and its action can be seen. This meter has been extensively used for upwards of 20 years.

321b. Half-inch Patent Water Meter, for the water supply for domestic and trade purposes on the constant supply system.
J. Tylor and Sons, London.

326. Water-meter, for cold water, for 26 mm. width of tube.
Dreyer, Rosenkranz, and Droop, Hanover.

327. Water-meter, for domestic use.
Dreyer, Rosenkranz, and Droop, Hanover.

---

331. Model of a Gas Meter of ancient construction, with glass sides.
School for Industry, Halle (Dr. Kohlmann, Director).

329. Apparatus for determining the capacity of Cartridge-cases as far as 20 cub. mm.
A. Bonsack, Berlin.

330. New Volumeter, consisting of A. Sauer's burette, a second glass piece, stands and tubes.
(Compare Fresenius, "Zeitschrift für analytische Chemie," xiv. heft. 3 and 4).
Berggewerkschafts-kasse, Bochum, Dr. Heintzmann.

V. MEASUREMENT OF MASS.

A. BALANCES.

333. Balance, with double column, 20-inch beam, fitted with steel knife edges working on agate planes, to carry 5 lbs. in each pan, and turn distinctly with ·01 grain. Fitted with apparatus for moving sliding weight without opening glass case. As made for the Warden of Standards, for comparison of standard weights.

L. Oertling.

334. Balance, with double column, very light beam, 10 inches long, fitted with agate knife edges and agate planes, to carry 30 grains in each pan, and turn distinctly with ·001 grain, with apparatus for moving sliding weight. *L. Oertling.*

335. Balance, with 14-inch beam, fitted with agate knife edges and agate planes, to carry 1,500 grains in each pan, and turn distinctly with ·001 grain. *L. Oertling.*

336. Balance, with 16-inch beam, fitted with agate knife edges and agate planes, to carry 2 lbs. in each pan, and turn distinctly with ·02 grain. *L. Oertling.*

337. Balance, with triangular beam, $6\frac{1}{2}$ inches long, fitted with agate knife edges and agate planes, to carry 3,000 grains, and turn distinctly with ·01 grain. *L. Oertling.*

338. Balance, with beam $6\frac{1}{2}$ inches long, fitted with agate knife edges and agate planes, to carry 2,000 grains in each pan, and turn distinctly with ·02 grain. *L. Oertling.*

339. Portable Assay Balance, with 6-inch beam, to carry 30 grains in each pan, and turn distinctly with ·001 grain.

L. Oertling.

340. Balance, constructed by H. Olland, of Utrecht, to weigh bodies up to 40 kilogrammes.

Prof. Dr. P. L. Rijke, Leyden.

This instrument is furnished with a double system of "fourchettes," directed by a rod 0·6 m. long. A difference of 1° in the pointer corresponded to a difference in weight of—

9·5 m. gr. when the weight was 20 kilogrammes.
10·5 ,, ,, 50 ,,
13·8 ,, ,, 73 ,,

With weights of about 50 kilogrammes, in a series of experiments under favourable conditions, between each of which the balance was set at rest, numbers not differing in the average by more than 0°·03 were obtained. When conditions were less favourable, the differences amounted to 0°·26, and only reached 0°·94 when the conditions were altogether unfavourable.

341. Analytical Balance, charge up to 500 grammes in each pan ; sensible to $\frac{1}{10}$ part of a milligramme with its full charge.

Beckers Sons, West Zeedyk, Rotterdam.

This balance is furnished with agate knife edges, and all bearings rest on agate planes ; it has a rest for pans and beam, and apparatus with adjustable shelf for taking specific gravities. The beam is divided in $\frac{1}{10}$ parts of a milligramme. Sets of weights from 500 grammes down to 1 milligramme. Three riders.

342. Analytical Balance, on plan suggested by Professor Dittmar, Andersonian University, Glasgow, for a charge up to 100 grammes in each pan.

Beckers Sons, West Zeedyk, Rotterdam.

This instrument shows a new method for displacing the centre of gravity of the beam, and for weighing up to 110 milligrammes by means of riders. The two riders form a part of the balance, with plunger for displacing exactly 10 grammes of water at 15° C. for taking specific gravities of liquids. Sets of weights.

343. Balance, with drawer and eccentric for lifting, movable pans, set screws and level, charge up to $1\frac{1}{2}$ kilos. in each pan, sensible for 20 milligrammes with its full charge.

Beckers Sons, West Zeedyk, Rotterdam.

344. Balance, charge up to 1 kilo. in each pan, sensible for 20 milligrammes with its full charge.

Beckers Sons, West Zeedyk, Rotterdam.

344a. New description of Balance of Precision, designed by M. Mendeleef, Professor of the University of St. Petersburgh, and constructed by Oertling. It is more particularly described in Appendix 10 to the Ninth Annual Report of the Warden of the Standards. *H. W. Chisholm.*

The peculiarity of this balance is that it has very short arms, and thus occupies very little room, and by its more rapid motion time is saved in weighings, whilst it gives results quite as accurate as those given by balances of precision with arms of greater length as ordinarily used.

Though constructed to carry a kilogram in each pan, the total length of the beam of this balance is less than $4\frac{3}{4}$ inches, whilst it is intended to give results within one tenth of a milligram. The balance beam to carry a kilogram is ordinarily 20 inches in length.

It can be used as a vacuum balance, as well as for weighings in air.

344b. Balance of Precision for minute weighings of 10 grains and under in each pan, constructed by Oertling.

H. W. Chisholm.

The beam is made as light as possible, and unusually so. The pans and suspending wire are of aluminium. The balance works upon fine points. A single action lowers the support of the beam and the supports of the pans.

348-9. Frerich's analytical **Balance,** capable of carrying 2,000 grms. with riders and a set of gramme weights.

F. Sartorius, Göttingen.

350. Analytical Balance, capable of carrying 500 grms., and a set of gramme weights. *F. Sartorius, Göttingen.*

351. Analytical Balance, capable of carrying 200 grms., with a set of gramme weights. *F. Sartorius, Göttingen.*

352. Frerich's Analytical Balance, with contrivance for weighing by means of torsion. *F. Sartorius, Göttingen.*

353. Pair of Russian Scales. *Bennet Woodcroft, F.R.S.*

354. Test Balance capable of carrying 20 grammes in each scale. *Edouard Sacré, Brussels.*

The bearings are taken off the knife edges when the balance is at rest. With 20 grammes the balance is affected by the 750th part of a milligramme. With 2 grammes it is affected by the 7,000th part of a milligramme.

354a. Model Balance, with two columns, specially intended for verifying the standard kilogram weights, mounting and tongue of aluminium bronze, tires and scales of aluminium, riders, carriage for shifting the weights from one scale to the other, rests for four weights, rules for the use of small sliding weights replacing the divisional series of the gramme, such as the decigramme, centigramme, milligramme, 10th of milligramme, hand of aluminium with double dial, parallel mirror for reading the oscillations at a distance, spherical level, two " Baudin" thermometers.

Messrs. Collot Brothers, Boulevard de Montrouge, Paris.

354b. Model Balance, with two columns, charge 300 grammes range, for chemical analysis, mounted on cast-iron tripod, mounting and tongue of bronze, platina tires, riders showing tenths of the milligramme, spherical level.

Messrs. Collot Brothers, Boulevard de Montrouge, Paris.

357. Printing Beam for Weighing Machine, admitting of the registration of each weighing. *M. Chameroy, Paris.*

This method of checking is applicable to all weighing machines of the nature of the steelyard. It would be found useful at custom houses, depôts, markets, railway stations, works, and other similar places.

Its advantages are : —

1. The affording of a record, by means of a printed impression on a special ticket, of the exact amount of the weight as determined by the machine itself.

2. The facilitating of the reading of the weights, either on the ticket or on the scale beam.

3. The preservation of an exact record of weighings, the authenticity of which is thus ensured.

358. Physical Balance, weighing up to five kilogrammes. *Hugo Schickert, Dresden.*

359. Physical Balance, weighing up to 200 grammes.

Hugo Schickert, Dresden.

363. Fine Assay Balance for weighing 20 grammes, turning with 1/100 mg.   *G. Westphal, Celle.*

364. Large Balance for determining the specific gravity of liquids.   *G. Westphal, Celle.*

365. Large Balance used in the **Manufacture of Sugar.**

G. Westphal, Celle.

366. Small Balance for determining the specific gravity of liquids.   *G. Westphal, Celle.*

367. Pharmaceutical Balance, for simple chemical operations.   *G. Westphal, Celle.*

370. Balance for chemical and physical purposes.

C. Staudinger and Co. (F. W. von Gehren), Giessen.

Balance of the exhibitors' construction; capacity of weighing, one kilogramme on each scale; sensitive at this weight, to 0·4 milligr. The balance is made of one piece of wrought (not cast) brass, and gilded. The centre and terminating knife edges are of steel, and all supports of hard stone. The weight of the beam with knife edges is = 793 grammes; deflexion of the beam at 1 kilogr. weight on each scale = 0·14 mm.; at 1·500 kilo. weight = 0·028 mm.; at 2·000 kil. weight = 0·042 mm.; at 3·000 kil. = 0·070 mm. A permanent deflexion has not been observed at such a weight.

375. Ten Plates of Rock Crystal for Balances.

Hermann Stern, Oberstein.

376. Chemico-physical Balance, executed by Ch. Jung, in Giessen.

Collection of Physical Instruments of the University of Giessen (Prof. Dr. Buff).

By shortening as much as possible the beam these scales offer the advantage of great sensitiveness and sufficient rigidity to weigh accurately from 250 grammes to $\frac{1}{10}$ milligrammes.

377. Analytical Balance, executed by Stollenreuther.

University of Munich.

379. Standard Weights in Glass, executed by Stollenreuther.   *University of Munich.*

381. Model of a Balance for determining the quality of grain, constructed according to the directions of the Imperial German Commission for Standard Weights and Measures, with a corn measure of 1 liter capacity.   *Reinhold Löhmann, Berlin.*

The manner of adjusting the several parts, as well as the successive series of applications of the same, is illustrated and facilitated by an explanation, with sectional and cross-sectional drawings, accompanying the model.

The practical employment and use of the apparatus for scientific and technical industries, in the first instance, and next for the solution of national-economical problems, will be demonstrated by two continuous memoirs, published by the Imperial German Commission on Normal Weights and Measures.

381a. Corn Balance, in box for showing the per-centage in value of corn by weight as a means of fixing the price for purchase or sale.   *L. Casella.*

382. Model of a Centesimal Weighing Machine, with glass platform.   *Dr. Kohlmann, Halle.*

384. Model of a Decimal Weighing Machine, with glass platform.
Physical Institute of the University of Halle (Prof. Knoblauch, Director).

386. Beam Balance with equal arms, sensibility 1 : 200,000.
Kleemann, Mechanical Engineer, Halle.

390. Beam Balance, for educational purposes.
Alex. Bernstein and Co., Berlin.

The beam, for educational purposes, has contrivances for demonstrating the different peculiarities of a scales-beam, or balance, namely, displacement of the centre of gravity, lifting and grinding of the principal bearings, unequal lengths of levers, non-parallelism of the knife edges, and position of *one* terminal knife edge out of the level of the two other knife edges.

388. Small Decimal Balance, for educational purposes.
Alex. Bernstein and Co., Berlin.

The decimal balance for instruction in schools has on each prism a scale, so that the influence of the weight on each prism can be shown by itself.

389. Analytical Balance.
Alex. Bernstein and Có., Berlin

This analytical balance is capable of carrying 500 grammes, and when fully weighted has a sensitiveness of $\frac{1}{10}$ mgr.; it has a perforated gilded brass beam with axes of agate and pans with arrangement for releasing all knife-edges, stop balance with pencil, and riders.

389a. New Balance for a Laboratory, carrying three kilogrammes in each pan, and turning with five milligrammes.
Deleuil, Paris.

When it is not in use, the beam is supported free of the knife edge, as in other accurate balances; vessels 25 centimetres in diameter can be placed on the pans, also vessels with long necks, and flasks of 1-2 litres capacity. By the aid of the second pan, the specific gravity of very bulky bodies can be obtained.

389b. Balance in mahogany case.
Universitäts Laboratorium, Berlin.

390a. Self-Acting Balance for Galvanic-plastic purposes.   *Alex. Bernstein and Co., Berlin.*

The balance for galvano-plastical purposes is so constructed that the conduction is interrupted automatically as soon as a deposit of a certain weight has been obtained.

391. Balance for Blow-pipe Experiments, in a case. with weights.   *Alex. Bernstein and Co., Berlin.*

The scales are for blow-pipe experiments; they have steel axes, and agate planes, two horn pans, two pairs of small gilded bowls, one bowl with hook for the determination of specific gravities, and a set of weights from 1 gr. to 1 centigr. of silver; from 1 centigr. to 1 milligr. of aluminium, and the fraction milligr. of quills.

392. Gold Assay Balance.
Alex. Bernstein and Co., Berlin.

The gold-alloy scales have a carrying capacity of 5 grammes, and are provided with bearings of agate, and indicate, when fully weighted, $\frac{1}{20}$ milligr.

378. Balance for Weighing in Vacuo, on von Jolly's principle.   *University of Munich.*

392a. Bullion Scales. The property of the Conservatoire des Arts et Métiers, 1866 ; constructed by Baron Séguier.
The late Baron Séguier, Membre de l' Institut.

419a. Spring Balance, with arrangement for suspending the lever and the scales on steel springs.
Physical Institute of the University of Halle (Prof. Knoblauch, Director).

428. Von Jolly's Spring-balance.   *University of Munich.*

B. Steelyards.

332. Roman Steelyard or Statera, of bronze. It was found in the year 1855, during building operations at Watermoor, a suburb of Cirencester, Gloucestershire.
Professor A. H. Church.

The beam, which is nearly 17 inches long, may be reversed, and it is consequently divided along both its upper and under edges. When the fulcrum nearer the head of the beam is employed objects can be weighed more than twice as heavy as those which can be accommodated when the beam is suspended by the other hook. To the head of the beam is attached a chain, branching below into two parts, each terminated in bold hooks adapted for grasping soft and bulky articles. This steelyard is a very good example of its kind. The locality which furnished it was the site of the Roman city of Corinium or Durocornovium, which has yielded an immense number of Roman remains, many of which are preserved in the local museum.

383. Model of a Roman Balance, with sliding weight of 75 grammes, and stand. *Physical Institute of the University of Halle (Professor Knoblauch, Director).*

C. WEIGHTS.

346. Five Standard Weights derived from the polar standard of length. *Prof. Hennessy, F.R.S.*

One of these weights is equivalent at 15° centigrade to a cube of distilled water whose side is the one hundred millionth part of the earth's polar axis. The others are submultiples of this weight, and the system is suggested in connexion with the polar standard proposed. *See* "Essay on a Uniform System of Weights, Measures, and Coins of All Nations," by Professor Hennessy.

346a. Series of Massive Copper Weights, standards of 1 gramme up to 20 kilogrammes, with subdivisions in platina. "Nécessaire" for inspector of weights and measures on his rounds, weighing about 10 kilos., and containing every requisite for testing scales, of weights, measures of capacity, and of length, and apparatus for stamping. *Messrs. Collot Brothers, Boulevard de Montrouge, Paris.*

346b. Ancient French Standard Weight of the city of Rouen, of brass, in the form of a series of cup weights in a closed box of ornamental shape, weighing altogether 8 lbs. of the old poids de marc de Charlemagne. Presented to the Standards Department in 1869 by Colonel Le Contens, Viscount of Jersey. *H. W. Chisholm.*

346c. Weight, wrought iron, ornamented with arabesques. flowers, and masks. Made for the old Mint at Madrid in the 17th century by the iron-master *Salinas.* *Archæological Museum, Madrid.*

360. Physical Weights. *Hugo Schickert, Dresden.*

361. Eight Sets of Weights, for analytical purposes. *G. Westphal, Celle.*

The first of these weighs from 1 kilogramme downwards, the second from 500 grammes downwards, the third from 100 grammes, the fourth, fifth, and sixth from 50 grammes, the seventh from 10 grammes, and the eighth from 0·5 grammes downwards.

362. Standard Weights. Each weight consists of one piece of solid metal adjusted by the gilding process. *G. Westphal, Celle.*

These consist of a 1 kilogramme weight, a set of weights weighing from 1 kilogramme downwards, and a set of standard weights with pin adjustment weighing from 500 grammes downwards.

362a. Box of Weights, containing two kilos. and fractions of a kilo. *Deleuil, Paris.*

362a. Case of Weights and Measures.   *T. Oertling.*
Containing—
Set of weights from 1 kilogramme to 1 milligramme.
Set of weights from 50 grammes to 1 milligramme.
Set of weights from 1,000 grains to $\frac{1}{100}$ grain.
Set of weights for assaying silver, $1,000=1$ gramme, in circular ivory box.
Sets of weights of rock crystal (one spherical set), from 50 grammes.
Set of measures from $\frac{1}{2}$ litre down.
Sikes' hydrometer, as supplied to the Honourable Board of Inland Revenue.
Bates' saccharometer, as supplied to the Honourable Board of Inland Revenue.
Set of petroleometers for testing liquids of 650 to 900 specific gravity.

262b. Iridio-Platinum Standard Kilogram.
Johnson, Matthey, and Co.

323. Standard Weights, according to natural principles.
Hans Baumgartner, Basle.

374. Sets of Weights and single Weights from 1 kilogramme, made of rock-crystal; amongst them some which have been examined and marked with an index error by the Imperial Commission for regulating Standard Weights and Measures at Berlin.   *Hermann Stern, Oberstein.*

The weights, as well as the measures of quartz or rock-crystal, were many years ago recognised as the best and most correct; but no one has, up to this time, executed them in such a manner as to afford institutes an opportunity of procuring them; which want has now been supplied by the exhibitor.

The other objects of agate are such as are produced by the Oberstein-Idar grinding and polishing mill, and can be employed in different kinds of machinery.

374a. Series of Spherical Standard Weights in Quartz, from 1 kilogramme to 1 milligramme.   *A. Hilger.*

374b. Standard Kilogramme Weight, quartz sphere, on brass stand.   *A. Hilger.*

374c. Standard Kilogramme Weight, of guaranteed value $1 \cdot 0000019$ kilo.   *A. Hilger.*

387. Set of Pharmaceutical Weights from 0·01 grammes to 200 grammes (19 pieces).
Kleemann, Mechanical Engineer, Halle.

Representation of the Weights and Measures used in the time of Henry VII., as well as the mode of punishment of fraudulent traders, copied from a painting on panel of the period (doubtless official).   *Gardner Collection.*

D. INSTRUMENTS FOR DETERMINING SPECIFIC GRAVITY.

347. Tangential Balance for measuring the density of liquids and solids by the angle of inclination, read on a divided circle to two minutes, thus giving the third decimal of specific gravity ; made by Oertling, of London.

Prof. Carl Wenzel Zenger, Prague.

355. Hydrostatic Balance, by Ramsden ; with **Weights,** by Robinson, presented to the Royal Society by Lady Banks.

Royal Society.

368. Xylometer (cylindrical form), with brass cylinder.

Zimmer Brothers, Stuttgart.

369. Xylometer of Glass, prismatic form.

Zimmer Brothers, Stuttgart.

These instruments are chiefly used in the management of forests, and for agricultural purposes.

They are employed for exact scientific examinations, especially for the cubing of irregularly-shaped pieces of wood, and for determining the specific gravity of wood.

Discs of wood which have been split out from the heart in the direction of the pith rays, and therefore contain proportionate parts of all veins of wood, can be quickly and exactly examined.

With both apparatus it is possible to read on the scale accurately down to 5 cub. centimeter (5 grammes water).

(See " Holzmessekunst," by Prof. Dr. Baur, Hohenheim.)

425. Hydrostatic Apparatus for ascertaining the specific gravity of woods.

Prof. Dr. Nördlinger, Hohenheim, Württemberg.

372. Densimeter of Major Bode's construction, for determining the specific gravity of all sorts of powder.

A. and R. Hahn, Cassel.

The densimeter is the only existing instrument with which the specific weight of all sorts of gunpowder (prismatic powder, powder-cakes, fine and coarse grained powder, &c.), can be easily determined, in quantities of 50 to 250 grammes, with the most perfect accuracy.

It is constructed by Major Bode.

This apparatus consists of a reservoir with bolt, two gutta-percha tubes, and a clamp.

1. The reservoir is formed by a steel capsule, with lid fitting air tight.

By means of the bolt the lid of the steel capsule can be screwed fast on this.

The contents of the reservoir are so measured in the clear that a prismatic powder grain can be easily placed in it.

Lid and steel capsule are vaulted, in order to accelerate the exhaustion of the air by pumping.

In the steel capsule in the upper part and in the lid in the lower part, there is an air-tight cock. The reservoir communicates with these cocks by means of two channels, which, for fine-grained powder, are shut off by a steel tinplate filter, the holes of which have a width of 0 3 mm. The two tubes, 1 and 2, are screwed air-tight on the plugs of the two cocks 1 and 2. At the upper ends of both tubes funnels of glass are squeezed in for more conveniently filling and emptying the mercury. The shorter tube, No. 1, of about 600 mm. length, carries in the centre a glass-tube, about 200 mm. long, divided into millims, and can, just above this tube, be closed air-tight by means of a screw cramp. The interior diameter of the tube No. 1 is about 9 mm., whilst that of tube No. 2, which is about 2,500 mm. long, measures only about 5 mm. The gutta-percha tubes are spun over on the outside.

Reservoir and tube 1 are fastened in a wooden frame; funnel 2 is in a wooden lining.

This precaution has been taken for the reason that the temperature of the mercury and of the apparatus should be altered as little as possible during the operation by the warmth of the hands. By means of two strings, which run over rollers fastened in the ceiling of the room, funnel 1 and funnel 2 can be pulled up or down at pleasure.

The auxiliary apparatus further required :

1. A thermometer, by means of which, previous to, during, and after the filling in of the mercury the temperature of the same will be ascertained.

2. A fine pair of scales indicating to 0·001 grammes, with a carrying capacity of 6 kilos. (each scale 3 kilos.)

3. A barometer for determining the pressure of the atmosphere at the place of operation.

4. A wooden scale, 1 m. long, with a steel point, and a slide for exactly measuring the difference in the levels of the mercury meniscus in the two tubes 1 and 2.

Theory of the Densimeter.

Let the weight of the reservoir, with the tubes screwed off, but inclusive of the connecting piece screwing the capsule and the lid together, = R grammes. After the reservoir has been exhausted, and filled with mercury, let its weight at temperature $t°$ of the chemically pure mercury = T grammes. Consequently the contents of the reservoir filled by the mercury amount to—

$$I = \frac{T-R}{13·59(1-0·00018\ t_0)} \text{ cub. centim.}$$

If now P grammes of powder be placed in the reservoir, and the air exhausted and the remaining space filled with mercury, the whole will weigh, at temperature $t°$, only T′ grammes, hence if Y is the volume of the P grammes powder, at $t°$ temperature,

$$V = \frac{T-T'+P}{13·59\ (1-0·00018\ t°)} ,$$

consequently the specific gravity of the powder to be examined

$$\frac{P}{V} = \text{specific gravity} = P\ \frac{13·59\ (1-0·00018\ t°)}{T-T'+P}.$$

The following examples will serve as illustration :

The specific gravity of chemically pure mercury amounts at

0° Cels. = 13·59 ; 10° Cels. = 13·5″   19° Cels. = 13·55 ; 27° Cels. = 13·53
5° Cels. = 13·58 ; 15° Cels. = 13·56   23° Cels. = 13·54.

Example 1.
T = the reservoir filled with mercury weighs at 19° Cels. = 1091·6
R = the empty reservoir - - - - = 329·6

$$Q = T - R.$$ Consequently mercury 762·0 grammes.
Consequently contents of the reservoir at 19° Cels.

$$I = \frac{T - R}{13·59\,(1 - t\,0·0018)} = \frac{762·0}{13·55} = 56·236 \text{ cub. centim.}$$

If now 60 grammes (= P) gunpowder is placed in the reservoir, and the remaining space filled with mercury, we obtain
611·6 grammes $(= T_1)$
Reservoir empty = 329·6
60 grammes powder = 60·0

R + P = 389·6 deducted,

remains Q = 222.0 grammes weight of the mercury filling the intervening space, occupying at 19° Cels.

$$I = \frac{222·0}{13·55} = 16·444 \text{ cub. centim.}$$

Consequently volume of the 60 grammes powder = 56·236 − 16·444
V = 39·792 cub. centims.

$S = \dfrac{P}{V}$ = specific weight of the 60 grammes gunpowder

$$\frac{60}{39·792} = 1·507$$

Example 2.
If a powder prism weighs 42·0 grammes at 190° C. and weight of reservoir, powder prism and mercury = 808·4 grammes.
Reservoir empty 329·6
42 grammes powder 42·0

371·6

consequently of 808·4
371·6 deducted,

= 436·8 grammes weight of mercury,

or $\dfrac{436·8}{13·55} = 32·236$ cub. centim. occupied by these 436·8 grammes.

Consequently volume of the 42 grammes weighing powder prism = 56·236 − 32·236 = 24·0 cub. centim.

Thus, specific weight of the powder prism $= \dfrac{42·0}{24·0} = 1·75$

General Formula.

$$S = \frac{P}{V} = \frac{P}{I - I^1} = \frac{P}{\dfrac{T - R}{13·59(1 - t\,0·0018)} - I^1} \text{ or } I^1 = \frac{Q}{13·59\,(1 - t^\circ\,0·0018)},$$

$$S = \frac{P\,[13·59\,(1 - t\,0·0018)]}{T - R - Q^1}. \qquad \text{Since } Q^1 = T' - R - P$$

$$S = \frac{P\,13·59\,(1 - t\,0·0018)}{T - T' + P}$$

The extension of the examples 1 and 2, therefore, is essentially facilitated :

Example 1.

Given $P = 60$ gr. $t = 19°$ C., consequently $13·59 \ (1 - 19 - 0·0018) = 13·55$

$T = 1091·6$ grammes, $T' = 611·6$ grammes.

$$\text{Thus } S = \frac{60·13,55}{540} = 1·507$$

Example 2.

Given $P = 42$ grammes $t = 190°$ C.,

consequently specific weight of mercury $= 13·55$.

$T = 1091·6 \ T' = 808·4$

$$\text{Thus } S = \frac{42·13,55}{325·2} = 1·75$$

Because the expansion coefficient of the reservoir made of steel is different for changing temperatures from that of the mercury, it will be necessary for determining once for all empirically the weights of Tn for ± 0, $+ 5$, 10, 15, 20, 25, 30, 35° Cels., to calculate the required cubical contents of the reservoir $= Vn$, to interpolate them graphically, and to embody them from degree to degree in a table.

DIRECTIONS FOR USE OF THE DENSIMETER.

The temperature of the mercury is determined and noted before, after, and during the period of operation. For that purpose it is advisable to employ a thermometer composed of a very fine glass tube, which admits of being inserted in the gutta-percha tube No. 1, which has been filled up to the aperture of the funnel.

The mercury in funnel 1 will show, on account of the friction and consequent heating, $1 - 3°$ more heat than that in funnel 2. This difference is, however, equalised in a very short time.

During the operation the apparatus must only be touched on the wooden lining, in order to avoid as much as possible any variations in the temperature which may be caused by the warmth of the hands. The powder to be tested must be of nearly the same temperature as the mercury to be employed, for which purpose it will be best to keep both before the testing operation for several hours in the same room. The reading of the barometer which indicates the pressure of the atmospheric air must be noted down.

1.

The two tubes 1 and 2 are screwed air-tight to the reservoir, the two cocks are opened, and the apparatus fastened in the wooden frame with vertical position of the tube No. 1.

Thereupon the funnel T'' is lifted to the level of T' (upon $+ 760$ mm.), and chemically pure mercury poured into the funnel T'', until both funnels are filled with mercury to a height of about 20 mm. The mercury will then stand 760 mm. high above the (± 0) point of the reservoir ; consequently the pressure upon the highest point of the reservoir will be altogether two atmospheres (1 atmosph. pressure corresponding in the mean to 760 mm. mercury.

Under this pressure the air in the reservoir will for the greatest part be already forced up, and in fact in the direction towards funnel 1.

2.

Now the cramp screw-piece is attached below the funnel 1, and above the glass tube in the centre of the hose 1 filled with mercury, and the latter shut off air-tight at a height of about + 600 mm. Then funnel 2 is sunk to about 1,000 mm. below the zero point of the reservoir. The mercury level will thereupon sink below the reservoir. In case reservoir and hose 1 were already exhausted of air, the difference of the level of both the mercury menisci will be exactly as much as indicated by the barometer, otherwise the difference will be smaller. All the mercury then flows back from the reservoir, &c. into the funnel 2, for which reason the same must have a sufficiently large capacity, and must be lowered carefully, not too quickly. Cock No. 1 is then shut, funnel No. 2 lifted above the zero point of the reservoir, and the latter, which in the most unfavourable case will contain only extremely rarified air, filled by the same; then cock 1 is opened, and the cramp at the hose No. 1, so that the mercury in the funnel No. 1 can rise again. This exhausting of the air is repeated a second time if necessary.

For testing whether the reservoir is entirely or sufficiently exhausted of air, funnel 2 is sunk so far until its mercury level has reached about 400 mm. below the zero point of the reservoir. Now occurs a Toricelli's vacuum in hose 1, and the mercury meniscus is seen in the glass tube.

If there were still air in hose 1, the level-difference in the hoses 1 and 2 will be smaller than the height indicated by the barometer for the day in millimeters. In this case the operation mentioned before is repeated. In all cases at a position of funnel 2 at about − 440 mm., the difference of the level of both the mercury menisci will be, at the utmost, smaller by 2 − 3 mm. than the indication of the barometer for the day. If the difference in the variation of the levels should show itself equal to the height of the barometer, which may be easily ascertained by the scale, by adjusting the slider fastened at the height indicated by barometer at the upper meniscus in the glass tube; the pressure at the upper part of the reservoir will then be—

$$\left. \begin{array}{l} 1 \text{ atmosph. air pressure} \\ - 1 \text{ atmosph. mercury pressure} \end{array} \right\} = 0 \text{ mm.}$$

consequently the reservoir is exhausted of air.

But in order to employ a further powerfully-acting means for exhausting the air, so far as this should not have been accomplished already, the funnel 2 is lifted as high as possible up to about $2\frac{1}{2}$, 760 = 1,900 mm., thereby the very small quantity of air still present will be forced into the hose 1 under $2\frac{1}{2} + 1 = 3\frac{1}{2}$ atmospheric pressure, and will ascend either towards the hose 1, or occupy only a small and practically insignificant place, of 0·001 to 0·0001 cub. cent. by shutting off the cocks 1 and 2.

The raising and lowering of hose 1 and 2 is performed by pulling or slackening the cords running over the rollers fastened into the ceiling of the room.

It may now be supposed that the air is completely exhausted from the reservoir, and that its vacuum is completely filled with mercury. At all events the hydrostatic air pump of 0 to 3·5 atmosph. pressure, attached to the apparatus in the simplest manner possible, will act much more powerfully than any other air pump.

Finally, the two cocks 1 and 2 in the mercury are shut off, the temperature of the latter being determined, the two hoses 1 and 2, which have previously been carefully emptied, are unscrewed, the reservoir cleaned of the mercury globules sticking to it (especially in the parts of the screw and the interior channel-openings of the cocks), and the weight of the reservoir, including bolt, determined with mercury.

DETERMINATION OF THE SPECIFIC WEIGHT OF THE POWDER TO BE TESTED.

The prisms or pieces of cake, or the coarse or fine-grained powder to be tested, are accurately weighed on a scale. The powder prisms of 25 mm. height, 40 mm. measured across the edges, 35 mm. on two sides, with channels each of 4·2—4 f mm. wide, or one channel 10 mm. wide, weigh, as a rule, at a specific weight of 1·6 to 1·8, from 36 to 44 grammes.

Of grained powder so much is weighed that the steel lid can be easily fixed on the reservoir, and fastened with the bolt, consequently about 50 grammes in case of a small reservoir with about 76 cub. cent. capacity, or 200 gr. in case of a larger reservoir of about 217—226 cub. cent. capacity. The quantity of powder weighed is filled into the reservoir; this is then closed, the two hoses 1 and 2 are screwed on, and the operation is thereupon proceeded with as detailed in the preceding explanations.

It must be observed that the operation is very simple and expeditious, excluding every personal error, so that, consequently, the method, being based on scientific principles, is a thoroughly rational one.

It should cause no surprise if the operations 1 and 2 must be repeated several times, when the powder has been filled in, in order to raise the difference of the level of the two mercury columns to the same height as the position of the barometer, as the large capacities of the coal of the powder for absorbing air is a notorious fact, and as also the moisture contained in the powder is to a very great extent evaporated in the form of aqueous vapour, or ejected by hydrostatic pressure and tension.

373. Mercurial Powder Balance, Major Bode's construction, for determining the specific gravity of prismatic and coarse-grained gunpowders.   *A. and R. Hahn, Cassel.*

The mercurial powder scale replaces the alcohol or so-called " volumetrical analysis method," and by means of this instrument the specific weight of the different sorts of powder, prismatic, powder-cake, and coarse-grained can be exactly determined with quantities of 40 to 50 grammes.

380. Balance for Weighing in Vacuo.
Paul Bunge, Hamburg.

The vacuum scale is a duplicate of a similar scale made by the exhibitor for the Physiological Institute at Kiel. For facilitating exhaustion it has been enclosed in a small receiver of 5 inches diameter and 10 inches in height, which was only possible by the exhibitor's system of employing a short beam. This is 69 millimeters long, and with a load of 50 gr. the balance turns with $\frac{1}{20}$ mgr.

The use of the scales is as follows : —After the body which is to be weighed in dried air or in a vacuum has been placed on the scales, and the exhaustion or the desiccation has been effected, the scale can be arrested and released by turning three studs fixed in the bottom plate. All weights of 20 gr. to 0·01 gr. can be placed on or lifted off the pan. Lastly, a rider can be traversed the whole length of a beam in the line of the axis.

VI.—MEASUREMENT OF VELOCITY.

A. Logs and Current Meters.

393. Patent Log, by Massey. For measuring speed at sea ; in use in H.M. Navy.
Hydrographic Department of the Admiralty.

394. Patent Log, by Walker. For measuring speed at sea.
Hydrographic Department of the Admiralty.

394a. New Ship's Log. *Benjamin Theophilus Moore.*

In this log a revolving cylinder, furnished with screw blades, is placed behind a tube, the front portion of which terminates in a solid pointed head. The cylinder turns upon a spindle, in bearings inside the tube. Within the revolving cylinder is a water-tight tube of glass containing the recording mechanism. This tube is drawn out for the purpose of reading the dials.

The log line is attached near the centre of gravity of the whole instrument, by which means the log is made to run horizontally, and *below* the surface of the water, and as steadily as an arrow in the air.

The mechanism, being entirely protected from contact with sea water, works smoothly, and cannot get out of order.

The glass tube is itself protected from any injury by the manner in which it is contained within the revolving cylinder.

This log will indicate any distance, from one-tenth of a cable to one thousand miles, with great accuracy.

394b. Massey's Patent Ship's Log. *E. Massey.*

Massey's Patent Frictionless Sounding Log.
E. Massey.

394c. Massey's Frictionless Propeller Conical End Log. *E. Massey.*

394d. Massey's Patent Self-registering Ship's Logs.
L. P. Casella.

These logs are constructed on the rotating system devised by E. Massey, and have registers and mile indices, to show distance run during the time the log is overboard towing astern of ship.

394e. Reynold's Patent Pendent Log.
J. Cohen & Co.

This log is composed of two parts ; first, the rotating log itself, and second, the apparatus for registering the distance run while the log is overboard.

398. Ramsten's Patent Ship's Log. *Elliott Brothers.*

396. Current Meter, for measuring the velocity of currents in rivers at different depths. *Elliott Brothers.*

An endless screw on a spindle turns two wheels at the same time, the one recording every revolution of the blades by moving one division ; the other indicating every complete revolution of the former.

397. Revy's Current Meter, constructed for measuring the velocity of currents in larger rivers.   *Elliott Brothers.*

The spherical boss is so determined that it will displace just as much water as will balance the weight of all the parts which are fixed to the spindle, so as to reduce friction to a minimum. Although the apparatus is covered with glass, it has to be filled, before using it, with pure water to establish similarity of pressure inside and outside. After every experiment the water is removed and the spindle thoroughly dried. This form of current meter was used by Mr. Revy in the survey of the Parana and Uruguay rivers.

397a. Darcy-Pitot Gauge or Current Meter, for determining the velocity of streams of water.   *Prof. W. C. Unwin.*

The velocity is obtained by a single measurement, and no time observation is required. Used in Darcy and Bazin's researches on the flow of water in pipes and canals.

399. Water Meter, based on the principle of measuring the volume of water by recording its speed.   *J. A. Muller, C.E.*

This water meter consists principally of an air and water-tight chamber or vessel, wherein moves a float, carrying two magnets of equal power, and fixed with their dissimilar poles in juxtaposition to each other: the whole combination of the float and its spindle, together with the magnets, is made as near as possible equal in density or specific gravity to the water. The water in passing through this measuring vessel is forced to take a rotary motion, by means of a screen or a tongue, being a metal piece, put at a certain distance from the inlet opening, and parallel with and lying along the inner circumference of the measuring vessel. The top cover of the measuring vessel is properly dished out, so as to allow of two small soft iron armatures, fixed to a thin metal arm or needle, to be brought outside the vessel, as near as can be to the poles of the magnets inside; the metal arm or needle is fixed to a light spindle, carrying an archimedean screw, which further gears with the registering parts of the apparatus. It is evident that the water in passing through the measuring vessel, or rather alongside the same, communicates its motion to the water inside the measuring vessel, which motion is also communicated to the float and magnets, and lastly to the needle and worm spindle and further gearing. It is plain that this meter really registers the true velocity of the water, and taking, moreover, into consideration the lightness of its different parts and the transmission of the speed of the float by means of magnets, it will be found to be a very correct and sensitive meter, of simple and durable construction.

399a. Current Meter, with electrical tell-tale apparatus, according to Amsler's latest construction. (*See* description.)
        Polytechnic School, Aix-la-Chapelle, O. Intze.

If the instrument makes 100 revolutions the electric current will be closed by a contact, and the chime work will be kept in motion during some revolutions of the instrument; it will not be necessary, therefore, in measuring the velocity in water-courses, to pull the instrument out of the water, but only to note the time which passes from one signal to the other. By experiments it must be ascertained what velocities of the current of the water correspond to certain intervals of time between the electric signals.

400. Patent Electric Velocimeter, invented by Francis Pastorelli, arranged for water currents, and for ascertaining the speed of vessels. It consists of three parts.   *Francis Pastorelli.*

1. Four hemispherical cups are fixed to the end of four strong metal arms that radiate at right angles from a central boss, mounted on a horizontal axis at right angles to a framework of metal, or other material, so that they may freely revolve when placed in the water. The horizontal axis has fixed to it a point or piece of platina; upon this work pressing points or surfaces, which can be made of any form, circular or otherwise; each revolution of the axis causes a contact to be made.

2. The same receiving instrument, as used for the mining instrument, No. 3388.

3. A Leclanché battery, as used for the mining instrument.

The receiving instrument can be placed in any convenient position on board.

N.B.—No. 1. This part of the instrument is intended to be fixed at any desirable and convenient part of the vessel, or it may be arranged to throw overboard; under such conditions it will give more accurate indications than the logs now in use, for it is not affected like them in their motion by depth, or the increasing density of water; assuming that corrections be applied for force and direction of currents, with respect to the course or line of motion, the errors would probably not be found to exceed 5 per cent.

402. Apparatus for indicating the **Speed** of a **Ship** by the aid of **Electricity.** *Bennet Woodcroft, F.R.S.*

409. Rhysimeter, without frictional parts, for measuring the speed of water or other liquids whether in pipes or open channels. *Alfred E. Fletcher, Liverpool.*

409a. Current Meter.
Benjamin Theophilus Moore, M.A.

This instrument will measure the velocity of running water at any depth below the surface, with accuracy and facility.

The frame is formed of thin brass bars united in front to a solid ogival head, and terminating, in the opposite direction, in a double vane or tail. It is suspended in water by a cord attached to a stirrup.

The frame supports a hollow cylinder which is provided with six screw blades, and is free to revolve upon fine pivots at its extremities. This cylinder contains a water-tight glass tube within which is a simple train of mechanism to record the number of revolutions made by the cylinder. This train of mechanism is suspended in such a manner, that it remains at rest while the cylinder revolves, and the dials which record the number of revolutions are seen through the glass without opening the cylinder. When the instrument is suspended in running water by a cord attached to the stirrup, the frame immediately takes up a position in which the axis of the revolving cylinder is parallel to the direction of the stream. The cylinder is set in motion, and stopped, at known instants, while under water, by means of a spring operated upon by a light cord.

For use in very deep water a simple automatic starting and stopping apparatus is placed inside the water-tight compartment of the cylinder, which operates in such a way that while the instrument is descending or ascending in water, the mechanism does not record the revolutions of the cylinder, but only while it is at the depth at which the velocity is required. In this case the spring is removed and one cord only is used.

409b. Deep Sea Current Indicator.
Benjamin Theophilus Moore, M.A.

This instrument is intended to be used for the purpose of ascertaining the direction of submarine currents.

It consists mainly of a water-tight globular shell of gun metal, pointed in one direction, and terminating in the other in a long double vane, and is carried by two pivots on a stirrup. Within the shell is a brass box, suspended by gimballs in the manner of a ship's chronometer, and containing a magnet with a graduated ring, and a train of clockwork. When the instrument is lowered into deep water, its principal axis takes the direction of the current, while the magnet settles itself in the magnetic meridian. After the magnet and the instrument have taken up their respective positions, the clockwork suddenly fixes the magnet at a known time. The instrument is then drawn up out of the water and opened, when the fixed magnet shows the direction, or bearing, of the current below.

This bearing is shown directly by the instrument when it is suspended from a fixed platform, or from a ship at anchor, or otherwise at rest. When the ship is in motion the instrument is to be used in combination with the deep sea current meter, by which means the velocity and direction of the submarine current can be determined simultaneously by a simple geometrical construction.

409c. Recorder for Indicating Speed, Pressure, &c.
W. H. Bailey and Co.

B. ANEMOMETERS.

408. Anemometer, without frictional parts, suited to measure the speed of air or gases, even when highly heated, or when contaminated with smoke or corrosive vapours. Used by H.M. Inspectors of Alkali Works.   *Alfred E. Fletcher, Liverpool.*

410. Lowne's Portable Air Meter, originally introduced by Casella.   *R. M. Lowne.*

The indications of this instrument are obtained by means of a light fan which communicates motion to indicating wheels; the dial of the instrument is placed at right angles to the fan, and is supported by three pillars on a base, which also carries the tube containing the fan. The works are extremely sensitive, the first centres running in jewels, and the indicating parts can be thrown in or out of gear with the fan.

410a. Lowne's Patent Magnetic Air Meter, especially adapted for measuring currents of air, gases, and fluids in positions where delicate instruments would be subject to corrosion.
R. M. Lowne.

The peculiarity of this instrument is, that the registering works are enclosed in an air-tight chamber, the connexion of the revolving fans with the works being made through a sheet of brass by magnetism. The fans carry a small bar magnet, and the first wheel of the indicating mechanism carries a piece of soft iron, so that when the fans revolve outside the plate of brass the soft iron revolves within by attraction and thereby moves the works.

410b. Lowne's Patent Colliery Air Meter, constructed expressly for use in mines.   *R. M. Lowne.*

The external aspect and form of this instrument is that of the well known "Biram's Anemometer." The improvements consist of.—1st, a strong, light, and anti-corrosive fan; 2nd, a large clear dial; 3rd, the indicating parts are perfectly protected from dust and smoke; and 4th, a lever is placed in a convenient position to enable the observer to throw the indicating wheels in or out of gear with the fan.

410c. Lowne's Patent Magnetic Anemometer and Current Meter, for measurement of velocity of currents of air, gas, and fluids.   *R. M. Lowne.*

In this instrument the registering works are enclosed in an air-tight chamber, the connexion of the revolving fans with the works being made through a sheet of brass by magnetism. This instrument is mounted on gymbals, with direction vane for use on board ship.

410d. Lowne's Patent Ventilation Anemometer, originally introduced by Stanley.   *R. M. Lowne.*

This instrument measures the air by means of a fan wheel placed in a clear opening, without any obstruction from the registering apparatus, which is in a separate chamber on the same plane as the fans, so that the instrument is quite flat for the pocket ; the whole of the works are of extreme sensitiveness, and the axes of the fans run in jewels, the indicating hands give the current that passes the fans in feet (after correction), and a lever above the dial throws the registering works in or out of gear with the fans.

410e. Mining Anemometer, for showing the velocity of currents in mines.   *Elliott Brothers.*

410f. Biram's Anemometer. Improved for Coal Mines.
Francis Pastorelli.

It consists of a broad brass ring ; fixed to it is a metal frame which carries three divided circles ; in the interior and centre of the ring is a spindle which carries eight vanes ; on one end is an endless screw ; this works a series of wheels, which give motion to the hands on the dials, which record the distance travelled by the air current every foot up to 100, 1,000, and 10,000 feet.

410g. Dickinson's Anemometer.
Joseph Casartelli, Manchester.

This anemometer consists of a disc, or plate, made of light material, suspended in a frame on delicate centres, having a balance weight attached to the top of the fan. To one side of the frame is fixed on pivots a quadrant opening out at right angles to the fan, and on it is marked the velocity of the current in feet per minute, as indicated by the angular rise of the fan upon which the current impinges. The advantage of the instrument consists in the fact that *it requires no timing* as required by every other instrument, and from actual experiment it is found as accurate as the most delicate instrument.

410h. Improved Biram's Anemometer.
Joseph Casartelli.

The improvement in this instrument consists in the fan being made of light material, thus greatly diminishing the friction, and rendering it a delicate and useful instrument.

410i. Biram's Patent Anemometer for ascertaining the current of air in mines, air flues, &c.   *John Davis and Son.*

This anemometer registers up to 1,000 feet. At the bottom there is a tube in which a stick may be inserted, so that the experimenter can stand at a distance from the instrument, otherwise the current of air would be deflected by the body of the experimenter.

The vanes may at will be disconnected from the indices by means of a stud at the side, thus rendering the process of timing more simple and exact.

410k. Biram's Patent Anemometer for ascertaining the current of air in mines, air flues, &c.   *John Davis and Son.*

The 4″ anemometer indicates up to 10 million feet. The size and angle of the vanes are calculated from theory and corrected by experiment, each instrument being corrected separately.

The registering apparatus consists in the 4 in. new anemometer of six small circles, marked respectively X, C, M, X M, C M, and M, the divisions on which denote units of the denominations of the respective circles ; in other words, the X index in one revolution passes over its ten divisions and registers 10 × 10 or 100 ft.; the C index in the same way 1,000 ft. ; and so on up to 10,000,000 ft.; so that an observer has only to record the position of the several indices at the first observation (by writing the lowest of the two figures on the respective circles between which the index points in their proper order), and deduct the amount from their position at their second observation, to ascertain the velocity of the air which has passed during the interval ; this multiplied by the area in feet of, the passage where the instrument is placed, will show the number of cubic feet which has passed during the same period.

The novelty in this anemometer is in its extreme portability and substantial workmanship; it is supplied with a lever which disconnects, at will, the vanes from the indices, thus rendering the process of timing more simple.

414. Edelmann's Anemometer with galvanic register.
          M. Th. Edelmann, Munich.

416. Anemometer for determining the velocity of the air, and other gaseous currents in pipes and air passages.
          Moritz Gerstenhöfer, Freiberg.

C. CHRONOGRAPHS.

401. Apparatus for measuring the velocity of projectiles, and capable of recording several measurements on one and the same trajectory and of the same projectile.
          Antoine Joseph Gérard, Liége.

403. Ballistic Chronoscope, with two pendulums, for ascertaining the speed of a projectile at any point of its trajectory, by measuring the time of flight of a portion of the trajectory; also for measuring portions of time between one tenth of a second and 25 seconds.   *Lieutenant-General Leurs, Brussels.*

404. Electric Chronograph, for measuring the initial velocity of projectiles.   *Le Boulengé, Liége.*

405. Electric Clepsydra, for measuring the time of flight of projectiles.   *Le Boulengé, Liége.*

405a. Electro-Ballistic Apparatus, for determining the velocity of a projectile, with description of experiments and additional apparatus.   *M. Navez, Paris.*

406. Electric Chronograph for the measurement of minute portions of time, &c., &c.   *Lieut. H. Watkin, R.A.*

This instrument consists of two upright cylinders resting on a base of wood ; between them, suspended by an electro-magnet, is a weight with projecting arms. The cylinders being connected with the secondary circuit of an induction coil, the circuit is complete with the exception of the small spaces on either side of the weight. When taking velocities of shot, the primary circuit is led through screens, constructed so that the current is broken and immediately made again during the passage of the shot. The gun being fired, the weight begins to descend; the shot in passing the first screen causes a spark to flash from one cylinder to the other through the weight which having been previously smoked registers by a white spot the position of the weight at that instant. As the weight continues to descend the same result is obtained at the next screen, and so on. Adjacent to the cylinder is a time scale divided into thousandths of a second, subdivided by a vernier of novel construction into hundred thousandths of a second, by which the absolute time taken by the shot between the screens is easily read off. For other uses to which the instrument may be applied, see Royal Artillery Institution Papers.

407. Clock-Chronograph, contrived for the purpose of measuring the time occupied by projectiles in passing over a succession of equal spaces, with a view to determine accurately the resistance of the air to their motion.   *Rev. F. Bashforth.*

If the fly-wheel be spun by hand, and the markers be brought down, they will trace two uniform spirals on the cylinder; each marker is, however, under the control of an electro-magnet. When the galvanic current is interrupted, a record is made by the corresponding marker being suddenly drawn aside. The circuit of the lower electro-magnet is interrupted once a second by a clock beating half-seconds, which gives a scale of time. The circuit of the upper electro-magnet passes along the tops of all the screens, as is shown in the case of one screen. When one or more threads are broken in any screen, a record is made on the cylinder. Thus, when an experiment is to be made, the fly-wheel is spun briskly by hand, the markers are brought down, and the gun is fired. The times of passing the screens are recorded on one spiral, opposite a scale of time on the other. This instrument was used in making all the experiments referred to in " Reports on " Experiments made with the Bashforth Chronograph to determine the " Resistance of the Air to the Motion of Projectiles, 1865-1870," published by authority. Generally 10 screens were placed at intervals of 150 feet, but in the experiments with the Whitworth gun (p. 162), 16 screens were placed at intervals of 75 feet; some of these records are shown. For a full description of the chronograph, see Proceedings of the Royal Artillery Institution, Woolwich, for 1866, which description is also published separately.

407a. Chronograph for projectile experiments with the recording apparatus of Deprez.   *Dumoulin Froment, Paris.*

407b. Electric Chronograph.   *Dr. Werner Siemens.*

This instrument, which was described in the year 1845 in Pogg. Ann. (Bd. 66, p. 435), serves for the measurement of high velocities, especially those of projectiles both along the barrel and in their further flight, and also that of electricity.

It is based on the circumstance that an electric spark leaves a sharp mark on polished steel, and that this mark can easily be discovered when the cylinder has been previously blackened. The cylinder is turned rapidly by clock-work, and each hundredth revolution is marked by the stroke of a small bell. By means of a regulator the rapidity of rotation is so arranged that the stroke of the bell coincides exactly with the beat of a second pendulum; the reading is made with a microscope with cross wires, the clockwork being stopped. The graduation of the micrometer head gives 0·0001 of a revolution of the cylinder or millionths of a second if the cylinder rotates 100 times a second.

The measurement of the velocity of projectiles is effected by the passage of an electric spark at the moment when the projectile touches an insulated wire which reaches to the inside of the gun, and thus is free from retardation caused by the inertia of matter or magnetism.

The same apparatus can be used for the measurement of the velocity of electricity in suspended wires.

The complete apparatus comprises a Leyden jar, induction coil, commutator, gun barrel, chronograph, and two batteries.

407c. Recording Cylinder, with original marks by which the speed of electricity in iron wires has been measured.

Dr. Werner Siemens.

Those marks which are surrounded by a halo or circle indicate the commencement of the discharge, the successive series of small marks has been formed by the electricity which has traversed the conductor; the angular distance between the first-mentioned mark and the first point of the series gives the measurement of the speed of electricity. By measurement of the time which the electricity requires to pass through lines of various lengths, the electrostatic retardation, which is proportional to the square of the length of the line, has been eliminated. By these researches it has been shown that electricity is transmitted in conductors with a constant velocity which is independent of the static retardation, and which for iron amounts to about 230,000 kilometres per second.

(See Monatsberichte der Kgl. Pr. Acad. der Wissenschaften, 6 Dec. 1875.)

411. Complete Apparatus for measuring the Velocity of Projectiles in the bore of a gun, and for measuring the speed of electricity.   *Siemens and Halske.*

412. Vibration Chronograph, for measuring the time of descent on an inclined plane, executed according to Beetz, by M. Th. Edelmann, at Munich. (A description accompanies the object.)   *Prof. Dr. Beetz, Munich.*

413. Edelmann's Apparatus for the descent of a falling Body accessory to Beetz's chronograph.

M. Th. Edelmann, Munich.

413a. Chrono-Goniometer, with magnifier.

Le Vicomte Duprat, Consul-General of Portugal.

D. STROPHOMETERS.

44. Counters and Speed Indicators.

T. R. Harding and Son.

(*a.*) Counters with reciprocating motion, as applied to marine and stationary engines.

Counters with rotary motion, suitable for shafting, printing, and other machinery.

Small counters with rotary motion applicable to spinning machinery and various other purposes.

Pocket counters for ascertaining the speed per minute of spindles or quick running machinery up to 10,000 revolutions per minute.

(*b.*) Counters actuated by pneumatic and electric apparatus at a distance from the motion to be indicated.

(*c.*) Speed indicators, showing by the height of a column of mercury the actual speed, at any moment, of engines and other machinery.

44a. Mercurial Indicator and Counter.

T. R. Harding and Son.

The above instrument is a combination of Harding's integrating counter and Brown's patent indicator, for which T. R. Harding and Son are sole licensees in Great Britain.

The mercurial indicator shows the speed per minute of the shaft above it; the counter records *half* the number of revolutions of the same shaft.

The principle of the mercurial indicator is very simple. The tubular arms of the rotating U tube are connected with the central glass tube, and when the instrument is at rest the mercury settles to a level in the glass tube and arrives at the zero of the scale. When the instrument is rotated, the mercury, owing to the centrifugal tendency, rises in the arms and sinks in the central tube to a greater or less extent according to the speed.

These indicators are of great importance for marine, stationary, and locomotive engines, as well as for various kinds of machinery.

By means of the counter and a watch, the accuracy of the mercurial indicator can at any time be verified.

47. "The Motometer," a machine to indicate the number
of revolutions made per minute, or other portion of time, by a steam engine or revolving shaft, or any body having intermittent motion, so that by simple inspection of a dial the rate of speed may be seen. *H. Faija.*

This instrument is constructed so as to indicate by a positive motion direct from the engine or other moving body to which it is attached, and is of purely mechanical construction independent of all centrifugal and other forces of an indirect nature. The indication is consequently absolute and not comparative.

The instrument is made in various forms to suit differences of speed, from the slow stroke of a pumping engine to the high speed of a locomotive, &c.

The skeleton machine exhibited is suitable to indicate the ordinary speed of a marine or stationary engine, while the one attached to the shafting is adapted for very high speeds.

395. Hearson's Patent Strophometer or Revolution
Indicator, an instrument for showing at a glance, by the position of a pointer on a graduated dial, the number of revolutions per minute an engine is at the time making. *Elliott Brothers.*

395a. Working Model of Revolution Indicator, for
engines and machinery, by J. Wimshurst. *J. F. Flannery.*

415. Mercurial Gyrometer, or "orbit meter."
Royal Polytechnic Academy (Prof. Reuleaux, Director), Berlin.

The instrument indicates directly the angular velocity of an axle, shaft, &c., in figures showing the rotations per minute. The reading takes place on an alcohol column, which shows on one side a millimeter scale, and on the other the rotation numbers. The instrument is so arranged that the scale of the rotations has uniform graduation.

532b. Reuleaux's Ball-Gyrometer.
H. Hädicke, Engineer, Demmin, Pomerania.

The object of the instrument is to indicate the rotations made per minute by any rotating body brought into connexion with the same. The number indicated will be read off a dial. As a peculiarity it may be mentioned that the scale of the dial shows a uniform division, although the position of the balancing balls moving the pointing hand depends—according to a complicated law—on the velocity of the rotation of the spindle.

The motion is worked by means of straps and pulley, and can, as a matter of course (on vessels, &c.), be effected by a fixed connexion with a shaft-movement.

The winch-handle, however, will enable the spectator to put the instrument in motion by the hand.

The accuracy of the indications of the instrument will be augmented if the pointing hand is turned off a little with the finger in the direction of the progressive numbers, and then allows it to jerk back freely.

A forcible turning of the pointing hand in the opposite direction, toward O, is not allowed.

415a. Revolution Indicator, to show the rate at which machinery is working.   *Frederick Guthrie.*

From the machinery an up-and-down motion is communicated to the piston of a pneumatic forcing pump. The compressed air escapes through a fine opening, and also exercises pressure on oil, water, or mercury, in a vessel provided with a manometer tube. The height of the liquid in the tube measures the rate at which the pump and machinery are working.

483d. Drawings of a Simple Counter, and of a Registering Counter.

(*See* Report of Baron Séguier to the Society of Encouragement, 1844.)   *M. Winnerel, Paris.*

VII.—MEASUREMENT OF MOMENTUM.

417. Model of the Ballistic Pendulum, erected in the Royal Arsenal in 1814, and transferred to the Royal Military Repository in 1836. Weight, 7,740 lbs. Centre of gravity below centre of suspension $10 \cdot 97$ ft.; centre of oscillation below centre of suspension $11 \cdot 88$ ft. Scale $\frac{1}{8}$th. *Major M. L. Taylor, R.A.*

418. Navez Electro-Ballistic Apparatus.
                          Major M. L. Taylor, R.A.

418a. Discussion on Electro-Ballistic Apparatus.

419. Model of Ballistic and Gun Pendulum, as erected at Shoeburyness in 1858. Oscillating system of gun pendulum, weighs 37 lbs. $10 \cdot 5$ oz.; that of the block pendulum weighs 31 lbs. $8 \cdot 25$ oz. Scale, $\frac{1}{8}$th. *Major M. L. Taylor, R.A.*

VIII.—MEASUREMENT OF FORCE.

421a. Attraction Meter. An instrument for measuring horizontal attraction.   *Dr. Siemens.*

This instrument consists of two horizontal tubes of wrought iron, terminating at each end in a horizontal tube of cast iron. The first-named horizontal tubes are partially closed at their extremities, and communicate with the transverse tubes below their horizontal mid-section. The transverse tubes communicate also by means of a horizontal glass tube of 2 millims. diameter at a superior level to the former.

The whole apparatus being mounted upon three levelling screws is filled to the level of the half diameter of the transverse tubes with mercury, which mercury also fills the whole of the longitudinal connecting tubes; the upper halves of the cast-iron transverse tubes and the glass connecting tube are filled with alcohol, comprising, however, a small bubble of air, which can be made to occupy a central position in the glass tube by raising or lowering the levelling screws.

If a weighty object is approached to either extremity of the connecting tube, an attractive influence will be exercised upon the mercury, tending to a rise of level in the reservoir near at hand, at the expense of the more distant reservoir; and this disturbance of level between the two reservoirs must exercise a corresponding effect upon the index of air in the horizontal glass tube, moving it away from the source of attraction. The amount of this movement must be proportional to the attractive force thus exercised. Variations of temperature have no effect upon this instrument, because the liquids contained on either side of the bubble of air are precisely the same in amount; and the total expansion of the liquids is compensated for by open stand tubes rising up from the centre of the connecting tubes through which the apparatus can be easily filled.

It is suggested that an instrument of this description may be employed usefully for measuring and recording the attractive influences of the sun and moon which give rise to the tides.

The instrument, which is of simple construction and not liable to derangement from any cause, would have to be placed upon a solid foundation with its connecting tube pointing east and west, records being taken either by noting the position of the index upon the graduated scale below, or by means of a self-recording photographic arrangement.

421b. Bathometer. An instrument for measuring the depth of the sea without the use of the sounding line. *Dr. Siemens.*

The total gravitation of the earth, as measured on its normal surface, is composed of the separate attractions of all its parts, and the attractive influence of each equal volume varies directly as its density, and inversely as the square of its distance from the point of measurement.

The density of sea water being about 1·026, and that of the solid constituents composing the crust of the earth about 2·763 (this being the mean density of mountain limestone, granite, basalt, slate, and sandstone), it follows that an intervening depth of sea water must exercise a sensible influence upon total gravitation if measured on the surface of the sea.

The bathometer consists essentially of a vertical column of mercury contained in a steel tube having cup-like extensions at both extremities, so as to increase the terminal area of the mercury. The lower cup is closed by means of a corrugated diaphragm of steel plate, and the weight of the column of mercury is balanced in the centre of the diaphragm by the elastic force derived from carefully tempered spiral steel springs of the same length as the column of mercury.

One of the peculiarities of this mechanical arrangement is, that it is parathermal, the diminishing elastic force of the springs with rise of temperature being compensated by a similar decrease of pressure of the mercury column, which decrease depends upon the proportions given to the areas of the steel tube and its cup-like extensions.

The instrument is suspended a short distance above its centre of gravity in a universal joint, in order to cause it to retain its vertical position, notwithstanding the motion of the vessel, and vertical oscillations of the mercury are almost entirely prevented by a local contraction of the mercury column to a very small orifice. The reading of the instrument is effected by means of a glass tube on the top, which connects the upper surface of the mercury with a liquid of less density. In this is enclosed an air bubble, whose position on a scale indicates the depth of water below the instrument.

Variations of atmospheric pressure have no effect upon the reading of this instrument; but a correction has to be made for variations of atmospheric density as affecting the relative weight of the mercury column, which correction might be avoided, however, by excluding the atmosphere from both the upper and lower surface of the mercury, and connecting the extremities of the column. The only necessary correction is that for the effects of latitude, which may be calculated as depths in fathoms, and tabulated for use with the instrument.

The readings of the instrument have been checked by actual soundings taken by means of Sir William Thomson's steel wire sounding apparatus; and the comparable results agree in all cases as closely as could be expected, considering that the sounding line gives the depth immediately below the vessel, whereas the bathometer gives the mean depth taken over a certain area, depending for extent upon the depth itself.

It is thought that the bathometer may render useful service to the mariner in warning him of changes of depth long before reaching dangerous ground; and the position of a vessel, when no astronomical observations can be taken, may be ascertained by means of the instrument, provided the contour lines of equal depths of oceanic basins were accurately laid down.

421c. Graphical Bathometer, after von Jolly.

University of Munich.

421d. Gravimeter. An instrument for the measurement of the variations of the earth's attractive force, invented by J. A. Broun, F.R.S., and constructed from his drawings by Dr. C. F. Müller, of Stuttgart.   *J. Allan Broun, F.R.S.*

The instrument consists of a weight suspended by two gold wires; a single wire fixed to the top of the weight and passing through its centre carries a cylindrical lever; when the lever is turned through 360° at the normal (say southern) station, the torsion of the single wire thus produced carries the weight round through an angle of 90°. The forces then in equilibrium are, the torsion force of the single wire and the attraction of the earth on the weight, which, as the two wires are no longer vertical, has been slightly raised and seeks to attain its lowest point.

On proceeding from a southern to a more northerly station the earth's attraction increases; the amount of this increase may be measured in two ways :—

1st. The lever will require to be turned through *more* than 360° in order to carry the weight to the height due to turning it through 90°. (Had the station been more southerly the lever would be turned through *less* than 360°.) The difference of the angle from 360° measures the increase (or diminution) of weight.

2nd. By removing a small portion of the weight, equal to that due to the increased attraction of the earth, the weight can be turned through exactly

90° by rotating the lever through 360°, as at the normal station. (On proceeding south weight has to be added.)

The following are the instrumental arrangements in order to make these observations :—

The weight has on each of three sides, at its base, a vertical mirror silvered, not quicksilvered) ; the middle mirror makes an angle of exactly 90° with the other two. The lever also carries a vertical mirror, which when there is no torsion in the suspension wire is immediately below and in the same vertical plane with the middle mirror of the weight. A telescope, having a glass scale at the focus of the eye-piece, is adjusted so that images of the scale can be seen (one higher than the other) reflected from the middle mirror of the weight and the lever mirror. When both of these mirrors are exactly in the same plane, the middle division on the scale seen directly with the eye-piece, coincides with the same division in the two reflected images.

By a wheel and pinion (with endless screw and clamp for delicate movement) placed below the instrument, a polished agate point can be made to act on a similar agate point fixed to the lever, so as to turn the latter through any angle. When turned through 360° the middle scale division again agrees with the image from the lever mirror. If the image reflected from one of the side mirrors of the weight does not agree also, the lever is turned through a greater (or lesser) angle than 360°, till this agreement is obtained; the difference of the angle through which the lever has been turned from 360° is obtained from the scale reading, as seen on the lever mirror.

The following apparatus is employed for very small increases or diminutions of the weight. Suspended to and vertically below the lever is a carefully *calibrated* glass wire (1 millimetre diameter), which enters a glass tube fixed below the instrument. At the lower end of this tube is a cistern containing a liquid (distilled water, or as at present, chemically pure glycerine). This liquid can be forced into the glass tube by a screw and piston (as in some barometer cisterns). The liquid is then raised till such a diminution of weight is produced by the immersion of the glass wire as to bring the mirror of the weight through exactly 90°, when the lever is turned through 360°. The length of glass wire immersed is read, by a micrometer microscope and scale, to a thousandth of a millimeter.

Though finely polished agate points have been employed for turning the lever so as to diminish the friction, there is an additional apparatus to ensure that vertical friction has no effect on the observation at last. The lever contains a magnet ; and two bar magnets, with rack-work adjustments for height, are placed one on each side of the instrument, so that by a pinion and rack movement they can be approached to the lever magnet till their force is exactly equal to the torsion force of the single wire, and the agate points are no longer in contact.

The instrument is made to serve for latitudes differing about 10° or 15°, but an auxiliary apparatus carries five platinum rings, which can be lowered upon the weight, so as to make the instrument serve from the equator to the poles, and to any height in the atmosphere.

There are special appliances for portability, by one of which the weight is fixed; another fixes the lever ; so that strain is removed from the suspension wires, and the suspended parts cannot be shaken from their places. Levels, a thermometer, and other details fit the instrument for the most accurate observations. The suspension wires are fixed at their ends in a special manner, so that the fixed points cannot vary. All the suspended apparatus is electro-gilt.

421d. Photograph of Automatic Bathometer.

A. Gérard, Liège.

In this instrument a lever of the second order supports a glass cylinder containing 100 grammes of mercury by means of a spiral spring attached to its extremity. The variation of the weight of the mercury, due to a change in the altitude of the point of observation, produces a corresponding variation in the length of the spring; this is rendered apparent by means of a pointer 30 centimeters in length, worked by a watch chain passing round a pulley on the axis of the pointer and attached to the lever.

421e. Photograph of Pendulum of Altitude.
A. Gérard, Liège.

The pendulum consists of a glass rod mounted on knife edges; the bob is a cylinder of mercury attached to the glass rod by a spiral spring, thus affording a perceptible variation in time of oscillation, due to variations of gravity.

425a. Drawing of a Registering Statical Gauge for Pressure in Guns. System of W. Paschkiéwitsh.
Captain W. Paschkiéwitsh, St. Petersburg.

2828. Dynamic Anemometer for obtaining the horizontal and vertical pressure of air in motion, upon inclined surfaces of different forms and angles. Manufactured by John Browning.
The Council of the Aeronautical Society of Great Britain.

This instrument is intended simultaneously to determine the component parts—*i.e.*, how much pressure is due to the horizontal, and how much to the vertical—of a current of air when directed against planes of different areas, and of different forms, at angles varying from 15° to 90°. The experiments are tabulated in the Aeronautical Society's Report for the year 1871 (Hamilton and Co.).

426a. Machine for measuring the slipping between hard surfaces rolling in contact. *Prof. Osborne Reynolds.*

This machine was constructed for the purpose of verifying the conclusions of the exhibitor respecting rolling friction, and the existence of a certain amount of slipping between two smooth surfaces of different curvature, or different hardness, when the one rolls on the other under pressure. It has also been used to measure the slipping between the surfaces, when the one is driving the other against various resistances and at various speeds, as well as the wear of the surfaces.

The large rolling surface is of cast-iron, supported so that it can rotate freely, but otherwise rigidly fixed. For the smaller surface various materials have been used, that exhibited being of steel; this cylinder is supported so that while its axis is always parallel to that of the larger cylinder, it can be pressed against the latter with various degrees of pressure by means of a lever acting through friction rollers. Arrangements are made for recording the number of revolutions of both cylinders; and connected with both spindles are driving pulleys and friction breaks on Appold's system, by means of which the force to be transmitted can be regulated.

An amount of slipping of not more than the one hundred thousandth part of the distance rolled can be measured with this machine.

The machine was constructed in Owen's College by Mr. Foster.

IX.—MEASUREMENT OF WORK.

429. Dynamometer, graduated up to 100 kilogrammes by intervals of 200 grammes, and showing dynams in kilogrammetres up to 981, each interval measuring two dynams nearly in absolute measure. *Prof. Hennessy, Dublin.*

430. Dynamometer graduated up to 10 kilogrammes, and giving absolute dynams in kilogrammetres up to 98, each interval measuring nearly one dynam in absolute measure.

Prof. Hennessy, Dublin.

Dynamometers similar to these are are employed at the Royal College of Science. Dublin, as referred to in the College Directory for 1876–77, page 17.

"The dynam or unit of force commonly employed throughout the course is one kilogramme moving through one metre in one second of time."

430a. Photograph of Electrodynamometer. Made by Professor H. A. Rowland, Johns Hopkins University, Baltimore, on the model of that of the British Association.

Cavendish Laboratory, Cambridge.

431. Drawing of a Dynamometrical Apparatus, constructed in 1844 by the exhibitor, to measure the real horse-power of steam-boats. *Prof. Daniel Colladon, Geneva.*

This apparatus, approved by the Academie des Sciences in 1843, was, in the same year, adopted in the Royal Dockyard at Woolwich.

644. Surface Spring Indicator.

H. Hädicke, Demmin, Pomerania.

The indicator was constructed by the exhibitor, and executed from his drawings by Messrs. Blanche and Co., in Merseburg. The piston has been replaced by a surface spring, and the construction aims at—

1. Avoiding the friction of the piston of the indicators hitherto in use.
2. Avoiding the slackness of the piston.
3. Avoiding the points of the diagram of machines in rapid motion produced by the mechanical momentum of the piston.

432. Richard's Patent Steam Engine Indicator, with Darke's Patent Detent and Cord Adjuster. *Elliott Brothers.*

By means of the detent, the paper cylinder is instantaneously set in motion or stopped by the movement of the pencil arm, as it is being applied or withdrawn, giving great facilities for taking a number of consecutive diagrams, also rendering its application to oscillating engines much more convenient.

433. Cooper's Patent Slide Valve Indicator. An instrument for ascertaining the relative position between the piston and slide valve of an engine at different points of the stroke.

Elliott Brothers.

434. Flexion Pandynamometer. An instrument designed to determine the work done by a steam engine, by means of the flexion of the beam.   *G. A. Hirn.*

On the upper edge of the beam is a rigid wooden bar of the same length, resting in the centre on a fork, which prevents it from swerving, fastened to one end of the beam with an iron rod, and free at the other end. To this extremity is attached an inelastic cord, which passes round a pulley, fixed at the head of the beam, and is carried thence towards the centre, where it is wound round the axis of a very light needle.

It is evident, from this arrangement, that when the beam moves in either direction, the end of the wooden bar which remains rigid approaches to or recedes from the head of the beam. The cord consequently winds itself round, or unwinds itself from, the axis of the needle, and the deviation of the latter indicates the degree of flexion of the beam, multiplied if desired. At the end of the needle is fixed a pencil, which works on a small board placed above the beam. This pencil, at each double stroke of the piston, traces a closed curve, of which the ordinates indicate the successive degrees of flexion of the beam during the work. To determine, once for all, the degree of flexion corresponding to a given load, the crank of the fly-wheel should be fixed at the dead point, and steam at a known pressure should be introduced into the cylinder.

435. Torsion Pandynamometer. This instrument is designed to measure the power supplied by an engine to a factory, by means of the torsion of the shafting through which the motive power is transmitted.   *G. A. Hirn.*

At the extremities of one length of the shafting are keyed two toothed wheels of equal diameter, which gear, one directly and the other by an intermediate wheel, into two smaller pinions. These pinions gear into the four bevelled wheels of an ordinary differential movement. The two intermediate wheels of this movement are loose on a shaft, which is prolonged in a vertical direction, and made of a light steel rod. The result of this arrangement is, that if the shaft twists, this rod deviates, and forms with a vertical line an angle proportionate to the torsion to which the shaft is subjected. At the upper and free extremity of the steel rod is secured, by means of a hinge, a horizontal very light wooden bar, carrying at its extremity a roller, to which is attached a recording apparatus. This roller, when the shaft is at rest, lies in the centre of a wooden disc covered with paper, and revolving uniformly on a vertical axis.

So soon as the steel rod deviates from a vertical line, in consequence of the torsion of the driving shafting, the roller leaves the centre of the disc, and begins to revolve. The turns registered by the recording apparatus are exactly proportional to the torsion of the shafting.

The mean torsion of the shafting being thus known for a day's work, two parallel levers, placed in contrary directions, are securely fixed at the extremities beyond the two toothed wheels, and the free extremities of the levers, so as to determine the deviation caused in the vertical bar by a given weight.

Simple proportion then gives the resistance, corresponding to the mean angle obtained during a day's work, and it becomes easy to determine the mechanical work which corresponds to this angle.

435b. Dynamometer Waggon, for marking and registering the tractive power, and the distances travelled.

Eastern Railway of France Company, Paris.

436. Theoretical Pressure Diagram for calculating the mechanical work in a steam cylinder.

H. Hädicke, Demmin, Pommerania.

X.—MEASUREMENT OF ANGLES.

437. 10-inch Protractor, by Ramsden. *Royal Society.*

438. Clinometer of Precision, employed.in 1865 by Professor Piazzi Smyth in the interior of the Great Pyramid.

Prof. Piazzi Smyth.

This instrument was made to order by T. Cooke and Sons, of York, in 1864, at the cost of Andrew Coventry, Esq., of Edinburgh, for measuring the interior slopes of the Great Pyramid. When thus used it was further mounted on a deep wooden beam, 120 inches long, armed with feet of gun metal.

The angle measuring portion of the instrument is a complete circle, provided with three pairs of opposite verniers, each reading to 10″ in order to eliminate errors of division as well as eccentricity, and the whole circle can be moved and clamped on its centre so as to repeat any required angle all round the circumference. On the voyage to Egypt a thermometer broke inside the box, and the mercury tarnished the divided rim in parts. The Pyramid angles thus obtained were printed in Vol. II. of " Life and Work at the Great Pyramid," by Professor Piazzi Smyth, in 1867.

439. Smaller Clinometer of Precision, with improved mounting, readers, and level. *Prof. Piazzi Smyth.*

This instrument was made to order by E. E. Sang, of Edinburgh, in 1869, and intended for measuring Great Pyramid angles of slope. It carries its own footbar, 25 inches long, has improved readers and illuminators, and a chloroform level, as being more quick and frictionless than either ether or alcohol. The circle can be rotated and clamped on its own centre for due repetition of the angles round the circumference ; the verniers read to 1 , and there are supplementary verniers for investigating errors of division.

435a. Method of ascertaining **Angles of Torsion** by means of instruments constructed by Professor Wischnegradski.

Laboratory of Mechanics, Technological Institute, St. Petersburg.

This is composed of a support fixed with two horizontal screws in the given section of the beam subjected to torsion. This support carries a horizontal axle, upon which is fixed an arc, bearing the teeth, whose pitch measures an angle of 2,440 seconds. This arc gears with an endless screw, the head of which bears a circle divided into 244 equal parts, and furnished with a fixed decimal vernier ; the arc also carries a very sensitive level, placed at the beginning of the experiment in a horizontal position.

The angle of torsion between the two given sections of the beam is calculated by two instruments exactly similar. The deformation of the twisted beam causes an inclination of the levels of both instruments; they are restored to their original position by means of the endless screws, and then is effected the reading of the angles described by the arcs of the two instruments. The difference between these angles is the angle of torsion required. In the Laboratory of Mechanics of the Technological Institute of St. Petersburg the well-known apparatus of Wöhler is used for the torsion of trees, the photograph of which, taken together with the instruments for measuring the angles of torsion, is exhibited. For demonstrating how to use the instruments, a provisional apparatus is exhibited, wherewith the torsion of the beam is effected by means of a simple lever.

442. Clinometers, devised by the Rev. Professor Henslow, one of which was used by Dr. Hooker in his Himalayan journeys.   *J. D. Hooker, M.D., P.R.S.*

442a. Three Clinometers.   *G. W. Strawson.*

443. Protractor, with scale, vernier, and magnifying glass. Reads to 1 min.   *Prof. Baron von Feilitzsch, Greifswald.*

443b. Instrument for the Measurement of Angles.
Dr. Fr. Holler, Selbo Drontheim, Norway.

440-1. Drawings and Photographs of Dividing Machinery.   *Messrs. Troughton & Simms.*

Fig. 1. General view of dividing machine.
- A. The circular table with racked circumference containing 4,320 teeth. each tooth, therefore, equal to five minutes of arc.
- B. The screw by which movement is communicated to Table A.
- C. A ratchet wheel attached to the screw shaft.
- D. A crank arm which during one half of a revolution gives a forward movement to the screw; during the remaining part of its revolution the screw is at rest. The axis which carries the crank arm has a bevelled wheel upon it, serving to communicate motion to the cutting apparatus.
- E. The cutting frame.
- F. A cam to give movement to the dividing knife or other tool by which the division is made.

The apparatus is so arranged that the division may be cut whilst the circular table is at rest, the tool being lifted by a second cam (not well seen in the drawing) when the table is in motion.

Fig. 2. Plan of cutting apparatus showing the relation it bears to the circular table and screw.
Fig. 3. Section of table and axis.
Fig. 4. Drawings of cams and cutting frame, the cam " h " for lifting the tool (just seen in Fig. 1) is here shown.

XI.—MEASUREMENT OF TIME.

Copy of the **Drawing representing the first idea of the Application of the Pendulum to the Clock; dictated by Galileo,** then blind, to his son Vincenzo and his disciple Viviani. Elucidated on the original of the Galilean manuscripts in the Biblioteca Palatina.

The Royal Institute of " Studii Superiori," Florence.

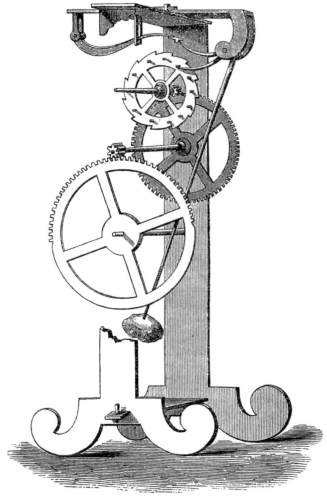

In an account which he gave Prince Leopoldo de' Medici, Viviani, after
having described Galileo's experiments on the pendulum, and the way in
which he applied it to the measurement of time, continues thus : " But as
" Galileo was most liberal in communicating his inexhaustible speculations,
" it frequently happened that the uses and newly discovered properties of
" his pendulum, spreading little by little, fell into the hands of persons who
" adopted them for their own ends or inserted them in publications, and
" by artfully passing in silence over the name of their true author, made
" such use of them that it was believed—at least by those who knew nothing
" of the origin of the discoveries--that the writers were the real authors of
" them.   He next speaks of the observations of the ' Stelle Medicee,' of the
" tables relating to them prepared by the Padre Renieri, of the offering made
" by Galileo to the States General of Holland of his method for determining
" longitudes by means of the eclipses of Jupiter's satellites, and of Galileo's
" determination to send his son Vincenzo and the aforesaid Padre to Holland,
" since he himself, being old and blind, was unable to travel thither."   He
then continues : " While, therefore, Padre Renieri was employed on the
" composition of the tables, Galileo gave himself up to meditations on his
" time-measurer ; and I remember, one day in the year 1641, when I lived
" near him in the Villa d'Arcetri, that the idea struck him that it would be
" possible to adapt the pendulum to clocks with weights or springs, and
" make use of it instead of the usual regulator, hoping that the perfectly
" equable and natural motion of the pendulum would correct all the defects
" in the mechanism of the clocks.   But as his blindness deprived him of the
" power of making plans and models of the designs he had formed in his
" mind, his son Vincenzo having arrived one day at Arcetri, from Florence,
" Galileo confided his ideas to him, and many times afterwards did they
" reason over the matter, and at last settled upon the method which is
" shown in the accompanying drawing, and then set to work at once
" in order practically to overcome those difficulties which for the most part
" it is impossible to foresee.   But Sig. Vincenzo intended to construct the
" instrument with his own hand, in order that by this means the secret
" of the invention should not be reported by the artificers before it had
" been presented to His Serene Highness the Grand Duke, his master, and
" to the States General (to be used for observing the longitude), but he
" put off the execution of his work so frequently that a few months later
" Galileo, the author of all these admirable inventions, fell sick, and on the
" 8th of January 1641, ' ab Incarnazione ' according to the Roman style, he
" died ; and consequently Sig. Vincenzo's energies so cooled down that it
" was not until the month of April 1649 that he actually began to make the
" present clock upon the idea explained to him by his father Galileo.   He
" then managed to obtain the services of a young man—who is yet living—
" named Domenico Balestri, a locksmith who had had some experience in
" making large wall clocks, and he made him construct the iron frame, the
" wheels and arbors, but the tooth-cutting and the remainder of the work he
" executed with his own hands, constructing on the highest wheel called the
" scape wheel (*tacche*) 12 teeth with as many pins (*pironi*) spaced between
" the teeth, and with a pinion of six leaves on the same arbor, and another
" wheel of 90 teeth which moves the above-mentioned.   He then fixed on
" one side of the support which is at right angles to the frame the detent
" (*scatto*) which rests on the scape wheel, and on the other side he fixed the
" pendulum, which was made of an iron wire screwed at the lower extremity
" for the attachment of a ball of lead, so that it could be lengthened or
" shortened for regulating.   When this much had been done Sig. Vincenzo
" wished me (as one who was in the secret of this invention and who indeed
" had urged him on to complete it) to see, by way of trial, the combined

" working of the weight and the pendulum. I observed the mechanism in
" operation more than once, and his workman was likewise present. When
" the pendulum was at rest it prevented the descent of the weight, but when
" it was raised and then let go, in passing beyond the perpendicular, with
" the longer of the two arms attached to the pivot of the pendulum, it raised
" the detent which fits into the scape wheel, which wheel drawn by the weight
" in rotating with its higher part moving towards the pendulum, pressed with
" one of its pallets on the other shorter arm, and gave it, at the beginning of
" its return, an impulse sufficient to cause it to swing to the height from
" which it had started, so that when it fell back naturally, and had passed
" the perpendicular, it returned once more to lift the detent, and immediately
" the scape wheel was set in motion and gave a fresh impulse to the pen-
" dulum, thus, the swinging of the pendulum was rendered continuous until
" the weight had reached the ground. We examined the operation together,
" connected with which, however, many difficulties arose ; but Sig. Vincenzo
" did not doubt but that he would be able to overcome them all, indeed he
" fancied that he would be able to apply the pendulum to clocks in a different
" manner and by means of other inventions ; but since he had got so far, he
" wished to finish it on this plan, as the drawing shows it, with the addition
" of hands to show the hours and even the minutes. For this purpose he set
" to work to cut another cog-wheel. But whilst engaged on this work to which
" he was unaccustomed, he was overtaken by a very acute attack of fever, and
" was obliged to leave it unfinished at this point, and on the 22nd day of his
" illness, on the 16th of May 1649, all his thoughts and aspirations, together
" with this most exact measurer of time, were for ever lost to him. He,
" their author, passed away to measure (let us hope) in the enjoyment of the
" Divine Essence, the incomprehensible moments of Eternity."

Dodecahedron, with eleven solar watches, made in Florence
in 1587. *The Royal Institute of " Studii Superiori," Florence.*

Horizontal Solar Watch.
The Royal Institute of " Studii Superiori," Florence.

Horizontal Watch, at the latitude of 43° 44', made by
Cammillo della Volpaja, Florentine, in the second half of the 16th
century. *The Royal Institute of " Studii Superiori," Florence.*

Vertical Watch, of boxwood, at the latitude of 43° 30',
made in Florence in 1590 by Girolamo della Volpaja.
The Royal Institute of " Studii Superiori," Florence.

Night Watch, at the latitude of 43° 30', made in Florence in
1568 by Girolamo della Volpaja.
The Royal Institute of " Studii Superiori," Florence.

483. Universal Dial, made in 1616 for Prince Charles.
The Royal United Service Institution.
Presented to the United Service Museum, in 1832, by Captain W. H.
Smyth, R.N., K.F.M., F.R.S., &c., &c.

485. Universal Dial, in use about 160 years ago.
The Royal United Service Institution.
Presented to the United Service Museum in 1838, by His Royal Highness
the Duke of Sussex.

484. Timekeeper, which was twice carried out by **Captain Cook.**   *The Royal United Service Institution.*

This timekeeper is thus spoken of in Cook's Voyage to the Pacific, 1776, Vol. I., p. 4 : " I had likewise in my possession the same watch or time-" keeper which I had in my last voyage, and which had performed its part " so well. It was a copy of Mr. Harrison's, constructed by Mr. Kendall."
This watch was taken out again by Captain Bligh, 1787, and when the crew of the " Bounty" mutinied it was carried by the mutineers to Pitcairn's Island. In 1808 it was sold by Adams to an American, Mr. Mayo Fletcher, who sold it in Chili, and in 1840 it was purchased for fifty guineas by Sir Thomas Herbert. It was repaired and rated at Valparaiso, and taken by Sir Thomas to China, and brought home in the " Blenheim " in 1843, having kept a fair rate with the other chronometers for the space of three years.
Presented to the institution by Admiral Sir Thomas Herbert, K.C.B.

484a. Two Hour Glasses. These were used in Spanish men-of-war at the beginning of the last century.
Ministry of Marine, Madrid.

484b. Chronometer, the fifth made by the English maker Arnold. It was used on board one of the Spanish ships at the battle of Trafalgar.   *Ministry of Marine, Madrid.*

491. Ancient Striking Clock.
H. M. Commissioners of Patents.

This clock is of Swiss manufacture, and supposed to have been made in the year 1348. It was obtained from Dover Castle, and had never been removed from there till the year 1872. It is interesting from the fact of its having the verge escapement, which was used many years before the pendulum.

444. Clock Dial. The hours, six only, are indicated by perforated Roman letters. The hand or pointer is formed of a revolving disc, painted in oil, with the subject of Aurora and the Hours; it must have gone round four times in 24 hours. The dial is fitted in the original carved door of the clock. *Italian.* 17th century.   *Rev. J. C. Jackson.*

445. Clock, in the shape of an orb of silver-gilt, covered with silver filigree, suspended from a ring which is surmounted by a cupid. The base of black marble is ornamented with beads enriched with silver-gilt filigree, enamels, and precious stones. *German (Hamburg).* Dated 1685.
Rev. J. C. Jackson.

446. Clock, in gilt ormolu case, engraved with figures of soldiers and festoons of flowers and fruit. It has a single hand, and strikes the hours. The present pendulum has been substituted for the old bob. *English.* Early 17th century.
Rev. J. C. Jackson.

462. Model of a Clock with four Faces, to be worked by water. *Major M. L. Taylor, R.A.*

463. Sir W. Congreve's Clock, in which the action of the pendulum is replaced by the motion of a small steel ball on an inclined plane, which it descends in 30 seconds.

Major M. L. Taylor, R.A.

486. Month Equation Clock, with double pendulum and dead-beat escapement by Quire, showing minutes and seconds both sidereal and mean time, also sun fast or slow, and containing an annual almanack, mentioned in Cooke & Maule's account of Greenwich Hospital in 1789. *Royal Naval College, Greenwich.*

447. Two Chronometers, by Arnold. *Royal Society.*

These chronometers were taken round the world by Captain Cook.

447a. Chronometer, with **Glass Balance Spring.**

E. Dent and Co.

This is the invention and handiwork of the late Frederick Dent, of the Strand and Royal Exchange, and the only specimen in existence. The spring requires far less compensation for any given change of temperature than a steel spring would, and the balance, which is composed of a glass disc, is compensated by the two small compensation laminæ mounted upon it.

448. Explanation of the principles of action of the **Compensation Balance,** with four models showing the various stages of construction. *James Poole & Co.*

"Cut open" for action of heat and cold; ordinary construction without auxiliary, in rough state from casting.

450. Silver Pocket Chronometer. *James Poole & Co.*

451. Diagrams and Models of the method of winding and setting watches without a key. *James Poole & Co.*

452. Keyless Watch, complete, with fuzee.

James Poole & Co.

452a. Dipleidoscope with Telescope. *M. Lutz, Paris.*

453. Keyless Watch, complete, with centre seconds and going barrel. *James Poole & Co.*

454. Ship Chronometer (2 day), complete.

James Poole & Co.

455. Ship Chronometer (2 day). Movement reversed, to show workmanship. *James Poole & Co.*

456. Chronometer Movement (2 day), as received from the factories in Lancashire. *James Poole & Co.*

456a. Six Chronometers with rates from the Geneva Observatory, stating records of trials of the years 1875 and 1876.

H. R. Ekegrén, Geneva.

No. 16,175, gold, open face, 21 lines, keyless.
No. 16,873,   „   „   18 „   „
No. 16,144,   „   „   Chronographs, 19 lines, keyless.
No. 16,534,   „   hunter, 21 lines, keyless.
No. 16,576,   „   „   19 „   „
No. 16,525,   „   „   18 „   „

457. Regulator Clock, filled with the Exhibitor's new patent gravity escapement, having no upward locking, and which cannot trip or slip teeth.   *Alfred John Higham.*

On the escape wheel there are two sets of teeth, one set longer than the other ; the teeth of each set are arranged, and the pallets are formed and placed, so that the shorter set of teeth only is used, except in case of tampering or other disturbance, when the longer set comes into action. This secondary locking entirely prevents irregularity in the clock rate, but the clock can be allowed to " run," if it is desired to be so set to time, by removing two extra stops which are adjustable. The action of the escapement is ordinarily exactly the same as that of the gravity escapements invented by Mr. Denison (Sir Edmund Beckett, Bart.), and the secondary locking can be applied to those escapements. For further description, see the Horological Journal, April 1876.

458. Regulator, with improved gravity escapement on the Bloxamic principle, as arranged and patented by Mr. Higham ; founded on the old pin-wheel escapement, and fitted with a galvanic interruptor for chronographical and other astronomical purposes.   *Charles Frodsham & Co.*

459. Marine Chronometer, fitted with a galvanic interruptor for chronographical and other astronomical purposes.
                          Charles Frodsham & Co.

459a. Two-day Marine Chronometer, fitted complete for ship's use.   *Charles Frodsham.*

460. Apparatus for demonstrating the application of the pendulum to the clock, at the same time serving for audibly indicating the minutes.
The Secondary Government School, Assen (Netherlands).

This apparatus is constructed by C. H. Van der Heyden, watchmaker, at Assen (Netherlands), in conjunction with Dr. A. Van Hasselt, teacher at the school for middle-class education, Assen. Price about 4*l*. The escapement may be pulled forward so as to allow the wheel to turn freely. In this manner it may be demonstrated, that clockwork without a pendulum will acquire an accelerated motion.
The escapement must be kept in a forward position, until the weight has reached the ground.
The apparatus, as it audibly indicates the minutes, may also be used for experiments to demonstrate the laws of the pendulum, the laws of hydrodynamics, &c.

460a. Wheatstone's Motor Magneto-Electric Clock (in teak case).   *The British Telegraph Manufactory, Limited.*

460b. Wheatstone's Sympathetic Magneto-Electric Clocks (4). *The British Telegraph Manufactory, Limited.*

460c. 18-inch Wheatstone's Sympathetic Magneto-Electric Clock.

The British Telegraph Manufactory, Limited.

461. New System of Electric Clocks.

Prof. Osnaghi, Imperial Central Meteorological Institute, Vienna.

In these electric clocks the uncertainty of the action of the greater number of electric pointers has been avoided by causing the electric current to flow with almost unabated force, as if there were no other clocks present. This is attained by giving the electro-magnets double coils of very unequal resistance. The spirals with great resistance serve for the attraction of the needle from a distance; the spirals with little resistance for retaining the already attracted needle. With every clock there is also a wire coil for the general return current, through which the electric current can circulate until the attraction of the needle is complete, when its course is diverted by certain mechanism, and is forced to pass over to the next clock.

464. Model of a Protomotive Clock.

The Committee, Royal Museum, Peel Park, Salford.

An apparatus consisting of a dial with hour and minute hands, and a gutta-percha tube 100 feet in length, the object of which apparatus is to demonstrate how a number of such dials in distant situations may be made, by means of a column of air at natural pressure, to indicate the same time as the clock with which they are connected. Invented and made by the late Richard Roberts, C.E., Manchester; about the year 1848.

465. Ley's Compensating Pendulum. *Henry W. Ley.*

An inexpensive pendulum compensation is to be obtained by the employment of zinc and flint glass.

466. Ley's Entirely Detached Gravity Escapement.

Henry W. Ley.

The object of this escapement is to cut off absolutely from the pendulum the clock train with its variations, so that there may be nothing whatever to disturb the arc of vibration. The arrangement of the escapement shown in the Figs. permits the motions of the various parts to be clearly followed. The scape wheel has six long teeth A, Figs. 1, 2, 3, and 4, by means of which it is "locked," and six "impulse" pins B near its arbor. The arbor carries a fly, not drawn. The pendulum receives its impulse at each alternate beat; at the beats from right to left in the Figures. The parts of the escapement are : (1), a pallet C; (2), a lever D, having the same axis as C, and resting normally against a fixed stop, from which it can lift, but below which it cannot fall; (it is in its normal position in Figs. 1, 3, and 4); (3), an arm E, of which one end can turn on a pin in D, and the other end, which is free, is lifted by the impulse pins, and rests on them successively; (it is resting on an impulse pin in Fig. 1); (4), a first detent F, against which C sets when at the top of its lift; (C is thus set against F in Fig. 1); (5), a second detent G; and (6), set on a spring, a stop H, against which the scape wheel locks.

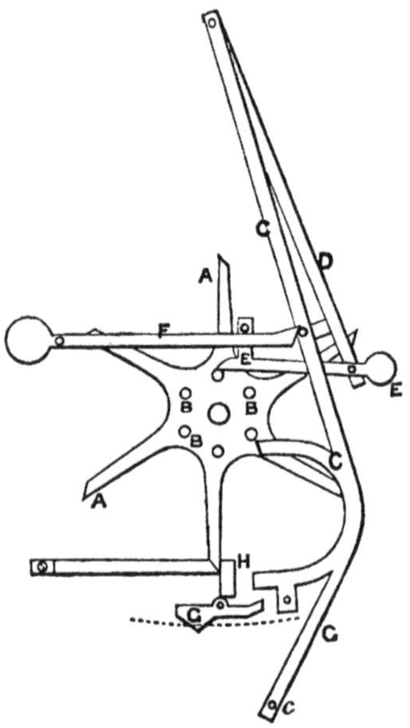

Fig. 1.—The pallet against   Fig. 2.—At the end of the pendulum
       the first detent.   swing to the right.

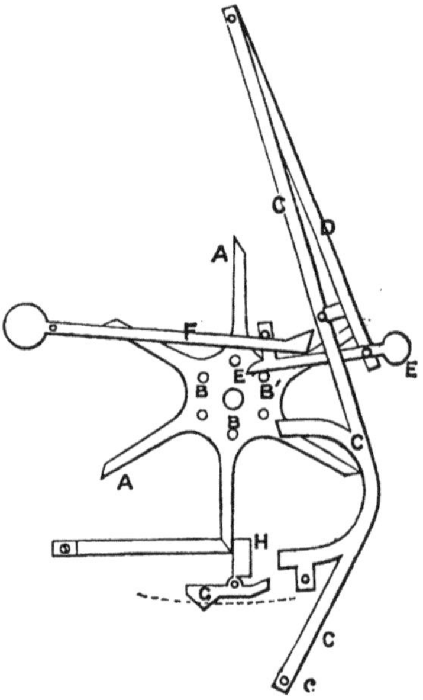

The action of the escapement is as follows:—Suppose (as represented in
Fig. 1) the scape wheel to be locked and that C has been lifted from its lowest
position through an angle $a + \beta$ to the top of its lift. Suppose also that
the pendulum is moving to the right from the end of its swing at the left.
First, a slot or a pin in the pendulum rod (a pin is supposed here for simplicity sake, and the path of the pin is shown by the dotted curve in each
Fig.) lifts G, idly, which falls back to its normal position, that of Fig. 1,
immediately the pin has passed; then the rod itself, towards the end of its
swing to the right, impinges against a "beat" pin c in C, and, still rising,
carries C with it through a further angle γ. In rising through γ, C takes
up D, and the free end of E, which was resting on the impulse pin by which
it was lifted, is carried clear of that pin (now see Fig. 2), and drops on to
B¹, the impulse pin next below, depressing F in its drop, and afterwards
holding F down (see E and F in same Fig.). The pendulum now returns,
from right to left, and C with D falls back through γ, D being arrested at the
fixed stop; the free end of its arm E still resting on B¹, and still holding F
down. The pendulum continuing its descent, C falls back through β (the
detent F being out of the way), as far as the detent G (here see Fig. 3),
where it stops. In this fall through β, C gives the impulse. The pendulum
now moves on by itself, until presently the pin in its rod once more lifts
G; not, however, idly now, but releasing C, which falling back further through
a to its lowest position (shown in Fig. 4), unlocks the scape wheel from
H. The position of H with respect to that part of C which acts upon it is

Fig. 8.—The pallet against
the second detent.

Fig. 4.—The pallet at its
lowest position.

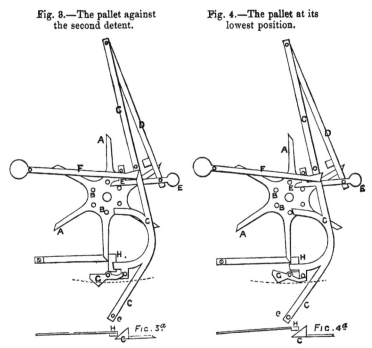

shown in plan in Figs 3a and 4a; in Fig. 3a just before, and in Fig. 4a just after the unlocking of the scape wheel. The pendulum having lifted G, continues its swing to the extreme left, whence it was supposed started. The scape wheel, free to move, lifts C and also E, the detent F, which is weighted so as to rise of itself, following E's motion, and being in position to hold C when the lifting is done, which is the case just before the next long tooth of the scape wheel coming round and setting against H (which returned to its normal position as C in lifting cleared it), the parts are again as represented in Fig. 1.

Such being the action, it is evident that the escapement is not only a detached one in the sense in which all properly so-called gravity escapements are so, that is, the pendulum is free from the scape wheel between the unlockings, but that the detachment is entire, seeing that the pendulum is *never in connexion with the clock-train at all*, being out of the way for the unlockings, as well as between whiles. There is therefore nothing whatever to affect its rate.

The pressure of the scape wheel against the stop by which it is locked varies. This variation, however, is altogether apart from the pendulum, as unlocking is the work of the pallet.

In the arrangement drawn the impulse is not given across the line of centres; it can, however, be so given; and other modifications of the escapement as here arranged can be made within the limits of its principles of action, which are (1) that the lifted pallet shall be held independently of any variations in the lift, and (2) that the unlocking shall be apart from the pendulum.

466a. Jamin's Compensator. *M. Lutz, Paris.*

466b. Diagram representing the Great Westminster Clock. *E. Dent and Co.*

This is by far the largest and most powerful clock in the world. The clock frame is 15 feet 6 in. long, and 4 feet 10 in. wide. The escapement is the double three legged gravity, and the pendulum which controls it weighs 685 lbs., is 14 feet 5 in. long, and vibrates once in two seconds. Its compensation is effected by zinc and iron tubes. The dials, four in number, are 22½ feet in diameter, and the bell on which it strikes, "Big Ben," weighs nearly 14 tons.

466c. Diagrams representing the New Standard Clock of the Royal Observatory, Greenwich. *E. Dent and Co.*

One of these is a view of the back of the movement of the Greenwich clock, showing the escapement, the galvanic contact springs, and the contrivances invented by Sir George Airy for altering the compensation of the pendulum and for altering the rate of the clock without stopping it. The other diagram shows the barometric compensation.

466d. Collection of Compensation Balances.
E. Dent and Co.

No. 1. An early form of balance.—Steel connexions are fastened near the root of the rims of a plain brass balance; the expansion or contraction of these being less than that of the central brass arm, the rims are by any change of temperature tilted towards or away from the axis of motion.

No. 2. An early form of balance.—Loops formed of brass melted on to steel are fastened upon each side of the axis of motion, in consequence of the greater expansion or contraction of the brass, these open or close with the change of temperature, and drag in or thrust out the small brass weights, to which they are attached by wires.

No. 3. An early form of balance.—The rims are of brass melted upon steel, the brass being outwards; with any change of temperature the rims open or close.

No. 4. An early form of balance.—A flat steel bar has soldered to its extremities underneath pieces of brass; the ends of the steel bar carry uprights bearing weights upon their summits, the brass pieces underneath having a different rate of expansion to the steel, bend it either upwards or downwards, and tilt the uprights carrying the weights towards or away from the axis of motion.

No. 5. A balance of similar design, but having brass melted upon the steel, instead of merely being soldered to its extremities.

No. 6. A balance of modern design, similar in its action to No. 5.

In order to obtain perfect compensation, it is found that for an increase of temperature the compensation weights must advance more rapidly towards the axis of motion, than for the same decrease of temperature they would recede from it. This peculiarity necessitates what is called secondary compensation. The following balances have been introduced to obviate this error:—

No. 7. Compensation pieces formed of brass melted upon steel receive such curves, that with any increase of temperature the compensation weights move towards the axis of motion more directly than they recede from it with any decrease of temperature. (Dent's balance.)

No. 8. A compensation bar is formed, as in No. 5, by brass being melted upon steel, and this bending upwards or downwards, with any change of temperature, tilts the weights carried by the staples towards or away from the axis of motion. But the staples are themselves compensation pieces, and they lift the weights higher with any increase, and depress them with any decrease of temperature, and in this manner increase the rate at which they approach the axis of motion, and diminish the rate at which they recede from it. (Dent's patent balance.)

No. 9. A balance of nearly the same form as No. 6, but the section of its rim is somewhat in the shape of a prism ; the form of the rim offers less resistance to the motion of the compensation weight inward than outward. (Dent's registered balance.)

No. 10. A balance similar to No. 5 is mounted upon the arm of a balance similar to No. 6. With any increase of temperature, the first balance can assist the second, but with any decrease of temperature its motion is checked. The whole combination, therefore, is more effective in the heat than in the cold. (Glover's form.)

No. 11. An experimental balance, contrived for the purpose of removing weight from the centre, both with an increase and decrease of temperature. (Wetherill's form.)

No. 12. An auxiliary compensation is added to a balance similar in form to No. 6. The auxiliary consists of two double compensation pieces, and the effect is to carry weight towards the axis of motion, both for an increase and decrease of temperature. The effect of the main compensation weights is therefore increased in the heat and diminished in the cold. (Dent's balance.)

No. 13. A balance of similar design to No. 8, but arranged so that the secondary compensation can be altered with greater facility. (Dent's balance.)

No. 14. A balance having the same general operation as No. 8, but the effect is obtained by straight bars only. The secondary compensation can also be altered without inconveniently.disturbing the main compensation, and both without producing any great alteration in the time of the chronometer. (Dent's balance.)

467. Drawings of Compensation Balances, Escapements, and other appliances connected with the construction of Clocks and Watches. *The British Horological Institute.*

Lever Escape Wheel.
Lever Escapement.
Double Roller Lever Escapement.
Two Pin Lever Escapement.
Chronometer Escapement.
Duplex Escapement.
Club Tooth Lever Escapement.
Verge Escapement.
Horizontal Escapement.
Double Roller Lever Escapement with Compensation Balance.
Marine Chronometer Escapement.
Compensation Adjustment by Sir G. B. Airy, Astronomer Royal, 1875.
Double Three-legged Gravity Escapement as used in the Westminster Great Clock.

Dead Beat Clock Escapement.
Pin Wheel Clock Escapement.
Zinc and Steel Compensation Pendulum as used in the West-
minster Great Clock.

467a. Compensation Balance arranged in Two Groups.

GROUP I.—Earnshaw's balance with circular rim (1795), and
modifications thereof to the present time.

Earnshaw's balance.
Modification of do.
 Do. do.
 Do. do., with extra adjusting screws.
 Do. do., with screws for weights.
 Do. do. do. do.
 Do. do., with screws for more minute adjust-
      ment.
 Do. do. do. do. do.
 Do. do., with double weights.
 Do. do., with variation of weights.
 Do. do., with auxiliary by Molyneux.
 Do. do., do. do.
Modification of Molyneux's auxiliary.
Eiffe's mercurial auxiliary.
Poole's auxiliary.
Example of recent auxiliary.
 Do. do. do.
 Do. do. do.
Compensation adjustment by the Astronomer Royal (Sir
G. B. Airy).

GROUP II.—Balances of a form distinct from Earnshaw's, from
Hardy (1805) to the present time.

Hardy's balance.
Arnold's do.
Dent's do.
Balance with laminated arm and rim.
Hartnup's balance.
 Do. do. cup.
Modification of Hartnup's balance.
Kullberg's flat rim balance.
 Do. double-flat rim balance.
Cole's balance.

The British Horological Institute.

468. Enlarged Model of Compensation Watch Balance.
The British Horological Institute.

469. Ordinary Marine Chronometer Compensation Balance.
The British Horological Institute.

471. Models (ten) **of Compensation Balances,** showing various attempts to overcome what is known as the "Error" of the ordinary Compensation Balance, by the late Thomas Hewitt.
The British Horological Institute.

472. Marine Chronometer by Earnshaw.
The British Horological Institute.

473. Marine Chronometer with Mudge's Escapement.
The British Horological Institute.

474. Grossmann's Micrometer.
The British Horological Institute.

475. Model of "Ferguson's Paradox."
The British Horological Institute.

476. Model of Cole's Resilient Escapement.
The British Horological Institute.

477. Callipering Engine, by the late Richard Roberts.
The British Horological Institute.

478. Models of English and French Repeating Motions for Watches. *The British Horological Institute.*

479. Watch Movement.
The British Horological Institute.

480. Marine Chronometer Movement.
The British Horological Institute.

481. Collection of Watch and Chronometer Balance Springs. *The British Horological Institute.*

482. Map showing allowance of time to be made for velocity of sound as applied to the **Westminster Clock Bell.**
The British Horological Institute.

482a. Working Model, for educational purposes, **of a Chronometer Escapement,** with a 6-inch compensation balance. *Ignaz Herrmann.*

482b. Working Model, for educational purposes, of a Lever Watch Escapement. *Ignaz Herrmann.*

Working Model, for educational purposes, of a Horizontal Watch Escapement. *Ignaz Herrmann.*

Model for demonstrating the law of the compensation balance or of any vibrating or rotating body. *Ignaz Herrmann.*

An arm carrying movable weights revolves about a vertical axis under the action of a spring. The time taken for the spring to run down is proportional to the square root of the distance of the weights from the axis.

484a. Eight-day Marine Chronometer.
Parkinson & Frodsham.

484b. Eight-day Marine Chronometer.
Parkinson & Frodsham.

484c. Chronometer, used by Captain Parry in the year
1819.   *Parkinson & Frodsham.*

This chronometer, used by Captain Parry in his voyage to the Polar Sea in 1819, was specially compensated for extreme cold. Others of a similar character were recently made for the Arctic Expedition under Captain Nares; also for Dr. Livingstone's Central African Expedition, the latter being compensated for a higher range of temperature.

487. Pendulum Clock for marking the time according to the time system of nature, thus forming the standard of a system of measurement including time and space together, with decimal subdivisions. The pendulum measures space by its length, and time by its period of oscillation.
Hans Baumgartner, Basle, Switzerland.

The pendulum has the exact length of a longitudinal unit of natural measure, that is to say, of the one hundred thousandth part of a degree, of which 540 go to a meridian, and measures the natural second, or the one hundred thousandth part of a mean day.

487a. Pendulum.   *Professor Dr. A. Krueger, Helsingfors.*

A barometer tube of about 350^{mm} length is attached to the pendulum rod in the plane of swinging; a little quantity of dry air is introduced in the upper part of the tube: height of the mercury column about 150^{mm}. The rising and falling of the mercury, depending on the variations of atmospherical pressure, will affect the length of the pendulum and the clock-rate. It will be very easy to calculate the distance from its centre, at which the tube is to be attached; then the barometrical variation in the clock-rate will be compensated. A pendulum of this construction has been used with success at the Helsingfors Observatory since 1866. *See* Astronomische Nachrichten, Vol. 62, No. 1482.

488. Clepshydral Escapement.
Prof. W. H. Miller, M.A., F.R.S.

By means of the fountain bottle of Berzelius, or Gay-Lussac's syphon washing bottle, or any similar contrivance, a current of water is directed into a capsule, from which it is transferred by a syphon to the mouth of an inverted syphon partly filled with fine sand, one leg being rather more than twice as long as the other. The upper end of the short leg is stopped with a cork, in which is inserted a short syphon about 0·29 inch (8^{mm}) in diameter. A compensated pendulum carrying near its upper end at a distance of 5·5 inches (140^{mm}) an inverted funnel about 0·63 inches (16^{mm}) long, 0·27 inches (7^{mm}) wide at its base, and about 0·04 inches (1^{mm}) at the upper end. The lower end of the upper syphon is supported at about 0·12 inch (3^{mm}) above the top of the funnel carried by the pendulum when at rest. A tube of about 0·08 inches (2^{mm}) in diameter, and 0·4 inches (10^{mm}) long, is supported with its upper end about 0·12 inches (3^{mm}) below the lower end of the funnel at rest.

The pendulum being made to vibrate through a small arc before reaching the upper syphon takes up a drop, and on arriving near its lowest point delivers a drop to be carried off. The time is thus measured without allowing the pendulum to come in contact with any solid body except the agate plane on which it is supported.

The drop given off by the lower tube at the end of every two seconds, may be used to record every alternate second of time by means of a timepiece having a very light pendulum timed in accordance with the pendulum the water clock.

No attempt has been made to exhibit the mechanism for counting the seconds.

488a. Model of Compensation Balance, applicable to watches and chronometers. (With a drawing.)

M. Winnerel, Paris.

488b. Model of Escapement, applicable to the model clock at the Paris Observatory. (With a drawing.)

M. Winnerel, Paris.

488c. Model of Escapement, with simplified suspension, applicable to clocks. (With a drawing.)

M. Winnerel, Paris.

488e. Two Movements for Chronometers, one eight days and one two days. *Victor Kullberg.*

489. Standing Pendulum Clock, in black wooden box with silvered dial. *Professor Buys-Ballot. Utrecht.*

This is one of the first clocks made after Huygen's principle (*i.e.*, provided with cycloidal pendulum). This peculiarity may be seen by opening the door.

490. Two Conical Pendulum Clocks, for determining short time intervals. *Professor Buys-Ballot, Utrecht.*

Each of these clocks is contained in a truncated wooden column covered by a circular brass plate, by lifting which the dial may be seen. The foremost part of the box can be removed to put the pendulum in motion, the spring being wound up. In this condition only one hand moves. By pressing on the button at the foremost part of the dial, the two other hands move until the finger is withdrawn. In this manner very short lapses of time can be measured. The instruments must be placed accurately horizontal. These two specimens were used by Moll and van Beelt on the heath near Amersfoort for determining the velocity of sound. They are constructed for the decimal division of time, and indicate the ten millionth part of a day (24 hours).

491a. Very curious Timepiece, designed by Mudge.

E. Dent and Co.

The escapement is a true remoutoire; two small pendulum springs are wound up at every beat of the scape wheel, and these give impulse to the balance. The balance is controlled by two pendulum springs, one above and the other beneath it; the first of these receives the action of the " compen-
" sation curb," the second is for ordinary regulation. The action of the

" compensation curb " is analogous to the ordinary regulation by curb pins, but the curb pins are advanced backwards or forwards along the spring by the operation of the compensation pieces, which, being constructed of brass melted upon steel, bend at every change of temperature. The whole time-piece has been designed and got up with a surprising degree of refinement.

492. Working Model of Chronometer Escapement, with two inch scape wheel.   *Philip John Butler.*

493. Small Electric Pendulum. Striking seconds on a bell, and thus capable of being used for astronomical studies.
Antoine Joseph Gérard, Liége.

494. Book containing plans of instruments, apparatus, and machines.   *Antoine Joseph Gérard, Liége.*

499a. Chronometrical Regulator, for putting in motion a registering cylinder.   *Mr. Yvon Villarceau.*

This system of regulator, the theory of which is due to Mr. Yvon Villarceau, is represented by the model included among the objects exhibited by M. L. Breguet.

500. Edelmann's Seconds Pendulum, with galvanic attachment.   *M. Th. Edelmann, Munich.*

501. Chronometric Comparateur, an instrument of coinci-dences, for determining the difference of time between two distant clocks.   *M. Redier.*

501a. Collection of Steel and Electro-gilded Pendu-lum Springs.   *E. Dent and Co.*

502. Clock employed in the Pantheon Experiment by M. L. Foucault.   *Conservatoire des Arts et Métiers, Paris.*

502a. Different applications of Metal Tubes of elliptical section to instruments for measuring pressure, temperature, weight, speed, and time.

1. Manometer for steam, air, or water pressure.
2. Barometer. Counterpoised barometer.
3. Thermometer.
4. Tacheometer, or speed indicator.
5. Balance for light and heavy weights.
6. Clock with pneumatic motor.
M. Eugene Bourdon, Paris.

502a. Motor Clock and case, with two electric dials, batteries, and fittings.   *T. Cooke & Sons.*

503. Electric Apparatus by M. Foucault, for keeping up continuously the motion of the clock.
Conservatoire des Arts et Métiers, Paris.

503a. Marine Chronometers for ships' use. Manufactured by Victor Kullberg, Liverpool Road, London, N.
South Kensington Museum.

503b. Two-day Chronometer, fitted complete. Two movements reversed, to show the compensation balances.
South Kensington Museum.

503d. Chronometer and case on stand with glass shade.
The Royal Observatory, Greenwich.

509. Marine Chronometer, regulated at sidereal time, used by the Scientific Commission of Noumea in observing the transit of Venus. *Messrs. Tondola and Co., Paris.*

510. Marine Chronometer, suitable for distributing the hour in quarter minutes to an unlimited number of electric receivers, going for one year without being wound up. (System applied for more than three years with complete success.)
Messrs. Tondola and Co., Paris.

511. Marine Chronometer, regulated to mean time. Specimen of ordinary construction. *Messrs. Tondola and Co., Paris.*

512. Geographical Clock, with revolving planisphere ; showing the time, longitude, and latitude, of all parts of the globe.
Messrs. Tondola and Co., Paris.

513. Astronomical Pendulum Clock.
W. Bröcking, Hamburg.

514. Wheel-work of Clock. *W. Bröcking, Hamburg.*

515. Astronomical Pendulum Clock with mercury compensation pendulum. *Th. Knoblich, Hamburg.*

517. Chronometer-escapement, model.
Th. Knoblich, Hamburg.

518. Anchor-escapement, model.
Th. Knoblich, Hamburg.

519. Pendulum Clock belonging to the tide-gauge of Mr. Reitz. *Th. Knoblich, Hamburg.*

519a. Clock worked by Water-power. *T. Hankey.*

520. Astronomical Pendulum Clock.
F. Dencker, Hamburg.

The Jürgens compensation pendulum system has an isochronous suspension spring, determined by calculation. It is, contrary to the formerly constructed repose pendulum, executed with sufficient stability. Not only is the expansion coefficient of the separate bars exactly determined by the pyrometer,

I

but likewise the whole pendulum is directly controlled in the pyrometer. The pyrometer employed is of quite a novel construction ; the observation takes place in a liquid, *without contact*, under two micrometer microscopes. As the observation through the microscope requires water of perfectly equable temperature, the uniformity of its temperature is ensured ; the bars likewise must be quite homogeneous, as otherwise a bending of them will take place by a change of the temperature, whereby the terminal ends will move out of the range of vision. With regard to compound pendulums, the centre of gravity is to be found by means of a balance the point of flexion of the spring of which is known to the exhibitor, and then the point of oscillation by calculation ; in a quite homogeneous and uniformly strong spring it is exactly in the centre. By means of a bar, which is adjusted exactly to the length of the pendulum,* the point of oscillation can be exactly fixed in the pyrometer, and the whole pendulum examined with regard to extension and stability. The working of the clock affords a check upon the accuracy of the measurement ; if no error be made, the pendulum thus adjusted and fitted to a clock will exactly vibrate seconds.

521. Model Escapement.   *F. Dencker, Hamburg.*

An anchor escapement, enlarged tenfold, with an impulse derived from the chronometer and acting like the same from fork upon balance. This requires no oil. The straight lines of the fork and the release stone render a quite exact execution possible, and consequently an effect which almost equals the direct impulse from wheel upon balance, without detracting from the great insensibility of the anchor escapement. The lifting which it is desired to give to the balance in this case is determined only by the length of the fork, independent of the length of the lever. Arranged for quarter seconds, it will be very useful for determining the time on journeys and on sea. The last seconds are regulated by a curb, permitting only little motion, but being securely guided by means of a screw. The last regulation by the balance screws always disturbs the equilibrium of the balance, and effects thereby a doubling of the errors at the change of the position. The flat spiral spring has an inner and an external curve.

522. Gold Watch.   *F. Dencker, Hamburg.*

The pocket watch has been executed exactly according to this model in the exhibitor's establishment at Geneva. It is provided with a flat spiral spring hardened in fire according to his invention. Up to the present time no flat spiral springs hardened in fire are employed, as far as the exhibitor knows.

523. Watch with spindle without spiral spring ; constructed in the East in the first half of last century, indicating month, day, and hour in Arabic figures. (Remarkable for its age and origin.) Property of H.H. Prince Pless, Fürstenstein.

The Breslau Committee.

523a. Watch, thickness of a crown piece, made for the late Sir C. Wheatstone by Mr. A. Stroh.

---

* The extension of the parts to be employed being known, a pendulum can be determined by calculations which swings in exactly one second.

SECTION 4.—KINEMATICS, STATICS, AND DYNAMICS.

WEST GALLERY, GROUND FLOOR, ROOM K.

I. SPECIAL COLLECTIONS.

COLLECTION OF APPARATUS USED BY 'SGRAVESANDE TO ILLUS-
TRATE HIS PHYSICAL RESEARCHES.

524. 'sGravesande's Apparatus to demonstrate the
Laws of Centrifugal Force.
Professor Dr. P. L. Rijke, Leyden.

(*See* 'sGravesande's "Physices Elementa Mathematica," 3rd edition, Vol.
I., p. 153.)

525. 'sGravesande's Apparatus to demonstrate the
Theory of the **Wedge.**
Professor Dr. P. L. Rijke, Leyden.

526. 'sGravesande's Apparatus, to show, by means of a
pendulum furnished with weights and springs, that the same
quantity of **Mechanical Work** produces the same quantity of
Vis Viva.
Professor Dr. P. L. Rijke, Leyden.

527. 'sGravesande's Apparatus to demonstrate the **Laws**
of **Falling Bodies.**
Professor Dr. P. L. Rijke, Leyden.

528. 'sGravesande's Apparatus for **Parabolic Motion.**
Professor Dr. P. L. Rijke, Leyden.

COLLECTION OF KINEMATIC MODELS, EXHIBITED BY THE KÖNIGL.
GEWERBE-AKADEMIE, BERLIN, PROF. REULEAUX, DIRECTOR.

The models in this collection are connected throughout with Professor
Reuleaux's treatment of the theory of machines. Their nature will be found
fully discussed in his " *Theoretische Kinematik* " (Vieweg und Sohn). The
English edition of this work (Macmillan), translated and edited by Professor
Alexander B. W. Kennedy, C.E., of University College, London, was published in June. The English names of the mechanisms here given are those
used by Professor Kennedy in his translation.

551. I.—Pairs of Kinematic Elements.

(*a.*) LOWER PAIRS.

1. Turning or cylinder pair, $R^+_= R^-$ or $C^+_= C^-$.

2. Sliding or prism pair, $P^+_= P^-$.

3. Twisting or screw pair, $S^+_= S^-$.

552. Pairs of Kinematic Elements.

(*b.*) HIGHER PAIRS.

4. Equilateral duangle in equil. triangle.

> These models of the higher pairs of elements can be inverted;
> that is, the movable element can be fixed, and the fixed element
> made movable. The centroids are shown in thick black or red
> lines; the roulettes, or point-paths, in thinner lines.

5. Expanded duangle in equil. triangle.
6. Equilateral curve-triangle in square.
7. Equilateral curve-triangle in rhombus.
8. Expanded isosceles curve-triangle (90°) in square.
9. Expanded equilateral curve-triangle (90°) in rhombus.
10. Regular curve-pentagon in square.
11. Symmetrical curve-pentagon in square.

**553. II.—Conic Axoids, with corresponding Spheric
Roulettes and Profiles.**

12. Spheric epicycloid.
> Ratio 1:3.

13. Spheric cycloid.
> A full cone rolling upon a plane cone (1:3).

14. Spheric hypocycloid.
> A full cone rolling in an open one (2:1).

15. Spheric hypocycloid.
> Ratio 1:3.

16. Spheric pericycloid.
> The curve upon the rolling cone passes always through the de-
> scribing point of the fixed one.

17. Spheric involute.
    Ratio 1:3.
18. Spheric involute.
    Ratio 8:9 ; the curve in red is a curtate involute.
19. Apparatus for describing spheric cycloids.
    Describes, among other curves, those here exhibited.

554. III.—Simple Kinematic Chains and Mechanisms.
20. Quadric cylindric-crank chain.
21. Slider-crank chain.
22. Quadric conic-crank chain.
23. Reduced slider-crank chain.
    The link c omitted.
24. Reduced slider-crank chain.
    Links a and c omitted.
25. Reduced conic-crank chain.
    The link c omitted.
26. Quadric crank chain with slot and sector.
27. Single crossed slide chain.
28. Double crossed slide chain.
29. Simple spur-wheel chain.
30. Simple spur-wheel chain with annular wheel.
31. Endless screw.
32. Stand for carrying the above models when in use.

555. IV.—Crank Trains.
33. Double slider-crank.
34. Slider-crank.
    With centroids.
35. Slider-crank.
    With centroids.
36. Double slider-crank.
    With centroids.   (The centroids are here Cardan's circles.)
37. Slider-crank.
38. Skew double slider-crank.
39. Double slider-crank with curved slide.
40. Double slider-crank with skew slide.
41. Lever-crank.
42. Double slider-crank with curved slide.
    With pin expansion.
43. Slider-crank.
    Link b is here a disc.
44. Slider-crank.
    With adjustable connecting rod.
45. Slider-crank.
    With adjustable cross-head.

46. Slider-crank.
    In the form of a marine engine.
47. Slider-crank.
    Slotted link gear.
48. Slider-crank.
    Double slide gear.
49. Slider-crank (Norman Wheeler).
    Three-fold.
50. Slider-crank, with pin expansion, 2 within 1.
51. Slider-crank, with pin expansion, 1 within 2.
52. Slider-crank, with pin expansion, 3 within 2.
53. Slider-crank, with pin expansion, 2 within 3.
54. Slider-crank, with pin expansion, 2 within 3.
    Annular expansion.
55. Slider-crank, with pin expansion, 1 within 2 within 3.
56. Slider-crank, with pin expansion, 3 within 2 within 1.
57. Slider-crank, with pin expansion.
    Adjustable stroke.
58. Swinging block (slider-crank).
59. Turning block (slider-crank).
60. Skew (turning) cross block.
61. Turning block (slider-crank).
    With pin expansion (can be used also as a turning slider-crank).
62. Turning block (slider-crank).
    With reduced centroids.
63. Double crank (drag-link coupling).
    With reduced centroids.
64. Turning block (slider-crank), Redtenbacher's "Maskirte
    Kurbelschleife."
    Quick return motion ; the stroke is adjustable.
65. Swinging slider-crank.
65a. Swinging double slider.
66. Swinging skew double slider.
67. Conic crank-train.
68. Isosceles double-crank (Galloway).
    Mean velocity, ratio 1 : 2.
69. Isosceles double-crank (Galloway).
    Mean velocity, ratio 1 : 2 ; arrangement for crossing dead points,
    by Reuleaux.
70. Anti-parallel cranks (Reuleaux).
    Special arrangement for crossing dead points.
71. Anti-parallel cranks (Reuleaux).
    With centroids, which are ellipses.
72. Anti-parallel cranks (Reuleaux).
    With centroids, which are ellipses and hyperbolæ.

73. Double parallel crank train, used as a coupling.
   For transmitting uniform rotation.
74. Double parallel crank train, used as a coupling (Reuleaux).
   For transmitting uniform rotation.
75. Crank train for transmitting uniform rotation (Heilmann).
76. Crank train for transmitting uniform rotation (Böhm).
77. Differential crank train (Römer).
   Numbers of teeth, 56 and 56, with apparatus for tracing diagrams.
78. Differential crank train (Römer).
   Numbers of teeth, 56 and 57, with apparatus for tracing diagrams.
79. Differential crank train (Römer).
   Numbers of teeth, 30 and 90, with apparatus for tracing diagrams.
80. Hooke's joint.
81. Universal joint (Blees).
82. Universal joint (Polhem).
83. Universal joint (Reuleaux).
84. Universal joint (Klein).
85. Universal joint (Klein).
   Simplified by Reuleaux.
86. Double Hooke's joint.
   The velocity ratio here can be made constant.

556. IVa.—Mechanisms for describing Straight Lines (exactly or approximately).

87. Roberts triangle, " parallel motion."
88. Triangle motion, inverted, by Reuleaux.
89. Elliptic linkwork (Nehrlich), 3rd form, inverted.
90. Elliptic linkwork (Nehrlich), 3rd form, inverted.
91. Hypocycloidal linkwork.
92. Hypocycloidal linkwork, inverted, by Reuleaux.
93. Epicycloidal linkwork, Reuleaux.
94. Elliptic linkwork, inverted.
   With the whole motion.
95. Tchebischeff's linkwork.
   Arranged so that it can be inverted.
96. Conchoidal linkwork, 1st form.
97. Conchoidal linkwork, 3rd form (Reichenbach).
98. Conchoidal linkwork, 3rd form (Reuleaux).
99. Lemniscoidal linkwork, 1st form (Watt).
100. Lemniscoidal linkwork, 2nd and 3rd forms.
101. Lemniscoidal linkwork, 1st form, inverted (Reuleaux).
102. Lemniscoidal linkwork, 2nd and 3rd forms.
   Steam engine model with Watt's planet wheels.
103. Sector mechanism (Reuleaux).
   Involute.

104. Sector mechanism (Reuleaux).
     Cycloid.
105. Sector mechanism (Reuleaux).
     Cycloid.
106. Cartwright's mechanism.
107. Maudsley's mechanism.
108. Tchebischeff's mechanism.
109. Harvey's mechanism.
110. Harvey's mechanism.
111. Pantograph.
     With elliptic linkwork, 1st form.
112. Pantograph.
     With prism guide.
113. Semi-pantograph.
     With prism guide.
114. Semi-pantograph.
     Steam engine model.
115. Rhombic linkwork.

557. V.—Apparatus for describing Curves.

116. Ellipsograph.
117. Ellipsograph, by Slaby, on Haman and Hempel's system.
     Describes also cycloids. Dr. Slaby's construction contains very
     essential improvements.
118. Elliptic chuck (Leonardo da Vinci).
119. Elliptic chuck (Delnest).
120. Sinoid and cardioid tracing gear.
121. Curve tracing apparatus.
122. Curve tracing apparatus.
123. Mechanism for describing Lissajous' figures.
     Describes also ellipses.
124. Hastie's conoid gear.
125. Tricentric gear.
     For the construction of three-grooved taps, &c.
126. (Form-) copying machine.
127. Rose-engine.
128. Rose-engine.
129. Rose-engine.
130. Rose-engine, special form.

558. VI.—Parallel or Translating Trains.

131. Parallel ruler.
     Single and double.
132. Parallel ruler.
     With crossed bars.

133. Complete lever parallel train.
     Weighing machine, of Roberval.
134. Incomplete lever parallel train.
     Weighing machine, of Milward.
135. Incomplete lever parallel train.
     Weighing machine, of Farcot.
136. Incomplete lever parallel train.
     Weighing machine, of Schwilgué.

559. VII.—Compound Parallel Trains.

137. Feathering paddle-wheel, of Buchanan.
     A combination of trains similar to parallel rulers. The floats
     remain always vertical.
138. Feathering paddle-wheel, of Oldham.
     The floats rotate about their axes as the wheel revolves.
139. Feathering paddle-wheel, of Morgan.
     With eccentric ring.

560. VIII.—Higher Couplings.

140. Uhlhorn's coupling.
141. Oldham's coupling.
142. Reuleaux's grooved disc coupling.
143. Köchlin's cylindric coupling.
144. Schürmann's cylindric coupling.
145. Conic coupling.
146. Pouyer-Quertier's coupling.

561. IX.—Toothed-wheel Trains.

147. Spur wheels (point-paths used for profiles).
148. Returning spur-wheel train.
149. Returning spur-wheel train, with annular wheel.
150. Returning spur-wheel train, with annular wheel.
151. Returning spur-wheel train, with two annular wheels.
152. Returning spur-wheel train, with two annular wheels.
153. Returning spur-wheel train, with intermediate wheel.
154. Returning spur-wheel train, with intermediate wheel.
     Reuleaux's so-called halving spur-wheel train.

155. Returning spur-wheel train.
     With Marlborough wheel.
156. Spur-wheel train.
     The centroids are Cardan's circles.
157. Beylich's universal wheels.
     " Pin-wheels."
158. Cylindric friction wheels.
     Held by axial pressure.

159. Screw wheels.
     Working as spur-wheels.
160. Screw wheels.
     Screw wheel and rack.
161. Bevel wheels.
     Plane- (face-) wheel and full wheel.
162. Mangle-wheel train.
     Automatic reversal.
163. Mangle-wheel train.
164. Mangle-wheel train.
     With internal teeth.
165. Whitworth's feeding gear for drills.
     The drill is under a constant pressure.
166. Reversing gear, claw coupling.
     With bevel wheels.
167. Reversing gear, bevel wheels.
168. Reversing gear, returning wheel gear.
     By Reuleaux.
169. Reversing gear, Sellers' arrangement.
     Open and crossed belts.
170. Reversing gear, with three pulleys.
171. Face-wheel and runner (Rupp).
172. Speed changing gear (Sellers).
     For lathes.
173. Speed changing gear with double pulleys.
174. Reversing and disengaging train (radial).
     Wheels of 103 and 53 teeth respectively.
175. Reversing and disengaging train (Fairbairn's).
176. Reversing and disengaging train (Brown's).
177. Engaging and disengaging train (Platt's).
178. Engaging and disengaging train (Curtis').

Globoid Gearing.

179. Globoid screw wheels ; spheric screw and wheel.
     Reuleaux.
180. Globoid ring, screw, and wheel.
     Reuleaux.
181. Skew globoid ring, conic screw and tooth.
     Reuleaux.
182. Globoid ring, cone, and wheel.
     Used by Stephenson in locomotive reversing gear.
183. Crossed globoid ring, screw, and tooth.
     Reuleaux.
184. Crossed globoid ring, screw, and tooth.
     Reuleaux.

185. Globoid screw and screw wheel.
   Endless screw.
186. Globoid screw.
   Applied in horse gins ; velocity ratio 1:12.

Parallel Wheels.

187. Parallel wheels with 24 teeth.
   Reuleaux. The teeth are revolutes.
188. Parallel wheels with 6 teeth.
   Reuleaux. Would work also with 3 teeth.   The teeth are pins.
189. Parallel wheels with 5 teeth.
   Reuleaux. One wheel annular.
190. Parallel wheels with 24 (pin) teeth.
   Reuleaux. The parallelism is destroyed by displacing the axes.

Planet Wheel Chains.

191. Planet wheel chain.
192. Planet wheel chain, $a=b=\infty$.
   With excanching wheels.
193. Planet wheel chain, $a=b=\infty$.
194. Planet wheel chain, with annular wheel.
195. Planet wheel chain, $b=c=\infty$.
196. Hyperboloidal endless screw.

562. X.—Belt-trains.

197. Returning belt-train.
   Shows the alteration of velocity due to the slipping of the belt.
198. Skew belt-train.
   Acts in one direction only.
199. Belt-train with crossed guide pullies.
   The necessary tension is given to the belt at the instant it is thrown into gear.

563. XI.—Slider-cam Trains.

200. Sinoidic cams.   Cardioids.
   Open cam with roller, pair-closure.
201. Sinoidic cams.   Cardioids.
   With second disc and centroid.
202. Sinoidic cams.   Cardioids.
   Pair-closure.
203. Sinoidic cams.   Polar sinoid.
   With centroid.
204. Sinoidic cams.   Polar sinoid.
   With centroid.

205. Cams with discontinuous profiles. Curve-triangle in skew
     curved slot.
     With centroid.

206. Cams with discontinuous profiles. Equilateral curve-quad-
     rangle.
     With centroid.

207. Cams with discontinuous profiles. Equilateral curve-penta-
     gon in straight slot.

208. Cams with discontinuous profiles. Equilateral curve-penta-
     gon in adjustable slot.
     Both parts are adjustable.

209. Cams with discontinuous profiles. Curved disc in curved
     slot.
     Both parts are adjustable.

210. Cams with discontinuous profiles. Curved disc.
     For the motion of a slide valve.

211. Cams with discontinuous profiles. Disc with looped slot.
     With shuttle, used in printing presses.

212. Slider-cam, two-lobed cylindric sinoid.
     Force-closure.

213. Slider-cam.
     Force-closure.

214. Slider-cam, cylindric sinoid.

215. Screw reversing train of Whitworth.

216. Steering gear of Scott and Sinclair.

217. Steering gear of Steel (Greenock).

218. Steering gear of McWilliam.

219. Steering gear of Reed.

220. Steering gear of Rogers.

221. Steering gear of Reuleaux.

222. Boring machine.

223. Boring machine (Stehelin).

224. Boring machine (Reuleaux).

225. Differential screws.

226. Differential screws, with wheel train.

227. Double screw train (Napier).
     Self-acting return.

228. Cam reversing train (Girard).
     For governors.

229. Leading screw with disengaging gear.
     With self-acting disengagement.

230. Screw disengagement (Whitworth).
     With pallet action.

231. Screw heckling machine (Houldsworth).

564. XII.—Ratchet Trains.

232. Click train.
  With two external and one internal clicks.

233. Centrifugal click train.
  If rapid rotation occur, the centrifugal force throws out the click.

234. Silent click train.

235. Pinching click train.

236. Click train of Wilbers.

237. Ratchet train of Langen.
  Used in gas engines.

238. Turning ratchet gear (Maltese cross wheel).

239. Turning ratchet gear (incomplete cross wheel).

240. Turning ratchet gear (spur-wheel).

241. Ratchet train.
  With automatic disengagement.

242. Ratchet train, with pin teeth.
  With automatic disengagement.

243. Ratchet train (Reed).
  With automatic disengagement and fall.

244. Ratchet train, with fast click.

245. Single acting ratchet train.
  The direction of motion can be altered.

246. Double acting ratchet train.

247. Reversing ratchet train (Francis).
  Used for governors.

248. Dividing machine (Nasmyth).

249. Lagarousse ratchet gear.
  With eccentric.

250. Crown wheel ratchet train.

251. Tooth ratchet train, with double acting free click.
  Model illustrating action of a force pump.

252. Escapement train (Mudge).
  Can be held in the stand, No. 32.

253. Throttle click train.
  Illustrates the action of the throttle valve.

254. Ratchet train, with pinching clicks.

255. Ratchet train, with several clicks.

256. Ratchet train, with free clicks.
  Directing gear can be added so as to illustrate the action of
  steam engine.

257. Ratchet train, with fast clicks.

258. Ratchet train, with Farey's director.

259. Ratchet train, with Watt's director.
  Watt's automatic valve gear.

260. Double acting ratchet gear.
261. Ratchet train, with cataract director (Hoffmann).
262. Double acting ratchet gear.
   (Reuleaux.) Shows that the steam engine is a ratchet train. The model can be worked with various forms of directing gear; it stands upon a large mahogany frame with columns.
263. Apparatus for using with Nos. 244 to 254 :—Column with fly wheel and two connecting rods.

Escapements.

264. Graham's anchor escapement.
265. Escapement of Reuleaux.
266. Pin escapement of Lepaute.
267. Escapement of Denison.
   With three-toothed escape wheel.
268. Gravity escapement of Denison (1860).
   As in the clock at the Houses of Parliament.
269. Gravity escapement of Denison.
270. Spindle escapement.
271. Cylinder escapement.
272. Anchor escapement.
273. Chronometer escapement of Jürgensen.

564a. XIII.—Chamber-crank Gears and Chamber-wheel Trains.

274. Chamber-crank gear, Simpson and Shipton.
   Steam engine.
275. Chamber-crank gear, Bährens, Napier, Bompard.
   Steam engine.
276. Chamber-crank gear, Wedding, Cochrane.
   Ventilator.
277. Chamber-crank gear, Ramelli.
   Pump.
278. Chamber-crank gear, Beale.
   Gas exhauster.
279. Chamber-crank gear, Cochrane.
   Steam engine.
280. Chamber-crank gear, Pattinson.
   Pump.
281. Chamber-crank gear, Minari, Stocker.
   Steam engine and pump.
282. Chamber-crank gear, Ramey.
   Steam engine and pump, with elliptic wheels.
283. Chamber-crank gear, Lemielle.
   Ventilator.
284. Chamber-crank gear, Cochrane.
   Steam engine.

285. Conic crank gear, Davies.
    Steam engine.
286. Parallel crank gear, Galloway.
    Steam engine.
287. Chamber-wheel train, Pappenheim.
    Pump.
288. Chamber-wheel train, Fabry.
    Ventilator for mines.
289. Chamber-wheel train, Fabry.
290. Chamber-wheel train, Root.
    Blower.
291. Chamber-wheel train, Root.
    Blower.
292. Chamber-wheel train, Payton.
    Water-meter.
293. Chamber-wheel train, Evrard.
    Pump.
294. Chamber-wheel train, Repsold, Lecocq.
    Pump.
295. Chamber-wheel train, Dart, Behrens.
    Pump.
296. Chamber-wheel train, Ganahl, Eve.
297. Chamber-wheel train, with three wheels.
298. Screw-wheel chamber train, Révillion.
    Ventilator.

564b. Kinematic Compasses with Three Branches, designed by Reuleaux.
Royal Academy of Industry, Berlin, Director, Prof. F. Reuleaux.

By means of this instrument it is easy when the orbs of two points of a system are given, to find all the other curves of the same system.

(See Reuleaux's " Kinematik," s. 24, and following.)

The third branch of these compasses may be altered longitudinally, and is, besides, provided with a joint, so that entirely obtuse angular triangles (amblygons) as well as entirely acute-angled triangles (oxygons) can be taken by the compasses, for which the three-branched compasses of older construction were not adopted.

These compasses were manufactured by J. Kern, mechanical instrument maker, at Aarau, Switzerland, by order of the exhibitor.

II. ELEMENTARY ILLUSTRATIONS.

528a. Parallelogram of Forces.
Dr. G. Krebs, Frankfort-on-the-Maine.

528b. Inclined Plane.
Dr. G. Krebs, Frankfort-on-the-Maine.

528c. Inclined Plane, constructed by Professor Dr. Bertram, Councillor of the Board of Education. *Ferdinand Ernecke, Berlin.*

The inclined plane is represented by two parallel iron rods, which can be placed at any angle with the horizontal bar.

Three distances can be measured as follows :—

1. The length of the inclined plane, that is to say, the distance from the joint to the perpendicular iron support bar, which maintains the plane in its proper position.

2. The base, that is to say, the horizontal distance from the joint to the support bar ; this is read on the horizontal pedestal.

3. The height, that is to say, the perpendicular from the terminal point of the first distance to a horizontal line drawn through the joint ; this is read on the support bar, the zero point of which is situated on a level with the joint.

The weight on the inclined plane can be balanced in two different ways : either parallel to the inclined plane, or in a horizontal direction. The carriage of the roller can be turned, and the pulling string can, therefore, be placed parallel to the inclined planes or horizontally. The double division on the slotted support bar serves for observing the horizontal position of the string.

At every experiment the weight carriage, that is, the two-wheeled axle with its scale, is balanced with the scale which is suspended on the string. This is effected by tare weights. Then the weight and the power of traction, that is to say, the weights which are balancing each other in the weight scale and the traction scale, are adjusted by means of the measured distances.

1. If the string remains parallel to the inclined plane, then the weight is to the traction in the proportion of the length of the inclined plane to the height.

For example, if the length be 80 and the height 40, then 20 grammes in the weight scale will be balanced by 10 grammes in the traction scale.

2. If the string remains horizontal, then the weight is to the traction in the proportion of the base of the inclined plane to the height.

For example, if the base be 40 and the height 40, then 4 grammes in the weight scale will be balanced by 20 grammes in the traction scale.

In order to make the difference in the two cases intelligible, such positions in the inclined plane are advantageous in which the three distances are indicated by round numbers, such as height, 20 ; length, 29 ; base, 21. The weight 20 in the traction scale balances with the string in a horizontal position, the weight 21 ; and with the string parallel the weight 29 will be balanced. With the height 30 and the base 40 the length will be 50, and 30 grammes in the traction scale will balance 50 grammes in the weight scale with a parallel direction, whilst the weight required in a horizontal direction will amount to 40 grammes.

528d. Parallelogram of Forces, constructed by Professor Bertram.   *Ferdinand Ernecke, Berlin.*

Apparatus for demonstrating the theorem of the parallelogram of forces.

If two adjacent sides of a parallelogram represent in magnitude and direction two forces acting at a point, the diagonal through the point will represent their resultant in magnitude and direction.

This theorem is illustrated by the apparatus. The angular point of the parallelogram is the (white) peg, over which a ring has been placed, on which are fastened the three cords ; the magnitude of the forces is determined by the weights in the pans, the directions pass along the three rails of which the one, AB, which is fixed, vertically ; the second AC, and the third, AE, movable around the peg A, always move in the diagonal of the parallelogram BADF, produced. The greatest of the forces is always taken in the direction of AB, and determined as equal to 100, and the value of the two others is read on the graduations of the lines BG and AF.

If, for example, the parallelogram is placed so that the lines are respectively AB = 100, BF = 70, AF = 80, the ring in that case will poise freely without coming into contact with the peg, if the weights in the pans amount respectively to 100, 70, 80 grammes; the weights of the pans, of course, must be adjusted previously by tare weights.

528e. Centrifugal Apparatus, complete.
Ferdinand Ernecke, Berlin.

529. Drawing. Experimental demonstration of the theory of the parallelogram of forces and velocities, used by the exhibitor since 1835. *Professor Daniel Colladon, Geneva.*

Two small pulleys are placed at some distance from each other on the edge of a table. On the opposite edge, held by the hand, is a small ball of the size of a musket ball, to which are attached, by one of their extremities, two helical springs of fine brass wire; the other extremity of these two springs is drawn parallel to the plane of the table by cords passing over the pulleys, and themselves stretched by the weights P and P'. The table being sprinkled with sand or seeds, two lines are traced upon it, marking in direction and intensity the tensions of the two springs which draw the ball, which are equal to the weights P and P'. On discharging the ball it traces on the table a straight line, which is in the direction of the diagonal of the parallelogram of the forces P and P' of the two springs.

III. PRINCIPLES OF MECHANISM.

529a. Models (14) of the various **Linkworks** for effecting the exact rectilinear motion of a point, commonly known as "parallel motions." *A. B. Kempe, B.A.*

Selected from a number described by the exhibitor in a paper published in the "Proceedings of the Royal Society," No. 163, 1875, and entitled, " On " a General Method of producing exact Rectilinear Motion by Linkwork," which points out the common principle on which the linkworks depend.
The points which have rectilinear motion are denoted by stars.

529b. The Sylvester-Kempe Parallel-Motion. Model of a linkwork for effecting the exact rectilinear motion of a point. *A. B. Kempe, B.A.*

Discovered simultaneously by Professor Sylvester and the exhibitor in 1875. The main portion of the apparatus consists of a linkwork of four bent rods called a "Quadruplane," which is such that four points, one on each rod, always lie at the angles of a parallelogram of constant area and angles. Two of the points consequently are situate at such distances from a third that the one distance is the inverse of the other. One of the points being fixed, another is made by means of a link to move in a circle passing through the fixed point, the other then describes a straight line. If the bent rods are made straight, the four points lie in a straight line, and the parallel motion becomes that of Mr. Hart.

529d. Model of a Linkwork, by which two rods may be made to rotate about the same axle with equal velocities in contrary directions.   *A. B. Kempe, B.A.*

This and the following linkwork are described by the exhibitor in the "Messenger of Mathematics," No. 44, 1874.

529e. Model of a Linkwork, by which rods may be made to rotate about the same axis with velocities proportional to 1, 2, 3, &c. The linkwork can also be used to divide angles into a number of equal parts.   *A. B. Kempe, B.A.*

529g. Model of a Parallel Ruler.   *A. B. Kempe, B.A.*

In this ruler the upper bar is constrained to move vertically up and down, and has no lateral motion.

529i. Link Motion.   *William Howe, Chesterfield.*

The sketch was made by W. Howe, in August 1842, which was the first sketch of the shifting link motion. The small rough model was begun by William Howe, in or about 1838, at the Vulcan Foundry, near Warrington, Lancashire, where the sectional cylinder, piston, valve, and foundation frame were made, but this was not for the purpose of applying the link motion, but a tappet motion. When the sketch referred to above was made, the link motion was applied to that model, and all the parts of the old model that could be brought in were used. The model of the twin bar link was designed in 1848 by William Howe, and made by Mr. William Usher, who was on a visit to William Howe at the time.

529j. Model of Mechanism invented by a pupil of the Royal School of Mines of Madrid, Dr. Horacio Bentabol, to ascertain the direct line of a given point by means of linkage.
Royal School of Mines, Madrid.

A memoir by the author accompanies this model, in which all the necessary details are given.

76c. Instrument, with joint, which makes its upper part movable in a horizontal plane.
Professor Tchebichef, University of St. Petersburg.

76d. Model of joint, which directly transforms a reciprocating into a circular motion.
Professor Tchebichef, University of St. Petersburg.

530. Drawing and Model of a connecting motion between two shafts turning in reverse directions.   *Charles Bourdon.*

530a. Four Models, for the description of tooth-profiles, and lines of contact.
Royal Rhenish Westphalian Polytechnic School, Aix-la-Chapelle.

No. 1 illustrates the construction of *general* toothing, according to Reuleaux's method.

No. 2 shows that by the describing point of a string which runs over two rollers, two evolutes constantly coming in contact are traced relatively to them, which for this reason are tooth-profiles correctly working together.

As the same profiles are described when the axes are removed from or brought nearer to each other, it follows that evolute-wheels may alter the distance of their axes, preserving, notwithstanding, correct contact.

The circles " K " and " K₁ " cut off from the " *contact line* " *a b*, the *contact space* PP.

No. 3 shows that at the *cycloid toothing*, the " *contact space* " consists of the curves (segments) cut off from the head circles, and that every normal placed on the tooth-profile in the common point of contact of two teeth always passes through the point of contact of the two dividing circles.

No. 4 evolves spontaneously the circumferential line of the smallest possible space between the profiles.

530b. Model of Weston's Differential Pulley, with weights complete.

Polytechnic School at Halle, Director Kohlmann.

IV. PENDULUMS AND GYROSCOPES.

531. Gyroscope. A mechanical contrivance to exhibit the phenomena of rotation, and to show experiments on the deviation of spherical projectiles.   *Elliott Brothers.*

532. Foucault's Gyroscope. Ordinary model.
Geneva Association for the Construction of Scientific Instruments.

532a. Gyroscope, by Foucault.   *College of France, Paris.*

532c. Electric Gyroscopes.   *G. Trouvé, Paris.*

Illustrating trials concerning the maintenance by means of electricity of the motion of the " Foucault " gyroscope. (Solution of the problem of gyroscopy, laid before the inventor by M. Foucault in 1864.)

533. Polytrope. A gyroscope mounted on circles so as to prove the laws of combined rotations about several axes. It may be used to determine the *meridian* or the *latitude* of a place, and to show the rotation of the earth on its axis, and for other experiments.   *The Council of King's College, London.*

534. "Soldier Experiment." Model designed to demonstrate the relative effects of revolution and of rotation, separate or combined, by the movements of soldiers.   *Henry Perigal.*

535. Compass Experiment, demonstrating that a magnetised needle does, but an unmagnetised needle does not, maintain its parallelism while revolving in a circle.   *Henry Perigal.*

536. Gyroscope, demonstrating the effects of revolution and of rotation, the two ways of turning round.   *Henry Perigal.*

537. Gyroscope, demonstrating that revolution alone will account for our always seeing the same face of the moon.
Henry Perigal.

538. Selenoscope, to demonstrate the kinematic effects of the three hypotheses of the moon's motion, as a satellite of the earth.   *Henry Perigal.*

539. Kinescopes, illustrating the laws of compound circular motion, by ocular demonstrations of their representative curves, shown by bright beads revolving with great rapidity.
Henry Perigal.

539a. Pendulum Apparatus, for the graphic representation of the combination of rectangular and non-rectangular vibrations, with illustrative plates.
Institute for Physical Science of the University of Halle, Professor Knoblauch.

An apparatus for the graphical representation of two simultaneous oscillations inclined to each other.

To a table are fastened the bearings of two pendulums, the oscillation-planes of which are permitted to change their angle of inclination. One of these pendulums transmits its motion to a horizontal bridge, the other to a writing pen which moves exactly above the bridge. The oscillatory directions of the bridge and the writing pen are the directions of the combined velocities. The curve the pin traces on the oscillatory bridge is to be considered as the trajectory of a point moving on a plane at rest under the simultaneous oscillations of the pin and the bridge.

This trajectory is recorded either by the motion of a steel pin over a piece of sooted paper, or on white paper by a narrow-pointed glass tube filled with aniline ink.

If the weights on the pendulums are displaced the proportion of their oscillatory movements is altered ; every difference of phase is obtained by an appropriate choice of the time, one pendulum beginning to move after the other. In this way the apparatus produces the greatest variety of geometrical figures.

The traces accompanying the apparatus may serve as specimens, which were drawn by the apparatus.

A more exact and scientific explanation is to be found in the " Zeitschrift " für die gesammten Naturwissenschaften von Dr. C. Giebel," Bd. XIV., October 1875.

The apparatus has been designed by P. Schœnemann in the Royal Seminary of Dr. Knoblauch, Professor of Physics at the University of Halle, and has been executed for the physical science cabinet of the University by Kleemann, mechanical engineer (Halle, Mauergasse 6).

539b. Compensation Pendulum.
Rohrbeck and Luhme, Berlin.

539c. Tisley's Compound Pendulum Apparatus, for combining two rectangular harmonic vibrations producing figures known as Lissajous' curves. See Report of the British Association 1873 ; Engineering, Feb. 6, 1874; also "Sound," 2nd ed., Tyndall. *Tisley and Spiller.*

539d. Donkin's Harmonograph, for compounding two parallel harmonic motions. See Proceedings, Royal Society, 1874. *Tisley and Spiller.*

V. VIBRATIONS AND WAVES.

540. Apparatus for the Composition of two parallel simple vibrations. *Dr. F. G. Groneman, Groningen.*

1. Principle. If the point C moves with constant velocity on the circumference of the circle HCG, it is known that its projection B upon the diameter GH performs a motion, which is called a simple vibration.

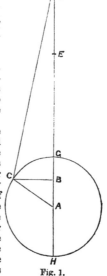

Fig. 1.

If to this variable point B a string is fixed, which is passed over the pulley D, and from this hangs down to E, so that BD + DE is the constant length of the string, it is easy to see that the point E will perform the same motion as B.

If this string is not attached to B, but to the moving point C itself, E will have a motion which is not strictly, but nearly that of B, the difference resulting from the difference of the two variable lines BD and CD. This difference can be diminished to any degree, by increasing the distance between the pulley and the centre A.

2. Two discs of the same diameter are placed in the same vertical plane, and can turn round horizontal axes, one of which has a handle. The motion of the first disc is communicated to the second by means of crown wheels and cog wheels. One of the latter can be fixed at any point of its axis, so that it may be made to gear with each of the six crown wheels of the second disc. By this arrangement the ratio of the velocity of the second disc to that of the first can be made successively equal to 1, $\frac{18}{17}$, $\frac{4}{3}$, $\frac{3}{2}$, $\frac{2}{1}$, $\frac{3}{1}$. For changing this velocity, or for changing the difference of the phases, when the velocities of the discs are the same, the washer at the end of the axis of the second disc must be a little loosened, and this axis pushed forward.

3. The discs have one knob each. If the distance of the knob of the first disc from the centre is called 1, that of the knob of the second disc can be made successively 1, $\frac{1}{2}$, or $\frac{1}{10}$. From the knobs proceed two strings, which pass through the pulleys on the top of the instrument, and are attached to the hooks A and B (Fig. 2).

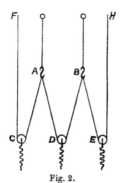

Fig. 2.

It is easy to understand that these hooks, when the handle is turned, perform two simple vibrations, the one being invariable, the other variable, in amplitude, phase, and velocity.

4. From A and B proceed two strings, passing through the pulleys C and E, and fixed in F and H. These pulleys have strictly the same motions as A and B, but reduced in the ratio one half. A third string proceeds from A to B, passing through D. At any moment the displacement of D will be the sum of the displacements of A and B, reduced one half. Its motion is therefore the resultant of the displacements of C and E.

The motions of the three pulleys are shown by the little white balls, placed on the front of the instrument.

5. With this apparatus an infinite number of combinations can be demonstrated, of which the following are examples :—

a. Two vibrations of the same amplitude and velocity, no difference of phase. The middle ball has double the amplitude.

b. The same, the difference of phase being 180°. The motion of the middle ball is nearly zero. The deviation that will appear results from the difference in the length of BD and CD (Fig. 1).

c. Combinations of a tone with one of its harmonics 2 and 3 with its quint $\frac{3}{2}$, or its quart $\frac{4}{3}$, as well as when they are of the same intensity or not (theory of the timbre).

d. Combinations of two tones of the same intensity, with the interval $\frac{1}{1}\frac{8}{7}$, showing the origin of beats.

541. Apparatus, of a new form, to illustrate **Wave Motion.**
C. J. Woodward.

This apparatus consists of a series of balls suspended from a horizontal beam by strings. These balls rest against a series of partitions in a wedge-shaped horizontal trough, which can be raised and depressed parallel to itself. The box, being drawn on one side in the plane in which the balls hang and then slowly depressed, the balls will be successively liberated, and a wave, similar to that of the sound wave, produced. If the beam be drawn aside prior to depressing the box, the balls will rest against one side of the trough and can be liberated in succession, causing them to oscillate in a plane at right angles to the beam, a vibration being produced similar to that of plane polarized light.

1576f. Wheatstone Wave Machine.
The British Telegraph Manufactory, Limited.

541a. Wheatstone's Apparatus, for illustrating the composition of rectangular vibrations.
The Council of King's College, London.

542. Drawings of new Apparatus for demonstrating the composition of **Vibrations.** *Dr. Leopold Pfaundler, Innsbruck.*

Plate I. Two blackened glass discs are each placed on a separate horizontal axis one before the other, in such a manner that the transparent curves apparently cut in their periphery, intersect each other nearly at right angles. A reflection of light is produced thereby, which, at the revolution of the discs, will generate figures such as are caused by the complex effect of the vibrations acting at right angles upon one another. By a simple mechanical contrivance the velocity of the rotary motion of each of the discs can be regulated according to equally simple relative relations, by which the figures of the different intervals are produced. By varying the tension of the strings more or less, an alteration in these places will be achieved. By changing one or the other of these discs, and replacing it by another with a different curve, the figures observed by Dr. Helmholtz, on oscillating strings with the vibration microscope, will be obtained, instead of those of Lissajous.

Plate II. Two thin rods of steel are fastened by screws at the two oblique corners of a strong bar of iron in such a manner that their further ends, one reaching above the other, vibrate vertically the one upon the other. To these ends small metal plates with incised slits are attached in a parallel position. The reflection of light produced by the intersection of these slits will show Lissajous' figures. A movable weight regulates the intervals.

B, the well known *tuning-fork apparatus* with mirrors is simplified by the tuning-fork being replaced by steel springs which are inserted in suitable movable wooden columns.

Plate III. Apparatus for simplified demonstration.

A. ramified tuning-fork.

The same produces a sound composed of two tones, and marks the corresponding musical note direct on a smoked glass plate.

B, resonator with monometrical flame without membrane.

A conical shaped Resonator, which is held with the large opening downwards and is filled from the top with illuminating gas, which is allowed to escape through a small tube attached, and then ignited. The flame will react on the tones in the usual manner.

C, an igneous Kaleidophon.

Mr. Tollinger has shown that by fastening with wax a glimmering candle to the end of a prismatic steel spring, a very admirable demonstration for large audiences of this well-known experiment can be produced.

542a. Apparatus for **combining waves** in one plane. The resultant shown is that of two sets of waves (superposed) that differ by half a wave-length. *Chas. Brooke, F.R.S.*

542b. Apparatus for **combining waves** in two planes perpendicular to each other. The resultant shown is a right-handed elliptic helix. *Chas. Brooke, F.R.S.*

1229e. Apparatus for showing the longitudinal vibrations of a row of particles, (1) stationary, and (2) progressive. *Chas. Brooke, F.R.S.*

The vibrations shown are those constituting the first harmonic subdivision of a pipe closed at one end.

543. Stationary Liquid Wave Apparatus and Sector Pendulum. *Frederick Guthrie.*

When such a system of stationary waves is formed in a deep cylindrical trough that the centre rises and falls as the edge falls or rises, the undulation is synchronous with a pendulum whose length is equal to the radius of the trough ; and the accelerations of motion of the wave and pendulum are identical.

544. Wheatstone's Wave Apparatus. A complete instrument, showing plane, circular, and elliptical waves, the phenomena of interference, &c.   *Elliott Brothers.*

544a. Apparatus to illustrate **Wave Motion.**
Rohrbeck and Luhme, Berlin.

544b. Apparatus to illustrate **Waves** and **Nodal Vibrations** of a row of **Mutually Influencing Particles.**
Sir William Thomson.

Each particle has only one degree of freedom, and is influenced by a force depending only on the relative positions of itself and of its next neighbours on each side. The shorter the wave-length the smaller the velocity of propagation of the wave.

The apparatus consists of a series of light rods loaded at each end, and strung transversely on two threads which form a bifilar suspension placed equidistant from the centre of gravity of each rod. The distance between the threads at any point is inversely proportional to the square root of the tension at that point.

545. Illustrations of Vortex Motion. Nos. 1 and 2 are vibrations ; Nos. 3–5 steady motion. (Proceedings of Royal Society of Edinburgh, 1 November 1875.) *Sir William Thomson.*

Series of 11 successive figures of a simple vortex ring, performing violent transverse vibrations of the first fundamental mode.

No. 2. Series of 11 successive figures of a simple vortex ring performing violent transverse vibrations of the second fundamental mode.

In Nos. 3–6 the motion is analogous to that of screw propellers backing, the vortex core being in each instance as it were the edge circumference of the screw propeller.

No. 3. Two-bladed screw.
No. 4. Three-bladed screw.
No. 5. Four-bladed screw.
No. 6. Trefoil knot described in Sir W. Thomson's papers on Vortex Motion (Transactions of Royal Society of Edinburgh for 1857 and 1858), and figured on the back of the "Unseen Universe," by Professors P. G. Tait and Balfour Stewart.

VI. FALLING BODIES AND PROJECTILES.

546. Drawing of a new Apparatus for demonstrating the lateral deflection of rotating conical projectiles.
Dr. Leopold Pfaundler, Innsbruck.

The conical projectile A turns within the horizontal frame B on its own horizontal axis, and can be put in rotation by pulling off the reel the string attached to *a*. Fastened, on the outside of the frame, on two little hooks, *b, b*, whose line of communication is perpendicular to the axis of rotation and passes through the centre of gravity of the entire body, are two threads, which join further up, and whose combined continuation is attached to a hook in the ceiling.

At the back there is a steering vane C with a counter-weight D, attached in such a manner that according to the position in which it is placed the line of the resultant of the atmospheric resistance will pass either above or below the centre of gravity of the projectile, without, however, altering the position of the centre of gravity itself. The vane can be turned on its axis, or be removed and replaced by a double vane C[1] with the two flat surfaces placed at right angles to each other.

The following experiments can be made with this apparatus :

1. Stability of the Axis of Rotation.

The apparatus is put in motion to swing in a curve of five meters length by taking hold of the vane and pulling it backwards, and then allowing it to drop. If the projectile does not rotate, it easily turns over and will deviate from its course by very slight causes ; if it rotates, it will remain parallel with its axis. The vane must be given a neutral position in regard to the atmospheric resistance.

2. Lateral Motion.

The apparatus is made to rotate to the *right* by the vane C being placed in an *upward* position, when the point in flying forwards will revolve to the *right*. If the direction of the rotation, or the position of the vane, be altered to the opposite course or direction, the point will revolve towards the *left*. If both are changed, the rotation will keep in the direction to the *right*.

3. Lateral Deflection.

The single vane C is replaced by the double vane C[1], the flat surfaces of which being placed in a vertical and horizontal position, the proceeding then is the same as described before.

Instead of the lateral motion, a parallel lateral deflection will be the result.

The latter experiment corresponds to the actual motion of the projectiles. The greater degree of velocity being equalized by the larger surface of the vane exposed to the atmospheric resistance.

682. Rotation Apparatus for determining the effect of Atmospheric Resistance on bodies of different shapes, particularly on projectiles. Constructed by Theodor Baumann, junior, Mechanical Engineer, Berlin.

Professor Dr. E. E. Kummer, Berlin.

(See " Abhandl. der Königl. Akademider Wissenschaften in Berlin, 1875 "; " Über die Wirkung des Luftwiderstandes," by E. E. Kummer.)

547. Simple and Inexpensive Form of Morin's Machine for demonstrating the law of falling bodies. It can be made by an ordinary carpenter, at a moderate price, made by the exhibitor.   *Dr. Stone.*

548. Apparatus by General Morin for the experimental demonstration of the laws of falling bodies.

M. Digeon, Paris.

549. Attwood's Machine with water clock attached.

The Council of the Yorkshire College of Science, Leeds.

The time is measured by a water clock, the orifice of which can be opened by means of a lever moving under the influence of an electro-magnet. The weights are supported by a thread grasped by a pair of iron pincers, which are kept shut by a spring, but can be opened by means of another electro-magnet included in the same circuit as that attached to the water clock, so that the water begins to flow and the weights to fall simultaneously. Another metal piece can be screwed on to the instrument. One of the binding screws with which it is furnished is insulated from it by a plate of ebonite pierced with a metallic rod in connexion with the binding screw, and on which rests one extremity of a lever in electric communication with the rest of the piece. This piece being included in the circuit the current cannot pass when the lever is raised, and the water clock is stopped as soon as this is effected by the falling weight.

549a. Attwood's Machine, with friction rollers and electro-magnetical release.   *Ferdinand Ernecke, Berlin.*

549b. Attwood's Machine, with pendulum attached.
                                   Rohrbeck and Luhme, Berlin.

VII. FRICTION.

550. Apparatus for determining the Friction between Water and Air.
                         Professor Viktor von Lang, Vienna University.

The above consists of a heavy stand with one fixed and three movable arms. The fixed arm bears a short glass tube, from which water is made to flow in a continuous vein. A crosspiece of four glass tubes is united, air-tight, by its longest arm to the water-delivering tube; the opposite arm is directed downwards, and closed by a caoutchouc mouthpiece passing over the "aspirating tube." This latter is supported by the two lowest arms of the stand, the remaining fourth arm securing the crosspiece. The vein of water passing through the "aspirating tube" moves the air, the quantity of which is determined by the motion of a soap lamina in the "measuring tube." This tube is joined to one of the horizontal arms of the crosspiece, the fourth arm bearing a water manometer.

550a. Machine for the **Examination** and **Measurement** of the **Sliding Friction** caused by the **Motion** and the **Variable Velocity** on **Rails.** Constructed by Herr Jung, University of Giessen.

This machine includes :—

    a. A board, with hooks for attaching a scale by means of a cord running on a roller, for the purpose of measuring the friction.

    b. Two pairs of iron and brass rails, to be fixed to this board.

    c. Three pairs of wooden rollers.

    d. A pair of iron rollers.

    e. A pair of brass rollers.
                         Physical Science Institute of the University of Giessen,
                         Professor Dr. Buff.

This apparatus was originally used for measuring sliding friction. It is at the same time a convenient appliance for demonstrating the friction of the steam engine on the railway line.

Illustration of Experiment with Magdeburg

Fig. IV.   Fig. V.

Fig. II.

D

N   N
N

Hemispheres, from Book described under No. 599.

SECTION 5.—MOLECULAR PHYSICS.

WEST GALLERY, UPPER FLOOR ROOM .

I. SPECIAL COLLECTIONS.

Hydrostamm (Hydrometer), with ballast of mercury, for liquids lighter than water.　　　*The Accademia del Cimento.*

Hydrostamm (Hydrometer), with ballast of shots, for liquids heavier than water.　　　*The Accademia del Cimento.*

The Hydrometer, No. 20, was probably used for determining the specific gravity of precious stones; by observing to what degree it was immersed without the precious stone, and with the precious stone placed upon the little metallic disc, suspended by the three little chains. This is also a fitting place to draw attention to the so-called "Palla d' Oncia" (ounce ball), of the Grand Duke Ferdinand, a glass globe which displaces very nearly an ounce of water. On the stem several rings were placed in order to make it sink, and then by the number of these rings, it was known what the specific gravity of the liquid was into which it had been immersed.

Besides the Hydrometer, No. 21, for determining simultaneously the temperature and the corresponding specific gravity, other ones of the Accademia del Cimento, which are true thermometers, having in addition the scale of graduation of hydrometers.

Hydrostamm (Hydrometer), with a small metallic balance, the first idea of Nicholson's areometer.

The Accademia del Cimento.

Photograph of the **Hydrostamm (Hydrometer),** enclosing a thermometer.　　　*The Accademia del Cimento.*

Ball of Metal, filled with water, and prepared by the Accademia del Cimento to serve in the experiments for the compression of water.　　　*The Accademia del Cimento.*

Ball of Metal, still containing some water that served for the experiments of the Accademia del Cimento on the compression of water.　　　*The Accademia del Cimento.*

Among the many experiments made by the Accademia del Cimento for the compression of water, either by means of rarefied air, by the pressure of mercury, or by the force of percussion, the last one which was carried out with a ball of very thin silver, has attracted the most universal attention. When beaten with a hammer, the water "sweated through all the pores of the " metal, like quicksilver spouting through some skin in which it was being " squeezed."

Two Tubes of Torricelli, one of which ends in a sphere. They served him for his experiments on atmospherical pressure in 1644. *The Accademia del Cimento.*

Galileo, by condensing air, had demonstrated its weight, and in his dialogue on the resistance of solids, he says of water, that in suction pumps, it does not rise higher than about 18 braccia, leaving the space above empty. Torricelli, pondering over this fact, was led to think of what would happen, if, in the place of water, mercury, which is so much heavier, were used; for he argued that by its means there would be much greater ease in obtaining a vacuum, in a much shorter space than that necessary for water. He then made a long glass tube of the length of about two braccia, which terminated at one end in a ball, likewise of glass, and remained open at the other; through this aperture he proposed to fill the tube and the ball completely with mercury; and then holding it with his finger, and turning it upside-down, submerge the orifice of the tube below the level of more mercury in a large vessel; and that being done, take away his finger and open the tube, thinking that the quicksilver would detach itself from the ball, and having glided down and remained suspended according to the various calculations, at about the height of 1¼ braccia, would in all probability leave a vacuum in the ball above and in part of the tube. He communicated this thought to his great friend Viviani, who, most anxious to see the result, agreed to the experiment, which he himself carried out, and was hence the first, about a year after Galileo's death, to see Torricelli's ingenious idea confirmed by the fact. He hastened to his friend, who, most joyful at the news of this evidence, was all the more persuaded that the weight of the air was really that which was in equilibrium with the column of water or mercury. Indeed, being asked by Viviani what would have happened if the experiment had been made in closed space, Torricelli, after having reflected for a short time, answered, the same thing; since the air is already compressed in it. This most important discovery was communicated by the author himself to Ricci in Rome, and by Ricci to Sig. de. Verdus, who, in his turn, made Padre Merseune acquainted with it, from whom Pascal learnt it, and made it famous, as everyone knows, in his celebrated Puy-de-Dôme experiment. Torricelli himself in a letter to Ricci, observes ; "that it would be possible, by means " of his instrument, to ascertain when the air was lighter or heavier ; and " that it might be the case that air, which is most heavy upon the surface " of the earth, becomes more and more light and pure as we rise higher and " higher to the tops of the loftiest mountains." And Carlo Beriguardi in his " Circolo Pisano," published in 1643, says ; "that the tube of quicksilver " leaves more space empty, when placed at the top of a tower or of a " mountain, than at the foot."

STANDARDS OF THE HYDROMETERS AND THERMOMETERS USED BY GOVERNMENT OFFICERS IN HOLLAND.

Exhibited by Dr. J. W. Gunning, Professor of Chemistry at the " Athenæum illustre," Amsterdam.

579. General Hydrometer (No. 1), with open stem and variable weight. Every degree has a bulk equal to $\frac{1}{100}$ of the part of the hydrometer below zero. When this instrument, having the arbitrary weight = W grammes, marks a degrees in a liquid, the

apparent specific gravity of that liquid at the observed temperature will be $= \dfrac{W}{(100 + a)1\cdot014608}.$

580. Instruments (Nos. 2 and 3) for ascertaining the strength
of alcoholic liquors. The degrees are the same as in No. 1, but the
weight is not variable, and the zero is the immersion point at
15° C. in a liquid, of which the specific gravity at that temperature
is equal to the specific gravity of pure water at 4° C. Tables are
added, those of Professor von Baumhauer (1863), used in Holland ;
those of the exhibitor (1873), used in the colonies. The latter are
based on the researches of Mendelejeff. Phil. Mag. (4) XXIX.
395.)

581. Hydrometer (No. 4), for liquids of a specific gravity
greater than 1. The construction is the same as in Nos. 2 and 3.
The Beaumé scale is added. Used for solutions of salts and sugar
juices.

582. Hydrometers (Nos. 5 and 6), for ascertaining the
specific gravity of seed oils and of petroleum, allowing immediate
application of correction for temperatures above or below 15° C.
(For description see Scheik. Bijdragen door J. W. Gunning,
Amsterdam, 1867.31.)

583. Densimeters (Nos. 7, 8, and 9), with flat stems. The
zero is the same as in Nos. 2 and 3.

584. Hydrometer (No. 10), for preparing a liquid having at
15° C. the specific gravity of pure water at 4° C. The instrument
is made in the following manner :—Through the open stem shots
are introduced till the instrument floats at the mark on the stem in
pure water of 15° C. The weight of the instrument is then
increased in the ratio 0·99915 : 1, in consequence of which it floats
at the same mark at 15° C. in a liquid having at that temperature
the specific gravity of pure water at 4° C.

585. Trough for comparing Hydrometers.

Thermometers. The instruments have the Celsius' scale.
By their mode of construction they possess the following advantages : (1.) They may be turned upside down and shaken in any
manner without breaking the column of mercury. (2.) Though
newly made, the zero is not subject to displacement.

The former advantage is obtained by filling the tube and the
upper space with perfectly dry air, free from dust, as highly compressed as possible.

The latter advantage is secured by placing the bulbs of the
newly made thermometers in a bath of paraffin, heated slowly to
100° C. and then allowed to cool slowly and in succession sixty to
a hundred times.

II. AIR PUMPS AND PNEUMATIC APPARATUS.

597. Air Pump, by Otto von Guericke, with stand.

Professor Dr. Lepsius, Berlin.

606. Otto von Guericke's Air Pump.

"*Collegium Carolinum,*" *Polytechnic School at Brunswick, Professor Dr. H. Weber.*

The earliest trustworthy information respecting *Otto von Guericke's* original apparatus is contained in a list of the physico-chemical apparatus of the *Collegium Carolinum,* at Brunswick, of the year 1816.   In this list it is stated that this apparatus was obtained from the legacy left by Aulic Councillor *Beireis,* in Helmstedt.   According to a special ordinance of His Highness *Frederich Wilhelm,* Duke of Brunswick, dated 8th October 1814, the collection of physical, mathematical, and astronomical instruments acquired from the legacy left by *Beireis,* physician in ordinary, at Helmstedt, was exhibited in the rooms, and by a later ordinance dated 9th March 1815 incorporated with the collections of the ducal *Collegium Carolinum.*   The air-pump has been preserved unaltered, with the exception of the lever and the piston attached to it, the former having been replaced by a new one of the same construction, and the latter by a wooden one, in the year 1864.   The pump has been described and faithfully represented by a drawing in a work published by *Otto von Guericke,* entitled "Ottonis de Guericke Experimenta nova (ut vocantur) Magdeburgica de vacuo spatio Amstelodani," 1672, cap. IV. p. 75, Tab. VI., in which work he also (p. 122) successfully refuted the assertion of *Augustus Hauptmannus,* doctor of medicine, in his "Berg-be-deneken, anno 1658, Lipsiae," "that it would not be possible to either angel or devil to bring about a vacuum."   This work is in the possession of the ducal library, at Wolfenbüttel.   But, previous to this air-pump, *Otto von Guericke* had constructed one more simple, consisting of only one cylinder and a piston, which is said to be in the library at Berlin (*see* No. 597).   The difficulty, however, connected with the motion of the piston, the resistance of the air against the free piston being so great that it required two strong men to pull it out repeatedly, p. 75, induced him to contrive their second improved construction. The receiver stand at present in use was unknown to *Otto von Guericke.* In order to produce a vacuum, he employed a hollow copper ball with stopcock,* which was placed in the axis of the barrel.   It was exhausted, and screwed on to other vessels, which, by the repetition of the process, were also exhausted.   The pail of tin-plate attached to the lower end of the barrel, as well as the copper bowl fastened to the upper end of the barrel, were filled with water or oil, in order to effect a greater tightness.

607. Two large Magdeburg Hemispheres of copper.

Professor Dr. H. Weber, Brunswick.

The two large hemispheres of copper are those of which *Otto von Guericke* states, in page 104, Tab. XI., that after their exhaustion they could not be separated by the united strength of 16 horses.   They have a diameter, as mentioned in the work alluded to above, of nearly ¾ Magdeburg ells, or, according to a second more exact statement, 0·67 Magdeburg ells.

---

* This ball (page 88, Tab. VIII.), as well as a pair of still larger Magdeburg hemispheres, nearly an ell, Magdeburg measure, in diameter, which 24 horses had not the power to separate (page 105), and, lastly, a copper boiler with inserted piston, which, when the air below the same was withdrawn, 50 men were not able to lift or to pull up (page 109, Tab. XIV.), are likewise enumerated in the list, but are no longer to be found.

Otto von Guericke's Air Pump, and Magdeburg Hemispheres, described under
Nos. 606 and 607.

[*To face page 159.*]

Title-page of Book described under No. 599.

608. Two smaller Magdeburg Hemispheres of brass.
Professor Dr. H. Weber, Brunswick.

The two smaller hemispheres of brass were used for experiments with weights (page 106, Tab. XII., of Otto von Guericke's " experimenta nova ").

598. Two Magdeburg Hemispheres.
Professor Dr. Lepsius, Berlin.

599. " Ottonis de Guericke Experimenta nova (ut vocantur Magdeburgica) **de Vacuo spatio."** Primum a R. P. Gaspare Schotto e Societate Jesu et Herbipolitanae Academiae Matheseos Professore : nunc vero ab ipso Auctore perfectius edita, variisque aliis Experimentis aucta. Quibus accesserunt simul certa quaedam de Aeris Pondere circa Terram, de Virtutibus mundanis, et Systemate mundi planetario, sicut et de Stellis fixis, ac Spatio illo immenso quod tam intra quam extra eas funditur.

Amstelodami apud Joannem Janssonium à Waesberge, Anno 1672. Cum Privilegio S. Caes. Majestatis.
Professor Dr. Lepsius, Berlin.

587. Air Pump, with double barrel (1662), by Boyle.
Royal Society.

622. Diagram of Von Guericke's Air Pump. Invented 1654. *A. Galletly, Edinburgh.*

It consisted of a globe of copper, with a stopcock, to which a pump was fitted.

The pump-barrel was entirely immersed in water to render it air-tight. This was the earliest of all air-pumps.

623. Diagram of the First English Air Pump, constructed in 1658–59 by Hooke and Boyle, but mainly by the former.
A. Galletly, Edinburgh.

As shown in the diagram, it had a single barrel, in which was a piston worked by a rack and pinion. In working it the valve G of the cylinder was shut, while the stopcock L of the receiver was open, during the descent of the piston. When the piston was driven home, L was shut and G kept open. In this air-pump a vacuum was produced slowly, and was imperfect at best.

624. Diagram of the Second English Air Pump, constructed by Boyle and Hooke in 1667.
A. Galletly, Edinburgh.

The piston was worked by a rack and pinion like the first English air-pump, but in this one the barrel was kept under water to keep the leather of the piston always wet.

The piston had an aperture at F and a stop-cock at I, which were worked as in the first English air-pump.

625. Diagram of Papin's Air Pump. 1676.
A. Galletly, Edinburgh.

The diagram represents only the working parts of the instrument, without the frame supporting them. This air-pump had the great advantage over earlier forms of having two barrels, in which case, as the air becomes exhausted, the resistance which it offers to the ascent of one piston is nearly balanced by the force with which it compels the other to descend. This air-pump was worked by moving the feet alternately up and down in stirrups.

626. Diagram of Hauksbee's Air Pump. 1703–9.

A. Galletly, Edinburgh.

This instrument, like Papin's, had two barrels, but the stirrup arrangement and pulley are replaced by racks on the piston rods, and a pinion moved by a handle, as in the modern double-barrelled air-pump. These diagrams 622–626 were made to illustrate a paper on the early history of the air-pump, published in the New Philosophical Journal, April 1849, by the late Dr. George Wilson of Edinburgh.

687. Diagram of Torricellian Vacuum. 1644.

A. Galletly, Edinburgh.

603. Air Pump, by J. van Musschenbroek, and two Magde-burg hemispheres.

The Royal Museum, Cassel, Dr. Pinder, Director.

1. *Air-pump, with obliquely placed cylinder, on Senguerd's system.*

This air-pump is, perhaps, the oldest of the kind existing, with a "four way" stop-cock. It was constructed in the year 1786, by *Jan van Musschenbroek*, in Leyden, and bears the Musschenbroek arms, namely, the "*Oosterche* lamp," and the crossed keys; the armorial bearings of the city of Leyden. The invention of the four way stop-cock notoriously originated with *Senguerd*, who, from the year 1675–1724, was a professor at Leyden. The air-pump, preserved in the Physical Science Cabinet of that town, which, according to the inscription, was constructed by *Samuel Musschenbroek*, in the year 1675, and which has a four way stop-cock, is there considered to be the first of the kind, but could not at first have been provided with this adaptation since it was invented at a later date. This appliance must have been adapted to it only at a later period. *Senguerd* first described this kind of cock in the second edition of his "Philosophia Naturalis," which was published at Leyden in the year 1685. In his work, "Rationis atque Experimentiæ Connubium," which was published at Rotterdam in the year 1715, he states that he made the invention in 1675, and that in 1679 he had the air-pump executed by a skilful artizan. That this artizan was *Samuel Musschenbroek* is proved by a notice in the handwriting of *Petrus van Musschenbroek*, which he wrote in a copy of his work entitled, "Beginselen der Naturkunde." However, *Senguerd*, in the first edition of his work, "Philosophia Naturalis," which appeared in 1680, and of which a copy is in the possession of the Library of the University at Utrecht, makes no mention at this time of the cock, but gives an illustration of an air-pump of the original construction of *Otto von Guericke*. That the air-pump was constructed is plain, from a record of *Uffenbach's*, describing the "Laboratorium Physicum" of the University of Leyden, dated the 19th January 1711 :

" In the centre stood an elevated table. Upon this table there stood an " 'Antlia pneumatica' of considerable size, of *Samuel Musschenbroek's* old " invention, inclined with a 'cista' to put water in." Although the words " from the invention of *Sam. Musschenbroek* " may also have reference to the air-pump, which is still preserved in the Physical Science Cabinet at Leyden, yet the description does not tally with it, inasmuch as its cylinder is perpendicular, and stands next to an elevated table which carries the plate, and is consequently united with it in one apparatus. It is possible that *Uffenbach*, who was not an adept in that science, was only induced to make use of the above-mentioned expression on seeing the arms of the firm of *S. v. Musschenbroek.* But how it happened that he did not see the other air-pump, since he mentions a third air-pump, of *Boyle's* construction, still in existence, is rather obscure, as *De Volder, Senguerd's* predecessor in the professorship of physics at Leyden, who himself laid the foundation of the establishment of the Physical Science Cabinet at Leyden, must have worked in the same laboratory. "There are given here (in the laboratory), as we were assured," continues *Uffenbach* in his narratives, "'Lectiones publicæ' four times a week "by *M. Senguerd; De Volder,* however, it was stated, has had larger "audiences, as he was more artful ('curieuser') in his experiments."

Senguerd did not deliver these lectures until after *De Volder's* retirement, in 1705. *De Volder* died in 1709, and it is possible that the air-pump was at that time in course of reconstruction. *Senguerd's* air-pump, therefore, as it is not to be found in the Physical Science Cabinet at Leyden, must, in all probability, have been private property; and it is possible that, like many old instruments, it is still somewhere in private possession in Holland. So long, however, as it is not discovered, the Cassel Museum may lay claim to the distinction of having in its possession the air-pump with the oldest four-way stop-cock. For although *Senguerd* had his air-pump made in 1679, the description of it did not appear until 1685, when it excited much attention. Christian Wolff, in 1718, had one constructed exactly according to the same pattern, by Leupold at Leipsic, which may still be seen in the Physical Science Cabinet of the University at Marburg. It is very singular that Musschenbroek, in his manifold writings, never as much as mentions Senguerd's air-pump. That, of course, this omission cannot be construed into an argument against the priority of *Senguerd* is plain from the subsequent remarks in the above-mentioned autograph notice, in which it is stated that "in the year 1679 *Senguerd* and the same *Samuel* "*Musschenbroek* had under consideration air-pumps of another kind, which "were afterwards improved by my father, *Jan van Musschenbroek,* and by "my brother" ("in den jaare 1679 heeft de Heer Senguerd met dienzelven "*Samuel Musschenbroek* lucht pompen van een anderen aart bedagt, welke "geduurig naderhand door mynen vader Jan van Musschenbroek, als door "mynen broeder verbeterd zyn.") It seems as if family partiality had influenced him.

The air-pump is accompanied by two powerful Magdeburg hemispheres with which the experiments of Otto von Guericke, of the Regensburg Diet, were repeated at Cassel.

604. Compression Machine, for throwing bombs.

The Royal Museum, Cassel; Dr. Pinder, Director.

This is one of the most ancient instruments in the possession of the Cassel Museum. On the 16th November 1709, von *Uffenbach* saw it in the Art Museum of Landgrave *Karl.* It was to serve for the purpose "of throw-"ing lighted grenades through the air to a distance of more than 100 yards, "with the usual effect." At that time some of its former defects had just been repaired. "Twelve grenades can be thrown one after another with great "rapidity. But, because the air is continually diminishing, the last, as may "be easily conceived, do not go as far as the others." This instrument, which acts on the principle of an air-gun, the brass ball carrying a funnel upon which the grenades were placed, and by which also red hot shots could be thrown without any danger to the men serving it, is of historical interest, as it is a surviving witness of the joint labours of *Papin* and the Landgrave *Karl,* and it even transports us to the midst of that catastrophe which put an end to Papin's activity. It was a similar machine, with which *Papin* contemplated throwing bomb shells by means of steam, which exploded in his

laboratory. " The other and greatest misfortune was that, having undertaken
" to shoot the same with steam as with powder, he might have provided a
" weapon of great destructive power. The machine prepared for that pur-
" pose, however, exploded, not only demolishing a great part of the labora-
" tory, and mortally wounding several men (one among whom had his jaw-
" bone carried away), but Landgrave Karl himself, who is a very curious
" lord, intent upon seeing and examining everything minutely, might have
" been hit, and have lost his life, had not His Highness been accidentally de-
" tained by other affairs and come to the laboratory later. On this occasion
" Papin was dismissed from his service."

614. Air Pump with two Barrels. The first pump of the
kind ever constructed. It is an exhausting and a condensing
pump, and was made for King George III. in 1761.
> *The Council of King's College, London.*

618. The Abbé Nollet's Air Pump.
> *Conservatoire des Arts et Métiers, Paris.*

586. Hand Pump, Regnault's System improved.
> *Geneva Association for the Construction of Scientific In-
> struments.*

Exhausting and forcing hand pump. The hand pump, constantly used in
laboratories, has now attained a satisfactory practical shape, from which not
much deviation is possible. Its general proportions are determined by the
consideration of how best use can be made of muscular power, with simplicity and facility of carriage of the apparatus. Silken valves, as being too
perishable, are excluded and replaced by cones of metal and leather. All the
movable parts, piston and valves, are equally accessible. Three cocks effect
the complete cutting off of all communication between the pump and the
exhaust and the compression receivers, as well as restore direct communication, either between both or one of these and the atmosphere. If
rarefaction is required in the compressing vessel, or *vice versâ*, without disturbing the tubes of communication, the relative position of the valves must
be inverted, which may be done in a few minutes, or else a change cock
must be joined to the pump.

For effecting all communication between the pump and its receivers, a
screw flush joint is always used, which is very safe, and can be fitted to
tubes made of india-rubber, copper, or lead. This avoids the deterioration of
the india-rubber, which is unavoidable on account of the constant action of
fastening the joints, especially as it is rarely that a thoroughly air-tight joint
can be made with india-rubber alone.

588. Sprengel's Mercury Pump.
> *Prof. Dr. R. A. Mees, Director of the Physical Labora-
> tory of the University of Groningen.*

This instrument differs from all others—(1) in the numerous curvatures of
the tube through which the mercury falls. These augment the exhausting
power of the pump, while the air bubbles, which are carried away with the
mercury, and which, when near vacuum, are so minute that they remain
hanging on the walls of the tubes before coming down, assemble in the curvatures, until their size is so far augmented that they are carried away by the
falling mercury. (2) The instrument is provided with a peculiar stop-cock
with two perforations, whereby the flowing of the mercury is regulated and

the acquired vacuum preserved. To effect this, the stop-cock is furnished with two iron rings, which float on the mercury, and, falling with it, close the opening of the stop-cock. By means of the second perforation of the stop-cock the instrument can be joined to an ordinary air pump, and the operation abridged by a partial exhaustion of the air. To produce a more complete vacuum than can be obtained by an ordinary air pump, the stop-cock is turned half round, and, when the vessel surrounding the stop-cock is filled with mercury, the Sprengel pump begins to act.

589. Aspirator, moved by clockwork with sulphuric acid ; U tube for determining accurately the amount of moisture in the atmosphere by the use of the balance. *Dr. Andrews, F.R.S.*

The amount of moisture at different periods of the day, or the average amount in 24 hours, may be determined with great precision by means of this arrangement, which the exhibitor proposed many years ago for use in meteorological observatories, instead of the present defective methods.

590. Spirator. Designed to get a constant current and known volume of air driven or drawn over a body.
Frederick Guthrie.

The principle resembles that of " Tantalus' cup." A constant current of water enters a flask, and (1) drives out its own volume of air through a mercury trap ; (2) when the flask is filled up to a certain point, a siphon acts, and, in emptying the flask, draws air in from another tube.

590a. Apparatus for collecting Gases without exercising upon them either pressure or rarefaction. *George Gore, F.R.S.*

612. Apparatus for Air Pump. An air-gun supported by two vertical mahogany pillars and cross-bar, by means of which it may be adjusted to any angle.
The Council of King's College, London.

613. Apparatus for Air Pump. Metal condenser with glass ends, large enough to take a pair of four-inch Magdeburg hemispheres with fittings, and a metal lever to which weights may be attached to measure the pressure of the air, either when compressed or at its ordinary pressure.
The Council of King's College, London.

592. Drawing of Mercurial Air Pump (1872).
J. P. Joule, D.C.L., F.R.S.

By alternately lifting and lowering the bulb attached to the flexible tube, the air being dried by the admission of sulphuric acid through a glass valve at the upper part of the perpendicular tube, a very excellent vacuum may be obtained in a short time.

592a. Sprengel's Mercurial Air Pump, improved form.
E. Cetti & Co.

592b. Modification of Sprengel's Air Pump, with air-trap and means of supplying sulphuric acid, in order to clean out the fall tube while the pump is in use. *Prof. H. McLeod.*

593. Mercury Air Pump.   *Dr. H. Geissler, Bonn.*

594. Photograph of an Air Pump, the exhibitor's construction (in frame).
        C. Staudinger and Co., F. W. von Gehren, Giessen.

This air-pump has a double-acting barrel of 10 cm. diameter and 32 cm. in height. The motion of the piston is effected by the rotation of a fly-wheel in one direction.

595. Hydro-dynamical Air Pump.
        Baron Dr. von Feilitzsch, Professor at the University of Greifswald.

An ordinary air-pump receiver communicates with a tube through which mercury is passed with great rapidity in such a manner as to flow in through a small aperture and to flow out again through a wide opening. A vacuum will thus be produced under the receiver.

596. Mercury Air Pump, with valve, on Mitscherlich's system.   *Prof. Dr. Mitscherlich, Münden, Hanover.*

(See "Poggend Ann. der Chem. und Phys.," vol. 150, p. 420.)

600. Air Pump.   *Charles Gustavus Pinzger, Breslau.*

This air-pump is provided with a double acting piston, and so arranged that it can be used as an exhausting and as a compressing pump. In the former case the barrel is placed in a diagonal position, and the pump provided with glass plate and stop-cock, but if the plate with the cock be screwed off, and the hose-screw attached to the wooden stand screwed on, and the barrel detached from its support and placed perpendicularly, the pump can in that case be used as a compression pump. Plate with cock are meanwhile screwed again on the wooden stand. The cock below the barrel, by reason of its parallel position, effects with this the connexion between barrel and plate, and in its upward position, the communication between barrel and the external air; and, lastly, in its downward position, the communication between the plate and the external air.

601. Mercury Air Pump, on Professor von Jolly's system,
        Prof. von Jolly, Munich.

This air-pump has been described by G. Jolly in Carl's Repertorium, 1865. As regards the alterations which have since taken place, and which have been introduced in the apparatus exhibited, a special description has been added to the object.

602. Apparatus for Demonstrating Mariotte's and Dalton's Law.
        Prof. Baron Dr. von Feilitzsch, University of Greifswald.

This apparatus enables the laws in question to be demonstrated when the pressure is above or below that of the atmosphere. A glass tube open at the upper end, and a glass gauge tube, which can be closed at the upper end by a stop-cock, communicate by means of a long gutta-percha tube, and can be displaced on a vertical scale. These are partly filled with mercury, the latter tube containing the gas to be tested, or the combination of gases and vapours, when by raising or lowering of the first tube, the level of the mercury can be changed as required, and the resulting volume of gas observed.

The apparatus can be easily adapted to all kinds of experiments, at pressures either above or below that of the atmosphere.

605. Suction Pump, on an iron stand.
Ferdinand Ernecke, Berlin.

605a. Suction Apparatus. Two small pieces of apparatus to illustrate the deficiency of pressure attending a high velocity in a stream of air. *Lord Rayleigh.*

By blowing through the electrotyped copper tube a suction of 6 inches of mercury may be realised at the narrow part.

In the second arrangement the novelty consists in the cap at the end of the brass tube, by which the efficiency is increased.

609. Double Cylinder Air Pump, with Babinet's stop-cock, cylinder 170 mm. high, 60 mm. wide, with manometer and glass plate of 200 mm. diameter, on a mahogany board, with the following auxiliary apparatus :—

1. Pair of Magdeburg hemispheres of glass.
2. Electric egg, consisting of two parts, with arrangement for carbon points.
3. Double mill.
4. Balloon, with bell.
5. Balloon, for the gravity of the air.
6. Fountain.
7. Quicksilver.
8. Ring.
9. Dasimeter.
10. Falling tube.
11. Heron's ball.
Warmbrunn, Quilitz, and Co., Berlin.

610. Apparatus for Lighting, on the system of Döbereiner, with contrivance for drying the gas.
Warmbrunn, Quilitz, and Co., Berlin.

610a. Water Vacuum Pump, invented by H. Sprengel in 1863. *Hermann Sprengel.*

610b. Air Pump and two receivers, by Staudinger, Giessen.
Chemical Laboratory, University of Berlin; Prof. A. W. Hofmann, Director.

611. Water Pump, either for exhausting or for compressing air. *Joseph Conquet.*

This apparatus can be used either in place of an air pump, or in place of bellows. Water at a pressure of two atmospheres is required for working it.

615. Electrical Apparatus belonging to Air Pump.
Vertical cylinder electrical machine with two leather rubbers.
The apparatus is so constructed that the cylinder may be enclosed
in a large glass receiver, which can be exhausted.

The Council of King's College, London.

616. Large Air Pump, arranged for producing the ordinary
vacuum or a dry vacuum of a millimetre of mercury, and also for
producing ice.   *M. Carré, Paris.*

616a. Small Air Pump, also capable of being transformed
at will into a compression pump.   *M. Carré, Paris.*

617. Apparatus for producing ice by vacuum and sulphuric
acid.   *M. Carré, Paris.*

619. Hydraulic Press of Copper and Glass (for demonstration).   *Luizard, Paris.*

620. Apparatus for Compression.   *Luizard, Paris.*

620a. Model of an Air Compressor, which may also be
used as a water meter.   *M. Eugene Bourdon, Paris.*

621. Air Pump for Double Exhaustion (Babinet's principle).   *Luizard, Paris.*

621a. Model of Pneumatic Machine.

M. Loiseau, jun., Paris.

627. Fick's Spring Manometer.   *Weber, Würzburg.*

630. Air Pump, made by Spencer & Sons, Dublin.

Prof. W. F. Barrett.

This air pump has no valves between the barrels and the receiver, hence
it is free from regurgitations of air, and is capable of making a vacuum nearly
equal to that obtained by means of Sprengel's air pump. The ends of the
pistons and barrels are conical, and, when they are in contact, no cavities
remain. All the air in the barrels is therefore expelled at each stroke of the
piston; the effect is the same when it is highly rarefied. The horizontal
portions of the pump admit of the barrel being brought into close connexion
with the plate, and leaves the valves accessible for cleaning or repairs.

630a. Single Barrel Air Pump.   *James How and Co.*

630b. Air-pump on Babinet's system.   *Dr. Stone.*

680a. Improved **Portable Spireometer.**   *E. Cetti & Co.*

III. OSMOSE, DIALYSIS AND DIFFUSION.

631. Electric differential Osmometer. Engelmann.
Prof. Engelmann, Utrecht.

Two glass vessels of quadrangular section, possessing in the plane ground surfaces facing each other a round aperture 30 mm. diameter, and each containing an electrode; a platinum disc 30 mm. diameter. Between the two vessels the cell is placed, a plane parallel plate of ebonite, furnished with a transverse perforation 30 mm. diameter, communicating with a short brass top, upon which a rise-tube, manometer, etc. can be screwed. The membranes or porous plates, whose electric osmotic permeability is to be compared, are placed between the cell and the glass vessels, the whole being secured by two brass vices. Both vessels and the cell are now filled with the same fluid. Inelastic partitions, such as clay plates, must, in order to prevent breakage and leakage, be provided, on each side, with an elastic ring (india-rubber or bladder).

Thin, very pliable, membranes, *e.g.* skin of a frog, are secured between sieve-like perforated plates of ebonite to prevent bending.

The apparatus is used to demonstrate :

1st, the fact of electric osmosis.

2nd, the specific influence of the membrane ; the rise and fall of the fluid in the cell shows which of the two membranes possesses the greater electric osmotic permeability. On removing one of the partitions, the apparatus becomes an ordinary osmometer.

(Onderzoekingen gedaan in het physiologisch laboratorium der Utrechtsche Hoogeschool. Derde Reeks, II., 1873, p. 365, etc.)

632. Osmometer, illustrating the transpiration of gases through capillary tubes. *Prof. W. F. Barrett.*

This is an apparatus constructed under the direction of Professor Sullivan for illustrating the law of the transpiration of gases. The diaphragm is made of a number of short lengths of capillary tubes.

IV. CONDENSATION AND SOLUTION OF GASES.

633. Addams' Apparatus, for condensing and solidifying carbonic acid. *William Sykes Ward.*

634. Apparatus used in 1857 for Liquefying Ice by Pressure at temperatures below −65° C.
Prof. Mousson, Zurich.

A small cylindrical chamber, hollowed out of a strong piece of steel, is filled with water containing a movable metallic index. When the water is frozen, the chamber is closed by means of a soft copper cone, subject to the action of a strong screw. The apparatus is then *reversed.* The chamber merges into a long, slightly widening cone of soft copper, on which a short steel rod presses by means of a powerful screw-nut acting on the principal piece, and a double lever a metre in length. If the chamber be opened, after the pressure is removed, the index is found to have been carried over to the other end, as a proof that the ice has been reduced to a liquid state. Soft copper is the only means of closing under excessive pressures.

The experiment can be successfully made at − 18° C.

635. Original Model of the Fluid Vein of Messrs. Poncelet and Lescros.   *Conservatoire des Arts et Métiers, Paris.*

636. Original Thilorier Apparatus for liquefying carbonic acid, 1857.   *Conservatoire des Arts et Métiers, Paris.*

637. Apparatus for the demonstration of Boyle and Marriotte's law to a class.   *Prof. W. F. Barrett.*

The mercury contained in the upper iron reservoir can be admitted to the tube through an aperture closed by a valve that is moved by pulling the string. The eye is kept level with the air chamber. Equality of pressure with the atmosphere is readily obtained by means of the stopper closing the lower bent tube or air chamber.

638. Compression Pump and Receiver. An apparatus for the liquefaction and solidification of gases.
          H. Lloyd, Trinity College, Dublin.

640. Apparatus employed in the researches of Dr. Andrews on the continuity of the gaseous and liquid states of matter, and on the properties of matter at high pressures and varied temperatures.   *Dr. Andrews, F.R.S.*

This apparatus consists of two cold-drawn copper tubes of great strength, communicating by a horizontal passage, and having massive end-pieces above and below, firmly bolted on with leather washers interposed, so as to be able to resist any pressure. The upper end-pieces are traversed by fine glass tubes, the junction between the glass and metal being made tight by a peculiar system of conical packing. The exposed parts of the glass tubes have a capillary bore, and, if sufficiently fine, will bear a pressure of 500 atmospheres or more without bursting. One of the tubes contains air or hydrogen, and serves as a manometer, the other contains the gas or liquid to be examined. The lower end-pieces carry well packed steel screws, which produce the pressure by compressing the water with which the apparatus is filled. The temperature of the air or hydrogen in the manometer, and that of the gas or liquid under examination, can be varied at pleasure by enclosing the glass tubes in outer cylinders of glass, or, where accurate readings are required, in rectangular vessels with plate-glass sides. With this apparatus, accurate measurements can be made to 500 atmospheres, or even higher pressures, and, with a slight modification, at any temperature which glass will bear without softening.

641. Single Apparatus, adapted to exhibit the properties of gases and liquids under different conditions of pressure and temperature, but without measuring the pressures employed.
          Dr. Andrews, F.R.S.

This apparatus is similar in construction to the compound form last described.

641a. Original Apparatus, by Despretz, by the help of which he proved the difference, shown by gases, in relation to the law of Marriotte.   *The Faculty of Sciences, Paris.*

641b. Original Apparatus, by Gay Lussac, for ascertaining the elastic force of gases and vapours.
Polytechnic School, Paris.

641c. Apparatus, by M. Dumas, for illustrating the density of vapours. *Conservatoire des Arts et Métiers, Paris.*

642. Sulphurous Acid Tube, to exhibit the flickering striæ which occur, from slight changes of pressure, near the critical point of temperature. *Dr. Andrews, F.R.S.*

643. Model, constructed by Professor J. Thomson, to illustrate the results of the experiments by which Dr. Andrews first established the continuity of the gaseous and liquid states of matter.
Dr. Andrews, F.R.S.

V. HYDROMETERS.

644a. Regnault's Apparatus for estimating the specific gravity of solid bodies.
Golaz, 24, Rue des Fosses, St. Jacques, Paris.

645. Compensation Salinometer.
H. Hädicke, Demmin, Pomerania.

The compensation salinometer is an areometer especially constructed for measuring the saliferous contents of sea water, or of the water in steam boilers, the readings of which are independent of the temperature of the respective fluids. The dimensions of the same have been so calculated that the increase of the volume of the instrument effected by the expansion of all its constituent materials (*i.e.*, aluminum, or iron and mercury) will avoid the deeper immersion which in the instruments of the ordinary construction would take place in consequence of an increase of the temperature of the fluids.

646. Apparatus for measuring the Density of Liquids.
Prof. Dr. Bohn, Aschaffenburg.

647. Apparatus for determining Vapour Density, on Dr. Frerich's system. *F. Sartorius, Göttingen.*

648. Oleometer. An instrument for ascertaining the density of oils. *Dring and Fage.*

649. Small Sykes's Hydrometer. *Dring and Fage.*

650. Barktrometer. Used by tanners for ascertaining the density of bark liquor. *Dring and Fage.*

The indication of the steam in degrees, and true specific gravity at 60 F.; the sliding rule accompanying the instrument enables indications at higher or lower temperature to be reduced to the equivalent at 60 F.

651. Small Barktrometer.   *Dring and Fage.*

The same in principle as the larger one, but it is not capable of indicating over so wide a range of specific gravities.

652. Salinometer, used on steam vessels to ascertain the amount of salt in the water in the boiler.   *Dring and Fage.*

On the stem of the instrument are the words " limit " and " blow." " Limit " indicates the maximum amount of salt in the water that can be used with safety, and " blow " when it is necessary to blow off a portion of the water in the boiler through there being too much salt in solution.

653. Atkins' Saccharometer.   (Obsolete.)
                                         Dring and Fage.

654. Four-weight Hydrometer on Sykes's principle. (Obsolete.)   *Dring and Fage.*

The stem is divided into 20 parts, and a part again divided into 2 parts, each of which is equal to one-half per cent. The sliding rule accompanying this instrument gives the equivalents of other hydrometers of the same period. About 1817.

656. Sykes's (Revenue) Hydrometer. For ascertaining the proof-strength of spirits.   *Dring and Fage.*

The instrument, used by the Revenue in collecting the duty on spirits, was first established under Act of Parliament in the year 1816; by this Act also proof spirit (which forms the basis of estimation in this instrument, as all strengths are indicated by their relation to this point) received a particular definition as that which weighed twelve-thirteenths of an equal bulk of distilled water at a temperature of 51° Fah. The stem of the instrument is divided into 10 parts, and each part again into subdivisions, which, with the nine weights (each of which is a multiple of the divisions of the stem), enables the instrument to measure all gravities from 67 over proof to just past distilled water at 60 Fah. The tables which accompany this instrument were compiled from the experiments of Gilpin, who carried them out with such accuracy that no error has ever been detected. The range of temperature given is from 30° to 80° Fah. These tables having been found inconvenient for use in hot climates, in the year 1851, Messrs. Dring and Fage compiled an extension from 80° to 100°, which meets all requirements for high temperatures.

657. Clark's Export Hydrometer (obsolete). Used for ascertaining the strength of spirits in the various stages of manufacture.   *Dring and Fage.*

The instrument used for determining the strength of spirituous liquors prior to the introduction of Sykes's hydrometer. It was the first of its kind established under Act of Parliament, and to which any definite kind of correction for temperature was applied. It is constructed to show the number of gallons of water, plus or minus, necessary to reduce to proof strength a sample of spirits under trial, and was only used for spirits in the various stages of manufacture.

658. Clark's Import Hydrometer (obsolete). Formerly used for ascertaining the strength of spirits imported.
                                         Dring and Fage.

The same in principle and construction as the one for export, only having nine extra or intermediate weights, called per cent. weights. This class of instrument was only used for determining the strength of spirits imported, and was adjusted slightly in favour of the importer.

659. Set of Six Twaddell's Hydrometers. Used for ascertaining the density of a solution ; principally used in trades for which no special instrument of the kind is constructed.

Dring and Fage.

The divisions on these instruments are so placed as to indicate equal differences of specific gravity. The specific gravity of a fluid is found from the indication of this scale by multiplying by 5, cutting off 3 decimal places, and prefixing unity.

659a. Hydrometer by Fordyce. *Royal Society.*

659b. Hicks's Patent Hydrometers, Urinometers, Salinometers, &c. *Hicks.*

These instruments have the scales and figures divided in black on a white enamel stem, thus avoiding all errors from shifting of scale, as with paper ; alteration of form, as with vulcanite ; or corrosion, as with metal.

651a. Tan Tester, for ascertaining the exact quantity of tannic acid in any substance by passing it through a piece of hide.

Thomas Christy and Co.

The solution having been gauged by a tannometer before being tested and after it has passed through the hide, the *difference* gives the exact value of any tannic matter, and a merchant knowing the price of oak bark can calculate the value at once of the substance he has tested.

655. Specific Gravity Instruments, for testing liquids from 650 to 900 sp. gr. *L. Oertling.*

659a. Fahrenheit's Metal Hydrometer.
The Physical Science Laboratory of the Technological Institute, St. Petersburg.

The theory of this instrument has been described and illustrated by an example by E. Lenz, Academician, in the "Bulletin physico-mathématique " de l'Académie des Sciences," Vol. XV., 1857.

660. Thermo-Dilatometer, by Baudin, showing by dilatation the per-centage strength of alcohol, from distilled water 0 to absolute alcohol 100. *M. Baudin, Paris.*

661. Thermo-Dilatometer, by Baudin, showing by dilatation the alcoholic strength of wines and other liquids, from 10 to 20 per cent.

(The last two instruments are the property of the "Conservatoire des Arts et Métiers.") *M. Baudin, Paris.*

662. Dring and Fage Saccharometer, for ascertaining the density of brewers' worts. *Dring and Fage.*

Used for determining the density (in pounds weight) of a barrel of wort, of 36 imperial gallons, in excess of the same quantity of distilled water at a temperature of 60° Fah. The rule accompanying the instrument shows the weight of the residuum if a barrel of wort were evaporated to dryness; also the amount of proof spirit to be obtained, and the specific gravity.

This instrument was the joint invention of Messrs. Dring and Fage, and perfected by the valuable experiments and calculation of Drs. Hope Coventry and Thomson.

663. Quin's Saccharometer. (Obsolete.)
Dring and Fage.

664. Dring and Fage Still, for ascertaining the original specific gravity of a wort from a sample of beer. *Dring and Fage.*

Used by the Excise and Customs when making the allowance for drawback on export beer.

664a. Apparatus by **Rakowitsch** for testing **Alcohols** and **Saccharine Matter** in **Liqueurs ;** likewise for ascertaining the fatness of butter, and for testing water.
R. Nippe, St. Petersburg.

665. Apparatus for determining the Specific Gravity of bodies, inclusive of a thermometer.
Ch. F. Geissler & Son, Berlin.

671. Piknometer.   *Dr. H. Geissler, Bonn.*

671a. Piknometer, for ascertaining the specific weight of small quantities of liquid, with thermometer attached.
W. Haak, Neuhaus, Thüringen.

672. Two Sets of Areometers.   *Dr. H. Geissler, Bonn.*

672b. Micromanometer.
Royal Mining Academy, Freiburg.

672c. Drawing of Apparatus for measuring low pressures of Gas.   *Prof. H. McLeod.*

The pressure is measured by compressing a known volume of the gas into a smaller space, and measuring the pressure under the new conditions. By dividing the pressure thus found by the ratio between the original volume and the reduced volume, the original pressure is obtained. *See* Proc. Phys. Soc. i. 30.

673. Manometer for Minute Observations, used by A. de la Rive in all his latest researches concerning the propagation of electricity in rarefied gases.
De la Rive Collection, the property of Messrs. Soret, Perrot, and Sarasin, Geneva.

This instrument, constructed by the Geneva Association for the Construction of Scientific Instruments, consists of two glass tubes dipped in a common mercury trough, of which one is a simple barometric tube serving as a point of comparison, the other communicating on its top with the quantity of rarefied gas by means of a pipe adjusted in the setting of the apparatus. The pressure, that is, the difference between the two mercury columns, is read by means of a small cathetometer, of which the millimetric graduation is turned towards the tubes and the lamp which lights them both. This graduation is reflected in the telescope by means of a total reflecting prism placed before the objective. Thus the level is at once read, and a micrometrical division placed at the focus of the eye-piece shows the $\frac{1}{25}$ of the millimetre.

673a. Delicate Pressure Gauge. *E. A. Cowper.*

The very light pressure required to produce a current through stoves can be shown by the delicate pressure gauge (enclosed in a mahogany case), which has a thin diaphragm exactly one square foot in area, and has a light lever and weights to cause it to bear the pressure brought on its surface, so that the whole pressure of the air on one square foot is exactly weighed with ease. As small a pressure as one hundredth of an inch of water can be measured.

673b. Multiplying Manometer, for measuring the force of the draught in chimneys and stoves, as well as for the pressure of gas. *Dr. K. List, Hagen, Westphalia.*

This apparatus was first constructed by the exhibitor in 1862 for the purpose of meeting the desire of several blast-furnace proprietors resident at Hagen, who wished to be able to measure more accurately the pressure of the blast in their furnaces than was possible with the ordinary manometer, consisting of a curved glass tube, as is mentioned in the description given in the "Zeitschrift des Vereins deutscher Ingenieure," (Journal of the Society of German Engineers) of 1863, vol. vii, p. 493. It consists essentially of a long narrow horizontal and two wider vertical glass tubes, the latter of which are filled to about half their length, and the first up to a long air-bubble, with coloured petroleum. If a suction is effected on one of the vertical tubes, the air-bubble contained in the horizontal tube must, if the vertical tube has a diameter n times as large as the horizontal tube, traverse a distance in the horizontal tube n^2 times as large as the liquid rises or falls, as the case may be, in the vertical tube; the sensibility, therefore, can be increased at pleasure. The space which the index travels is $\frac{n^2}{2}$ times as large as the difference in the respective heights of the liquid in the two vertical columns.

The apparatus has been constructed for some years by the exhibitor himself; the scale indicating in water-millimeters the difference of pressure was prepared by him in the manner described in the Journal of the Society of German Engineers. The apparatus has answered well everywhere. It has been recommended for measuring the pressure of gas by authorities such as Schiele in Frankfort-on-the-Maine and Schilling at Munich.

673c. Level Manometer, by Gallaud. *M. Breguet, Paris.*

673d. Two-fold Control Manometer.
C. D. Gäbler, Hamburg.

1. The advantage here consists in two independent manometers, acting under quite similar conditions, being combined in one case, whereby not only a greater guarantee of control is obtained, but by the close proximity of the two scales the simultaneous reading of the divisions is facilitated. The upright position of the tubes prevents the collection of impurities out of the

water, as the latter flows freely away after use.   The instrument may be connected by a simple screw arrangement with the boiler.

673e. Four-fold Control Manometer.
C. D. Gäbler, Hamburg.

This manometer, consisting of two manometers of the above construction,
combined one with the other, offers the greatest possible safety in controlling ;
because, if only two-fold action be required, the one pair of manometers
can be shut off by closing the stopcock, and may thus be used as reserve or
control of the manometer thence used.   The connexion with the boiler
flange is performed by means of the accompanying two thumb nut-screws.

673f. Control Manometer, showing the inner construction.
C. D. Gäbler, Hamburg.

Shows the extremely simple construction of the inner mechanism of the
manometers described above.

674. Alcoholometer, consisting of two cylinders of ebonite and brass, keyed together.   *G. Recknagel, Kaiserslautern.*

This instrument is not liable to break, and answers all requirements of
accuracy in reading.

675. Areometer of ebonite and brass, with adjustable cylinder.
G. Recknagel, Kaiserslautern.

This not being liable to be broken can be used for educational purposes as
well as for practical application.

The level upper terminal plate is adapted for the addition of small weights
by means of which the value of the divisions can be demonstrated.   The
instrument, moreover, is arranged with a *cylindrical slide*, which can be
extended to the double volume of the divided spindle.   If the scale is over-
run, the slide is to be pulled out a volume more, and the scale is then again
at disposal for use.

For instruction and practice the most suitable is the uniform scale of the
Gay Lussac areometer.   As, however, there is space for four scales, the
others can be arranged for direct indication of specific weights, or, like the
present models, for alcoholometry.

676. Areometer Case, containing three standard areometer-cylinders, for determining the specific gravity of all kinds of liquids, with indicator scale fused into them.   *W. Zorn, Berlin.*

677. Areometer. The indicator scale is not fused into the glass, but fastened only with sealing-wax.   *W. Zorn, Berlin.*

Each of these areometer-cases contains three glass spindles, which, loaded
at the top in a similar manner to Nicholson's metal spindles, with weights,
indicate with the greatest accuracy the specific weight of *all* liquids.

The liquid to be tested must be brought to a temperature of 15° Celsius ; if
one of the spindles is inserted in the liquid, so many weights must be placed
on the glass plate as are required to make the spindle sink as far as the black
mark on the milk-white glass line in the neck of the spindle.

The lightest spindle embraces all liquids from 0·650 to 1,000 ; if this
has been used, 0·650 must be added to the weight placed on the spindle.   If
the medium (1,000) spindle has been employed, 1,000 must be added to the
weight, and at the heaviest (1,400) spindle (1,400) 1,400 must be added to the weight
placed on the same.   The sum obtained will give the specific weight of the
liquids with an accuracy extending a little over the third decimal.

The proof of the correctness is simple ; it is only necessary to have a good pair of scales and some distilled water :

650 weight = spindle 0·650 ;
spindle 0·650 + 350 weight = spindle 1,000 ;
spindle 1,000 + 400 weight = spindle 1,400.

The construction of these areometers is *new*. For many years the exhibitor had constructed similar areometers, consisting, however, of two spindles only, which were very much liked, under the name of Wittstock's areometer, on account of their accuracy ; but it was too fragile, owing to the light spindle (there being but two of them) having to carry too much weight. Moreover, the late Privy Councillor, Dr. Wittstock (apothecary to the Royal Court at Berlin), had devised a peculiar proportionate weight to the same, from which the spindles had derived their name, but which is now no longer in use.

The areometers constructed by the exhibitor, the normal weight of which has been recommended by Mr. Hirsch, apothecary, at Giessen, and which corresponds to the gram scale, have the following advantages, compared with other similar instruments :

They can be easily tested as to their accuracy (as shown before); they will not be affected by the liquids ; they can be easily cleaned, and will sink slowly and uniformly in any liquid, and are not surpassed, nor even equalled, by most similar instruments, in respect of accuracy.

One of the exhibited areometer-cases contains three spindles, which are perfectly equal in point of accuracy with those in the second case. The milky glass lines at the neck are, however, fastened only with sealing-wax, which, in case of great carelessness, may dissolve, although persons of experience have used these spindles for years without injury ; besides, every particle of sealing-wax can be easily supplied as soon as any defect has been noticed. But, in order to avoid this, the milky glass lines on the three spindles in the second case have been melted together with the glass plate.

Besides the weights, there are added to the two cases a cylinder for weighing the liquids, a pair of forceps for placing the weights, and a thermometer for determining the temperature of the liquids, which latter is adapted in its form and the strength of the glass to be used as a stirring rod.

The exactness of the indications of Zorn's spindles affords also the advantage of testing the correctness of other liquid scales, such as Tralles, Baumé, and others made on a scientific basis.

679. Alcoholometer.
Siemens Brothers and Co., Charlottenburg.

An apparatus for measuring simultaneously the quantity of spirit flowing through it, and the per-centage of proof spirit contained in that spirit. The quantity is measured by a revolving drum imparting motion to a counter under the control of a hydrometer containing spirit such as is the average production of the still. The measurement is unaffected either by the velocity at which the spirit enters or by the friction of the bearings of the spindles of the drum.

679a. Metallic Alcoholometer, on Tralles's principle.
W. Gloukhoff, St. Petersburg.

The metallic alcoholometer, with additional weights, newly adapted by the Russian Government, is made on the principle of Sykes's hydrometer, but its scale is adapted to the system of Tralles, legalized in Russia.

VI. MISCELLANEOUS.

680. Apparatus, serving to illustrate the **Mechanical Effect** of the **Expansion** of **Liquids.**   *T. A. Snyders, Delft.*

Consisting of gun-barrels closed by a lead plate 3 millimetres thick, which is retained by a perforated screw-plug. The expansion of the liquid causes a cylinder of lead to be forced out through the aperture of the plug.

681. Apparatus, constructed by **Leschot** and **Thury,** to **suppress Friction** by the interposition of a stratum of air.
Geneva Association for the Construction of Scientific Instruments.

The apparatus is composed of two plates, superposed, and perfectly adjusted, between which is introduced air or gas under pressure. The air, spreading from the centre between the two plates, exerts upon these a total pressure measured by its elastic force multiplied by the extent of the surfaces brought into apparent contact. When this pressure equals or exceeds the weight of the plate acted upon, it supports the plate, and the friction is reduced to its smallest limits.

683. Two Apparatus for measuring the Transpiration of Air at different Temperatures.
Dr. O. E. Meyer, University of Breslau.

(Described in "Poggend Ann., 1873," vol. 148, p. 203.)

684. Apparatus to illustrate to a large audience the fact that the **pressure** exerted by a **curved liquid** film increases with the curvature.   *The Yorkshire College of Science.*

The apparatus consists of a glass tube communicating with two others, each of which is furnished with a stop-cock. Bubbles of different diameters are blown at the ends of the tubes, one stop-cock being closed while the other is open, and then communication with the outer air being cut off, and the cocks both opened, the smaller bubble is seen to diminish and the larger one to increase, thus proving that the air inside the smaller bubble was the more compressed. The tubes are bent, so as to bring their extremities close together, and the experiment can be shown to a large audience by throwing a magnified image of the bubbles on a screen.

685. Apparatus for demonstrating **Leidenfrost's phenomenon** of **drops.**   *Dr. J. Hoogewerff, Rotterdam.*

This apparatus was constructed by Mr. Kellenbach, curator of the Batavian Society, Rotterdam, and belonging to the academy of plastic arts and technical sciences at Rotterdam. The apparatus is used as follows : a Grove pile is connected with the instrument, a galvanometer being placed on the conducting wire, and the copper tray or platinum capsule into which the drops of water have been put, are heated by means of a gas lamp. Every time Leidenfrost's experiment is repeated, no current is indicated by the galvanometer, contact having been interrupted by the layer of steam ; when on the contrary the water, the surface of which is in contact with the copper wire, comes into immediate contact again with the metallic surface of the tray or capsule, the galvanometer distinctly indicates the current.

686. **Apparatus** for determining the **Tension** of the **Vapours** of different liquids at the boiling point.   *Prof. W. F. Barrett.*

This is a useful and strong form of the usual apparatus. The liquids whose tensions are to be measured are inserted in the barometer tubes, and either steam or the vapour of some liquid having a lower boiling point, such as alcohol, is sent through the larger tubes.

670. Parabolic Diagram of the relation between tension and volume of the saturated steam.

H. Hädicke, Demmin, Pomerania.

The area of this parabolic tension diagram shows that it is justifiable to replace the area of Mariotte's oder Pambour Navier's tension-lines of the saturated steam by a parabola. This method gives for the ratio of the mean pressure (p_0) on one side of the piston to the absolute pressure at the commencement (p), and the cut off (ϵ) the elementary formula : $p_0 = p - \frac{1}{6}$ $(1 - \epsilon)$ $^2(4\,p + 1)$, or

$$\epsilon = 1 - \sqrt{\frac{6(p - p_0)}{4\,p + 1}}.$$

(See "Practical Tables and Rules for Steam Engines," by H. Haedicke Kiel, 1871.)

687a. Independent Bed Plate for **Pneumatic Machines.**

Geneva Association for the Construction of Scientific Instruments.

This plate is made entirely of cast-iron, as well as its stand, so as not to become damaged in experiments where mercury is employed.

SECTION 6.—SOUND.

WEST GALLERY, UPPER FLOOR, ROOM $\left(\text{Q}\right)$

I.—SOURCES.

688. Apparatus used by M. Rijke to cause a tube to emit sounds when wire gauze placed in its interior is heated.
Prof. Dr. P. L. Rijke, Leyden.

689. Whistles for producing shrill notes, within and beyond the limits of ordinary audition. *Francis Galton, F.R.S.*

These whistles were designed for testing the limits of the power of men and animals of hearing very shrill notes. The plugs that close the whistles can be screwed up and down, and the length of the whistle can be ascertained by the attached graduations, whence the number of vibrations per second may be calculated. The whistles are of three forms : (1) a small cylindrical tube, which gives a pure note, but of small power ; (2) a flat, wide and narrow whistle, of which the plug is a broad thin plate of metal ; (3) an instrument which is externally a cylinder of 2½ inches in diameter, but of which the effective part is merely an annulus ; the plug of this is a cylindrical sheet of brass ; it gives a powerful note, but not a pure one.

690. Brass Tube to sound the constant proper tone of the mouth, characterising the vocal sound.
Prof. Donders, Utrecht.

This consists of a brass tube terminating in a broad slit at one end and at the other end in an india-rubber tube to be placed on a blower (" *souffleur* "). (Donders.) The blast, directed by the slit on the borders of the lips, sounds during the time a vocal sound is sung in different tones, the constant proper tone of the mouth characterising the vocal sound. (Compare Donders, Uber die Natur der Vocale, Holl. Beit. zur Nat. u. Heilk. 1846.)

691. Set of Vowel Forks and Resonance Globes.
Frederick Guthrie.

692. Set of Organ Pipes. *Frederick Guthrie.*

693. Set of Tuning Forks. *Frederick Guthrie.*

926o. Acoustical Instrument, illustrating harmony and discord. *S. F. Pichler.*

This instrument is constructed to demonstrate the relations between musical sounds. By its means, harmony and discord, gravity and acuteness of pitch, beats, waves, and amplitude of notes, may be rendered visible as well as audible.

Two metallic vibrators, each with a small speculum, are fixed at right angles to each other, and sounds are produced by a current of air acting on one or both of them at pleasure. The perpendicular vibrator is tuned to a given note, and the horizontal vibrator is fitted with a mechanical arrangement, whereby its pitch can be graduated to any degree of nicety within the compass of two octaves. An apparatus is also provided with a means of concentrating a pencil of light upon the speculum of the perpendicular vibrator, whence it is reflected to the speculum of the horizontal vibrator. For educational purposes, artificial light may be used, which can be further reflected and magnified upon a screen so as to be visible to a number of spectators.

When musical sounds are produced by the vibrations, various luminous geometrical figures are formed on the horizontal speculum, which, being thence thrown on the screen give the curves described by the pencil of light, by the single or joint action of the vibrators, and the form and motion of such figures demonstrate the exact relation to each other of the musical notes produced.

Sounds which harmonise to the ear, produce regular figures to the eye, as for example, segments of the circle, ellipses, ovals, or straight lines, and if the amplitude of each vibrator be equal, these luminous figures will appear on the speculum or screen with an apparent steadiness. If the sounds do not harmonise, the figures are confused, unsteady, and complicated.

The mathematical relations of musical notes can also be demonstrated by this instrument; regular simple forms being produced by combinations of those notes which result from vibrations bearing a definite numerical ratio to each other, while irregular and unsteady figures are caused by notes which have no such ratios.

This apparatus has advantages over those in which tuning forks are used for a like purpose; (1.) In being able to prolong the effects to any period desired by the operator; and (2.) In its capability of representing combinations and graduations otherwise unattainable.

694. Photograph of a Chemical Harmonica of glass for gas-flames, with eight pipes (major-scale from d^1 to d^2 inclusive), with double regulating cocks, and key-board for playing. With a copy of a few melodies executed on the same for two and three voices.

Prof. J. Joseph Oppel, Frankfort-on-the-Maine.

The glass tubes can be turned by means of metal mountings, and can be adjusted according to the height of the flames, which being regulated by the taps, and the acoustic effect of the major and minor accords (particularly when carefully tuned) are astonishing.

695. Diapason Tuning Fork. *F. Ernecke, Berlin.*

695a. Diapason of 200 double vibrations, arranged for continuous action. *T. Hawksley.*

695b. Model of a Scientific and Musical Instrument called the Pyrophone, invented by the exhibitor, composed of a series of glass tubes of different lengths and dimensions, in which gas jets, when ignited and moved by a very simple apparatus, produce the most perfect musical notes, and establish the new scientific principle of the " interference " of melodious flames.

Frédéric Kastner, 43, Rue de Clichy, Paris.

M 2

The great pyrophone, which this model represents, has three octaves, like those of a piano, and is composed of a series of glass tubes, similar to organ pipes, of different lengths and dimensions, in which gas jets burn. A very simple mechanism causes each key to communicate with the corresponding supply pipe of the flames in the glass tubes. On pressing the keys, the flames separate and the sound is produced as soon as the little jets in each tube are separated from each other; when re-united and becoming one flame, the sound immediately ceases.

In this instrument it is shown that when two or more flames are introduced in a tube they vibrate in unison, and produce a musical sound when they are placed one-third the length of the tube, and a new law has been demonstrated, as an application of which an instrument which approaches the human voice more nearly than any yet made has been constructed.

A contrivance opens and shuts like the fingers of a hand, at the extremities of each of which a small jet of gas is lighted. When the fingers are separated, the sound is produced; when they are closed or approach each other, the sound ceases, and the numerous jets become one silent flame.

The new principle is as follows :—" If two or more flames of a certain size " be introduced into a tube made of glass or other material, and if they be " so placed that they reach to the third part of the tube's height (measured " from the base), the flames will vibrate in unison. This phenomenon con- " tinues as long as the flames remain apart, but as soon as they are united the " sound ceases."

The pyrophone gives sonorous and penetrating tones, and may be constructed from one octave to a most extended compass, and for the *cabinets de physique*, with one, two, three, or more notes to show the principle of the " interference " of singing flames. The exhibitor has also invented a gaselier (*lustre chantant*), lighted by electricity, which can be placed in the centre of a room, and by unseen electric wires be made to produce powerful and melodious sounds, and on which any kind of music can be performed.

696. Tube for Singing Flames, according to Schaff-gotsch's system.   *Albrecht, Tübingen.*

697. Set of Glass Tubes for illustrating singing flames, with paper sliders for adjusting the pitch of any tube.
Prof. W. F. Barrett.

The only novelty here is the paper sliders, which enable the pitch of the note to be readily adjusted, and were originally suggested by the exhibitor.

697a. Stand and Burner for Sensitive Flames.
Prof. W. F. Barrett.

The sensitive flame is an illustration of resonance. The vibrations accepted by the flame are those which the flame itself would emit when roaring. A flame to be sensitive must be brought to the verge of roaring by a proper adjustment of pressure on the gas supply. The flame is then in unstable equilibrium, and a feeble sympathetic vibration will then produce the same effect on the flame as a slight increase in the gas pressure.

The flame to be extremely sensitive must be fed with gas which flows smoothly and freely to the orifice. A bell gas holder is far better than a gas bag for obtaining the necessary pressure. The gas cocks must be fully open and the pressure adjusted by altering the weights on the gasholder. The stand shown allows the gas to flow smoothly, and the best burner is a steatite " jet photometer " burner carefully enlarged till it gives the *tallest possible* flame under a pressure just short of roaring. Such a flame with good gas can be had 2 feet high, shrinking down under the influence of a sound to less than one half this height. With similar burners and pressures the quality of different specimens of coal gas is accurately determined by the degree of sensitiveness of the flame. See Phil. Mag., March and April 1867.

697b. Suitable source of Sound for experiments with Sensitive Flames. *Prof. W. F. Barrett.*

This is simply a loud ticking watch enclosed in a padded case with a movable cover, and mounted on a sliding stand.

697c. Practical application of Sensitive Flames.
Prof. W. F. Barrett.

By using a suitable burner a sensitive flame can be made to spread out sideways into a fish-tail flame under the influence of sound. Under such conditions the flame can be made to touch a compound metallic ribbon, which curves by unequal expansion, closes an electric circuit, and rings an electric bell. An arrangement of this kind could be adapted to detect burglars, or to act as a self-recording phonoscope. First exhibited by Prof. Barrett at the Royal Dublin Society, January 1868.

698. Set of Resonant Tubes. *Prof. W. F. Barrett.*

By suddenly and successively withdrawing the stopper, the resonance of the air within the series of tubes will give the notes of the common chord.

699. Apparatus for experiments with singing gas flames.
Yeates & Sons.

The above, consisting of glass tubes of similar size, with the assistance of a revolving mirror, will illustrate most of the phenomena of interference, harmony, &c.

700. Inferior Limits of Audibility. An apparatus to show the lowest number of vibrations that will produce sound.
Elliott Brothers.

701. Sir Charles Wheatstone's Resonating Tube.
W. Groves.

By moving a piston up and down, and thus diminishing or enlarging the resonator, a perfect two-octave scale of aliquot parts may be produced from a spring, which would otherwise sound but one note. The scale may thus be played as rapidly as by the fingers upon a pianoforte. Set the spring in vibration by a twang with the forefinger of the left hand, and draw the piston with the right hand.

701a. Sir Charles Wheatstone's original " Magic Lyre," for rendering vibrations audible at a distance through wires. *Robert Sabine.*

704. Set of five Steel Tuning-forks on a new system.
Dr. Stone.

II.—MEASUREMENT.

705a. Apparatus, by M. Regnault, for ascertaining the velocity of sound.   *College of France, Paris.*

706. Acoustic Apparatus, for ascertaining the velocity of transmissio of sound through water, used in 1826 on the Lake of Geneva at a distance of 13,487 metres, and subsequently in 1841 at a distance of 35,000 metres.

Prof. Daniel Colladon, Geneva.

This apparatus was used in 1826 and 1841, during a series of experiments upon the transmission of sound through water, and upon the direct measurement of the velocity of sound in the water of the Lake of Geneva.

Mémoires de l'Académie des Sciences de l'Institut de France, savants étrangers, sciences mathématiques et physiques, vol. 5, p. 267 and following pages. Comptes-rendus de l'Institut, vol. 13, p. 439, séance 23rd August 1841.

N.B.—With this instrument it is possible in calm weather to hear, at the distance of more than a hundred kilometres, the blows struck upon a bell of about half a ton weight immersed in the water, which may thus be used as a submarine telegraph, or for transmitting signals in foggy weather.

715. M. Le Roux's Apparatus, for determining the velocity of sound.   *Conservatoire des Arts et Métiers, Paris.*

707. Apparatus for registering Tuning-fork Vibrations.   *Prof. L. von Babo, Freiburg, Breisgau.*

707a. Original Apparatus, by Duhamel, for registering vibrations.   *Polytechnic School, Paris.*

708. Helmholtz' Double Siren.
H. Lloyd, Trinity College, Dublin.

705. Revolving Drum, for determining pitch of note.
Frederick Guthrie.

The styles attached to two vibrating forks mark sinuosities on the blackened surface of the drum when it turns and advances on its screw axis. The pitch of the notes is thus compared.

709. Double Siren, such as was employed by Professor H   .1z in his researches on sound.
Frederick Guthrie.

710. Siren, of card-board, with four circles of holes, 64, 80, 96, and 128, giving major chord. Made by Yeates and Sons.
Prof. W. F. Barrett.

The above is provided with an air-chest and four keys, so that any or all the circles of holes can be made to sound at pleasure.

711. Siren, an instrument for showing the number of vibrations corresponding to a given note.   *Elliott Brothers.*

712. Savart's Toothed Wheels. A set of wheels, of different sizes and numbers of teeth, to produce a succession of notes. *Elliott Brothers.*

713. Sonometer, with sound-post on the principle of the violin. Also adapted for passing the galvanic current through strained wires. *Dr. Stone.*

714. Metronome, invented by Francis Wollaston.
G. H. Wollaston.

716. Phonometer. *Prof. Lucae, Berlin.*

The phonometer, "speech-measure," is intended to determine accurately the intensity of speech, that is to say, the pressure of expiration employed in speaking.

The apparatus consists of a short metal tube, one end of which expands in the shape of a funnel to a kind of mouth-piece, the rim of which is coated with india-rubba. The other end of the tube is attached to a contact lever oscillating in an axis, the lower section of which is formed by a round aluminium plate, which closes the tube when the lever is in a vertical position and in repose, whilst the upper end of the contact lever, terminating in a point, indicates on a quadrant the oscillations of the pendulum. By any word which is spoken into the mouth-piece, the plate will be pressed outwards according to the pressure of the air employed. When speaking is discontinued a spiral spring attached to the axis has the effect of suspending the action of the lever at the maximum of the motion transmitted to it, and its inclination can be read on the quadrant. The practical use of the instrument in the first instance is, to determine when speaking in a loud or a low voice the relative intensity of one and the same word, or, rather, the preponderating sound prominent in the word uttered, and imparting to it the greatest colour. This object the apparatus perfectly accomplishes, since the force of utterance is proportional to the density of the air effected in the tube. The apparatus consequently affords, among other things, a more exact test of hearing with persons slow of hearing than has been the case hitherto with ordinary speaking.

716a. Phonoptometre, by M. Lissajous.
M. J. Duboscq, Paris.

716b. Experimental Windchests, for measuring the effect of heat on reeds. *Dr. Stone.*

III.—ANALYSIS AND SYNTHESIS.

717. Series of Chladni's Figures. *Frederick Guthrie.*

Sand being scattered on a square brass plate, clamped in the middle and horizontal, the plate is bowed at various points of its edge, while various other points are touched with the finger. The sand is accumulated in the lines of least motion or nodal lines. Gummed paper is then pressed upon the figures so formed.

718. Five Wire Figures for representing **Lissajous' Figures.** *Prof. Buys-Ballot, Utrecht.*

On a horizontal wooden rod are placed five figures of wire, so bent that their shadows or lens images form the figures of Lissajous. Each interval has its own wire figure, and the changes · produced by various phase-differences are shown by turning the figures round their vertical axes.

719. Wooden Board for constructing **Lissajous' Figures.** *Prof. Buys-Ballot, Utrecht.*

The instrument consists of a white-painted wooden board, on which a circle is traced. Horizontal and vertical lines cut the circle in points corresponding to the angles of a regular inscribed icosagon. At the inter-section of each pair of lines a hole is made in the board for fixing pins, which can be joined by threads for showing the figure. Along one of the horizontal and vertical sides of the board two rods can be fixed, provided with ciphers corresponding with the horizontal and vertical lines. Each of these lines may be figured to represent the phase of a vibrating particle either by sound or light for each twentieth part of a vertical and horizontal direction.

For instance, the figure exhibited by the interference of two notes of the same pitch is constructed in the following manner:—The horizontal rod indicates the vertical chords on which the particle is at a supposed moment by a horizontal vibration, then the perpendicular rod indicates in the same manner the horizontal on which it would be at the same moment by a vertical vibration. By fixing pins at the intersections of those chords indicated by the same ciphers, and joining them by a thread, the desired figure originated by both motions is obtained.

If the oscillations differ in phase, another rod not beginning with the same cipher, but with another differing as many twentieth parts of the oscillation as required is found.

If the two interfering notes are not of the same pitch, but one figure is the octave of the other, for the higher note a difference of phase double as great as for the other is taken, and so on.

719a. Tonophant, a simple arrangement for showing **Lissa-jous' Figures.** *See* Phil. Mag., Sept. 1868.

Prof. W. F. Barrett.

719b. Apparatus for exhibiting the **combination of rectangular Vibrations,** made by Yeates & Son, Dublin.

Prof. W. F. Barrett.

By turning the handle the two mirrors have a vibratory motion imparted to them in planes at right angles to each other; the relative rate of vibration of the mirrors can be adjusted by a simple mechanical contrivance. A beam of light reflected from one mirror to the other and projected on a screen thus enables a large audience to see the whole of Lissajous' figures.

1576a. Wheatstone Kaleidophone.

The British Telegraph Manufactory, Limited.

1576b. Wheatstone Kaleidophone.

The British Telegraph Manufactory, Limited.

720. Melde's Universal Kaleidophon.

Ferdinand Süss, Marburg.

(*See* Poggendorff's Annalen, vol. 114, p. 117.)

This apparatus is too well known already to require in this place a more detailed description as regards its capabilities. It will therefore be necessary to give only a short explanation of the reading indications on the Lamellæ, and to refer, as to all specialities respecting this apparatus, to Prof. Melde's publications. Poggendorff's Annalen, Vol. CXIV., page 117, " Lehre von den Schwingungs-Curven," by Dr. Franz Melde, p. 25.

The great Lamella has on each side a line as a mark, and on its upper end the figures I. and II. ; the smaller Lamellæ have the same figures I. and II., and on each side three lines.

If the great Lamella is placed upon the mark of side I., and the small Lamella on the line of side I., with the indication $\frac{1}{4}$, then the vibrations of both Lamellæ to each other are as 1 : 4. The indications of side II. naturally correspond to the mark of side II. on the great Lamella.

The Lamella with the little mirror is used when the curves are to be shown to a whole audience, for which purpose the apparatus ought to be so placed that the rays of the sun, or electric light, falling on the mirror are thrown either on the ceiling or on the wall of the room.

The round bars serve for the production of oscillating curves of two elliptical vibrations.

Finally, it is to be observed that the apparatus may be used equally well for fixing the vibration curves, for which purpose a phonautograph is employed, the cylinder of which is covered with a paper, *as smooth as possible*, on which soot is lightly scattered.

A small piece of the top of a feather fastened with wax upon one of the smaller Lamellæ will be sufficient to describe the curves. In this case, the oscillating surfaces ought to be parallel.

721. Atlas, belonging to the same, illustrative of the theory of oscillation curves, by Dr. F. Melde.

Ferdinand Süss, Marburg.

722. Melde's Tuning-fork Apparatus for producing stationary waves on a thread.

(*See* Poggendorff's Annalen, vol. 109, p. 193 ; and vol. 111, p. 513.)   *Ferdinand Süss, Marburg.*

In Poggendorff's Annalen, Vol. CIX., p. 193, and Vol. CXI., p. 513, a detailed description is given of this apparatus, and of several experiments made with it; also an account of the theory of the oscillation curves ("Lehre von " den Schwingungscurven "), by Dr. Franz Melde, p. 94.

By means of the small sliding rod of glass which is screwed into one of the prongs of the fork, and rubbed with wet fingers, the tuning-fork is brought into a state of vibration. (The small glass rod, owing to its fragility, must be inserted reversely into the wooden frame, when not used, so that only the brass neck of it is visible.) On the lower part of the tuning-fork there is a peg for tuning, which takes up one end of the thread, and serves for stretching it. From this point the thread passes through the neck of the other prong of the fork to a clamp admitting of adjustment to any point of the bar, which is about a meter in length, and thus allows of any length of the thread up to this limit. In order to read the length of the thread, there is on one of the narrow sides of the bar a division indicating half centimeters.

The tuning-fork can be turned about its vertical axis in such a way that its surface of oscillation falls in a parallel line with the longitudinal direction of the thread, perpendicular to it, or at any other selected angle.

The bar can be turned about a horizontal axis, and arrested in any desired position by means of the set screw fixed at the back, so that the thread makes any desired angle with the longitudinal axis of the tuning-fork.

It requires but little practice to effect such a tension of the thread that the required number of waves always appears.

723. Melde's Apparatus for the Combination of Two Thread-vibrations.

(*See* Melde's Lehre von den Schwingungscurven, p. 99.)

Ferdinand Süss, Marburg.

723a. Melde's Apparatus for the production of **Oscillation Curves** of two **Rectilinear Vibrations** on a **Strained Thread.**   *Ferdinand Süss, Marburg.*

The apparatus consists of two principal parts, one of which is a Lamella, oscillating vertically, fastened on a pedestal, and which is set in oscillation by an electro-magnet. The Lamella itself forms the interference. The other part consists also of a Lamella, which is brought into vibration by an electro-magnet, and the surface of vibration of which can be placed at any desired angle to that of the first Lamella.

In order to put the apparatus in motion for experiments both parts must be firmly fastened, either to a long, or on two separate solid tables, at such a distance (about 8–12 feet) from each other that the thread when strained measures 3 to 3·5 meters in length. (The thread can, of course, be shortened at pleasure, or replaced by one a few meters longer; but for class experiments the length stated appeared to be most practical.) The thread is fastened to Lamella I., and passed through the hole in Lamella II. towards the peg by which the correct tension is effected.

The white points on the (red) thread serve for facilitating the better observation of the curves, and, in order to render them more prominent, the black screen is placed behind the oscillating thread. The screen has on one side small holes, which are hinged in the hooks of the frame of Lamella I. ; when the screen is expanded, the foot of it is screwed firmly by a vice-pin to the opposite table.

Two (large-sized) chrome elements, the strength of which may be easily regulated, will serve best as electro-motors. In the present case two chrome elements were employed, each with two carbon plates, and a zinc plate, of 18 cm. in height, and 6 cm. in width, joined together one behind the other.

When newly charged, it was only necessary to dip the zinc plate from 1 to 1·5 cm. into the acid, in order to obtain the requisite strength of electric current. Both elements and the wire spirals of both magnets are, of course, united to form *one circuit.* As already observed, Lamella I. serves at the same time for interference, for which purpose an attachment screw is fixed at its reverse end, which is intended to be connected with the wire leading to the battery. To the attachment screw, which is in connexion with the mercury bowl, a wire is fastened, which is to be connected with the first attachment screw of the wire spirals, if the proportion of the numbers of the vibrations of the two Lamellæ are from 1:1 to 1:2 (accord, octave). If Lamella II. is to perform three or four vibrations in the same time in which Lamella I. makes only one vibration, then only one wire spindle of Lamella I. is set in motion by fastening the wire coming from the mercury bowl in the centre attachment screw. The current must always be so powerful that the interference, whose rough displacement is effected by raising and lowering the platinum pin, and the more minute displacement by turning the mercury bowl, can be so placed that the shutting off of the current is effected as quickly as possible, as upon this the purity of the figures chiefly depends.

The intensity of the vibrations on Lamella I. is regulated by raising and lowering the electro-magnet, which is kept in its position by a set screw, and on Lamella II. by screwing in and out the iron cores, and by the shifting the whole magnet.

With a powerful current the vibrations will be most regular when the magnet is entirely moved back, and the cores are screwed so closely to the Lamella that the contact just ceases.

Lamella I. has only one mark on which it always remains accurately adjusted. If it is required that it should oscillate uniformly with Lamella II. the latter is moved to the back, marked with the annotation $\frac{1}{1}\frac{\alpha}{2}\frac{1}{2}$ and upon the spot marked with × the weight is placed with the annotation $\frac{1}{4}$. If the weight is removed further from Lamella II. the oscillations of the two Lamellae are as 1: 2. If Lamella II. is moved to the mark indicated by $\frac{1}{3}\frac{\alpha}{4}\frac{1}{4}$ the oscillations of both Lamellæ are as 1:3 (S fifth of the octave).

By placing the weight with the annotation $\frac{1}{4}$ upon the spot indicated with × on Lamella I. the oscillations are as 1: 4 (double octaves).

The two smaller weights of different size serve for the more exact regulation in case great changes in the temperature, or simultaneous oscillation of either one or the other of the tables, should be of injurious influence.

Lamella II., as has been observed at the commencement, can be turned about an horizontal axis, so that its vibrations can be adjusted either rectangularly, parallel, or in any angle whatever to Lamella I. In case it is necessary to ease and screw fast the hexagonal nut, the key is added for the purpose. The small key is for unscrewing and fastening the screws which fix the Lamella.

724. Melde's Wave Apparatus, for showing the production of Chladni's figures of sound, according to the theory of Wheatstone.

(*See* Poggendorff's Annalen, Jubelband, p. 101, and the accompanying description.)   *Ferdinand Süss, Marburg.*

This apparatus has been more particularly described by Professor Melde, in Poggendorff's Annalen, Jubelband, page 101.

The present apparatus is distinguished by several differences from that described in the above-mentioned work, by which the manipulation is considerably facilitated, but before entering into this, it must be mentioned that in the construction of this apparatus it was kept in view that two systems of waves of equal length and equa intensity pass swiftly through a square plate.

The upper system of waves is formed of 33 rows of 17 pins each, which together make two wave lengths. The saddle upon which the pins are placed corresponds to a length of 1·5 of wave lengths; it has on two sides (above and below) wave-systems, which in so far differ from one another, as the one is like a system of two mountains with one valley in the middle, and the other like two valleys intersected by a mountain. On the longitudinal sides of this saddle rectangular zinc plates have been screwed, which are provided with divisions for adjusting the desired phases.

O denotes the centre position (middle between mountain and valley). The lines marked on the black board correspond to the waves formed by the pins. The longitudinal line serves as index upon which the lines of the saddle are indicated.

When the saddle has been adjusted in the desired position, the board is lifted by the handles fastened on the sides—as far as possible perpendicularly—

so high, that the wires running parallel under the board catch into the hooks, and so prevent the board from descending.

In order to let the board and the saddle down again, a slight pressure with the thumb on the pegs protruding from the board next to the handles, while holding the handles themselves firmly in their position, will be sufficient to prevent the too rapid, or sideways, sliding down of the board.

Finally, it is to be observed, that this apparatus can also very well replace Eisenrohr's Interference Apparatus in so far as two waves will be sufficient; it would only be necessary to add a number of bars, or if entire surfaces are desired, some saddles, the waves of which are of different and certain lengths.

Great care is required by placing the one part of the apparatus upon the other.

725. Wave Disc, by Professor J. Müller.
J. Wilhelm Albert, Frankfort-on-the-Maine.

The wave-disc by *Prof. Dr. Joh. Müller,* Freiburg (Breisgau), is an adaptation of the well-known stereoscopic disc (Phenàkistoscope, or Wunderscheibe), for demonstrating the wave undulation. The eight drawings are for the purpose of illustrating the water, rope, sound, and air waves, in covered as well as in open pipes.

See "Lehrbuch der Physik," VII. Edit. Vol. I., § 155).

726. Telephon, on Reis' system, for the reproduction of sounds by galvanism.
J. Wilhelm Albert, Frankfort-on-the-Maine.

The telephon is based on the experiments of Wertheim and others regarding galvanic sounds. *Philipp Reis,* at Friedrichsdorf, made use of these with a view of reproducing by means of galvanic action the musical sounds produced by singing (or by pipes, &c. played upon), by employing an elastic membrane and an interference apparatus constructed by him.

(*See* Jahresbericht des physikalischen Vereins zu Frankfurt a Main. Jahrgang 1860-61 ; also, "Müller's Lehrbuch der Physik, VII. edit. Vol. II., § 135.

727. Crank Apparatus, for showing the production of progressive waves in water, &c.
Prof. J. Joseph Oppel, Frankfort-on-the-Maine.

The rotating liquid molecules are represented by white wooden balls on a black background, and are all put into motion by a crank attached at the back.

The whole shows two wave lengths.

728. Cylinder Apparatus, for showing directly and comparatively, progressive and stationary waves of sound, and the essential difference between them.
Prof. J. Joseph Oppel, Frankfort-on-the-Maine.

Contains 1½ corresponding wave lengths on both cylinders. Other drawings likewise, for example, with different wave lengths (for illustrating the musical intervals, &c.), can be mounted on the cylinders, which are of a somewhat conical shape.

729. Two Wave Discs of paste-board for stroboscopic illustration of a progressive and a stationary (water) wave.
Prof. J. Joseph Oppel, Frankfort-on-the-Maine.

Would be best fastened to a rotatory apparatus such as are used for colour spindles, &c., and placed in front of a mirror. All waves, with the exception of a horizontal one, should be covered by a black screen.

730. Brass Tube, with Gas-burner, for Intonation. (Compare Poggendorff's Annalen, vol. 129, 1866.)

Albrecht, Tübingen.

731. Rotating Mirror, movable towards all sides.

Albrecht, Tübingen.

This mirror has not yet been described, but several specimens have already been executed by Mr. Albrecht. That position of the mirror in which its normal forms a moderate angle with the rotatory axis, is peculiarly adapted to change the reflected image of the sonorous flame into a beautiful elliptical crown.

731a. Barlow's Logograph, an instrument for recording pneumatic effects of speech, showing the consonant actions and the vibratory effects accompanying vowel sounds.

W. H. Barlow, F.R.S.

731b. Apparatus for Synthesis of Vowel Sounds.

Prof. Clifton, F.R.S.

732. Apparatus for the projection on the screen of the curves produced by the combination of rectangular vibrations.

Yeates & Sons.

787a. Electro-Diapasons, showing the composition of vibratory movements by producing fixed acoustic figures.

M. Mercadier, Paris.

IV.—INTERFERENCE.

733. Apparatus for demonstrating, by the aid of flames, the interference of two musical sounds.

Prof. Dr. R. A. Mees, Director of the Physical Laboratory of the University of Groningen.

The apparatus consists of two curved movable cross tubes, narrowed at one end; their narrow openings, being near each other, are tightly fixed in a longer tube, also ending in a narrow opening; this is placed close to a strong flame, and also a very small flame, which can be seen in a rotating mirror. A small movable burner for this small flame is attached to the apparatus. When a sound-wave proceeds from the opening of the instrument, the strong flame diminishes abruptly in length and begins to roar, while the small flame rapidly vibrates, its motion being visible in the rotating mirror. The open ends of the two curved tubes can be placed before the mouths of two unisonant organ pipes, or above two different segments of a vibrating plate. When only one of the pipes is sounded, the flames show the vibrations, but, when both pipes are sounded, there is no agitation of the flames, the two sound-waves counteracting each other. When the two openings are above segments of the plate, which are in the same phase of vibration, the flame is agitated, but, when they are above segments in the opposite phase, the flame remains at rest.

734. Apparatus, of simple character, for demonstrating by the aid of flames the interference of two musical sounds.
Prof. Dr. R. A. Mees, Director of the Physical Laboratory of the University of Groningen.

This apparatus is designed for the same purpose as that last described, but is of more simple construction.

735. Quincke's Interference Tube, to demonstrate by the action of a flame the diminution and increase of sound by interference.
Prof. Dr. R. A. Mees, Director of the Physical Laboratory of the University of Groningen.

This apparatus is furnished with a supplementary brass tube having a narrow opening. It can be used with the same flames as those provided for the instruments last described.

735a. Interference Apparatus, by Jamin.
M. J. Duboscq, Paris.

735b. Interference Apparatus.   *Prof. W. F. Barrett.*

This is a circular modification of the usual trombone apparatus. When used with a pitch pipe, the extinction of the sound is made evident to a large audience, the two resonant columns being rendered in opposite phases by adjustment of the movable circular tube.—Phil. Mag., Aug. 1874.

V.—ABSORPTION, REFLECTION, AND REFRACTION.

736. Apparatus for showing Approach caused by Vibration.   *Frederick Guthrie.*

A suspended card or mass of cotton wool, or an air ball floating in water, approaches a resonant fork ; and the latter, when free to move, is also urged towards neighbouring matter.

737. Apparatus for showing the Expansion of Gases by Sound.   *Frederick Guthrie.*

One prong of a tuning fork is enclosed, air-tight, in a glass tube provided with a capillary exit tube, in which water stands at a certain height. On bowing the free prong of the fork the water level is seen to fall about a quarter of an inch.

738. Apparatus for the Reflection of Sound by heated air and vapours.   *Prof. Tyndall, F.R.S.*

Sound of high pitch from a vibrating reed is passed through a long rectangular chamber, and caused to agitate a sensitive flame. Air, saturated with the vapour of a volatile liquid, is gently driven through six narrow openings into the chamber, at right angles to the direction of the sound. The atmosphere within the chamber is thus immediately rendered heterogeneous, and the sound waves being reflected, the agitated flame is rapidly stilled. The removal of the heterogeneous medium instantly restores the flame to its former agitation.

For the action of heated air, the rectangular chamber is turned upside down upon its support, and the heated air from six gas jets allowed to stream in through six narrow openings across the sound waves. The air thus rendered heterogeneous has the effect of immediately rendering the sensitive flame quiescent.—Phil. Trans., 1874.

739. Diagrams and Apparatus illustrating the reflection and refraction of sound-bearing waves, as exhibited to a class by means of a sensitive flame.　　　　　　　　　*Prof. W. F. Barrett.*

The arrangement shows a suitable source of sound, a good form of gas pillar for yielding a tranquil flow of gas to the burner, and the best shape and size of steatite burner for the flame, together with a useful form of gas-holder for giving steady pressure of gas larger than that usually given by the street mains.

These experiments were first shown by the exhibitor in connexion with a paper read by him before the Royal Dublin Society in Jan. 1868. See Quarterly Journal of Science, Jan. 1870.

VI.—RESONATORS.

740. Resonator of adjustable pitch.　　　　*Lord Rayleigh.*

Resonator, whose pitch can be rapidly adjusted to the various notes of a harmonic scale—A$\flat$, a$\flat$, e$\flat$, a'$\flat$, e''. The smallest hole is made first and adjusted until the resonator responds a$\flat$. The second hole is then made and adjusted, until, *with both holes open*, the note is e$\flat$. Similarly with three holes the note is a'$\flat$, and with four holes e''. When the note A$\flat$ is sounded on the piano or harmonium, and the resonator is suitably fingered, the various overtones are heard with great distinctness, and the phenomenon is more marked than usual in consequence of the contrast afforded by the rapid transition.

740a. Sonorous Tubes, by Dulong.
Polytechnic School, Paris.

741. Six Resonators of glazed cardboard, for the sounds:
c' (256 vibrations), e' (320 vibrations), g' (384 vibrations),
c'' (512 vibrations), e'' (640 vibrations), g'' (768 vibrations).
Gustav Schubring, Erfurt.

VII.—MUSICAL INSTRUMENTS.

742. Enharmonic Harmonium, with generalised keyboard; 84 keys in each octave; compass, 4½ octaves.
R. H. M. Bosanquet.

This instrument is tuned according to the division of the octave into 53 equal intervals, a system sensibly identical with a system of perfect fifths.

References.—Proceedings Royal Soc. XXIII. 390.
Philosophical Magazine, XLVIII. 507, L. 164.
Proceedings of the Musical Association, 1874–5.
Novello's Dictionary of Musical Terms, Article Temperament.
Ellis's Helmholtz, pp. 692–699.
Elementary Treatise on Temperament; Macmillan, 1876.

742a. General Thompson's Enharmonic Organ. Built by Messrs. Robson, London, 1856.   *John Curwen.*

It is an improvement upon a similar instrument he exhibited in Hyde Park, in 1851, which also was an improvement on the first organ built for him in 1834. It is capable of being played in 21 keys, with their minors of the same tonic. The organ is fully described in General Thompson's pamphlet, " The prin-" ciples and practice of just intonation with a view to the abolition of tempera-" ment." Effingham Wilson, Royal Exchange. On the middle finger-board the keys of C, G, D, A, E, B, with their minors, can be played perfectly; on the lowest finger-board there can be played, besides the keys of C and G, those of F, B♭, acute E♭, acute A♭, and D♭. The fingering is mainly the same as in other instruments. The red shows the principal key tone of the board. The black shows the fourth and sixth of that scale, as well as the grave second, with which they make a true chord. The white shows the fifth and seventh of the scale, as well as the acute second of the same scale, with which they accord truly. The small oblong quarrils and the flutals (finger keys of a flute) are always a komma shriller or deeper than the digital in which they are embedded. The buttons are always a diatonic semitone deeper or shriller than the adjacent digital. The serrated edges are for the blind.

742b. Harmonium, with double key-board.
                                                M. Guéroult, Paris.

This instrument, of which the two key-boards are tuned in fifths, has a *comma* at $\frac{81}{80}$ interval one from the other, which serves to verify the theories of musicians and natural philosophers upon the melodic or harmonic gamut.

743. Patent Double Trumpet, called Bi-Clairon.
                                          Franz Hirschberg, Breslau.

The double trumpet (Bi-Clairon), constructed by the exhibitor, is described by him equally as an interesting and practical invention. As the instrument consists of two bell-mouths of different measure and construction, into which the air can be admitted or from which excluded at pleasure through the valve, it has been rendered possible to produce by the same two kinds of sounds, which, according to their sonorous colour, are equal to at one time the bugle horn, at another to the piston (or, also, to the cornet and the trumpet). The instrument is particularly adapted for being used in concerts, inasmuch as by the different sonorous colours more, " light and shade," consequently more variation, is imparted to the execution, and, therefore, no band of musicians should be without it. In a weak orchestral band, in which both bugle horn and piston (respectively cornet and trumpet) are not always represented, this instrument supplies the place of both. Its pitch is high C with B, and A low, and is so constructed that the smaller bell-mouth can be screwed off, in which case the instrument can be used as a common bugle horn (or cornet).

744. Model of the Action of Grand Pianofortes.
                                                Messrs. Erard.

745. Molineux's Patent Self-acting Escapement and Check Repeater Action for upright Pianofortes. It combines extreme simplicity, a firm and elastic touch, with freedom from friction and great durability. *Thomas Molineux.*

746. The first of the now generally adopted **obliquely strung upright Pianofortes,** patented Robert Wornum, of the firm of Wilkinson & Wornum, in 1811.

Messrs. Wornum & Sons.

The large factory in Oxford Street, in which this instrument was made, is shown by an engraving within the lid. This factory was burnt down in October 1812, and the partnership was then dissolved. In the following year Robert Wornum made the first successful "Cottage" pianoforte, with *vertical* stringing, to which he gave the name of "The Harmonic Pianoforte." He accomplished this by discarding entirely the use of brass wires, and adopting the closely-spun copper-covered strings in their stead.

747. Model of the Elastic Tie Action for the Piccolo Pianoforte, invented and patented by Robert Wornum in 1826.

Messrs. Wornum & Sons.

The principles of this mechanism are very generally adopted in France and Germany, as well as in England.

748. Model of Robert Wornum's method of returning the hammer, in his **down-striking Action** for Horizontal Grand and Square Pianofortes, patented in 1842.

Messrs. Wornum & Sons.

This action greatly economises the cost of manufacture. The usual actions are up-striking.

749. Model of Alfred Nicholson Wornum's new **Patent Action** for Grand Pianofortes (1875), in which the heads of the hammers are reversed, and now face the wrest plank, this being effected by an entirely new method. *Messrs. Wornum & Sons.*

By this invention longer strings may be used, relatively to the external dimensions of the case, than in an instrument of the ordinary construction.

750. Model of the action of **Ancient Great Hydraulic Organ,** from Mr. Chappell's description of the instrument.

Dr. Stone.

750a. Model of a Keyboard for an Organ.

Colin Broun.

751. Marimba or Balafo, from South-eastern Africa. Modern. Given by Captain J. Stuart.

South Kensington Museum.

The instrument has twelve slabs of a sonorous wood, beneath which are fastened, by means of a dark-coloured cement, twelve gourds, to increase the sound. In each gourd are two holes, one of which is at the top, and the

other at the side. The latter is covered with a delicate film, to promote
the sonorousness. Several African travellers have noticed this curious
acoustic contrivance. „Du Chaillu says that the film consists of the skin of a
spider; Livingstone mentions spiders' web being applied to instruments of
this kind used by certain native tribes in Southern Africa. The marimba is
a favourite instrument of the negroes as well as of the Kafirs.

752. Glass Harmonica. Modern. Made by E. Pohl, in
Bohemia.   *South Kensington Museum.*

The glass harmonica consists of a series of glass bells, which are affixed in
regular order to an iron spindle lying horizontally in a case, and which by
simple machinery are set in motion by the feet. The sound is produced by
the performer moistening his fingers and pressing them on the bells while
these are rotating.

753. Sol-Fa Harmonicon, invented by Miss Glover. Intended as an assistance in learning singing, and the theory of
music.   *South Kensington Museum.*

754. Organ Pipes, a selection in illustration of their manufacture, showing the middle C pipe of each stop.   *H. Speechly.*

755. Chromatic Harmonium, peculiarly constructed key-
board, " showing the twenty-four progressions. The common
method is seen at the back of the instrument in connexion with the
keys."   *Mrs. Read.*

756. Chromatic Pianoforte, peculiarly constructed key-
board, in which the keys are distinguished by different colours.
Intended to facilitate the playing in the different major and minor
keys.   *Mrs. Read.*

757. Models of several **Ancient Egyptian Pipes,** the
originals of which are in the Egyptian Museum at Turin or in the
British Museum. Those from Turin are copied in brass, and
those from the British Museum in cane.
                                      W. Chappell, F.S.A.

The original pipes were found in Egyptian tombs, some examples being as
old as the fourth or fifth dynasty of Egypt. They were played upon by
means of a cut, or split, piece of reed, or of straw, inserted in the end, as was
usual with ancient shepherds' pipes, and much in the manner of the modern
hautboy, or bagpipe, reeds. Parts of the ancient reed or straw remain within
one of the pipes in the British Museum, and another at Turin. Usually a
fresh long piece of reed or straw was laid in the tomb by the side of the pipe,
and it may be assumed that the object was to supply the dead man with a
stock of those perishable inciters of tone, in order that he might play continuously upon his pipe when he awoke. Examples of the straws or reeds, so
deposited, are included in the Museum at Leyden and in the British Museum.

The pipes selected for copying were those which varied in length and in the
number of finger holes, so as to obtain varieties of pitch, and varieties of the
prehistoric scales. Through the kind assistance of Dr. W. H. Stone, himself
an accomplished player upon reed instruments, the following have been
ascertained.

1. Pipe from Turin 14⅛ inches long, with six finger holes. Scale—

This is the scale of E major, but only extending to six notes. It lacks the addition of D ♯ and E to complete the octave. The fundamental note, or tone of the whole pipe, was not obtained.

2. The longest pipe from Turin, 23⅜ inches, but with only three holes. The scale is—

This forms a Diatessarōn, or Fourth, from B-flat to E-flat, therefore one note below our C D E F in pitch. In order to obtain the notes from this pipe, it was found necessary to lower the reed into the pipe, as in the drone of a bagpipe. It extends three inches down the tube.

3. A pipe in the British Museum, copied in cane. It has four holes, and is 8¾ inches in length. The scale is a Diatessarōn, or Fourth, exactly one note above the last, but with an F sharp added to it at the top. Possibly this F sharp may have been intended for a G to make a Fifth; or as F sharp to lead into the key of G upon a treble pipe.

4. Also from the British Museum, copied in cane. It has four finger holes, and the entire length is 10½ inches. This pipe has a hole bored through it near the mouth end. It would have been absolutely impossible to produce sound from the pipe if this hole had been left open. It may therefore be assumed that it was once covered with thin bladder, such as that of a fish. The intention of placing bladder there would have been to give a tremulousness to the tone of the pipe so as to assimilate it to the human voice. The old English pipe called the Recorder, referred to by Shakespeare in Hamlet, and in A Midsummer Night's Dream was of the same kind, and differed in no other respect from the English flute, both being blown at the end. It is curious to find that such an appreciation of the difference of tone that might be produced has been anticipated by the ancient Egyptians. A film of gutta percha is now tied over the hole. The scale is—

This is a peculiar scale, a pentaphōnic major scale, such as is popularly entitled the Scotch scale. It is suitable for playing tunes upon the black keys of the pianoforte.

It is remarkable that no one of these pipes gives any indication of a minor key, and they seem therefore to be older than the introduction of the minor scale, inherited by us from the Greeks, by the junction of two tetrachords. For this an astronomical or theological reason is assigned, that, as there were but seven planets (according to ancient supposition), and seven days in the week, or quarter of the lunar month, &c., so there should be but seven notes instead of eight in a musical octave. Therefore a scale of two tetrachords, each of four notes, was reduced to seven, through uniting them by one note common to both. Hence the intervals of our B C̆ D E—E F G A.

757a. Indian Vina, with resonating gourds.

W. Chappell, F.S.A.

757b. Patent Comma Trumpet, producing approximately correct intonation, by means of a valve, which raises the pitch the interval of a comma. *H. Bassett.*

757c. Marimba from **Angola,** on the principle of the musical box. *H. Bassett.*

757d. Wooden Trumpet from **Angola,** made from the root of a tree. *H. Bassett.*

757f. Double Bass, with heavily covered fourth string, going down to C C C. *Dr. Stone.*

758. Series of Acoustic Models.   *M. Lancelot, Paris.*

759. Savart Violin.

Conservatoire des Arts et Métiers, Paris.

759a. A new Orchestral Musical Bass Instrument, with concertina fingering, full tone, and expressive for part and solo performance. *S. F. Pichler.*

759b. Acoustical Instrument, illustrating Harmony and Discord. *S. F. Pichler.*

759c. Violin fitted with Tension Bars.   *Dr. Stone.*

759d. Viol d'amore, illustrating the principle of consonating springs. *Dr. Stone.*

759e. Tenor Bassoon, or **Alto Fagotto.**   *Dr. Stone.*

VIII.—SPECIAL COLLECTIONS.

DETAILED LIST OF INSTRUMENTS MANUFACTURED BY
M. LANCELOT OF PARIS.

Exhibited by Auguste Bel & Co., London.

760. Eight Pieces of Wood, giving a scale.

761. Mouth of a Flute Pipe, showing the inside of the air-chamber.

762. Wertheim's Apparatus.

763. Pipe with glass side, and membrane.

764. Pipe carrying a slide, enabling holes of different sizes and shapes to be opened in the same situation.

765. Four Pipes, all containing the same mass of air, one cylindrical, one cubical, one tetrahedral, one spherical.

766. Three Pipes containing the same mass of air, one prismatic, one in the shape of a right cone, one in that of an inverted cone.

767. Mons. Bourbouze's Pipe with glass side, bearing three membranes, with mirror, one placed at the node of the first sound, the two others at the two nodes of the second sound.

768. Six Plates, five of different woods, one of brass.

769. Two Flat Rods of brass for transverse vibrations. Support for fixing these.

770. Sonometer on Mons. Barbereau's principle.

771. Circular Membranes, with varied tensions.

772. Two Tuning Forks on resonance box, to give four beats a second.

773. Two Tuning Forks, mounted between the poles of electromagnets, with contact-breaker.

774. Duhamel's Vibroscope.

775. Apparatus for transmission of vibrations through liquids.

776. Instrument for showing the quality of vowel-sounds.

Various Scientific Instruments illustrating the Phenomena of Sound, invented and made by the late John Henry Griesbach.

777. Monochord, with apparatus for printing and registering the vibrations of strings, with a view to ascertaining the number of vibrations per second.

778. Phonometer, by means of which the equally tempered 12 fixed sounds in the octave may be produced.

779. Monochord, giving the positions of the Nodal Points in vibrating strings.

780. Monochord, which affords the means of measuring the 200th part of an inch, with a view to ascertaining the number of vibrations given by that length of string.

781. Apparatus for producing musical sounds, mainly consisting of notched wheels of different diameters, which, being set in motion at a given speed, and duly prepared pieces of cardboard brought in contact with their teeth, produce the notes of the common chord; the number of notches in the wheels corresponding with that of the vibrations required to produce those notes.

782. Apparatus for showing the relative positions of the vibrations of two strings or tubes under the operation of altering the ratios; the strings first tuned to coincidence as unisons, the ratios then altered by lowering the pitch of one of the strings, as from 81 to 80, 82 to 80, or any other numbers within the scale of the apparatus.

783. Set of Flute Pipes, with bellows attached, some of which have too high sounds to be heard singly, but which together give the resultant tone.

784. Large Set of Coloured Diagrams for illustrating lectures on sound.

784a. Collection of Tuning Forks.
One tuned to Handel pitch.
Two in a box.
Two normal diapason.
One tuned to Big Ben.
One C. 528.

784b. Paper printed by J. H. Griesbach to Handel's pitch, and one to the true diapason.

-----

785. Collection of Acoustic Apparatus.
George Appunn and Sons, Hanau.

785a. Three Acoustic Wind-Chest Tables. These three tables are required for placing all the following apparatus :—
George Appunn and Sons.

1 *Overtone Apparatus,* consisting of 64 lingual tones, the first 64 part tones of the fundamental or key tone (primary sound) C $\frac{-2}{}$ = 32 vibrations in a second with reed pipes.

2. *The same Apparatus*, consisting of 32 lingual tones, the 32 part tones of the fundamental tone C^{-2} = 32 vibrations.

On the overtone apparatus there result quite plainly the corresponding difference-tones of all phases that may be chosen at pleasure; also, up to a certain point, the corresponding resultant tones. By means of this overtone apparatus not only the waves and the quality of the sound can be demonstrated, but likewise the different degrees of harmony of the various musical proportions (rhythm), and of poly-accords in different keys and transpositions. The latter, in particular, in combination with the *tone limit apparatus* for low tones, mentioned hereafter.

3. *Tonometer*, consisting of 65 lingual tones; every subsequent tone higher by 4 vibrations (waves) than the previous one, from $\tilde{c}$ = 256 to $\overset{=}{c}$ = 512 vibrations in a second; with reed pipes.

4. *Tonometer*, consisting of 33 tones; each succeeding tone being 4 vibrations higher than the preceding one, from c = 128 to $\tilde{c}$ = 256 vibrations in the second.

(*Note.* By the number of vibrations given, double vibrations (waves) are always to be understood.)

5. *Tone-limit Apparatus*, for low (pitch) tones; consisting of 57 lingual tones, with reed pipes, from c = 128 vibrations downwards to 8 vibrations in a second, namely, from c^0 = 128 to C^{-1} = 64, every subsequent tone in the descending scale lower by 4 vibrations than the next preceding tones from C^{-1} = 64 to C^{-2} = 32, each two vibrations lower, and, lastly, from C^{-2} = 32 vibrations downwards, each one vibration lower as far as 8 vibrations.

6. *Tone-limit Apparatus*, for high (pitch) tones; consisting of 31 small tuning-forks, the diatonic Durton scale c; d; c; f; g; a; h; c;

$$24 : 27 : 30 : 32 : 36 . 40 : 45 : 48.$$

through $4\frac{1}{4}$ octaves, namely, from c^{iv} = 2048 vibrations (double) to e^{viii} = 40,960 double vibrations, with two bows.

In order to be able to observe, and to perceive better and more distinctly with the tone-limit apparatus for high tones (tuning-fork apparatus of 31 tuning forks), the ascending scale up to the highest pitch, it will be advisable to intonate with two bows the scales in octaves, one after the other, and thus to sound every tone with its octave simultaneously or one directly after the other.

7. *Two large Gedact Pipes* (stopped mouth pipes), whose pitch can be lowered an octave, with small wind chest and wind trunk, for illustrating the interference, waves, resultant tones, &c.

   a. Two smaller Gedact Pipes, with small wind chest and wind trunk, for the same purpose.

   b. Very powerful-sounding *Lingual (reed) Pipe*, accompanied by a large number of overtones, with bell-mouth : C^{-2} = 32 vibrations in a second.

   c. Twenty-nine Resonators to the same, from the 4th to the 32nd overtone. (The resonators are conical, and made of zinc.)

   d. Reed pipe, with bell-mouth, C^{-1} = 64 vibrations.

   e. Four open and four stopped *Labral Pipes*, with small wind chest and valves, producing the C major accord : $\overset{-}{c} : \overset{-}{e} : \overset{-}{g} : \overset{=}{c}$.

$$8 : 10 : 12 : 16$$

   f. Accord Reed Pipe, producing the C major accord $c : e : g : \overset{-}{c} : \overset{-}{e} : \overset{-}{g} : \overset{=}{c}$ (The two apparatus *e* and *f* are for demonstrating the quality of sound.)

786. Stand of Apparatus illustrating the progress of Œolian Principles.   *J. Baillie Hamilton.*

    1. Primitive Æolian types.
        The rod.
        The bow.
        The harp.
    2. Modern Æolian harp.
    3. Wind concentrated upon a string, and applied to its entire length. By Professor Robinson.
    4. Wind applied to a portion of the string, as by Wheatstone, Green, Isoard, &c.
    5. Further modifications of the same.
    6. Use of a free-reed. By Pape. The connexion with the string being effected by a silk thread.
    7. Julian's mode. A metal string flattened into a tongue at the part where the wind impinges.
    8. Farmer's mode. A reed-tongue substituted for the flattened portion.
    9. Farmer and Hamilton's mode. A rigid connexion between reed-tongue and string allows the reed to be used as in reed-organs.

SUBSEQUENT INVESTIGATIONS BY HAMILTON IN CONJUNCTION
WITH HERMANN SMITH.

    10. Improved string-organ note, in which a sympathetic resistance is offered to the string, the vibration transmitted to the soundboard, and constancy of pitch preserved by a spring-bow. The reed tone is modified by a tube, and the connexion effected through the tube by a "purse," the latter suggested by Hermann Smith.
    11. Further modification by Hamilton. The necessity for the "purse" abolished by setting both reed and string inside the register. The economy of space effected by Hamilton's spiral spring, and the use of a short metal spring-bow.
    12. Application of these improvements for use in a wind-viol. Also a conical string, invented by Hamilton, for obviating the following difficulties peculiar to reeds and strings in combination.
      a. The difficulty caused by the string breaking into segments, owing to the constraint on the reed, and the scarcity of notes obtainable.
      b. By the irregularity of intervals which, in a cylindrical string, are crowded together near the reed, and are far apart when remote from it.
      c. By the irregularity of tone in different portions of the string's length. When an ordinary string is used in short lengths the reed's motion is confined, and the tone is consequently

pure; as the string lengthens, the tone becomes loose and coarse.

   d. By the reducing of the string's bulk by stretching when tuned, as the reed remains constant the intervals would be changed when the string is thus attenuated.

The conical string meets these difficulties thus,—

   a. The bulk of the string lies in the part used for playing upon, and thus no intervals are wasted.

   b. The bulk is increased where the intervals would otherwise be wider.

   c. The increasing bulk controls the reed equally at all points.

   d. If tuned from the smaller end the string does not become attenuated, as more bulk is brought over the other bridge.

13. Apparatus for studying the relative amount of tone contributed through the string.

   a. By the action of a soundboard and bridge.

   b. By the reciprocal action of a spring-bow.

The spring-bow can also be rendered rigid, and the tone is then due merely to the constraining effect of the string on the reed's motion.

14. Æolian effects produced by percussion. For investigating the causes of the Æolian tone.

15. Apparatus investigating the most effective modes of—

   a. Constraint upon a reed.

   b. Transmission to a soundboard.

An intermediate mass is here used for transmission.

16. Apparatus showing how far quality of tone can be now affected by reaction from the soundboard. By placing a weight on different parts of the soundboard any quality of tone can be produced.

17. Further modifications for reducing these principles to practical use by altering the relation of levers and spring resistance, and substituting for the intermediate mass the structure of the register, which, as in No. 18, is itself the conductor to the soundboard.

18. Register embodying the foregoing improvements into a form for practical use, and illustrating the different forms of constraint applicable to reeds.

19. A new form of vibrator applicable to all solid bodies as well as to columns of air. Invented by James Baillie Hamilton, April 1876.

20, 21, 22. Apparatus for studying the laws of strings combined with reeds.

IX.—EDUCATIONAL.

787. Apparatus for illustrating lectures. *Auguste Bel & Co.*

788. Graphical Representations of Musical Scales, on paste-board.   *Gustav Schubring, Erfurt.*

The logarithms of the numbers of vibrations and their differences where first recognised by *Euler* (Leonhard Euler, tentamen novae theoriae musicae, 1739). These logarithms were at a later period used by *Opelt* ("Natur der "Musik," 1834, "Theorie der Musik") for the graphical representation of musical scales. *Herbart*, likewise, has employed them in his psychological speculations, and, lastly, *Prof. Drobisch* (Leipsic) has extensively applied them in his calculations of the musical intervals (Abhandlungen der fürstlich Jablonowskischen Gesellschaft, 1846).

The exhibited plates are especially intended to illustrate the musical scale calculations of Prof. Helmholtz; they agree in their annotations with those employed in the older edition of the "Lehre von der Tonempfindungen" (Part III., sections 15 and 16). A reconstruction of these plates, according to the annotations used in the new edition, is in course of execution.

789. Model for the Higher Tones.
Gustav Schubring, Erfurt.

Prof. Mach (Prague) has made use of the high-tone scales, drawn according to logarithmic scales, in order to produce a model for the high-scale tones. By means of the same not only can the higher tones of any sound be ascertained directly, but the higher tones of several sounds can also be compared with one another, and the theory of the musical consonance and dissonance advanced by Prof. Helmholtz can be demonstrated in the most excellent manner.

Prof. Mach's model had a length of three octaves, and contained the high tones according to the tempered free-balancing scale.

The model exhibited is more than four octaves long, and contains the high tones in the pure (natural) key.

789a. Model, similar to the two preceding, but not on paste-board.   *Gustav Schubring, Erfurt.*

790. Sir Charles Wheatstone's Mechanical Illustration of the Vibration of Strings or Rods.
The Council of King's College, London.

SECTION 7.—LIGHT.

West Gallery, Upper Floor, Room

DETERMINATION—VELOCITY.

791. Original Apparatus, by M. Fizeau, for measuring the velocity of light. *Polytechnic School, Paris.*

792. Apparatus, by M. Foucault, for measuring the velocity of light. *Polytechnic School, Paris.*

I.—DISTRIBUTORS.

a. Lenses.

794. Burning Glass (German), made probably by Count von Tschirnhausen in the 18th century. The property of Prince Pless Furstenstein. *The Breslau Committee.*

798. Early Stereoscope, made by the late Sir David Brewster, the inventor of this instrument. *John Maclauchlan.*

798a. Early Stereoscopic Pictures, prior to the application of photography. From the collection of the late Sir C. Wheatstone. *Robert Sabine.*

798b. Early Stereoscopic Pictures. Daguerreotypes of 1, Biot; 2, Bequerel ; 3, Foucault, from the collection of the late Sir C. Wheatstone. *Robert Sabine.*

798c. Seven Earliest designs for the Stereoscope printed in black and white. From the collection of the late Sir C. Wheatstone. *Robert Sabine.*

798d. Early Stereoscopic Pictures. Two Daguerreotypes of Faraday, from the collection of the late Sir C. Wheatstone. *Robert Sabine.*

799. Double Opera Glass. An early example, probably made in Holland about 1700. *South Kensington Museum.*

801. Binocular Field Glasses, Nos. 17, 29, 37. *Voigtländer and Son (Chevalier von Voigtländer), Brunswick.*

804. Telescope, No. 14.

Voigtländer and Son (Chevalier von Voigtländer), Brunswich.

807. Iceland Spar Ball, for showing double axis.

A. Hilger.

810. Series of Metrical Glasses. The dioptric unit is a lens of one metre focus ; the lens 0·50 to two metres focus is a semi-dioptric value of the unit ; the lens 2 to 0·50 focus is a dioptric value of double the unit. The same rule applies to all the other lenses in the collection.   *M. Crétès, Paris.*

810a. Globe made of Spar.   *M. Lutz, Paris.*

811. Early form of Stereoscope.
The Council of King's College, London.

812. Early form of Stereoscope.
The Council of King's College, London.

812a. Polistereoscope.—Apparatus which serves as tele-stereoscope, pseudoscope, iconoscope, &c., &c.

Augustus Righi, Professor of Natural Philosophy, Royal Technical Institute, Bologna (Italy).

This apparatus consists of two plane mirrors, one of which (on the left in the figures) can turn about an horizontal and a vertical axis ; the other mirror, besides these movements, can be fixed at different distances from the former. The eyes must be applied at two cylindrical tubes fixed to a diaphragm, which can take different positions. One of the eyes sees directly the objects, while the other sees the same object but apparently in a different position. This virtual position can be determined by forming the image of the eye, given by the left mirror, and afterwards the image of the point so determined in relation to the other mirror. If objects not too near are observed the illusion succeeds equally, though the image in the eye which sees by reflection is smaller than the other. According to the inclination which is given to the mirrors, it is possible to make any determinate point of the observed objects appear in the true position.

Fig. 1.—If the apparatus be placed as in Figure 1, it produces the effect of the telestereoscope.

Fig. 2.—Placed as in Figure 2, it acts as a pseudoscope. According to the distance between the mirror, diminution or augmentation of relief can be obtained, together with the inversion of relief. Some curious effects (which cannot be obtained with a Wheatstone's pseudoscope) are observed by looking at rotating geometrical solids, constructed with metallic wire, or by looking at these solids while the observer moves round them.

Fig. 3.—With the apparatus placed as in Figure 3, the effects of an iconoscope are obtained. A very narrow mirror is substituted for the large mirror.

Fig. 4.—In Fig. 4 the apparatus is placed so that the eyes see the objects as if they were in a same plane perpendicular to the right line which joins the eyes ; the relief in objects made with vertical wires then disappears. If the diaphragm which bears the two tubes is kept fixed, and the instrument turned slowly round the left tube, very curious apparent motions occur in the objects under observation.

For the mathematical theory of this apparatus, *see* the "Nuovo Cimento," 2[d] ser. t. xiv.

813. Jewel Lens (Ruby) of $\frac{1}{60}$ inch focus. Made by Andrew Pritchard, at the suggestion of Sir David Brewster. (*See* Brewster's "Optics," 1831, p. 337.) *John Spiller, F.C.S.*

813a. Vertical Apparatus for Projections.
M. J. Duboscq, Paris.

813b. Projecting Apparatus, for all phenomena of double refraction and polarization. *M. J. Duboscq, Paris.*

813c. Support, with Reflector, by Fresnel.
M. J. Duboscq, Paris.

b. LANTERNS, CAMERAS, &c.

814. Oxy-hydrogen Lantern, of new form, suitable for lecturers. *C. J. Woodward.*

The lantern is mounted on a "Willis's stool," so that supports of various kinds may readily be attached. The body of the lantern is swung between two uprights, and can be clamped so as to send a beam of light at any angle ; this, combined with a rotatory motion in a horizontal plane, enables the lecturer to direct a beam in any required direction. A rod carries lenses and a mirror when it is required to throw the light vertically upwards on an object, as, *e.g.*, for cohesion figures.

816. Camera Obscura. An early example, said to have belonged to Sir Joshua Reynolds. *South Kensington Museum.*

This camera when closed has the form of a large folio leather-bound book. It is recorded to have been given by Sir Joshua Reynolds to Lady Yates, by whose great grand-daughter, Mrs. J. R. Harrison, it was, in May 1875, presented to the Museum.

816a. Camera, by Colonel Laussedat. *M. Lutz, Paris.*

816d. Camera Lucida, invented and used by Dr. W. H. Wollaston. *G. H. Wollaston.*

816g. Camera Lucida, with slight magnifying power.
A. Nachet, Paris.

816h. Camera for Landscapes, by M. Govi.
A. Nachet, Paris.

816i. Camera Lucida, invented and used by Dr. W. H. Wollaston. *G. H. Wollaston.*

816j. Wollaston's original periscopic Camera Lucida.
See Tilloch, xxvii. (1807), p. 343 ; Phil. Trans., 1812, p. 370.
Wollaston Collection, Cavendish Laboratory, Cambridge.

816k. Wollaston's Camera Lucida adapted to Telescope.
Wollaston Collection, Cavendish Laboratory, Cambridge.

817. Improved Electric Lamp and Lantern for Lecturer's use. *John Browning.*

The lamp is automatic, the carbon poles being drawn asunder in proportion to the strength of the battery power used; this is effected by drawing iron rods into a hollow coil of insulated copper wire. The lantern has two nozzles, one intended for exhibiting screen experiments in spectrum analysis, polarisation, refraction, reflection, diffraction, &c., the other for exhibiting diagrams on the same screen, without altering the arrangement of the apparatus for physical experiments.

818. Lithographs for the Stereoscope, from drawings by J. Müller, Hessemer, Oppel, Nell.

J. Wilhelm Albert, Frankfort-on-Maine.

The coloured drawing marked with A (upon grey cardboard) is the original made by the late Prof. *Müller.* Images 2, 3, and 4 serve for a stereoscope without glass. The other images refer to stereometrical, astronomical, and optical subjects (colour combinations). Some images appear, by a slight change of position, in low or high relief.

819. Edelmann's Spectral Lamp, for the projection of spectra, with printed description. *M. Ph. Edelmann, Munich.*

821. Duboscq's Lantern. To be used in connexion with the following apparatus :—

1. Top with illuminating lens ; is to be employed with spectral slit and polariscope.

2. Spectral slit ; can be regulated by means of a fine screw.

3. Stand with convex lens; serves for the projection of the rays in spectrum experiments.

4. Two hollow prisms.

5. Prism plate for two prisms, arranged for being turned and put higher or lower.

6. Stand with holder for a prism ; can be regulated.

7. Polariscope.

8. Lens system, with four-inch illuminating lenses and achromatic objective; serves for the projection of photographic and other images of about 3 inches in diameter.

9. Microscope. This can be screwed to the end of the preceding system of lenses, after the objective has been removed.

10. Regulator for producing electric light.

11. Hydro-oxygen gas lamp.

A. Krüss, Hamburg.

Apparatus for the use of Lecturers, to show sketches, drawings of instruments, anatomical preparations, &c., by means of hydro-oxygen illumination, arranged for the projection of opaque objects. *A. Krüss, Hamburg.*

821a. Photogenic Lantern. *M. J. Duboscq, Paris.*

822. Double demonstrating Oxy-hydrogen Lantern, with triple condensers, consisting of two 10-inch plano-convex lenses, and a 7-inch meniscus, next to the source of light. The last lens is made to slide backward and forward between guides, so as to increase or diminish the cone of rays, and enable large or small diagrams or pictures to be exhibited without material distortion. *Dr. Stone.*

823. Magnesium Lamp, provided with brass cylinders and reflector. *A. Herbst, Berlin.*

824. Brewster's Patent Kaleidoscope (with case). The original form of the instrument made by Bate, of London, in the year 1815. (Position of the reflectors capable of adjustment, non-central eye piece, and rotating terminal disc, or box containing the coloured glass.) *John Spiller, F.C.S.*

II.—SELECTORS.

a. SPECTROSCOPES, PRISMS, &c.

825. Photographic and **Spectro-photographic Specimens** and **Apparatus** of Sir John Herschel and Sir William Herschel. *Prof. A. S. Herschel.*

1. Original fragments and complete photographs on glass with chloride of silver of the forty foot telescope at Slough. Produced in 1839 by Sir John Herschel, as a new modification of the process of Daguerre. Paper wrapper of the specimens inscribed in autograph by Sir John Herschel with the above description of the plates.

2. Prismatic apparatus designed and used in researches on the photographic action of the different rays of the spectrum, by Sir John Herschel, Slough, 1839. Original description, and notes of experiments with the instrument extracted from MS. journal. Specimens of photographed spectra obtained with the instrument by Sir John Herschel, at Collingwood, in 1859.

3. Heliostatic mirror (used by fixing outside to aperture in a window shutter, and turning screws by hand inside to direct the sun's rays horizontally or in a required direction). Glass prism to receive and bend downwards the reflected ray to a table on which thermometers were exposed to action of the different rays of its spectrum. Constructed and used (with other apparatus not preserved) by Sir William Herschel, in experiments on thermal radiation in the solar spectrum, described in the Philosophical Transactions, 1800–1801.

A plate of light blue cobalt glass, mounted on cardboard diaphragm pierced with an eye-hole; used with the prismatic photographing apparatus to examine test papers submitted to the solar spectrum (before sensitizing), and to mark with pencil on the test paper the exact position of a certain yellow colour of the spectrum. When the paper had been thus fixed and marked, the sensitive solution to be tested, if not already present in the paper, was applied to its exposed surface with a brush, and the time of exposure and intensity of the direct sunlight was at the same time recorded. · Two small card-leaves have, for the purpose of examination, been attached to the cardboard diaphragm, by closing which upon the glass, the "fiducial" yellow ray transmitted by blue cobalt glass will be observed with the accompanying eyepiece of a small pocket spectroscope placed with the plate. Through the narrow slit left between them the selective absorption of the glass can easily be distinguished if white light is examined through it, and spectroscopically analysed by means of the dispersion of the prisms.

825a. Fluorescent Eyepiece, by M. Soret, for adaptation to the **Spectroscope.**

Geneva Association for the Construction of Scientific Instruments.

It consists of a plate made of a fluorescent and transparent substance (uranium glass or different liquids contained between two glass plates), which is placed in the focus of the objective of the spectroscope. The ultra-purple spectrum projected upon this plate becomes visible, and is observed through an eyepiece which is movable upon the axis of the observing telescope. A very intense light is necessary. (For description of this apparatus, *see* Poggendorff's Annalen, 1274. Jubelband, p. 407 ; Archives des Sciences physiques et naturelles, 1874, vol. 41, p. 338 ; and 1875, vol. 54, p. 255.)

825a. Combination of Three Prisms of different dispersive power, giving by one combination deviation without dispersion, and by another dispersion without deviation. Formerly the property of Dr. Priestley.   *Mrs. Parkes.*

825b. Apparatus used by **Sir C. Wheatstone** in his early researches in **Spectrum Analysis.**   *Robert Sabine.*

825c. Arrangement of Apparatus for Experiments on the Assay of Gold Alloys, by means of the Spectroscope, in the manner suggested by Mr. Lockyer.

W. Chandler Roberts, F.R.S.

The apparatus consists of: —

(1.) An induction coil, capable of giving a 10-inch spark in air, which is provided with a Foucault contact breaker in order that the spark may be perfectly continuous.

(2.) A frame on which the portions of metal under examination can be so arranged as to be easily brought in succession under a fixed pole of aluminium. Accompanying this frame is a fixed microscope provided with cross wires in the eyepiece, and the table bearing the assay pieces can be adjusted by a micrometer screw, so that the image of the apex of each assay piece can be brought to coincide with the cross wires, thus ensuring that the striking distance remains constant.

(3.) A lens to throw an image of the spark on the slit of—

(4.) A large spectroscope in which the spectra of the alloys are examined. It is provided with a micrometer, the wire of which is horizontal and moves in a vertical plane.

When a spark from the induction coil (the two terminals of which are also connected with the coatings of a Leyden jar) passes between the aluminium pole and one of the alloys, an atmosphere of the vapours of gold and copper is formed round the lower pole which does not extend to the upper pole, and therefore in the spectrum observed the lines due to these metals will not cross the field of view. Mr. Lockyer observed, that, when all other conditions remain the same, if the composition of the alloy be slightly altered, the relative heights and intensities of the lines of the two metals vary. For these comparisons the gold line having a wave length of 5,230 tenth-metres, and the copper line 5,217, are the most convenient. If a series of known alloys varying slightly in composition is examined, a curve may be constructed, the ordinates of which represent the ordinary assays, and the abscissæ the micrometer readings for the points at which the above two lines are equally bright, and then, theoretically, if an unknown alloy of about the same composition be examined, this curve enables us to determine its exact composition when the micrometer reading is known.

In practice, however, it is found necessary to vary the striking distance with the composition, and the amount of this variation is still under investigation.

826. Spectroscope for determining the smallest displacement of spectral lines, and for measuring the velocity of motion of the object. *Professor Carl Wenzel Zenger, Prague.*

This new instrument gives double images, two spectra produced by an additional prism of quartz or calcspar, giving two dark lines in parallel directions, *e.g.*, the D. line, and of constant distance, if there be no motion towards or from the luminous body. The motion of heavenly bodies producing, therefore, the displacement of both D lines, and an accurate micrometer measuring it, gives the amount of velocity.

827. Hermann's Haematoscope, for examination and demonstration of absorption bands in fluids by the spectroscope. *Professor Dr. L. Hermann, Zürich.*

The fluid is poured into the little chamber, and the thickness of the layer is regulated by sliding the inner tube until the bands appear.

828. The Collection of Prisms of crown and flint glass used in the construction of refractors and spectroscopes by Steinheil and Merz at Munich, and by Hofmann at Paris, whose refractive indices for 50 lines in the solar spectrum were determined by Prof. Van der Willigen. *Teyler Foundation, Haarlem.*

Steinheil No. I. flint, No. II. flint, No. III. crown glass.
Merz No. I. and No. II. both of the same heaviest flint, No. III. crown, No. IV. crown, No. V. and No. Va. both of the same ordinary flint glass.
Hofmann No. I. heavy flint glass.
See " Archives du Musée Teyler, " Vol. I. p. 31, 64 and 205, and Vol. II. p. 183.
See the chemical composition of crown Steinheil No. III., and Merz No. IV., and of flint : Steinheil No. II., Merz No. I. and No. II., and No. V. and No. Va., and Hofmann No. I., given by the late Prof. P. J. van Kerckhoff, " Archives du Musée Teyler," Vol. III. p. 117.
Steinheil No. II. and No. III., Merz No. I. and No. II., No. IV. and No. V. and No. Va., and Hofmann No. I. are accompanied by parallelopipeds and plates of the same glass and by pieces or powder for chemical analysis.

829. Powerful Spectroscope, with Browning's automatic action, for adjusting the prisms to the minimum angle of deviation of the ray under examination. *John Browning.*

In this instrument the ray can be made to pass four times through the six prisms, and a dispersive power of 24 prisms thus obtained can be used, or that of any lesser number of prisms at pleasure. The instrument is fitted with a new reflecting bright line micrometer; when measuring with this contrivance no light is visible in the field of view, but the wires of the micrometer are seen faintly illuminated.

830. Universal Spectroscope, with Browning's automatic action, giving a dispersive power of from 2 to 12 prisms.

John Browning.

791. Rotating Tube Holder, a contrivance for containing a number of Pflücker's tubes, and obtaining their spectra successively without loss of time.   *John Browning.*

792. Rotating Metal Holder, suggested by Mr. Lockyer, for holding specimens of all the principal metals, and obtaining their spectra successively, or for the purposes of comparison.

John Browning.

831. Direct-vision Spectroscope, with apparatus - for registering observations.

Geneva Association for the Construction of Scientific Instruments.

This direct-vision spectroscope is distinguished from others of the same description, in that the distance of the lines of the spectrum is measured, not by the superposition of the spectrum upon an illuminated micrometric scale, but by measuring the angle formed by the eyepiece in moving the cross wires of the telescope from one line to the other, and then comparing it with that formed by the telescope when directed successively to two lines of known distance. A tangent screw effects the angular displacement of the telescope; this screw is graduated on the head to read angular displacements of less than 20 seconds. A recorder, formed of a movable pencil which acts upon a counter, serves to make series of observations in the dark. The collimator telescope possesses also an angular motion, which serves to bring any portion whatever of the spectrum to the centre of the field of view.

832. General Apparatus for Spectroscopy, Polarisation, Reflexion, Refraction, and for various experiments in **Fluorescence.**

Geneva Association for the Construction of Scientific Instruments.

This apparatus has been constructed with the object of carrying out, with one and the same instrument, all, or nearly all, experiments in spectroscopy, rotatory polarisation, reflection, and refraction. The divided circle is movable around an axis, and serves to bring the rays of light at any angle upon the eyepiece of the line of collimation. For experiments in spectroscopy, a table of one, three, or six prisms may be set up. The prisms are raised above the divided circle sufficiently to allow of their being heated from below if required. For determination of the line of collimation suitable arrangements have been provided, and for experiments in polarisation, eyepieces fitted with divided circles and nicols; a Babinet compensator may also be adapted to them.

The first apparatus of this kind was constructed for the use of Professor Wiedemann, of the University of Leipzig.

833. Large Spectroscope, according to Meyerstein's system for determining the relations of refraction and dispersion of different media, with contrivance for polarisation.

Schmidt and Haensch, Berlin.

834. Smaller Spectroscope, of the exhibitors' own construction. *Schmidt and Haensch, Berlin.*

835. Spectroscope, according to Abbe's system, with divided circle of 20 cm. diameter, repeating circle, and micrometer apparatus for determining the dispersion. Also a hollow prism with metal body. *Carl Zeiss, Jena.*

The spectroscope has only one telescope, which serves at the same time for collimation and observation. The adjustment brings about automatically the minimum deviation for every ray. The measurement of the refracting angle and that of the deviation takes place without change in the instrument. The determination of the dispersion is effected, independently of ascertaining the absolute refractive index of one colour, by micrometric measurement.

836. Two Prisms of Glass, for observing the **dispersion** of coloured liquids ; constructed by Steinheil, of Munich.

Prof. Kundt, Strassburg.

The refracting edge of the hollow prisms is so sharp that they show the dispersion of even highly coloured liquids.

837. Spectroscope, with five-inch circle, according to Dr. Meyerstein's system, for measuring the refraction and dispersion of different media, for chemical and optical analysis, as well as for all kinds of goniometrical measurements.

There are belonging to this :—

  a. Small circle with pivot and plate.
  b. Telescope with stand.
  c. Slit tube (Spaltrohre).
  d. Scale tube.
  e. Crystal stand.
  f. Prism.

Aug. Becker (Dr. Meyerstein's Astronomical and Physical Workshops), Göttingen.

The telescopes, when required, are screwed on to the places marked "Fernrohre," "Spaltrohre," &c. The small circle is put into the middle of the larger one. To determine the refracting angle of a prism, the telescope is attached to that part of the instrument which is marked "Zur Bestinmung, &c.," it is left here for all goniometric measurements, but the smaller circle is removed and the crystal-holder put into its place.

838. Spectroscope, of the latest construction, according to Dr. Meyerstein's system, arranged for the relations of refraction and dispersion of different media, for the reflection of polarised light at the free surface of liquids, as well as for the reflection of solid bodies, for all kinds of goniometrical measurements, and for chemical and optical analysis.

There are belonging to this :—
a. Small circle with plate.
b. Two telescopes.
c. Two slit-tubes (Spaltrohre).
d. Scale-tube.
e. Babinet's compensator.
f. Two weights for balancing telescope and slit in a vertical position.
g. Crystal stand.
h. Flint-glass prism.
    Aug. Becker (Dr. Meyerstein's Astronomical and Physical Workshops), Göttingen.

For attaching the telescopes, collimators, &c., the same rules apply as in the previous case. The larger telescope and collimator serve for determining the refraction and the refracting angle; the small tubes are for the polarisation. Determination of refraction and of refracting angles are effected as with the larger instrument, except that the telescope is put into the place of the micrometric tube. Solid bodies, when submitted to polarisation, are fixed with some wax against the plate which is put up in the black dish. When liquids are to be investigated, the small circle with its clamps is entirely removed, the screw, which maintains the principal circle in horizontal position, taken out, and the instrument turned down until the main circle stands vertically, when it is fixed again by the screw. For all polarisation experiments a *Babinet* compensator is fixed by means of the two screws upon the bearer of the telescope.

839. Hollow Prism, according to Dr. Meyerstein, which is used for optical analysis with the spectroscope.
    Aug. Becker (Dr. Meyerstein's Astronomical and Physical Workshops), Göttingen.

840. Rigid Spectroscope by Browning, constructed for Mr. Gassiot on the design of Dr. Balfour Stewart, with the view of determining whether the position of the D lines of the spectrum is constant, whilst the co-efficient of terrestrial gravity is made to vary.   *Kew Committee of the Royal Society.*

It is described in the Proceedings of the Royal Society, Vol. XIV., p. 320.
Observations were made by it from 1866–1869, on board H.M.S. "Nassau," during a voyage to and from the South Pacific, and subsequently at the Kew Observatory until 1872, the results of which are not yet published.
It consists of a train of three prisms, the last of which is silvered on one side, so as to return the light which falls upon it. Close to the slit another prism is placed, which reflects the rays on their way back into a micrometer, by which the position of the D lines is measured."

841. Vierordt Spectroscope, adapted for the measurement of absorption spectra and for quantitative chemical analysis.
    Schmidt and Haensch, Berlin.

This apparatus is described in Vierordt's work on the "Application of the Spectroscope to the Photometry of Absorption Spectra, and in Quantitative Chemical Analysis." Tübingen, 1873.

843. Mitscherlich's Stand for use in the spectroscopic examination of coloured flames.

Prof. A. Mitscherlich, Münden.

843a. Direct-vision Spectroscope of great dispersive power, provided with a specially constructed micrometer for measuring intervals between lines to ·0001 in. *A. Hilger.*

843b. Two powerful Direct-vision Spectroscopes, 7 in. in length. *A. Hilger.*

844. Pocket Spectroscope, showing sodium line in a simple burning candle, of great use in chemical and meteorological observations. *A. Hilger.*

845. Hilger's Pocket Spectroscope, showing all the principal Fraunhofer lines, dividing D easily with nickel line between; when on sun, a sliding slit with division adjustable to position. besides a limb in front of eyepiece. *A. Hilger.*

846. Pocket Spectroscope of the simplest and cheapest kind. *A. Hilger.*

847. Rhomboid of Iceland Spar. *A. Hilger.*

Iceland **Spar Prism** of 60°, 1½ in. surfaces, single refracting. *A. Hilger.*

Spar Prism, 1¾ in. surfaces. *A. Hilger.*

Flint Prism, 1⅛ in. surfaces. *A. Hilger.*

Flint Prism, 1 in. surface. *A. Hilger.*

Two Half Prisms, 1 in. surface. *A. Hilger.*

Lateral Flint Prism, 1¾ in × 3 in. *A. Hilger.*

Three compound Direct-vision Prisms, one of a new form receiving light at right angles. *A. Hilger.*

Four compound Prisms without colour. *A. Hilger.*

Two dull Glass Prisms for photometric use. *A. Hilger.*

Nicol Prisms for polarisation. *A. Hilger.*

Double Image Prism. *A. Hilger.*

Parallel Plane Glass Plate, surface 3 in. × 2½ in. *A. Hilger.*

848. Nicol and a Double Image Prism. *A. Hilger.*

848a. Large Prism of 60°, with perfectly worked surfaces 3 in. by 3⅝ in. of the finest Thalium glass, on brass stand. *A. Hilger.*

848b. Two Nicol Prisms mounted for polarisation. *A. Hilger.*

849. Spectroscope made by Yeates, of Dublin, fitted with a diaphragm instead of cross threads for measurement of position of lines. *Prof. Jos. P. O'Reilly.*

The diaphragm, above, being perfectly opaque, is always visible against even the faintest lines; moreover, it dispenses with the introduction of an extraneous light which may by its brilliancy interfere with that of faint lines. This spectroscope is specially adapted for the examination of fluorescent minerals, the prisms and lenses being of quartz.

850. Spectroscope, with bi-sulphide of carbon prisms and lens, arranged for projection. *Yeates & Sons.*

The prisms and collimating lens are so proportioned that no light is lost by passing outside the prisms or otherwise.

851. Spectroscope, with compound prism and angular scale. *Yeates & Sons.*

852. Spectroscope, with two prisms. *James How & Co.*

853. Three Foucault's Prisms.

Iceland-spar, cut in three directions.
Small rhombohedron of Iceland spar.
Cone and pyramid (black) for Guérard's apparatus.
Cone for the projection of annular spectrum.
Pyramid for the projection of four spectra.
M. Mascart's prism.
Prism of crown glass.
Two polyprisms (glass).
One polyprism (quartz).
Collection of nine prisms.
Boscowich prisms.
Fresnel's tri-prism.
Fresnel's Parallelopipeds, mounted in brass.
Laurent, Paris.

854. Prisms for direct Vision in 3 pieces.

„ „ „ **5** „
„ „ „ **7** „.
Rochon prism.
Wollaston prism.
„ „ (small).
Achromatized Iceland- spar prism.
Cube of fluor-spar.
Fluor-spar lens.
Cemented cells for spectroscopic work (9).
Tubes with platinum wire (6).
Collection of quartz and Iceland spar prisms.
Spectroscope with one prism, one and two burners.
Object-glass for projecting ray spectra.
Laurent, Paris.

854a. Collection of Prisms, for optical purposes.
Laurent, Paris.

854b. Spectroscope. *M. J. Duboscq, Paris.*

854c. Prisms, by Arago. *Paris Observatory.*

854d. Prisms, by Borda. *Paris Observatory.*

854e. Collection of various kinds of **Glass** for optical purposes. *Feil, Paris.*
  1. Disc of Crown Glass, 4 inches.
  2. Disc of Flint Glass, 4 inches.
  3. Plate of Crown Glass (heavy English).
  4. Parallelopiped, of ordinary flint glass.
  5. Crown Glass Prism.
  6. Flint Glass Prism.
  7. Flint Glass Prism, (sp. gravity 4·4.)
  7a. Prism manufactured by Fraunhaufer Guinand.
  8. Flint Glass Prism (very heavy), sp. gr. 4·4.
  9. Flint Glass Prism, sp. gravity 4.
  10. Flint Glass Prism, sp. gravity 5·2.
  11. Flint Glass Prism, sp. gravity 5·5.

Series of Assays of the Metallic Earths.

  11a. Silicate of Potassium and Calcium with Titanium.
  12. Aluminate of Silicium and Magnesium.
  13. Crystallisation of Alumina and Magnesia.
  14. Crystallisation of Fluosilicate of Magnesium and Calcium.
  15. Alumina and Magnesia Crystallised by Fluosilicate of Potassium.
  16. Crystallisation of Fluosilicate of Aluminium.
  17. Crystallisation of Barosilicate of Aluminium.
  18. Crystallised Aluminate of Magnesium and Silicium.
  19. Manufacture of Adamantine Boron.
  20. Plate of Crystals of Aluminium.
  21. Blue Obsidian.
  22. Obsidian coloured by Cobalt.
  23. Samples of Crown Glass, extra white.

854f. Objective of Rock Crystal, 10 centimetres in diameter. *M. Lutz, Paris.*

854g. Astronomical Glasses, for cabinets of physics.
M. Lutz, Paris.

855. Micro-spectroscope, with prism for comparing two spectra, and with Abbe's measuring apparatus for the direct estimation of the wave lengths of dark or bright lines in a spectrum.
C. Zeiss, Jena.

This spectroscope gives the position of the bright or dark lines by means of a scale on which the spectrum is thrown, and which, by means of its peculiar graduation and numbering, allows the wave length at any place to be read off in micro-millimetres.

856. Apps' Improved Gas and Electrode Holder for Spectrum Analysis.   *Alfred Apps.*

857. Improved Automatic Chemical Spectroscope, invented and made by the contributor. The object glasses and prisms by Chas. Owen, Optician, Strand.

Rev. Nicholas Brady.

Prisms with a circular face are cemented to the object glasses of the collimator and telescope, the circular face of the prisms being of the same size as the object glass. The base is rectangular to the surface of the object glass, and the refracting angle about 30°. The beam of light rendered parallel by the collimating lens passes through the first half prism perpendicularly and suffers no refraction, but is refracted on emergence from its exterior face; the refracted and dispersed beam is received on the external face of a second object glass prism, suffers more refraction and dispersion, and, emerging at a perpendicular incidence, is taken up and brought to a focus by the object glass. In all positions the ray under examination passes parallel to the base of the prisms, and therefore at the angle of minimum deviation. The variation of the angle between the two prisms by the motion of the tangent screw of the telescope merely brings one ray after another into the field of view. This automatic arrangement gives a dispersion equal to one dense glass prism of 60°. Should greater dispersion be required a prism of 60° has been arranged in the centre of the instrument, which by a very simple automatic contrivance of one lever and a slot is moved by the arm carrying the telescope, so that any ray is still preserved at the angle of minimum deviation. A further advantage of this new principle is, that with these object glass prisms the field is completely filled with light, which has not usually been the case, unless the prisms are extra large, and therefore expensive; and if a train of prisms be inserted their faces only require to be equal for them still to entirely illumine the field : thus much light is gained, and comparatively little is lost by absorption and reflection, as the surfaces are fewer, so that the violet end of the spectrum is very extensive, and the lines beautifully defined.

858. Gas Lamp Apparatus for placing before the slit of the spectroscope with Bunsen burners, &c., insuring the proper adjustment of the platinum wires carrying the substance under examination in the flame without displacing the eye from the ocular of the telescope; and also an arrangement for quickly and efficiently exchanging one or both burners for either one or two vacuum tubes.   *Rev. Nicholas Brady.*

Two photographs accompany the instrument, showing its use in two different positions.

859. Ordinary Spectroscope, arranged for the exhibition of diffraction phenomena, with apertures and gratings, &c. of various kinds, under common and polarised light, and with the means of observing the spectra of the diffracted beam. Designed and made by exhibitor.   *Rev. Nicholas Brady.*

971c. Curve for obtaining **Wave-lengths** of **Spectra;** and **Map** of the absorption spectra of bromine and iodine monochloride.
Professors Roscoe and Thorpe.

857a. Automatic Motion for the Spectroscope.
Walter Baily.

The apparatus consists of an axle and four parallel discs, the outer pair being fixed, and the inner pair rigidly connected together, and capable of turning on the axle. Between the inner discs are four arms, which also turn on the axle.

Taking the centre of the axle as origin, each disc has 4 slits, the equation to their middle lines being $r = \mathrm{F}(\theta)$, $r = \mathrm{F}\left(\dfrac{\theta}{2}\right)$, $r = \mathrm{F}\left(\dfrac{\theta}{3}\right)$, $r = \mathrm{F}\left(\dfrac{\theta}{4}\right)$. The discs in each pair are placed with their slits parallel, but the inner discs are turned over. The arms have straight slits radiating from the axle. A pin is passed through the first slits of the outer discs, the 4th of the inner discs, and the slit in one of the arms. The remaining arms are connected with the discs by three other pins inserted in a similar manner. The first and last prisms are fixed on the initial lines of the outer and inner pair of discs respectively, and the four other prisms are carried by the arms. Motion is given by moving the inner discs. The angles at which the slits cross are constant and differ least from right angles if $\mathrm{F}(\theta) = a e^{-\theta\sqrt{5}}$, which is the form adopted in the model.

859a. Photo-Spectra of Metals and Gases, obtained with a Browning Direct-vision Spectroscope.
John Rand Capron.

These spectra were photographed with wet plates by J. R. Capron and G. H. Murray, for use and reference in connection with auroral investigations.

The optical apparatus consisted of a Browning direct-vision prism, with, in the case of metals, a collimating lens of 6-inch focus, and a projecting lens of 9-inch focus. For the gases a similar prism was used, with collimating and projecting lenses, each of 4-inch focus. The photographs are enlarged to twice the originals. The electrical apparatus consisted of a 4½-inch spark Ruhmkorff coil, worked by four half-gallon bichromate cells, a condenser of four glass-coated plates, and spark terminals.

The spark was placed about half an inch from the slit. Mr. Lockyer's plan of interposing a lens was tried in some cases, but given up, as the long and short lines were found to be equally well observed in the spark itself. The gases were mostly in Geissler tubes. It is proposed in further experiments to get rid of the air lines by taking the spark between the metal electrodes in glass bulbs, through which a gas passes.

One of the spectra represents the spark between platinum electrodes taken in a current of coal gas passing through a tube.

The dispersion of the direct-vision prism is as follows:—A=17·62; B=23·62; C=30·42; D=50; E=77·35; F=103·90; G=159·73. With the instrument as used for the gases, it is found possible to photograph faint spectra satisfactorily.

b. Polariscopes, &c.

861. Jellett's Saccharometer, for the measurement of the rotation which certain fluids are capable of producing in the plane of polarisation of the transmitted ray. *Trinity College, Dublin.*

This rotation is measured by the method of compensation, the original position of the plane being restored by transmission through a column of fluid possessing an opposite rotatory power. This fluid is contained in a vessel closed at the bottom with glass, and the length of the column is regulated by means of a tube, also closed with glass, which is capable of moving in the direction of its axis, the amount of this movement being read off on a scale. A full description of the instrument and of the analysing prism used in its construction is given in the "Transactions of the Royal Irish Academy," Vol. XXV., pp. 373–82.

861a. Laurent Polarimeter and Saccharometer, with two divisions on the plate, with inversion tube and one thermometer.   *Laurent, Paris.*

861b. Saccharometer, by Soleil, with penumbra (large model).   *M. J. Duboscq, Paris.*

861c. Large Circle, by Messrs. Jamin and Sénarmont.
                                        M. J. Duboscq, Paris.

861d. Large Circle for measuring the elements of elliptic and rotatory polarisation in solids and liquids, and reproducing all impressions of polarisation and refraction. (This apparatus belongs to the School of Photography.)   *M. Lutz, Paris.*

862. Polarising Apparatus, according to Dove's system, complete, with polyoscope and dichroscope.
                                        Schmidt and Haensch, Berlin.

863. Simple, handy Polarising Apparatus, according to Carl's system.   *Schmidt and Haensch, Berlin.*

864. Melde's Models, for illustrating the **colours of thin leaves** in polarised light.   *Ferdinand Süss, Marburg.*

865. Melde's Model, for illustrating **circular polarisation** by means of gypsum and scales of mica.
                                        Ferdinand Süss, Marburg.

866. Paste-board Models, according to J. Müller's system, for illustrating the **colour phenomena** in **polarised light,** and the **uni-axial** and **bi-axial crystals.**
                          J. Wilhelm Albert, Frankfort-on-Maine.

Ten models of cardboard, together with a treatise on them. Described in *J. Müller's* "Lehrb. der Phys., 7 Aufl., I. Bd., 3tes. Buch, caps. 9 and 10."

867. Polarising Apparatus, for projection with rotatory analyser, according to E. Mach's system, with quartz plate and ¼ undulation plate. (Comp. Poggendorff's Ann., 1875, No. 12.)
                          J. Wilhelm Albert, Frankfort-on-Maine.

The ray of sun or electric lamp falls through a Nicol, which is protected with a shade, upon a press, in which the object is fastened by means of spring clamps, and passes thence through a tube which can be rotated with great velocity. This tube is provided at one end with a shade capable of rotating with the tube, and the analysing Nicol over which there is a slit or a square aperture.

At the end of the tube is a deflection prism of crown glass, to which, for some investigations, a direct vision prism is added. The ray, as it issues from the tube, is received by a lens, which throws upon a screen a sharp image of the slit or square aperture. This image moves in a circle as the azimuth changes, and thus shows by quick rotation all the phenomena which, in ordinary polarising instruments, appear successively *side by side*.

868. Twelve Plates with Pictures, of gypsum and mica, for polarised light. *Prof. Karsten, Rostock.*

The form of images has been chosen to represent the different colours of thin plates in polarised light. Any kind of polarising apparatus may be employed for these observations.

869. Nörremberg's Polarising Apparatus, small size. *W. Apel, Göttingen.*

870. Nörremberg's Polarising Apparatus, large size ; according to the design of Professor Listing. *W. Apel, Göttingen.*

The apparatus serves not only for purposes of lecture demonstration, but also for accurate measurements. The advantage of the instrument over the ordinary polarising microscopes lies in the circumstance that in the *Nörremberg* apparatus the polarised light passes to and fro through the same crystal plate, when placed on the horizontal mirror. The movable glass plate of the middle table serves for measuring the angle of the optical axes by means of a graduated semicircle.

870a. Large Apparatus by Norremberg, improved by Wheatstone. *M. Lutz, Paris.*

870b. Apparatus used for observing the **Polarisation** of **Light** in **Water.** *J. Louis Soret, Geneva.*

It is formed of a telescope tube closed on the objective side by a glass plate. The eyepiece is formed of a " Nicol " prism.

The observer, placed in a boat, immerses the objective end of the tube and looks through the "Nicol." He then finds the light of blue coloration reflected by the lower strata under the surface of the water, and by turning the Nicol ascertains if it is polarised.

See "Notes sur la Polarisation de la Lumière de l'Eau." Archives des Sciences physiques et naturelles, 1869, Vol. 35, p. 84, and 1870, Vol. 39, p. 352.

871. Apparatus for the Observation and Measurement of the cyclopolar double refraction of Quartz in the direction of the optical axis. Designed by Professor Listing, executed by R. Winkel in Göttingen.

Royal Mathematical and Physical Institute of the University of Göttingen, Prof. Listing.

The telescope can, before being put into the holder of the apparatus, be adjusted for distant objects, or for an object of but 2-3 meters distance from the object glass. The *Fresnel* triple quartz prism is fixed in upright position in the support below the telescope, and protected by a cardboard shade against side lights. By means of the achromatic lens, situated below the prism, a virtual image is produced of an appropriate object (line cut by a diamond upon glass, &c.), placed upon the black table, which image, when seen sharp and double in the telescope, will be just as distant from the object glass of the telescope as the latter has been adjusted for. The angle of the two images is read off on the micrometer in the eyepiece, and from the number obtained the diversion of the two rays after their passage through the triple prism is calculated. The ocular can now be provided with tourmaline and $\frac{1}{4}$ mica plate, which may be used singly or combined. The tourmaline alone shows, on turning, in all azimuths, the double image without alteration of intensity in the component parts; the two rays undergo, therefore, neither linear (plain) nor elliptical polarisation. The tourmaline with mica plate below it, shows, as well known, that both rays are circularly polarised, the one right, the other left; the tourmaline must in this case be so adjusted that its line of principal action be azimuth 0° or 90°, and the main section of the interposed mica plate is turned to form with that line $\pm 45°$.

The aim of the measurements is to determine the refractive indices of the two rays of opposite circular polarisation, propagated with unequal velocity along the optical axis of the quartz.

872. Stauroscope, according to the design of F. von Kobell, executed by Wiedemann.

Prof. Dr. Franz von Kobell, Munich.

873. Analysing Prism of Iceland Spar, made by the inventor, the late William Nicol, in his 80th year.

Edinburgh Museum of Science and Art.

The Nicol prism is so constructed that only one polarised ray can pass through it.

873a. Nicol's Prism for Polarising Light, by C. D. Ahrens.
W. Spottiswoode, F.R.S.

This, which is one of the largest ever constructed, has a clear field of $3\frac{1}{4}$ inches in diameter. With a view to saving bulk and weight, the acute angles have been cut off, and the whole reduced to an octagonal form. The advantages of this will readily be seen by comparing this instrument with that by Tisley and Spiller, the field of which is greater by only a quarter of an inch.

873b. Nicol's Prism for Polarising Light, by Tisley and Spiller.
W. Spottiswoode, F.R.S.

This, which is the largest and purest ever constructed, has a clear field of fully $3\frac{1}{2}$ inches in diameter.

874. Soleil-Ventzke Polarising Apparatus, with several improvements.
Franz Schmidt and Haensch, Berlin.

Soleil having introduced compensation by the use of rock-crystal, Ventzke subsequently improved the colour-giving power, and Scheibler made further improvements, principally in the manner of inserting the observation tubes. Messrs. Schmidt and Haensch, besides a few minor changes, succeeded in making improvements which greatly facilitate the use of the instrument, by a change in the construction of the wedges, and have thus reduced the irregularities frequently observed in the polarisation of diluted solutions to from one to two tenths per cent. in each part of the scale. They have thereby done away with the principal cause of the variations which so frequently occur in the observations of different analysts.

875. Jellett-Corny Half-shade Polarising Apparatus, provided with wedge compensation.

Franz Schmidt and Haensch, Berlin.

This apparatus differs from the foregoing in having the double plate replaced by a double Nicol's prism. In using it both fields of the apparatus are adjusted to equal half darkness, instead of equal colour, as in the " Soleil. The double Nicol prism was first proposed by Professor Jelett, of Dublin, and employed by Professor Corny in Duboscq's polariscope for circular polarisation, known as saccharomètre à pénombre. The improvement in the instrument exhibited by Messrs. Schmidt and Haensch consists in combining with it their wedge-compensation, so as to obtain the advantage of lineal readings. The instrument recommends itself for dark solutions ; it is indispensable for colour-blind operators, and prevents the colour-weariness to which the healthy eye is liable. Its sensitiveness perceptibly exceeds that of a Soleil.

876. Wild's Polari-Strobometer.

Franz Schmidt and Haensch, Berlin.

877. Jellett-Corny Polarising Apparatus, constructed for circular polarisation. *Franz Schmidt and Haensch, Berlin.*

877a. Handy Polariscope with Nicol prisms to show rings in bi-axial crystals. *W. Previte Orton.*

This instrument was designed by the exhibitor and made for him by Pastorelli. Its object is to utilize the Nicols of a small microscope so as to show the effect of polarised light in a biaxial crystal ; and further, to do this in a handy manageable and inexpensive form.

878. Mica-preparations of mono- and bi- axial mica, for polariscopes. (See Mineralogy.) *Max. Raphael, Breslau.*

879. Mica-preparations of foliaceous mosses ("Laubmoosen "), Algal, &c., for microscopes. *Max. Raphael, Breslau.*

879a. Quartz Axis Plates. *M. Lutz, Paris.*

879b. Amethyst cut parallel to the axis. *M. Lutz, Paris.*

880. Dichroscopic Lens.
Four Nicol's prisms.
Prazmowski prism.
Two Tourmalines parallel to the axis.
Iceland spar of M. Bertrand's arrangement.
Quartz and mica for compensating the refraction of crystals.
Heated crystals, felspar, gypsum, carbonate of lead.
One blue glass, red glass, and green glass.
Billet lens on stand.

Laurent.

881. Ellipsometer. Before the eyepiece of the glass a double refracting prism is made to turn until a wire, moving perpendicularly to the principal section of the prism, passes through the two intersecting points of the two reflections of the ellipse. An index shows at the moment the position of the prism.

882. Table Polariscope, made by the exhibitor when a youth.   *Rev. Nicholas Brady.*

The interest of this instrument consists in showing with what simple materials a student can construct a fairly useful apparatus, the divided circles being common stamped protractors; the clamping screws, teapot thumb-screws, and the mountings of the lenses ordinary simple microscope frames.

882a. Original Apparatus for Rotatory Polarization, by Biot.   *College of France, Paris.*

883. Airy's Polariscope, with appliances for approximately measuring the angle between the planes of polarisation and analysation, and for determining the angle between the optic axes of bi-axal crystals in air or in a fluid medium. Modified and arranged by the contributor when a student.   *Rev. Nicholas Brady, M.A.*

883a. Large Polariscope for Projection, by Ladd.
                                            W. Spottiswoode, F.R.S.

A pair of Nicol's prisms, by Ladd, the first of a large size ever constructed. They are furnished with a system of lenses for showing the crystal rings, as well as with other contrivances for the various phenomena of polarised light.

883b. Revolving Analyser for Polariscope, constructed by Tisley and Spiller.   *W. Spottiswoode, F.R.S.*

A revolving analyser, consisting of a double image prism, furnished with wheel-work, whereby it may be caused to revolve with such rapidity that the eccentric image may remain upon the retina during a complete revolution, and thus give the appearance of a ring of light. By this means all the phases of polarised light as seen successively in ordinary polariscopes may be seen simultaneously. The instrument is adapted to show all the phenomena of chromatic polarisation, both plane and circular. An instrument for a similar purpose was invented independently by Prof. Machs, of Vienna.

883c. Portable form of Polariscope, comprising a Nicol's prism, a double-image prism, a plate of tourmalin, a Savart's wedge, a bi-quartz, a dichroscope, and a quarter undulation plate. These various parts may be used either separately or in any combinations at pleasure; and are consequently adapted either to illustrate the general laws of polarised light, or for actual observations of atmospheric or other polarisation not involving actual measurements. It will be observed that the tourmalin plate gives the opportunity of using convergent as well as parallel light. The instrument is fitted in a case 2 inches long and $\frac{3}{4}$ inch in diameter, but its size might be considerably reduced below the dimensions of the specimen here exhibited.   *Mrs. W. Spottiswoode.*

883cc. Spottiswoode's Pocket Polarising Apparatus, consisting of Nicol's prism, Savart's polariscope, tourmaline, double-image prism, bi-quartz, dichroscope, and ¼-undulation plate. The whole is packed in a leather case, 2 inches long, by ¾ inch in diameter. *W. Ladd & Co.*

883d. Large Circle for measuring the **Azimuths** of **Elliptic** and **Rotatory Polarisation,** and reproducing all experiments of polarisation and reflection.
School of Pharmacy, Paris.

883d. Polariscope for detecting faint traces of Polarisation independently of its direction. The wedges are right and left-handed quartz, with their axes parallel to that of the instrument. The eye should be placed in the focus of the lens, with the Nicol interposed. Designed by the exhibitor and made by Messrs. Tisley & Spiller. *R. H. M. Bosanquet.*

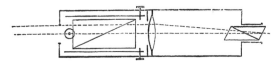

883e. Arago's Polariscope. *M. Lutz, Paris.*

883f. Savart's Polariscope. *M. Lutz, Paris.*

883g. Tourmaline Plates. *M. Lutz, Paris.*

883h. De Sénarmont's Polariscope. *M. Lutz, Paris.*

884. Wheatstone's Polar Clock. To determine the true solar time by the polarisation of light reflected from the sky.
The Council of King's College, London.

885. Latest form of Wheatstone's Polar Clock. To determine the true solar time by the polarisation of light reflected from the sky.
The Council of King's College, London.

Soleil's Compensator. *S. Laurent, Paris.*

887. Norremberg's Polarising Apparatus, with Wheatstone's improvements. *H. Lloyd, Trinity College, Dublin.*

888. Duboscq's Polariscope, for determining the inclination of the axes in bi-axial crystals.
H. Lloyd, Trinity College, Dublin.

889. Wheatstone's Apparatus to illustrate the laws of interference of polarised light.
H. Lloyd, Trinity College, Dublin.

889a. Ladd's Polariscope, consisting of a bundle of glass, selenite design, and Nicol's prism, &c.   *W. Ladd & Co.*

2531. Photograph of a **Wild's Polari-strobometer,** for determining the rotation at different temperatures.   The tube containing the liquid is surrounded by a jacket, through which water of a given temperature flows.   The apparatus is manufactured by Messrs. Hermann and Pfister, Bonn.

Prof. H. Landolt, Aix-la-Chapelle.

2532. Photographs of a simple **Polari-strobometer** with two Nicol's prisms, constructed for holding tubes, one meter in length, containing the liquids which can be placed in a water-bath; and a blow-pipe lamp, over which is suspended a platinum gauze cage for holding the sodium salt which is used for producing the monochromatic sodium flame.

Prof. H. Landolt, Aix-la-Chapelle.

The lamp is manufactured by Dr. Meyerstein of Göttingen, and by Feldhausen, philosophical instrument maker, Aix-la-Chapelle.

2533. Photograph of the same apparatus, provided with a short tube for holding the liquid.   A bottle containing a solution of potassium dichromate is interposed between the tube and the sodium flame, to ensure a purer monochromatic light.

Prof. H. Landolt, Aix-la-Chapelle.

III.—PHOTOMETERS.

891. Great Atmospheric Photometer.   De la Rive model, designed by M. Thury, and constructed by the Geneva Association for the Construction of Scientific Instruments.

De la Rive Collection.   The property of Messrs. Soret, Perrot, and Sarasin, Geneva.

This apparatus is particularly intended to measure the transparency of the atmosphere.   It is used for simultaneous observation, with one eye, through two similar eyepieces of two similar objects placed at different distances. The difference of brightness and of tint between the two reflections indicates the effect of the intervening stratum of air.   The computation of this difference is arrived at by equalising the two images by means of diaphragms of different aperture, and of glass plates variously tinted.   The instrument is composed of a telescope with a single eyepiece, and two objective tubes, of which the angular distance can be varied between 0° and 29°.   A system of four total reflection prisms unites the two divergent cones in the eyepiece. The apparatus is movable round three different axes, and may be worked in the most varied directions.   Graduated circles measure these different angular motions.   It is a general photometer, and can be specially used as an astronomical photometer.   De la Rive has effected with it a long series of observations on the transparency of the air.   (See Comptes Rendus, vol. 63, p. 1221.)

892. Photometer, according to Glan's system, for photometrical determination of the absorption spectra for homogeneous light. *Schmidt and Haensch, Berlin.*

892a. Photometer, fitted with clock, governor, pressure gauge, and all necessary apparatus complete, as adopted by the Government of Canada. *William Sugg.*

893. Photometer by **Bunsen,** simplified by Professor Bohn. *Physical Collection of the University of Giessen, Prof. Buff.*

The standard for comparison is a pure stearine candle of known weight. The measure wound upon the cylinder serves to determine the distance at which the oil spot on the paper, when viewed from the second flame, is made to disappear. First the standard candle, and then the flame, to be measured, are thus investigated. The intensities of the two lights are to one another as the squares of their distances from the oil spot.

894. Photometer, for ascertaining amount of daylight. *Scottish Meteorological Society.*

The light is reduced by turning round the graduated milled head at the side, which works simultaneously and by equal degrees the two shades which thus reduce the area of the aperture. At the opposite end of the box a printed page is looked at through the eye-piece till it ceases to be legible, when the result is read off in revolutions of the milled-head. Designed by Thomas Stevenson, C.E., F.R.S.E., Honorary Secretary, and described in Society's Journal, vol. iii., page 292.

895. Selenium Photometer. *Siemens and Halske, Berlin.*

It being the property of selenium that its electrical resistance is diminished by the action of light, the diminution being dependent on the intensity of the light, this apparatus is constructed with a plate of selenium forming part of an electric circuit which is brought by rotating the cylinder containing the plate alternately under the action of a normal candle sliding on a scale and of the light to be measured. The normal light is adjusted on the sliding scale until the electrical resistance of the selenium remains constant under the action of the two sources of light, and the intensity of the light to be measured is calculated from the relative distances of the lights from the selenium plate.

896. New Optometer, with double-refracting lens of calcspar, giving double readings, and greater precision in determining the distance of sight. *Prof. Carl Wenzel Zenger, Prague.*

IV.—RADIOMETERS.

899. Collection of Radiometers of different construction, with lamps and screen for making experiments. *Prof. Adolph Weinhold, Chemnitz.*

The apparatus serves to perform the radiometer experiments, described by the exhibitor in *Carl's* "Repert. der Experimental Phys., 1876, Heft. 2." Compare also the description annexed to the apparatus.

900. New Radiometers. *Dr. H. Geissler, Bonn.*

901. Radiometer. *John Browning.*

These instruments are set in motion by either light or heat; they consist of four small discs on two arms at right angles to each other ; the discs may be of pith or mica; those exhibited are made of mica, as they appear to be the most sensitive to minute traces of light. One side of each disc has a dead black surface. The action of light or heat repels the black surfaces, and continuous motion is obtained so long as any light or heat falls on them.

902. The late Prof. T. T. Müller's Apparatus for illustrating the influence of the intensity of light on its rapidity of propagation (Poggendorff's Annalen, 1872, cxlv. p. 86.)
Prof. A. Mousson, Zurich.

Use is made of Newton's rings, produced between a plane glass *a*, and another glass *b* (the latter very slightly convex), which may be separated in a known manner so as to produce differences of progression up to 50,000 waves. At this distance, the convex glass *b*, on which *a* rested at first, radiates in the centre of a square iron vessel *c* on mercury. The three screws of the support *d* which surrounds the glasses are fixed apart. Then, the mercury having been allowed to flow out (through the cock *b*), *c* is brought down and fixed also, by means of three other screws *y*. The distance can be calculated with great precision by means of the weight of the mercury, and the known area of the vessel *c*.

The luminous point used is the small image at the opening of a collimating tube *e*, lighted by a monochromatic sodium flame, upon the hypothenuse surface of a small prism *f*. From this point the rays diverge and fall on the lens *g*, placed on the glass *c*, which makes them parallel. These rays return, with interference, from the two reflecting faces, towards the point *f*, where the eye is placed close to the prism.

Now, if the intensity of the light be lessened by interposing absorbing glasses, it will be seen that the greater the difference in the number of waves the more the lines change place, the increase of rapidity being proportionate to that of intensity.

V.—REFLECTION, REFRACTION, AND DIFFRACTION.

903. Total Reflection Apparatus, for the projection of objects placed in a horizontal position.
J. and A. Molteni, Paris.

904. Small Prism for double reflection. *Laurent.*

905. Coloured Rings on an 80 millimetre tripod.
Laurent.

905a. Fresnel's Parallelopiped. *M. Lutz, Paris.*

908. Apparatus designed to exhibit Double Reflection, which arises when a ray of light traversing a uni-axial or bi-axial crystal reaches the surface of contact of the crystal with the surrounding medium. *Arthur Hill Curtis.*

The incident light passes through a small orifice in the cap terminating one of the tubes. If the eye be applied to the other tube, as the stage on which the crystal rests is turned round its vertical axis, four, three, or two images of the orifice will be seen formed by the two rays which, refracted at the upper surface, are (in general) *each* doubly reflected at the lower surface. A Nicol's prism is added, which, though not essential to viewing the phenomena, may be introduced into *either* tube to polarise the incident light, or to examine the planes of polarisation of the reflected rays.

I. Sphere of Calcite, 3¾ inches in diameter.

II. Polyhedron of Calcite, cut from a large rhombohedron of that mineral, so as to represent the optical characters of the crystal in directions perpendicular—

1. To the pinakoid, and along the optic axis.
2. To a prism plane, and perpendicular to the optic axis.
3. To the cleavage planes (of the rhombohedron) {100}.
4. To the plane {ī22} correlative to the cleavage rhombohedron. (These were made by Mr. Ahrens.) *Prof. N. S. Maskelyne.*

909. Dichroic Apparatus. *A. Hilger.*

910. Iceland Spar Prism, of 60°, showing single refraction for any line in the spectrum. *A. Hilger.*

910a. Two Nicol's Prisms. *South Kensington Museum.*

911. Prism, with Double Reflector, of Dr. de Wecker. Two triangular prisms are joined together at their hypothenuse; while the observer looks directly through the cube formed by the union of the two prisms, an observer looking obliquely sees in the hypothenuse the reflection of the former as though in a mirror. The lens serves to show the reflection smaller and reversed. *M. Crétès, Paris.*

912. Prism, Movable, by Crétès. Two prisms of 15° each are placed in a setting. When placed base on edge, their refraction becomes annulled : (15—15=0). When placed base to base, their effect becomes added : (15+15=30). Between these two extremes, an ascending scale of 0 to 30° can be obtained. The prismatic axis remains fixed, because the glasses move equally in reverse ways. *M. Crétès, Paris.*

912a. Three Rectangular Prisms, crown glass. *M. Lutz, Paris.*

912b. 32 Rectangular Prisms, flint glass of various sizes. *M. Lutz, Paris.*

912c. Prisms for Camera.   *M. Lutz, Paris.*

912d. Prisms with Compartments.   *M. Lutz, Paris.*

912e. Prisms with Compartments.   *M. Lutz, Paris.*

912f. Bi-refracting Spar Prisms.   *M. Lutz, Paris.*

912g. Rhomboids of Spar.   *M. Lutz, Paris.*

912h. Collection of 12 Nicol Prisms.   *M. Lutz, Paris.*

912i. Four Large Nicol Prisms.   *M. Lutz, Paris.*

912j. Prisms for M. Desain's Experiments.
                                              M. Lutz, Paris.

912k. Uranium Glass Prisms.   *M. Lutz, Paris.*

912l. Uranium Glass Cube.   *M. Lutz, Paris.*

912m. Three Quartz Prisms.   *M. Lutz, Paris.*

912n. Prisms for Spectroscope (direct vision).
                                              M. Lutz, Paris.

912o. Prisms for Spectroscope (direct vision).
                                              M. Lutz, Paris.

912p Two Pyramidal Prisms.   *M. Lutz, Paris.*

912q. Three Flagon Prisms.   *M. Lutz, Paris.*

912r. Four Polyprisms.   *M. Lutz, Paris.*

912s. Four Prisms for Bisulphide of Carbon.
                                              M. Lutz, Paris.

912t. Two Isosceles Prisms of Flint Glass.
                                              M. Lutz, Paris.

912u. Three Foucault's Prisms.   *M. Lutz, Paris.*

912v. Two De Sénamont's Prisms.   *M. Lutz, Paris.*

912w. Three Braszinowski and Hartnack's Prisms.
                                              M. Lutz, Paris.

912x. Six Rochon's Prisms.   *M. Lutz, Paris.*

912y. Boit's Prism.   *M. Lutz, Paris.*

912z. Large Flint Glass Prism, extra denticulated, set in
a peculiar manner.   *M. Lutz, Paris.*

913. Instrument to show the phenomenon of conical refraction, with models of Fresnel's wave surface. (Soleil, Paris.)
                        H. Lloyd, Trinity College, Dublin.

914. Jamin's Optical Bank of Diffraction.
H. Lloyd, Trinity College, Dublin.

914a. Jamin's Apparatus with Parallel Mirrors.
M. Lutz, Paris.

914b. Large Steel Mirror. *M. Lutz, Paris.*

914c. Series of Barton's Iris Buttons, consisting of gold and steel faces engraved upon which are numbers of very fine lines, illustrating most beautifully *iridescence* or decomposition of light from ruled surfaces. The lines on the large steel button are 100, 200, 400, 500, 1,000, 2,000, and 4,000 to the inch.
Robert C. Murray.

915. Optical Bank, improved by Professor Clifton, to observe the interference and diffraction of light and measure the bands. *Elliott Brothers.*

915b. Collection of Six Gratings (*réseaux*) by Nobert of Barth, and Steeg of Homburg. *Teyler Foundation, Haarlem.*

Nobert B. of 1,801 lines in six Paris lines.
   ,,   C. of 3,001 ,,   ,,   ,,
   ,,   D. of 10,801 ,,   in one Paris inch.
   ,,   E. of 2,001 ,,   ,,   ,,
   ,,   F. of 3,001 ,,   ,,   ,,
Steeg A. of 3,201 lines in five millimètres.

Nobert B and C were used by Prof. Van der Willigen for the determination of the wave-lengths of fifty lines in the solar spectrum.

915c. Specimens of **Circular Gratings** (*Réseaux*) photographed on **Glass.** The rays of the successive circles limiting the opaque and transparent parts are in the proportion of 1 to $\sqrt{2}$, $\sqrt{3}$, $\sqrt{4}$, $\sqrt{5}$, &c.

915d. Large **Circular Grating** on **smoked Glass,** transparent traces of equal width, having rays in the proportion of $\sqrt{3}$, $\sqrt{7}$, $\sqrt{11}$, $\sqrt{15}$, $\sqrt{19}$, &c. *J. Louis Soret, Geneva.*

See " Mémoire sur les Phénomènes de Diffraction produits par les Réseaux circulaires " (Archives des Sciences physiques et naturelles, 1875, Vol. 52, p. 320. Poggendorff's Annalen, 1875, No. 9).

915e. Telescope with Circular Gratings. Constructed by the Geneva Association for the Construction of Scientific Instruments. *J. Louis Soret, Geneva.*

1st arrangement. The smoked glass grating is used for objective with a common eyepiece. Looking at a gas jet (for instance) at seven metres distance, the distance of the objective and the eyepiece being from 34 to 41 centimetres, then pulling out one of the tubes the image of the jet is seen reversed, and coloured more or less. By pushing in the tube as much as possible, the second image is seen green-coloured.

2nd arrangement. A common objective is used, and for eyepiece the small photographic grating. The distance from the objective to the eyepiece being of 50 centimetres (maximum length of the telescope), the image of the gas jet, reversed, is obtained as in an astronomical glass. By pushing in the tube to 31 centimetres, the direct image' is got as in the Galileo telescope. (*See* Mémoire sur les Phénomènes de Diffraction produits par les Réseaux circulaires, Archives des Sciences physiques et naturelles, 1875, vol. 52, p. 320.)

915f. Diffraction Grating on speculum metal. *A. Hilger.*

916. Refractometer, according to **Abbe's** system, for determining the refractive indices, and the dispersion of any kind of liquids. *Carl Zeiss, Jena.*

The refractometer enables the determination of the refractive index of a liquid to be effected up to four decimals with a single drop of the substance. The readings refer to line D, and are read off from a graduated sector.

917. Procentum Refractometer, for determining the percentage of solutions and mixtures by optical means.
Carl Zeiss, Jena.

The instrument is designed for liquids whose index lies between 1·3 and 1·4. The determination takes place at a numbered scale in the field of view of a small telescope. Besides the scale for the absolute index of refraction, there is another scale, which gives directly the per-centage strength of saccharine liquor.

917a. Refractometer by M. Jamin.
Polytechnic School, Paris.

917b. Jamin's Interference Apparatus with two Spars. *M. Lutz.*

917c. Refraction Goniometer, constructed by the Rev. Baden Powell, and used in some of his experiments, and afterwards by the Rev. T. Pelham Dale and Dr. Gladstone in their earlier researches on refractive indices of liquids at different temperatures. *Mrs. Baden Powell.*

919. Abbe's Refractometer, for determining the power of refraction of different liquids as far as the fifth decimal, with direct reading of the refractive index, without calculation. (Comp. Abbe : " Neue Apparate.")
Franz Schmidt and Haensch, Berlin.

921. Apparatus on J. Müller's Principle, for Experiments on the Refraction of Rays of Light in Fluids.
Warmbrunn, Quilitz, and Co., Berlin.

921a. Apparatus by M. Mascart for studying the Refraction of Gas.
M. Mascart, Professor at the College of France.

921a. Drawing of the **Apparatus** used in 1842 by Prof. Daniel Colladon to show the total refraction of light in the interior of a vein of water. *See* Comptes Rendus, vol. 15, p. 800.

Prof. Daniel Colladon, Geneva.

This remarkable experiment was made on a large scale by the French Government on several occasions at public festivals.

921b. Apparatus for determining the Refractive Index of Solids and Liquids.

C. Czechovicz, Teacher of Physics at the Gymnasium, Belostok, Russia.

Consists of a horizontal board and vertical divided pillar with movable support for a telescope. The body under examination is put on a glass plane attached over a slit in the board, through which a light beam is reflected by an inclined mirror. A linear mark made on the upper surface of the glass (if the body is solid), or on the upper surface of the vessel (if the body is liquid), is brought in coincidence with a movable wire which touches the upper plane of the body. The distance of this wire and the height and inclination of the telescope give the necessary data for calculating the index with sufficient approximation.

VI.—FLUORESCENT BODIES.

922. Fluids showing the **Phenomenon** of **Fluorescence.**

Charles Horner.

A. Soda salt of anthracene in water.
B. Fluorescine in water.
C. Eosine in water.

923. Fluids showing the **Phenomenon** of **Fluorescence.**

Charles Horner.

Small Tubes in Stand.

D. Turmeric in castor oil.
E. Harmaline in water.
F. Magdala red in alcohol.
G. Ebony wood (*Amerimuum ebenus*) in castor oil.
H. Induline in chloroform.
I. Esculine in water.
K. Camwood (*Baphia nitida*) in castor oil.
L. Esculetine in water with alkali.
M. Fraxine (*Fraxinus excelsior*) in water.
N. Fustic (*Maclura tinctoria*) in solution of alum.

924. Selection of eight fluorescent liquids.

Dr. Th. Schuchardt, Görlitz.

Eosine.
Magdala Red.
Saffranine.
Fraxine.

Extract of Cuba-wood.
Aesculine.
Bichloranthracine.
Bisulphurous acid.

VII.—PHOTOGRAPHY.

a. PHOTOGRAPHIC PROCESSES.

925. Frame containing glass negative, gelatine "relief," leaden "mould," and impression from the latter, showing the various stages in the production of a "Woodbury" permanent photograph.

Woodbury Permanent Photographic Printing Company.

926. Frame containing transparencies for the magic lantern, printed by the "Woodbury" process.

Woodbury Permanent Photographic Printing Company.

927. Specimens of Willis's Aniline Process. Printed from tracings by Vincent Brooks, Day and Son.

William Willis.

This is a method of photographic printing differing greatly from all other kinds in the chemical actions involved and in the manipulations required. The blacks of the picture are produced by the action of aniline vapour on free chromic acid ; the paper having been first coated with the latter substance, and exposed to light under the drawing to be copied. On placing this exposed sheet in a chamber filled with aniline vapour the yellow unaltered chromic acid becomes speedily blackened, and produces a permanent print. No negative is required, but a positive print is obtained by one operation from a positive original. The principal application of the process is the copying of engineers' and architects' tracings.

928. Willis's Platinum Printing Process.

Wm. Willis.

This is a method of photographic printing by which the picture is made to consist of platinum black instead of silver. The reduction of the platinum salt, with which the paper is coated, is effected by the action of light on a persalt of iron, which forms an additional coating to the paper, followed by a floating of the print on a solution of potassic oxalate.

928a. Illustrations of the Heliotype Process.

B. J. Edwards and Co.

The photographs are printed in printers' ink, at an ordinary printing press, from a film of gelatine, to which the photographic image has been transferred by the action of light.

1. Gelatine film ready for exposure to light under the negative.
2. Film laid down upon a metal plate and inked up ready for printing from.
3. Proof in permanent ink from the same film.

928b. Photograph Reproductions.

1. Embroidered stuff on velvet ground.
2. Faience dish, after Bernard Palissy.
3. Blue framed enamel, pâte tendre de Solon.
4. Louis XIV. shield, repoussé copper.
5. Incense-burners, silver filigree, enamel and stones.

6. Mirror, silver-gilt and precious stones.
7. Tobacco jar, gold and silver.
8. Holy-water vessel, Limoges enamel, with frame and precious stones.
9. Hunting knife handle, silver and stones.
10. Byzantine dish, repoussé copper.
11. Aliotide shell, from nature.
12. Cup, silver and rubies.
13. "The First Fable," after the oil painting by Simonetti (Salon de 1875).
14. "After Action," after the oil painting by Marchetti (Salon de 1875). *M. Léon Vidal, Paris.*

928c. Specimens of Dallastype. *D. C. Dallas.*

928d. Specimens of Dallastint. *D. C. Dallas.*

928e. Specimens of Chromo-Dallastint. *D. C. Dallas.*

928f. Specimens of Carbon transparencies.
Col. Stuart-Wortley.

928g. "Cleopatra," a solar enlargement on salted paper, by R. Fenton, from a photograph of the original in the National Gallery. *Robert Sabine.*

928ee. Chromo-Woodbury Type, combination of Woodbury type with chromo-lithography. *Walter B. Woodbury.*

928ef. Kaleidoscopic Photograph of Ferns.
Walter B. Woodbury.

928h. Seven Photographic Prints on Salted Paper, from Waxed Paper Negatives.

Two views of Windmill for reflecting stereoscope, by B. B. Turner.

Two views of the First Post on the road from Kief to Moscow by R. Fenton, for reflecting stereoscope.

Two views of a Russian Cottage, for reflecting stereoscope, by R. Fenton.

View of ruined Interior, by Le Gray. *Robert Sabine.*

928j. Two Sheets of Photographic Prints, by the Iron and Uranium process. M. Nièpce de St. Victor, 1857.
Robert Sabine.

928k. Five Sheets of Photographic Prints, from waxed paper negatives on albumenized paper, by R. Fenton.
Robert Sabine.

928l. Ten Photographic Prints from waxed paper negatives on albumenized paper, 1853. *Robert Sabine.*

928m. Two Sheets of Talbotype Prints from waxed
paper negatives.   *Robert Sabine.*

928n. Two Views for the Reflecting Stereoscope;
the varnish used to render the prints transparent, having preserved
the details of image from fading.   *Robert Sabine.*

929. Specimens of Dujardin's photo-engraving process.
Dujardin, Paris.

930. Specimens of Photo-type Printing on Zinc.
Capt. Abney, R.E., F.R.S.

931. Gillot's Photo-type Process.   *Veuve Gillot.*

932. First known Photograph on Glass, taken on precipitated silver chloride, by Sir J. Herschel (Slough, 1839.)
Prof. A. S. Herschel.

" Having precipitated muriate of silver in a very delicately divided state from
water very slightly muriated it was allowed to settle on a glass plate; after 48
hours it had formed a film thin enough to bear drawing the water off very
slowly by a siphon, and drying. Having dried it I found that it was very
little affected by light, but with washing with weak nitrate of silver and drying
it became highly sensible. In this state I took a camera picture of the telescope on it. Hyposulph. soda then poured cautiously down washes away the
muriate of silver, and leaves a beautiful delicate film of silver representing the
picture. If then the other side of the glass be smoked and black varnished
the effect is much resembling daguerreotype, being dark on white as in nature,
and also right and left as in nature, and as if on polished silver."—*Sir J.
Herschel* (MS. Journal of Experiments).

937b. Original Book of Experiments made by **Sir J.
Herschel** on the **Metallic Salts** sensitive to **Light.**
Prof. A. S. Herschel.

933. Second Daguerreotype Proof, obtained by Daguerre
in 1839.   *Conservatoire des Arts et Métiers, Paris.*

Lithographic Stone of Poiteven, with a proof on paper.
Conservatoire des Arts et Métiers, Paris.

935. Photographs by Daguerre.   *M. Fizeau, Paris.*

**937. Specimens illustrating the History of Photo-
graphy.**   *French Photographic Society, Paris.*

937a. Daguerreotype full-length Portrait, taken in Paris
in **1840,** by special appointment, on the roof of a house in the
open air, at 6 a.m. Exposure 20 minutes, in June sun.
James Martin.

938. Instantaneous Photograph. Waves breaking on the
shore of Britain in 1876.   *James Martin.*

939. Engravings with the Aid of Photography.
MM. Goupil et Cie., Paris.

Proofs obtained by impression with fatty ink on copper plates engraved by hand, the lines on which are obtained by means of a photographic negative, with the use of chemical substances sensitive to the action of light.

1. " Pollice verso," after the painting by Gérôme.
2. " L'eminence grise,"   do.   do.
3. " Rembrant dans son atelier " (Rembrant in his studio), after the painting by Gérôme.
4. " Il Décamerone " (The Decameron), Sorbi.
5. " La rentrée au Convent " (The return to the Convent), Zamacois.
6. " Le premier coup de canon " (The first cannon shot), Berne, Bellewurt.
7. Part of the colonnade of the Louvre, from nature.
8. Reproduction of a mineralogical fragment, from nature.
9. Frame containing the two plates from which proofs Nos. 7 and 8 were printed.

940. Photographic Prints, in ink.   *Thiel Ainé, Paris.*

CLASS THE 1ST.

1. Reproduction of water-colour painting—A cow, after Troyon.
2.   ,,   chalk drawing—Park of the Marquis de Mégrigny, after Lalanne.
3.   ,,   chalk drawing—An old courtyard at Colombes, after Lalanne.
4.   ,,   oil painting—The Resurrection, after Lazerges.
5.   ,,   chalk drawing—Borders of the Rhine at Mulhouse, after Niederhausen.
6.   ,,   chalk drawing—Ruins of a temple, after Lalanne.
7.   ,,   oil painting—Presenting the Bride, after de Boucher ville.

CLASS THE 2ND.

8. Reproduction of chalk drawing—In the park at Plombières, Vosges, after Allongé. (Salvon, 1875.)
9.   ,,   of chalk drawing—Rocks and lake, after Appian.
10.   ,,   from nature—Church of St. Augustin, Paris.
11.   ,,   ,,   —Opera-house, Paris.
12.   ,,   ,,   —Notre Dame, Paris.
13.   ,,   of chalk drawing—Borders of the Lake of Arandou, after Appian.
14.   ,,   of chalk drawing—Borders of a pond, after Allongé.

CLASS THE 3RD.

15. Reproduction of terra-cotta—Marguerite with the jewels, after Carrier-Bellenze.
16.   ,,   of oil painting—The imprisoned loves, after Chaplin.
17.   ,,   from nature—Study of foreground, stereotyped on paper of Mr. H. Le Secq.
18   ,,   from nature—Cathedral of Rheims, stereotyped on paper of Mr. H. Le Secq.
19.   ,,   of chalk drawing—The Bois de Boulogne, after Lalanne.
20.   ,,   from nature—Cathedral of Rheims, stereotyped on paper of Mr. Le Secq.
21.   ,,   from nature—Study of foreground, stereotyped on paper of Mr. Le Secq.
22.   ,,   of oil painting—The death of Asala.
23.   ,,   of drawing—Head of Christ, after Lazergues.

24. Reproduction of oil painting—Diana's toilet, after Devedeux. (Imitation of photography with salts of silver.)
25.   ,,   from nature—A hawk ; still life. (Imitation of photography with salts of silver.)
26.   ,,   from nature—Sèvres vases. (Imitation of photography with salts of silver.)
27.   ,,   from nature—Naumachy in the Park Monceaux. (Imitation of photography with salts of silver.)
28.   ,,   of chalk drawing—The borders of the Yères, after Allongé.
29.   ,,   from nature—The bridge of Solférino, Paris. (Imitation of photography with salts of silver.)
30.   ,,   from nature—Sèvres vases. (Imitation of photography with salts of silver.)
31.   ,,   of a print—Musings, after de Moussy.
32.   ,,   of terra-cotta—The Mountebank's, after Deca.

943. Photo-lithography Process of Simonan and Toovey. *Veuve Simonan and Toovey.*

1. Plan of the town of Liège.
2. Portrait of Archbishop St. Lambert, after an old engraving.
3-7. Topographical plans, photographed by Capt. Hanot.
8-13. Six drawings of the " Campagnie des bronzes " at Bruxelles.
14-15. Two reproductions from a line drawing by Licot de Nivelles.
16-19. Four archæological drawings.
20. Frontispiece of an ancient MS.

This is a photo-lithographic process, and depends on the fact that if gum be mixed with potassium dichromate, and when dry be exposed to the action of light, it becomes insoluble. A paper is coated with gum and potassium dichromate, and exposed under the negative of a line subject, or under an etching on glass, having a non-actinic ground. When light has sufficiently acted, the paper which has a faint impression of the lines is placed under a pile of damped paper on the surface of a polished lithographic stone, and submitted to pressure for about an hour. The paper is then removed from the surface of the stone, the insoluble part forming the lines coming away with it. The lines of the engraving are thus left ungummed on the stone. A little olive oil is brushed over the surface, when the gum on the stone has been allowed to dry in a dark room. The surface is next washed, which dissolves away the gum, leaving the lines of the picture only. The stone is then rolled up with a lithographic roller, and is ready for giving impressions.

944. Specimens of Paul Pretsch's Photo-typography. *Warren De La Rue, F.R.S.*

945. Electro-chemical Process for reproducing lithographic impressions on copper. *M. Erhard.*

A proof freshly pulled from an autograph, lithograph, auto-lithograph, or a copper-plate which is intended to be reproduced, is, by this process, transferred to a copper-plate, and furnishes in a few minutes an intaglio copy of the plate, as clean and good as the original, which is in no wise injured by the operation.

By means of this process : firstly, it is unnecessary to preserve the cumbrous and fragile lithographic stones ; secondly, a plate in use may be reproduced so as to ensure repeated impressions ; thirdly, corrections may be made on the copper, which could not be made on the original plate, worn by repeated working. The cost of reproduction on copper by Erhard's process is small, and may be estimated at about 3 to 5 centimes per square centimetre.

1. Album containing 36 maps and plans reproduced by this electro-chemical process.

2. 10 copper plates obtained by this process, the impressions from which are shown in the album.

946. Photoglyphic engravings, 1853.
H. Fox Talbot, F.R.S.

947. Silver prints of views in **Knoll Park.**
School of Military Engineering, Chatham.

947a. Proof by Papyrotype Process.
School of Military Engineering, Chatham.

947b. Specimens of Enlarging Process.
School of Military Engineering, Chatham.

948. Second Proof of Photographic Engraving, obtained by M. Fizeau, without retouching, in 1843, and printed in greasy ink. *M. Fizeau, Paris.*

949. Daguerreotype Proof, fixed by M. Fizeau's process with chloride of gold, by Hubert, in 1840. *M. Fizeau, Paris.*

949a. Daguerrean Print, obtained by the continuous action of red rays, without mercury. *M. E. Becquerel.*

950. Photochromic Proofs (selection). *M. Vidal, Paris.*

951. Early Talbotypes.
The Council of King's College, London.

951a. Specimens of Enamel Process. *Wm. Mayland.*

952. Table of Specimens of Historical Records of Photography. *French Photographic Society.*

952a. Application of Photography to Cartography.
The Topographical Department of the Imperial Russian General Staff, St. Petersburg.

In the application of the negative process, Rupell's drying system, with tannin, has been employed. The positive prints are either black silver copies or blue iron pictures : preference is given to the latter if the photograph is to be traced over with Indian ink, and the photographical ground to be removed afterwards by etching for the purpose of producing a clean drawing.

1. Reproductions of Central Asiatic surveys and maps.
2. Copies of plates of survey sheets in European Russia.

952b. Photolithography.
The Topographical Department of the Imperial Russian General Staff, St. Petersburg.

Transfer on stone of a printing picture, well covered with ink, which has been produced on a gelatine ground rendered primarily sensitive by double chromate of potash.

Reproduction of a Hebrew manuscript of the 10th century, belonging to the Imperial Russian Public Library.

952c. Helio-Engraving.   Sediment of galvanic copper on a photographical gelatine relief.
The Topographical Department of the Imperial Russian General Staff, St. Petersburg.

Copies of a heliographical edition of the survey of Bessarabia, on the scales of 1 : 100,000, and 1 : 126,000, and the survey of Finnland, scale, 1 : 42,000; map of Khiva, scale, 1 : 580,000, transferred on stone, and prepared as colour print.

952d. Examples of Heliographic and other Processes.
Imperial Establishment for the preparation of official papers, St. Petersburg.

1. Portfolio of heliographic copper-plate and mezzo-tint engravings by the process of G. Scamoni (manager of the Heliographic Department of the Establishment), containing :—
   27 reproductions of historical portraits ;
   10 reproductions of fine engravings ;
   12 reproductions of etchings ;
   8 reproductions of drawings executed with pen and ink, water-colour, and crayon ;
   17 reproductions of pen and ink drawings ;
   6 reproductions of wood engravings.
2. Heliographic plate in electrotyped copper.
3. Heliographic plate in electrotyped iron.
4. Typographic printing form in electrotyped iron, from type.
5. Typographic printing form in electrotyped iron, net-work from type.
6. Typographic printing form in electrotyped iron, net-work from relief.
7. Typographic printing form in electrotyped iron, guilloched net-work.
8. Typographic printing form in electrotyped iron, id-annealed.
9. Glass plate, with the surface irregularly broken up into floral and other forms through a coating of gelatine ($\frac{1}{2}$ millimeter thick), springing up from it when dried at a temperature of about 70° C., thereby producing a form from which an inimitable printing-plate can be made.
10. Handbook of heliography, by G. Scamoni.

b. Photographic Apparatus.

953. Photographic Lenses for Landscape, Architecture, and Copying, showing progressive improvements from the original single meniscus lens:—

(*a.*) Single meniscus lens, used from 1851 to 1861.

(*b.*) Triplet, consisting of front combination. Double convex crown and plane concave flint ; middle combination.

Double convex crown and double concave flint; back combination. Double concave flint, and double convex crown. Used from 1861 to 1864.

(c.) Doublet, consisting of front combination. Double convex crown and double convex flint; back combination, meniscus crown and concavo-convex flint. Used from 1864 to 1874.

(d.) Symmetrical lens, introduced in 1874, and consisting of front combination. Concavo-convex and meniscus lenses; back combination, exactly similar, the denser element being on the outside in both cases.   *Ross & Co.*

954. Photographic Lenses for Portraiture, showing the progressive improvements from 1839 to present date :—

(a.) Original compound portrait lens. The first lens made in England by Andrew Ross for daguerreotype portraiture in 1839.

(b.) Compound portrait lens, with Waterhouse diaphragms, in front. Date 1851.

(c.) Compound portrait lens, with Waterhouse diaphragms, giving a flat field. Date 1858.

(d.) Compound portrait and group lens, giving a flat field, and straight marginal lines. Date 1874.   *Ross & Co.*

954a. New Tourists' Photographic Apparatus for taking Wet Plates without the use of Dark Tent, all baths and chemicals being placed in water-tight compartments under body of camera.   *Harvey, Reynolds, and Company.*

954b. Photographic Lens with which the pictures of Mr. Fox Talbot's Pencil of Nature were taken. (This is the first publication of photographs printed from negatives on paper.) Presented.   *B. B. Turner.*

955. Photographic Apparatus. " Poor man's photo graphy," for wet collodion negatives of the smallest possible size, but rapid and well defined. Twelve examples of negatives, 1 inch square, two framed and magnified positive copies, and the bath-holder in which these negatives were taken.

Prof. Piazzi Smyth.

These negatives are on microscope slide glasses, and were taken with a lens of rather less than 2 inch solar focus by Professor Piazzi Smyth, in Egypt, in 1865. They represent scenes inside the Great Pyramid by magnesium light, and outside it by daylight, including, in one of them, camels in motion. The two positive copies on glass, each 10 in. high, are exhibited to show to what extent magnifying may be carried without definition being lost to any sensible degree.

The peculiar bath-holder in which the small negatives were taken is also shown. It has been described in the " British Journal of Photography," in 1866, with improvements, in the almanacs of the same journal for 1874 and 1876.

955a. Photographic Apparatus, by M. A. Chevalier, for plan-drawing.   *M. J. Duboscq, Paris.*

956. Lichtpaus Apparatus, for photographing maps, plans, &c.   *Romain Talbot, Berlin.*

957. Press used in printing a " Woodbury " permanent photograph, with leaden mould in position.

Woodbury Permanent Photographic Printing Company.

958. Photographic Portrait Camera, D.
Voigtländer and Son (Chevalier von Voigtländer), Brunswick.

958a. Photographic Portrait Camera, No. 3.
Voigtländer and Son (Chevalier von Voigtländer), Brunswick.

959. Photographic Lens for Astronomical Photography.   *John Henry Dallmeyer.*

A double combination lens (rapid rectilinear) of 4″ diameter and 30″ focal length, consisting of two symmetrical combinations, each having a focus of 63″, and composed respectively of a crown and flint-glass lens, united by a permanently transparent cement to avoid reflection at the contact surfaces. The flint lens occupies the exterior position in each combination. It is concavo-convex, convex side external. The crown lens is a meniscus, the convex side of the same radius of curvature as the flint lens.

The radii of curvatures are so apportioned between the lenses that the spherical,˙ and chromatic, aberrations are destroyed ; or, in other words, the combination is aplanatic, and this for aperture $\dfrac{f}{7}$ to $\dfrac{f}{8}$.

The lens described (one of a series) was constructed and used for photographing the sun's corona.

959a. Photographic Lens for Self-Recording Instruments, *i.e.,* Barographs, Thermographs, &c.

John Henry Dallmeyer.

A double combination lens (No. 2 C) of $2\frac{3}{4}''$ diameter and $4\frac{1}{4}''$ focus. It consists of a cemented front and an open back-combination, having a large angular aperture, *i.e.,* possessing great intensity. The front, or cemented combination, is composed of a double convex crown lens, of unequal curvatures, the shallow side occupying the outer position ; and a double concave flint lens is united to the deep side of the crown, the adjacent surfaces being identical. At an interval equal to the diameter of the front is placed the back combination, composed of a concavo-convex flint lens, convex side facing the front combination, and a crown lens nearly plano-convex, with the more convex side nearest the flint lens, but of different radii of curvature, and therefore not cemented.

The ratio of foci between the front and back combinations is 2:3 nearly, and the radii of curvatures are calculated to produce well defined images for aperture $=\dfrac{f}{2}$, and angle of picture $= 30°$.

959b. Photographic Lenses for the reproduction of maps, plans, &c., as employed at the Home and Foreign Government Topographical Establishments.   *John Henry Dallmeyer.*

1. A triple achromatic lens of 18″ focal length for copying on plates 15″ × 12″. As its name implies, this lens consists of three achromatic, or actinic, combinations, two of which, *i.e.* the front and back, are positive or converging lenses, and a negative or diverging lens placed between them. The positive combinations, the front of 2″ and the back of 3″ diameter respectively, with focal lengths of similar proportion, are composed each of a double convex crown, and a plano-concave flint-glass, lens; the adjacent surfaces being identical and cemented. The convex or crown lenses occupy the external position in both, and the combinations are separated by an interval equal to the diameter of the largest or posterior lens. Between the two, and proportionate to their diameters or foci, is placed the diaphragm aperture, or stop ; this is also the position of the negative achromatic combination 1½″ diameter, the crown of which is in this case a double concave, and the flint a plano convex nearly. This combination limits the aperture to $\frac{f}{10}$; it does not affect the direction of the pencils as refracted by the front and back positive combinations, which produce an image free from distortion, though too much curved to admit of its reception on a flat screen ; but its action (that of central pencils only) is confined to the proportionately greater prolongation of the oblique or marginal pencils, in virtue of its negative or divergent power. In other words, it lengthens these pencils and produces the required amount of flatness of field, the *sine quâ non* for copying purposes.

This lens, one of a series, was introduced in 1860, and is reported upon by the jurors of the International Exhibition of 1862: "As the first aplanatic " non-distorting view lens placed within the reach of photographers, and the " best lens extant for copying purposes, &c." It was first used in the Ordnance Survey Office at Southampton.

2. A 3″ symmetrical combination of 24″ focal length for plates 18″ × 16″. This lens, constructed for copying, consists of two combinations identical in all respects ; each has a focal length of 46″, and is composed of a convexo-concave flint, and a convexo-concave or meniscus crown-glass lens ; the radii of curvatures of the concave of the one and the convex of the other are identical and they are cemented.

The combinations, with their convex or flint elements external, are mounted in a tube a given distance apart, in this case about ⅓ of the compound focal length, and midway between them is a perforation for the insertion of diaphragms or stops. Both combinations being identical in form and focal length, the direction of the finally emergent pencils is parallel to the incident ones ; hence the lens is free from distortion.

Flatness of field and correction of aberrations are obtained, for the qualities of glass employed, by the forms and foci of the component elements, and in order to suit particular requirements the construction is modified for varying angular apertures of from $\frac{f}{4}$ to $\frac{f}{10}$, including proportionate angles of pictures of from 35° to 90°. This lens was introduced in 1866.

960. Photographic Lenses for Portraits and Views.
John Henry Dallmeyer.

1. & 2. These portrait lenses, introduced in 1866, are constructed on a new formula, as compared with the first portrait lens, the invention of Professor Petzval of Vienna, and which appeared in 1841.

Unlike the lenses used for astronomical photography, copying, &c., required to produce a sharp image of an object situated in *one* plane, and necessitating perfect correction for aberration, a portrait lens must produce a presentable picture of the human face, head, and body, situated in *different* planes ; or

40075. Q

possess what is called "depth of focus." A *perfectly corrected lens,* of sufficient angular aperture or rapidity of action to make portraiture possible, has no "depth of focus." An instrument, possessing a certain residual amount of spherical aberration, solves the difficulty to some extent. Unfortunately, however, a lens so constructed works at its best only for a given size of image or distance of object. If this be removed to a greater distance, as for a smaller image, the aberration is in excess, and if placed nearer, the converse obtains.

The lens now to be described surmounts this difficulty. Its aberrations, both spherical and chromatic, are perfectly corrected when used intact, or as sent out by the maker; and by the simple turn of a screw, separating the component elements of the posterior combination, this correction can be modified at will, *i.e.* positive spherical aberration, or "depth of focus," is obtained, proportionate in amount to the separation of the posterior lenses.

The front combination, composed of a double convex crown, and a double concave flint glass lens, is cemented, to prevent loss of light from reflection. The convex or crown lens occupies the exterior position. At a distance equal to the diameter of the front is placed the back combination; an uncemented compound of the same diameter as the front, but of greater focal length, *i.e.* as 2 : 3. The crown element of the back combination is a meniscus with its concave surface facing the front combination. The flint is concavo-convex, the convex side external : this lens is mounted in a cell, the screw of which affords the means of approach to, or separation from, the companion crown lens. An index and division registers the amount of separation. The converging cone of rays refracted by the front is incident upon the crown element of the back combination ; on emergence it is more strongly convergent when it meets the concave surface of the posterior flint lens. It is evident that any alteration in position or distance at once reduces or increases its effective diameter, or, in other words, its aberration (y^2). The aberrations being perfectly corrected when the posterior flint lens is screwed home, the index pointing zero, the smallest amount of unscrewing or increased separation at once introduces positive spherical aberration.

In Petzval's lens the position of the lenses composing the back combination is the reverse of the above; *i.e.* the flint, or convexo-concave element, faces the front combination, the rays when refracted by it are parallel or nearly so, and any alteration of distance of the companion crown lens is without effect upon the state of aberration of the objective as a whole.

The portrait lens constructed on the new formula is free from disturbing reflex images ; it produces pictures approximately free from distortion, and illuminates a larger angle of field. It is made of three descriptions: viz. the A series, aperture $= \frac{f}{4}$, B $= \frac{f}{3}$, and D $= \frac{f}{6}$.

3. A single combination landscape lens of 2″ diameter and 12″ focal length for pictures on plates 12″ × 10″. It is composed of three lenses, two of which are of crown glass, but of different optical properties, and between these two is the correcting flint glass lens. The crown lenses are deep menisci, ratio of foci of front to back as 1:3; the flint is concavo-convex. The adjacent surfaces of the crowns and flint being identical they are cemented, and externally the combination is a deep concavo-convex, the concave side facing the view or landscape. Its aperture is limited by a diaphragm plate placed at a distance of ¼ the focal length in front of the lens; the stops provide apertures from $\frac{f}{12}$ to $\frac{f}{40}$. Correction of chromatic and spherical aberration is obtained by the foci and forms of the lenses employed.

The angle of picture included exceeds of 70°, and though marginal distortion of image is not entirely avoided, it is perhaps counterbalanced by the increased brilliancy of image and equality of illumination, due to the absence of all disturbing reflex images.

4. A wide angle rectilinear lens of $1\frac{1}{2}''$ diameter, and $7''$ equivalent focal length for $12'' \times 10''$ pictures.

This lens is constructed upon the same general principle as the symmetrical $3''$ combination already described, with this difference, that the lenses composing it, though similar in form, are of proportionately smaller diameter, *i.e.*, they are thinner, and being placed nearer together, they transmit more oblique pencils, and a larger angle of picture is included, as much as 90°, with apertures $\frac{f}{10}$ to $\frac{f}{40}$.

This instrument is designed for photographing objects in confined situations, such as interiors of buildings, monuments, &c., where the camera cannot be removed to more than a given distance. It produces images free from distortion and flare.

962a. Apparatus for producing photographs in permanent pigments : consisting of,—

Registering pressure frame.

Plates of zinc, porcelain and opal glass (the plates having upon them : *a.* The sensitive tissue ready for development. '*b.* The supporting paper stripped off. *c.* The picture partly developed. *d.* The picture ready for transfer).

Reservoir for hot water, with means of keeping it warm, and grooves for the plates.

Zinc trays for development washing, &c.

Wooden stool and squeezer for the mounting of the exposed sensitive tissue. *The Autotype Company.*

If a negative in half-tones be placed under a paper coated with gelatine pigments of any kind, and potassium dichromate, and then be exposed to light, the gelatinous film is rendered insoluble, to depths varying according to the intensity of the light passing through the various portions of the negative. If the paper were at this stage exposed to the action of hot water, it would be found that the soluble portions lay between the exterior surface of the film and the inner surface of the paper. In order to develop the picture, it is transferred by simple atmospheric pressure to a zinc plate or other temporary and impervious support. The original paper peels off, and the development takes place by the simple application of hot water, the shades of the image being formed of different thickness of gelatine and pigment. The image may be retransferred from the temporary support, or if developed on a paper support may be left as it is, in which case, unless the negative be reversed, the picture will appear reversed.

962c. Sir John Herschel's Actinometer, by Robinson, marked I and K.

The Meteorological Committee of the Royal Society.

977a. Actinometer. *Prof. Balfour Stewart.*

962b. Actinometers, or instruments for ascertaining the intensity of the action of the light, and by which the exposure of the sensitive pigment paper under the negative is regulated.

Johnson's, Vogel's, Spencer's, Lambert's, Sawyer's, Burton's, Vidal's. *The Autotype Company.*

Chemicals Employed.

Granulated bichromate of potash.   Chrome alum.

Colours.

Indian ink, vegetable black, Paris black, plumbago, Indian red, Venetiau red, vermillion, purple madder, brown madder, vandyke brown, indigo, lac de vin, various kinds of gelatines.

962c. Pigmented Papers, in various colours, with subjects printed to show the various shades.

962d. Transfer Papers (*i.e.*, papers prepared with insoluble gelatine, and upon which the pictures finally rest).

962e. Sawyer's Pàtent Flexible Support (being paper prepared with insoluble gelatine and an aqueous solution of lac, upon which the picture is first developed previously to its transfer to the final support.   *The Autotype Company.*

962f. Reversing Mirror, being a piece of plate glass polished to a perfectly true plane, and silvered by a chemical process used to produce *reversed* negatives, enabling them to be printed by the single transfer process of permanent pigment printing.   *The Autotype Company.*

962g. Wave Bath, for nitrate of silver solution (being a new and convenient form for sensitizing large plates with a comparatively small quantity of solution.   *The Autotype Company.*

962h. Sawyer's Collotype Process.
          The Autotype Company.

Glass plates prepared with gelatine and isinglass, potassium dichromate, &c., hardened by a spirituous extract of gum resins, upon which the photograph is impressed by the action of light; after which they are placed in a type printing press, damped, and inked by lithographic rollers.
(*a.*) Plate in first stage of preparation.
(*b.*) Plate ready for exposure under negative.
(*c.*) Plate after exposure under negative.
(*d.*) Plate after being inked up in the press.
(*e.*) Plate with the same picture partly showing on the paper and partly on the plate.

988i. Four Prints from Photographs, by Paul Pretsch's mechanical process.   *Robert Sabine.*

VIII.—EDUCATIONAL.

963. Frame with 163 Photographs on Glass, for projection by the lantern, for instruction in the natural sciences.
          Romain Talbot, Berlin.

964. Four simple Models, for instruction in the use of the telescope and the microscope.

J. Wilhelm Albert, Frankfort-on-Maine.

The four models for optical instruction are open instruments with lenses and shades. They show the course of the rays and illustrate in a simple manner the Galilean, astronomical, and terrestrial telescopes, and the compound microscope.

These models are largely used in German and foreign educational institutions.

964a. Sciopticon and three Manuals.

W. B. Woodbury.

965. Coloured Chalks for lectures, with a Black Board.

C. Blattner, Munich.

966. Interference Apparatus by Fresnel, executed for educational purposes by Ch. Jung, Giessen.

Physical Collection of the University of Giessen, Prof. Buff.

918. Apparatus for demonstrating the **Refraction of Light** in liquids, according to J. Müller.

J. Wilhelm Albert, Frankfort-on-Maine.

The semicircular plate of the refraction apparatus is of glass, ground on its outside and having the scale burnt into it in black. The ray refracted from the liquid appears therefore on the outside of the graduated plate and can thus be viewed by large audiences.

967. Educational Apparatus, for elementary experiments on **Refraction.** *Prof. Dr. J. J. Oppel, Frankfort-on-Maine.*

This apparatus is only a modification of the old experiment of viewing a coin at the bottom of a basin full of water. The coin is replaced by a white line upon black ground, and the water by a movable cube of glass. The position of the eye is fixed by a dioptric plate.

968. Educational Apparatus for illustrating **astronomical refraction** and its effect on measured heights of stars.

Prof. Dr. J. J. Oppel, Frankfort-on-Maine.

The eye looks through a dioptric plate over a wooden globe ("earth") towards a few stars and the rising moon, which appears above or below the horizon, according as the piece of glass, which represents the refracting atmosphere, is lifted up or removed by means of a contrivance attached to it.

969. Two Colour-tables, for illustrating "colour blindness," with greenish glass (absorbing the ends of the spectrum) belonging to them. *Prof. Dr. J. J. Oppel, Frankfort-on-Maine.*

Both tables, each with 10–12 colour-couples upon black ground, have reference to the most frequent form of achromatopsy, the so-called *red,* and *green,* blindness. The one table contains colour-couples which are most generally mistaken for one another; the other those that are never mistaken. The green glass, added to the tables, enables a normal eye to get an idea of that peculiarity of vision.

970. Apparatus for illustrating the Colours of double refracting Bodies, in the form of a Gothic window, composed of gypsum plates systematically arranged according to the colours; with a blackened glass plate belonging to it.

Prof. Dr. J. J. Oppel, Frankfort-on-Maine.

971. Some characteristic Drawings for illustrating the **Stroboscopical Principle,** with manifold movements, as,—forwards and backwards, centripetal and centrifugal, undulatory, oscillatory, and quite irregular.

Prof. Dr. J. J. Oppel, Frankfort-on-Maine.

Is a collection of those principal forms of periodic movements which can be represented stroboscopically.

IX.—MISCELLANEOUS.

897. First Heliostat, invented by 'sGravesande.

Prof. Dr. P. L. Rijke, Leyden.

(*See* 'sGravesande's "Physices Elementa Mathematica," ed. III., Tom. I., page 715.) 'sGravesande was an eminent Dutch geometrician, b. 1688, d. 1742.

898. Heliostat, G. Johnstone Stoney's modification, made by Spencer and Sons, Dublin, a cheap and useful form.

Prof. W. F. Barrett.

971aa. Self-adjusting Heliostat, mounted on a declination bar magnet. *Dr. Stone.*

971ab. Mounted Prism of Quartz. *Dr. Stone.*

971ac. Mounted Prism of Iceland Spar. *Dr. Stone.*

971ad. Equatorial Heliostat for maintaining a reflected beam of solar light in a constant position, applicable to places whose aspect will admit of a beam of light from the north only being obtained.

By changing the dial wheel of the clock, it may be used from the south as a polar instrument. *Conrad W. Cooke.*

It consists of a plane mirror, adjustable in declination, revolving upon a polar axis and driven by a clock. The clock has two interchangeable dial wheels whose circumferences are respectively one-half and one-fourth that of the wheel upon the polar axis of the mirror. By using the smaller, the mirror is driven round once in 48 hours, and if the instrument be placed due north of the spot to which the reflected beam is to be directed, and in the same equatorial plane with it, and adjusted in declination, the beam will be maintained in that plane and in a fixed meridional direction.

By substituting the larger dial, the mirror will rotate once in 24 hours; and if the instrument be placed due south and in such a position that its polar axis passes through the spot to be illuminated, the beam will be equally constant in direction from the south.

Heliostat and glass prism.   *Prof. A. S. Herschel.*

971a. Wheatstone's Apparatus.   *Paris Observatory.*

971b. Interferential Apparatus, by Arago.
             Paris Observatory.

3699. Reflecting Stephanoscope.
           Prof. Dr. Lommel, Erlangen.

(1.) The little apparatus is intended for the observation of interference phenomena produced by a dimmed (trübe) mirror. The mirror is at one end of a short brass tube, which has a lateral excision, and contains a second tube, whose one end is cut off at an angle of 45° to the axis, and is there closed with a plane glass. If the light of a candle is allowed to fall through the lateral apertures upon the plane glass, so that it is reflected vertically to the mirror, it will, on being viewed through the tubes, appear surrounded by *Newton's* rings, which pass, when the tube is slightly inclined, into the so-called, *Whewell's* bands. If a pencil of solar rays is caused to enter by the lateral opening, and a lens of large focus is put on to the end of the second tube, the above images can be thrown upon a screen.

3700. Erythroscope.   *Prof. Dr. Lommel, Erlangen.*

The erythroscope is an eye-glass of combined blue cobalt glass with red copper-suboxide glass, which gives passage only to the ultra red before line B. Since chlorophyll does not absorb this colour, the green parts of a plant appear, when viewed through the erythroscope, quite light coloured ; the foliage of a tree in sunshine, for instance, will appear as light as a white cloud, and form a light spot upon the dark ground of the sky.

3701. Melanoscope.   *Prof. Dr. Lommel, Erlangen.*

The melanoscope, a combination of dark red copper glass and light violet glass, allows chiefly the red rays between B and C to pass, which are absorbed with avidity by plants. Viewed therefore through the melanoscope plants will look very dark, almost black. This contrivance and the preceding one show in an instructive manner that plants absorb eagerly the middle portion of the red part of the spectrum, but not at all the ultra red.

3702. Erythrophytoscope I.
           Prof. Dr. Lommel, Erlangen.

This combination of *Simmler,* consisting of blue cobalt glass and dark yellow iron oxide glass, transmits ultra red up to B, and also to yellow-green, blue-green, and blue. Leaves, looked upon through it, appear coral red, the sky blue, the cloud reddish-violet, the soil violet-grey.

3703. Erythrophytoscope II.
           Prof. Dr. Lommel, Erlangen.

The erythrophytoscope II., combined of blue cobalt and light red copper glass, allows, besides ultra red, only blue-green and blue to pass. The effect is similar to that of the preceding combination, only more striking.

It should be remarked that, only vegetable and anilin green appear, as described under these combined glasses ; mineral green would look dark blue-green.

3704. Coloured Gelatine Leaflets as objects for the spectroscope.   *Prof. Dr. Lommel, Erlangen.*

To demonstrate the absorption phenomena of soluble colouring matters, plates of isinglass, which had been dyed with the colouring matter, are recommended, instead of the solutions. To protect these plates against dust, &c., they may be placed between two glass plates. By placing feebly coloured plates in a fan-shaped way upon one another, the increase of absorption by increasing thickness may easily be illustrated.

972. Mixoscope (colour-mixer), executed according to the directions of the exhibitor by M. Ph. Edelmann, Munich.

Prof. W. von Bezold, Royal Polytechnic School, Munich.

This apparatus gives the true colour resulting from the combination of two colours by actual trial with the brush, and thus a correct colour table can, by its means, be made with greater facility than with the revolving disc. The achromatised calcspar prism is so to be adjusted that on looking through the apparatus only six squares should be visible. It is easy to find this position through moving the telescope and turning the prism. On bringing the two colours to be mixed under two of the square openings, these colours will be right and left of the mixed colour, which fills the middle part of the three contiguous squares. If, now, the other two apertures contain the true optical compound colour, the two central squares will appear in the same tint. Compare the description annexed to the apparatus.

973. Clockwork for colour discs, for lecture purposes, according to Kühne's and Becker's principles.

Rud. Jung, Heidelberg.

The clockwork, provided with a strong spring, is capable of rotating coloured discs of 28 centim. diameter with such velocity that a disc covered with the tints of the spectrum will appear white. By a simple contrivance the one or the other of the spectrum colours can be excluded, and thus its complementary and contrast colour produced. By inserting a second axle, the movements of the clockwork can be retarded.

1065a. Colorimeter.   *M. Laurent, Paris.*

3659. Three Diaphragms and lamp stand for magic lantern.
Laurent.

3659a. Diaphragms (4), various.   *Laurent, Paris.*

974. Apparatus for demonstrating the **Glory on bedewed Meadows,** consisting of glass globes filled with water.
Prof. Dr. Lommel, Erlangen.

975. Apparatus for demonstrating the **Glory on bedewed Meadows,** with solidified drops of Canada balsam.
Prof. Dr. Lommel, Erlangen.

These glass globules serve to demonstrate why dew drops shone upon by the sun appear so brilliant just round the shade of the head of the observer. The explanation is that each dew drop is a lens concentrating the rays upon the ground below the drop, which latter sends them back diffused.

If the glass globules are placed upon a sheet of white paper, and the shadow of the observer's head is permitted to fall into the circle, the globules will appear very brilliant; as soon, however, as the white reflecting sheet is removed, and the globules rest on a dark dull ground, the brilliancy at once disappears.

The preceding experiment can also be well executed with solidified drops of Canada balm.

975a. Apparatus of M. Bravais, for producing halos, parhelia, and other kindred phenomena. *W. G. Lettsom.*

It consists of a hollow equilateral prism made with plates of parallel glass. The prism is filled with water at an orifice on the top. It is fixed on a vertical axis rotating by clockwork. With 100 revolutions in a second it reproduces the varied series of positions of the vertical prisms of ice producing parhelia. The prism is to be illuminated by a beam of sunlight, or by a lamp at a distance of 8 or 10 yeards. The apparatus reproduces also the Anthelion; by substituting for the prism a quadrangular plate of glass turning round on one of its vertical edges. If striæ are traced on the surfaces of this plate, the anthelion becomes traversed by two symmetrical arcs, arranged in the form of a St. Andrew's cross.

See M. Bravais's *Mémoire sur les Halos*, Paris, 1847, p. 266, 4to.; or *Journal de l'École Royale Polytechnique*, XXXI. cahier.

976. Four Absorption Cases, all in glass, 5 and 10 mm., and two pieces for spirits and water.
Warmbrunn, Quilitz, and Co., Berlin.

977. Bunsen's Apparatus for Experiments in Spectrum Analysis, a battery of four Leyden jars, a stand with arrangements for producing a spark spectrum, with holder for spectrum tubes. *Keiser and Schmidt, Berlin.*

978a. Apparatus for measuring the magnitude of Gas Jets at a distance. Invented by Mr. C. Wolfberger.
Geneva Association for the Construction of Scientific Instruments.

This apparatus, invented by the civil engineer Wolfberger, is intended for the comparison of gas lamps from the street level.

Founded on the principle of the sextant, it is composed of two mirrors : the one fixed, the other movable parallel to the first along a graduated scale.

In order to measure the magnitude of the gas jet, the operator holding the instrument by its handle on a level with the visual ray, and looking through the sight, observes simultaneously, in a direct line, the jet to be measured, above the fixed mirror, and indirectly the image of the same jet by double reflection. The screw-head placed below the handle is then turned, until the right edge of the jet, seen in a direct line, coincides with the left edge of the jet, seen by reflection. The magnitude of the flame, ascertained, is then shown on the metallic scale which is read in millimetres.

978b. Patent Illuminating Power Meter, for showing the illuminating power of gases in the terms of the Parliamentary sperm candles and the standard quantity of five cubic feet of gas per hour by the observation of one minute. *William Sugg.*

979. Trepiscope. An optical apparatus made by the late Richard Roberts, C.E., of Manchester, and first shown at the meeting of the British Association at Dublin in 1835.

The Committee, Royal Museum, Peel Park, Salford.

On being turned by hand or by power the card on the disc is caused to revolve from 6,000 to 40,000 times a minute; on viewing the revolving disc *through the eye-hole,* the printing on the card can be read with ease and distinctness.

The time given for one view of the card may not exceed the 150,000th of a second when the disc is revolving at the highest speed.

980. Radiograph.

The Committee, Royal Museum, Peel Park, Salford.

A small apparatus to show that the spokes may be counted whilst the wheels revolve at a very high velocity. Invented by the late Richard Roberts, C.E., of Manchester.

979a. Phosphoroscope, by Becquerel.

M. J. Duboscq, Paris.

979b. Phosphorescent Tubes, set in the form of writing. These tubes preserve their brilliancy in the darkness long after being exposed to solar light. *Alvergniat Frères, Paris.*

981. Newtonian Disc, for rotating movement transmitted by caoutchouc bands. *Luizard, Paris.*

981a. Newton's Apparatus. *M. Lutz, Paris.*

983. Pair of Reflectors, for the demonstration of the laws of reflection. *Elliott Brothers.*

983a. Reflector of 22 mm. diameter, for Foucault's telescope. *M. Lutz, Paris.*

983b. Apparatus illustrating Persistency of Vision. *S. F. Pichler.*

983c. Magic Mirror. *Robert von Tarnow.*

This mirror is a curiosity, and consists of a brass concave disc with finely polished surface. At the reverse rough side there are several Arabic characters in relief. By exposing the polished surface to the rays of the sun in such a way that they reflect them on the wall, the Arabic figures of the reverse side of the disc become plainly visible in the reflected light on the wall.

983c. Selenium Eye. *C. W. Siemens, F.R.S.*

994. Government Safety Magazine Lamp. *J. Gardner & Sons.*

Constructed at the request of the Home Office. The aim was to invent a lamp which would burn in a powder magazine and other dangerous places in perfect safety, and would exclude the powder which, it is well known, is found floating in powder magazines and stores in the form of fine dust. After a series of experiments it was ascertained in the most conclusive manner that all safety lamps wholly or partially constructed of gauze are useless for this purpose, the gauze failing to exclude the powder dust, which, collecting inside the lamp explodes. Now the flame of a powder explosion is so much more violent than that of gas that it instantly penetrates the gauze and carries through with it incandescent particles of powder.

This patent safety magazine lamp has, therefore, been constructed on the following principle :—

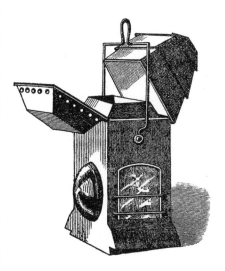

To prevent risk of explosion from the entry into the lamp of the fine powder dust which may be present in a powder magazine or store, the air passages which supply this lamp, and also the exit passages for the burnt air from the flame, are constructed so that the air must pass under and over a series of screens. Air to support combustion enters the lamp under an inverted outer ledge, and then passes through holes made in the casing to a narrow space formed by an inner lining ; so that the air must first pass up to reach the holes in the casing, then down the inner space, and finally up another narrow space before entering the lantern. The top part of the lamp is constructed on substantially the same principle, that is, the exit air passages are made zig-zag ; but, in case they ever become clogged with soot, two out of three parts which form the passages are hinged to the casing, and are secured in place by a spring lock. When these parts are unbolted they can be turned back on their hinges and easily cleared of any soot that may have become deposited therein.

The bottom and sides of the lamp are the only parts fixed to one another, and the burner is dropped in through the top of the lamp, which is then secured with a spring lock as already mentioned.

Every detail of the outer casing of the lamp has been carefully considered, and there are no projecting parts where dust can accumulate and settle. The lamp has a bull's-eye lens in front ; the side lights are glazed with glass one-eighth of an inch thick, protected by strong copper wire. The handle moves on a pivot. The burner is a ¾-inch flat wick, and a reflector is added to increase the brilliancy. The lamp and lantern are made of copper, bright tin, or tin japanned.

The highest temperature of the outer casing of the lantern has been 126° the exploding temperature of powder being 600°.

Spectacles, for shooting.   *G. W. Richter.*

SECTION 8.—HEAT.

I.—SOURCES OF HEAT.

984. Double-Chambered Lamp and Reservoir for heating water or air, or both. There is no blast-pipe, or communication between the chambers. *J. L. Milton.*

The heat is generated by the combustion of methylated spirit, and applied on the principle of driving a ring of flame from the holes in the top of the outer chamber against the flame issuing from the inner compartment, thus securing a great and continuous heat. The spirit in the inner chamber alone requires to be lighted. A chambered and tubed reservoir accompanies the lamp. The consumption of 1 oz. of methylated spirit in the outer, and ⅓ oz. in the inner chamber, will produce and maintain, for from 10 to 15 minutes, as great a heat for a vapour-bath as most persons can bear.

985. George's Patent Gas Calorigen, for warming and ventilating apartments. *John F. Farwig.*

The peculiarity of construction in this gas stove, which diffuses heat principally by convection, consists of an outlet so arranged with regard to the inlet (both being external to the apartment) that only so much air passes either way as is required to support and carry off the products of combustion.

The heat generated by combustion warms a thin coil of sheet iron in the interior of the stove, the coil being in communication at one end with the external atmosphere, and at the other with the apartment; thus a stream of fresh air, which is warmed in its passage, is drawn into, and equally diffused throughout, the apartment.

986. Bunsen Burner, improved form, with air jet to increase the temperature of the flame to any required extent without re-adjustment of height or position. *Thomas Fletcher.*

In the above, the blow-pipe flame obtained with the blast tube, when confined by the loose cap, is compact and very powerful, owing to the partial mixture of air before the blast begins to act.

987. Injector Gas Furnace, with blower, for the treatment of refractory substances at very high temperatures. *Thomas Fletcher.*

This furnace will burn perfectly, in the same space, any available gas supply from 10 to 50 ft. per hour, or more, giving temperatures in exact proportion. With ½ inch gas supply, day pressure, starting with a cold furnace, silver can be melted in three minutes, cast iron in eight minutes, and cast steel in 25 minutes. With a $\frac{1}{8}$ bore gas-pipe the furnace can be constructed to give the same results in half the time, and so on in proportion, the power being limited only by the gas supply and the fusibility of the refractory clay jacket. A small foot blower only is necessary to create a current of air in the burner tube, the jet of air from the blower acting as an injector, and drawing in the air required for combustion from the atmosphere. By closing the air slide to a greater or less extent the same burner acts equally well in proportion with any gas supply from 10 cubic feet per hour. The texture of the refractory casing is cellular throughout, and the loss of heat by radiation is practically nil.

988. Low Temperature Gas Burner, to dispense with drying closets, sand and water baths, and adapted for drying, evaporating, boiling, &c. *Thomas Fletcher.*

This burner gives a range of temperature from a gentle current of warm air without visible flame to clear red heat, and is so perfectly under control that a common glass bottle may be placed on tripod, and heated to required temperature, without risk of fracture.

For very low temperatures, the ring must be lighted through the lowest opening. This gives a steady current of heated air through the gauze above For boiling, &c., a light must be applied on the surface of the gauze, thereby providing a large body of blue flame, which can be urged by the blast-pipe until it gives a clear red heat.

989. Hot Blast Blowpipe, for temperatures up to the fusion of platinum. *Thomas Fletcher.*

The air jet in the above is coiled round the gas pipe in a spiral form, and both are heated by three Bunsen burners underneath, which are controlled by a separate tap. By this arrangement the power is double that of the ordinary blow-pipe. When the jet is turned down to a small point of flame it will readily fuse moderately thick platinum wire.

990. Gas Crucible Furnace, for temperatures up to white heat, and requiring neither blast nor attention. *Thomas Fletcher.*

991. Gas Muffle Furnace, requiring neither blast nor attention; for temperatures up to fusing point of cast iron.
Thomas Fletcher.

992. Diagram of the Porcelain Furnace at Sèvres, part of it being open to show the interior construction. Painted by Mr. Hubertus Sattler.
Dr. Alexander Bauer, Professor, Polytechnic Institute, Vienna.

993. Diagram of the Bottom of a Blast Furnace for Smelting Iron.
Dr. Alexander Bauer, Professor, Polytechnic Institute, Vienna.

995. Gas Furnaces. Perrot System.
Geneva Association for the Construction of Scientific Instruments.

These may be set up wherever gas is laid on ; the slightest draught is sufficient ; and in default of a chimney in the workroom it is enough to let out the
funnel through a window pane.   The shape of these furnaces varies according
to their intended use.   There are two principal models, the melting furnace,
and the muffle furnace ; the latter is advantageously used for assaying copper,
gold, and silver, for roasting minerals, and for melting metals for analytical
purposes.

Temperatures up to 1,300 and 1,400 degrees can be obtained rapidly and
with economy, and once obtained can be maintained unchanged during any
length of time, and may be reduced at will.

996. Cowper's Regenerative Fire-brick Hot-blast Stoves.
E. A. Cowper.

The diagrams and model of regenerative fire-brick hot-blast stoves are
illustrative of the progress of science as applied to heating the air supplied as
a blast under pressure to blast furnaces for smelting iron ores and iron stone.
In early times the air was always used cold, but a blast heated in cast-iron
pipes was introduced by Mr. J. B. Neilson in 1829, and from that time till
the year 1857 the temperature of the blast was generally only about 600°
Fahrenheit, but by the application of regenerative fire-brick hot-blast stoves
the temperature of the blast has been raised to 1400° and 1500° Fahrenheit,
and has been accompanied by a very large saving of fuel, amounting in some
cases to 7 cwt. 3 qrs. 14 lbs. of coke per ton of iron made, whilst at the same
time a largely increased make of iron has been produced, varying from 20 to
30 per cent. of the original make of iron.   The entire wear and tear of cast-
iron pipes is avoided, as the air only passes over fire-brick surfaces previously
heated by the combustion of the waste gases obtained from the top of the
blast furnaces.   Two stoves are used alternately, one heating blast whilst the
other is being heated.   Two or three stoves will heat the blast for two or
three blast furnaces.

996a. Drawings of Bull's Patent Semi-continuous Brick Kilns.
Hermann Wedekind.

These kilns have been worked with great success in India and are now
being introduced into this country.

They are the cheapest kilns with regard to cost of building and effect a
saving of nearly two-thirds of the expenditure usually involved, and are especially suitable for temporary works.

The effective mode of feeding invented by Hofmann, has also been applied
by Bull to his kiln with great success.

996b. Model of Hofmann's Circular Kiln.
Hermann Wedekind.

Hofmann's Patent Annular Ovens for the continuous burning of bricks and
tiles, limes and cements, at a saving of from two-thirds to three-fourths of the
fuel usually employed.

It is the joint invention of M. Fred. Hofmann, of Berlin, and M. A: Licht,
of Dantzig.

It is now well known and appreciated by the trade in this country, and has
been extensively employed by the Government on the extensive works at
Portsmouth Dockyard, where five of the kilns have been working at a great
saving to the country.   About 1½ millions of bricks are daily burnt in them
in England alone.   Taking coal at an average price of 10s. per ton, this invention will give a saving of at least 60,000l. per annum.   In burning lime
equally favourable results have been realised as in burning bricks.

997. Gas Lamp, consisting of four Bunsen burners, and provided with an air-regulating system.

S. Hoogewerff, Dr. Phil., Rotterdam.

This lamp, which is intended for heating tubes, was constructed by Mr. Verkerck, mechanical engineer, Utrecht, under the directions of Dr. Hoogewerff, and belongs to the middle school at Rotterdam.

998. Gas Lamp (Bunsen's system), intended for heating porcelain vessels of large size.

S. Hoogewerff, Dr. Phil., Rotterdam.

It was constructed, under the directions of Dr. Hoogewerff, by Mr. Verkerck, mechanical engineer, Utrecht, and belongs to the middle school at Rotterdam.

999. Apparatus for showing **the liberation of heat** during solidification. *Will. Haak, Neuhaus am Rennweg, Thüringen.*

II.—EXPANSION.

"1088a. Apparatus, by **Peter von Musschenbroek,** a Dutch mathematician (born 1692, died 1761), to determine the relative values of the **Coefficients** of the **Expansion of Solid Bodies.** *Prof. Dr. P. J. Rijke, Leyden.*

1097b. Original Apparatus, by **M. Dulong** and **Petit,** for measuring the **Dilatation of Mercury** with overflow thermometer. *Polytechnic School, Paris.*

1097c. Apparatus for measuring the expansion of Bodies by Heat. *Physical Institute, Freiburg.*

1097d. Apparatus to demonstrate the mechanical effects of the expansion of Fluids by Heat. *Prof. Boscha, Royal Polytechnic School, Delft*

1090b. Original Apparatus of **Regnault** for the **Dilatation of Gases.** *College of France, Paris.*

1075. Apparatus for demonstrating the **Expansion of Solids by Heat.** *A. Steger, Kiel.*

1090. M. Fizeau's Apparatus for measuring the **Coefficient of Dilatation.**

Conservatoire des Arts et Métiers, Paris.

1082a. Photograph of M. Fizeau's Apparatus for determining the **Coefficients** of **Dilatation.**

M. Laurent, Paris.

1082b. Microgoniometer. An instrument for measuring the expansion of metals by heat. *Prof. Dr. F. Pfaff.*

1087. Model of Circular red hot Copper Railway, for causing a metal ball to rotate by means of unequal expansion by heat.   *George Gore, F.R.S.*

Model in wood of circular railway, which when formed of copper heated to redness, and a thin cold ball of German-silver placed upon it, the ball rotates by the influence of unequal expansion produced by the heat. (*See* Philosophical Magazine, August 1859.)

III.—THERMOMETRY AND PYROMETRY.

Air Thermometer, in the form first given by **Galileo.**
The Royal Institute of " Studii Superiori," Florence.

Padre Benedetto Castelli, one of Galileo's disciples, writes about the thermometer as follows :—" I recollect an experiment shown me, about the year " 1603, by our Sig. Galileo. He took a glass bottle of about the size of an " egg, having a neck nearly two palmi iu length and as thin as a wheat " straw, and having warmed it well with his hands, he then turned its mouth " upside down into a vessel placed underneath in which there was a little " water. When he had removed his hands from the bottle the water began " immediately to rise in the neck, and mounted higher, by more than a palmo, " than the level of the water in the vessel. Sig. Galileo made use of this " effect to construct an instrument for examining the degrees of heat and of " cold, concerning which much might be said."

And then Vincenzo Viviani, another disciple of Galileo's, in the life of that great man, which he wrote between 1593 and 1597, states that Galileo invented the thermometer.

1828. Registering Thermometer of Fontani.
The Royal Institute of " Studii Superiori," Florence.

Thermometer, cinquantigrade, with spherical bulb. The freezing point corresponds to 13°·5. The academicians used this thermometer for the meteorological observations instituted first by them in 1654.   *Accademia del Cimento.*

Thermometer, cinquantigrade, with cylindric bulb. The point of ice melting corresponds to 13°·5.
Accademia del Cimento.

Thermometer, settantigrade. The freezing point is at 23°·5.
Accademia del Cimento.

Thermometer for bath. When immersed in the bath it must indicate 49°, as is written in the upper little ball.
Accademia del Cimento.

Thermometer, with foot divided in 470°, corresponding to 26° centig. It served for the experiment made to ascertain whether the cold of ice reflects itself from the mirror like the heat of burning coals and light.   *Accademia del Cimento.*

Thermometer, with elaborate foot. An object of art.
Accademia del Cimento.

Thermometer in winding form, very sensitive, height 32 cm. ; the spiral tube is 2·30 meters long. *Accademia del Cimento.*

Thermometer, with balls (thermoscope). The alcohol dilating by heat, the little balls fall one after another according to their weight. *Accademia del Cimento.*

The Accademia del Cimento was founded by Prince Leopoldo de' Medici, brother of the Grand Duke Ferdinand II., who also greatly favoured it. The first assembly was held at the Palazzo Pitti in Florence on the 18th of June 1657 ; and it chose for a device the celebrated motto " Provando e Riprovando." After ten years of existence, Prince Leopold having been made a Cardinal, the academy was dissolved.

1075a. 'sGravesande's Ball and Ring Pyrometer, for showing expansion. *Harvey, Reynolds, and Co.*

1076. Musschenbroek's Pyrometer made in the first half of the 18th century, with five different metal bars, and an autograph. Property of His Highness Prince Pless, Schloss Fürstenstein.
The Breslau Committee.

The apparatus has been constructed after the description and drawing given on p. 12 and Table XXX. of Musschenbroek's "Tentamina Experimen- " torum Naturalium Captorum in Academia del Cimento, Lugduni, 1731, " Pars. II." The orthography of the French, on an annexed slip of paper, is that of the beginning of the last century. The instrument, which is in capital condition, may therefore be considered as one of the oldest of its kind.

1027. Original Spirit Thermometer, of the Florentine Accademia del Cimento (17th century).
The Royal Institution of Great Britain.
Presented to the Royal Institution by Sir Henry Holland, Bart., F.R.S.

1003. Photographs of Old Thermometers ; a small al- cohol thermometer, with Florentine scale, and four larger ones by Michelo du Crest (1754).
Prof. Hagenbach-Bischoff, Director, The Physical Insti- tute in the Bernoullianum, Basle.

In these alcohol thermometers zero indicates the temperature of the cellar under the Observatory of Paris, and 100° the boiling point of water.

1070. Wedgwood's Pyrometer, invented in 1782.
Edinburgh Museum of Science and Art.
Dry clay when exposed to high temperatures contracts uniformly, and Wedgwood believed that by the amount of contraction the temperature which produced it could be measured. The instrument, however, is not trustworthy. This specimen was made by Josiah Wedgwood, and presented to the Edinburgh Museum by his grandson Mr. Godfrey Wedgwood.

1067. Wedgwood's Pyrometer, consisting of pieces of clay, contracting according to the heat to which they are exposed. These are afterwards slid along the gradually diminishing and graduated groove in the brass plate, and so indicate the degree of heat to which they have been exposed. *Robert Garner, F.R.C.S.*

1078. Early Pyrometer (by Funiey).
The Council of King's College, London.

1079. Daniell's Pyrometer, employed in researches by
Professor Daniell.   *The Council of King's College, London.*

1052a. Original Air Thermometer, by M. Regnault.
College of France.

1000. Thermometer. "The Great Pyramid temperature
scale, and its standard reference point of 50° P." With a map of
the world to illustrate the advantages of this standard.
Prof. Piazzi Smyth.

This consists of a large table thermometer, graduated according to the indications of the Great Pyramid system of standards ; firstly, by colours into
fifths of the distance between freezing and boiling of water, and then each
fifth into 50, or 250 for the whole distance.

A map of the world on an equal-surface projection accompanies the
thermometer, and exhibits the mean temperature of the whole earth's surface
according to the Great Pyramid scale; illustrating also the territorial and
international advantages to all civilised nations of adopting the mean temperature standard of the Great Pyramid, viz., 50° Pyr. or 68° Fahr., as the
temperature reference standard for all human purposes, scientific, social,
and commercial.

1001. Legible Spirit Thermometers, with line at above
and below the proper temperature of a room, so that the degree
can be read off at a long distance, at the opposite side of a large
room, or at the ceiling, for experiments in ventilation.
Peter Hinches Bird, F.R.C.S. Lond.

1002. Apparatus for determining the **Boiling Point of a
small quantity of Fluid.**
The Secondary Government School, Assen (Netherlands).

In this simple apparatus, constructed after the design of Dr. A. Van Hasselt,
(teacher at the school for middle-class education at Assen), the small tube is
filled for the greater part with mercury ; the remaining space with the fluid.
The tube is then turned upside down into a small beaker-jar, which is also
filled with mercury ; part of this must be removed until the quantity left rises
about one or two centimetres above the bottom of the jar. When the apparatus has been placed in the large beaker-jar, water or oil is poured into the
latter, so that the tube is quite immersed. The jar is then heated, agitating
the fluid meanwhile with a moving apparatus.

The millimetre scale serves to determine the height of the fluid in the larger
jar, together with the difference between the position of the mercury in the
tube and that on the outside of it. In determining this difference, the pressure
of the fluid in the larger beaker-jar and the barometric height must be taken
into consideration.

To know whether a fluid is homogeneous, two experiments must be made,
one with a fluid which is partially evaporated. In both cases the results
must be the same.

When the vapour of the fluid in the tube has the pressure of one atmosphere
the boiling point of the fluid must be observed.

1375a. Air Thermometer after Riess.

Warmbrunn, Quilitz, & Co., Berlin.

1004. Trough for comparing Thermometers, provided with in- and out-flow tubes for water, and stirring apparatus.

Dr. J. W. Gunning, Professor of Chemistry at the " Athenæum illustre," Amsterdam.

The thermometers are placed in a movable frame, in which they may be transported from one trough into another containing water of another temperature. Two sheets of glass, placed on either side of the frame, prevent the inner portion of the water from cooling.

1005. Dial Thermometer, designed by L. R. Brühne, Leyden.

Dr. D. de Loos, Director of the Secondary Town School, Leyden.

This thermometer is intended to admit of a large number of students seeing, from a distance, the change in the volume of the mercury as the temperature varies.

In the mercury is a small glass tube, balanced by another similar tube, both being joined together by a thread, which is suspended over a small copper box, at the extremity of which is a needle moving over a dial.

1006. Mercurial Dial Thermometer, adapted for class experiments on specific and latent heat.

Prof. W. F. Barrett.

The expansion of the mercury in the bulb of the thermometer lifts a small iron piston which communicates its motion to the index hand. Small variations of temperature are thus readily seen by a large class, e.g., the expansion of the glass of the bulb causing a momentary retreat of the index hand, is seen to be the first effect produced by heat on the bulb. By making the scale on glass the dial can be projected on a screen and determination of specific and latent heat made before a large class; electric contact can also be made by the hands, and thus a self-registering thermometer constructed. The instrument was made by Mr. Yeates, of Dublin.

1011. Various Thermometers, of different kinds, in metal, ivory, porcelain, glass, and wood. *Elliott Brothers.*

1012. Standard Thermometer. *Elliott Brothers.*

1013. Insolation Thermometer, for determining the intensity of the rays of the sun (maximum thermometer), with holder.

Ch. F. Geissler & Son, Berlin.

1014. Eight Normal Thermometers, executed by Greiner and Geissler, Berlin.

Imperial Admiralty Hydrographical Bureau at Berlin, and Deutsche Seewarte, Hamburg.

These thermometers are employed in the stations of the Naval Observatory and in the Imperial Navy.

R 2

1015a. Thermometer, with corrected Freezing Point.
W. Gloukhoff, St. Petersburg.

This thermometer is constructed on a principle much used in Germany. To it is added only a contrivance to render the scale more steady, and to correct the error of *freezing point*, by raising or lowering of the scale. By unscrewing the upper metallic cap of the thermometer, this contrivance becomes visible.

1016. Reaumur's Scale.   *Dring and Fage.*

Formerly much used in Germany and Russia, now mostly in Norway and Sweden, and some parts of Denmark. The zero of this scale is at the melting point of ice. The interval between this and boiling point is divided into 80 degrees.

1017. De Lisle's Scale.   *Dring and Fage.*

This scale is seldom used; zero is fixed at boiling point; the interval between this and freezing point is divided into 150 degrees.

1018. Six's Thermometer on a porcelain scale (named after its inventor, Mr. Six of Canterbury) for registering extremes of temperature.   *Dring and Fage.*

The indices are little pieces of steel coated with glass which are enabled to retain their position in the tube by means of a hair fastened round them, and by this means the highest or lowest temperature is recorded.

1018a. Six's Thermometer with a very flat bulb which renders it as sensitive as an ordinary mercurial thermometer.
S. G. Denton.

1018b. Six's Thermometer with mercurial wet bulb thermometer attached, thereby combining four instruments in one, namely, maximum, minimum, hygrometer, and present temperature.
S. G. Denton.

1019. Long Brass-Cased Thermometer. Showing the difference in length of the mercurial column after being pointed and divided with the whole length of the tube immersed in water at the various temperatures between 32° and 212°; the same with the bulb only in the water.   *Dring and Fage.*

1020. Very delicate Spiral Bulb Thermometer. Extremely sensitive, capable of indicating small variations of temperature.   *Dring and Fage.*

1022. Standard Thermometer, calibrated throughout.
Dring and Fage.

A standard thermometer divided on the tube, used for purposes where great accuracy is required. The tubes used for these thermometers are selected with great care, particular attention being paid to the uniformity of the bore. The method of ascertaining this is usually performed as follows:— A portion of mercury is introduced into the tube, and the length it occupies is noted ; it is then carried a little further on, and its length compared with the former length. So on all down the tube; if the length has decreased from the first measurement, it shows that the bore of the tube has increased, and vice versâ. The process is known as calibration.

1023. Four Thermometers. Showing the different scales principally in use. Fahrenheit, Celsius or Centigrade, Reaumur, and De Lisle. *Dring and Fage.*

Fahrenheit's scale is used principally in Great Britain, its Colonies, and the United States. The zero of this scale is obtained from a mixture of salt and snow; thirty-two degrees is the point at which ice begins to melt, and 212, or boiling point, from boiling water, when the barometer stands at 29·905. One advantage of this scale is that temperatures may often be expressed in whole degrees, whereas in other scales fractions of degrees are frequently necessary.

1024. Celsius or Centigrade Scale. Generally used on the continent. *Dring and Fage.*

The zero of this scale is that point at which ice begins to melt, and 100 the point at which water boils when the barometer stands at 760mm. Celsius is the name of the inventor of this scale; it is called Centigrade from its being divided centesimally.

1025. Becquerel's Thermo-Electric Thermometer.
Conservatoire des Arts et Métiers, Paris.

1025a. Becquerel's Thermo-Electric Pyrometer.
Conservatoire des Arts et Métiers.

1026. Hodgkinson's Actinometer. Described in the Proceedings of the Royal Society, vol. XX., p. 328.
Kew Committee of the Royal Society.

This is a large thermometer, filled with alcohol coloured blue, and having a bore much contracted for a great part of its length, in order that the scale may be very open; at its top it opens out into a large chamber, which receives the superfluous fluid at the time of observation.

A tin case, capped with glass at both ends, prevents the access of extraneous rays to the bulb of the instrument at the time of observation.

1028. Drawings, various, partly new, of constructions of Differential Atmospheric Thermometers.
Dr. Leopold Pfaundler, Professor of Physics, Innsbruck.

This plate presents a general view of all possible forms of construction, which partly appear as modifications of Berthelot's atmospheric thermometers, partly are based on independent principles.

For further details, see Transactions of the Imperial Academy of Sciences at Vienna, Vol. LXXII., 1875.

1029. Melloni's Thermo-Electric Apparatus.
M. Ruhmkorff.

1029a. Line Pile for Spectrum and Galvanometer.
M. Ruhmkorff.

1286. Nobili's Thermo-Electric Pile, of 54 pairs of bismuth and antimony bars, soldered alternately together; the smallest temperature between the two faces of the pile develops a current, readily indicated by a suitable galvanometer. *Elliott Brothers.*

1033. Thermometer Electric Alarum, for giving notice when a given temperature is reached. *Dr. Letts.*

The apparatus consists of an open thermometer with large bulb and wide tube. A platinum wire is sealed into the bulb, and another wire passes down the tube. The latter can be so adjusted that at a given temperature its end is touched by the mercury in the tube. The two wires being connected with an electric bell and battery, as soon as the mercury touches the wire, contact is made and the bell rings.

The apparatus was used in experiments with the glass digester, and served to give notice at some distance from the room in which the latter was being heated when the desired temperature had been reached, thus rendering an actual observation of the temperature unnecessary, and so preventing all danger in case of an explosion.

1034. Normal Thermometer, divided in tenths of a degree from 0° to 105° C.
Will. Haak, Neuhaus am Rennweg, Thüringen.

1035. Normal Thermometer, in a narrow glass cylinder with a small mercury bulb; divided in tenths of a degree from 0° to 105° C. *Will. Haak, Neuhaus am Rennweg, Thüringen.*

1036. Normal Thermometer, divided to tenths from —35° to +50° C. *Will. Haak, Neuhaus am Rennweg, Thüringen.*

1038. Two Thermometers, for chemical work, from —10° to +360°. *Will. Haak, Neuhaus am Rennweg, Thüringen.*

1039. Two Thermometers, from 100° to 360°.
Will. Haak, Neuhaus am Rennweg, Thüringen.

1040. Two Thermometers, from 0° to 150°.
Will. Haak, Neuhaus am Rennweg, Thüringen.

1041. Two Thermometers, with divisions etched on the tube. *Will. Haak, Neuhaus am Rennweg, Thüringen.*

1045. Various Pressure Thermometers, on the plan of Mitscherlich. *See* accompanying pamphlet.
Prof. Mitscherlich, Münden.

1046. Metallic Thermometer for Lectures, on the plan of W. Beetz, constructed by Sauerwald, of Berlin. Consult the adjoining description and scheme. *Prof. Beetz, Münich.*

1047. Model of an **Apparatus for measuring Temperatures by means of Thermo-batteries.**
Dr. J. Pernet, Breslau.

The apparatus permits the quick and trustworthy determination of the temperature of the several soldered places, as well as the differences of temperature of any two soldered places.

In the first case the resistances of the single conductors (except the resistances of the two soldered places, which are exposed to constant temperatures, and show always the same resistance) may be different. In the other case the resistances of all soldered places must be capable to be made equal.

1048. Diagrams, illustrating the application of the **Apparatus for the Measurement of the Temperature of the Earth** and of metal tools. *Dr. J. Pernet, Breslau.*

1049. Normal Thermometer, divided in tenths of a degree from — 5° to + 105° C. *Ch. F. Geissler and Son, Berlin.*

1050. Chemical Thermometer from — 10° to +360° C.
Ch. F. Geissler and Son, Berlin.

2582. Collection of **Thermometers.**
Dr. H. Geissler, Bonn.

1050a. Thermometer for Cooking, range to 600° F.
Harvey, Reynolds, and Co.

See Mrs. Buckton's book "Health in the House," page 153.

1051. Apparatus for determining the **Temperature of Fusion.** (Compare the adjoined description.) *Prof. Dr. Himly, Kiel.*

1052. Air Thermometer on the plan of Jolly. Compare Poggendorff's Annalen, Jubelband, 1873.
University of Munich (Prof. v. Jolly).

1053. Thermopile on the plan of Melloni of 64 bismuth antimony elements. *Wesselhöft, Halle.*

1054. Thermometer Stick for measuring temperatures at some depth. *Ludwig Meyer, Berlin.*

The instrument, adapted for depths down to 3 feet, is chiefly distinguished by the strength of its construction.

The bulb is in the nickel clamp, which latter stands by means of mercury in thermal connexion with the bulb. This mercury serves also as buffer to the thermometer bulb.

The horn clamp is replaceable by an iron screw, which facilitates the introduction of the thermometer into the ground.

Care is taken that only the clamp be thermo-conducting, not the whole tube.

1055. Milligrade Thermometer. The milligrade scale is one in which the interval of temperature between the freezing and boiling points of mercury is divided into one thousand degrees. *John Williams, F.C.S.*

According to Dulong and Petit, mercury freezes at —39·44° C., and boils at +360° C. For convenience, assuming that the freezing point is —40° C., the interval is therefore 400 degrees C., thus it follows that $2\frac{1}{2}$ degrees milligrade are equal to 1 degree centigrade. Upon this scale the following results are obtained. Water freezes at 100° M. and boils at 350° M., the interval 250° being just one-fourth of the interval between the freezing and boiling points of mercury. Many other substances also show a curious relation in the interval between their freezing and boiling points to that of mercury, facts which are not obvious upon other thermometric scales.

The practical advantages of this system of graduation consist in the comparative smallness of the degrees, thus avoiding in many cases the necessity of the use of fractions to express the boiling point of substances; also that the zero point being so low the scale is a continuous one, all numbers under 100° M. representing temperatures below freezing water, but avoiding the necessity of the use of the minus sign, and at higher temperatures as 1,000° is approached, giving a clear idea that the heat is arriving at the extreme limit of thermometric registration.

In practically graduating this thermometer reference is not made to the freezing or boiling points of mercury, but the freezing point of water is marked as 100°, and the boiling point as 350°, and the scale carried upwards or downwards as required.

The conversion of centigrade degrees into milligrade degrees, or *vice versâ*, is extremely simple. A centigrade degree multiplied by $2\frac{1}{2}$, and 100 added, gives the milligrade degree, thus 40° C. multiplied by $2\frac{1}{2}$ is 100, and 100 added gives 200, the degree on the milligrade scale. The correspondence between the Fahrenheit and the milligrade graduation is not so simple, as the interval on the Fahrenheit scale between the freezing and boiling points of water being 180° F., higher numbers are required to be used in the calculation. The following are the lowest common numbers for the scales:— 25° milligrade, equal to 10° cent. and equal to 18° Fahr.

Thus it follows that the following rules can be applied to calculate one scale with the others—

To convert centigrade into milligrade degrees - n × 5 ÷ 2 + 100.
To convert milligrade into centigrade degrees - n − 100 × 2 ÷ 5.
To convert Fahrenheit into milligrade degrees - n + 40 × 25 ÷ 18.
To convert milligrade into Fahrenheit degrees - n × 18 ÷ 25 − 40.

1055a. Thermometer, with 19 differently graduated scales, traced on a silvered metal plate ; the centre is occupied by the thermometer-tube and bulb. This instrument was made in 1754.

Prof. Buys-Ballott, Utrecht.

1055b. Four Registering Thermometers.

E. Cetti and Co.

1055c. Siemens' Pyrometer.   *E. Cetti and Co.*

1055d. Metallic Thermometer, indicating two temperatures.   *Francis Pizzorno, Bologna, Italy.*

The movements of the index are in this instrument produced by the dilatation of two zinc blades which in the figure are seen edgewise. Along the graduated arc can be fixed two sliding pieces ; if the index touches one of them, it closes an electric circuit and rings a bell. Two small pearls carried by a thread stretched between the extremities of the graduated arc, and which are displaced by the index in its movements, serve to indicate the maximum and minimum temperature.

1055e. Five Thermometers, various.   *E. Cetti and Co.*

1055f. Mathiessen's Differential Thermometer.

E. Cetti and Co.

1068. Pyrometer, of iron and copper, for lecture illustration.
Yeates and Sons.

The above consists of a compound bar of iron and copper, bent into the form of ∪, one arm of which is firmly attached to the stand ; the other arm is free, and carries a long index. If the compound ∪ be immersed in a beaker of boiling water, the index will move over several degrees of the scale.

1069. Reflecting Pyrometer, for showing by projection the difference of expansion of different metals. *Yeates and Sons.*

1074. Pyrometer. *O. Schütte, Cologne.*

The pyrometer for determining the temperature of the heated blast-current deserves particular attention on the part of proprietors, overseers, &c. of foundries, on account of the simplicity of its construction, which is proof against deranging influences, although constantly exposed to a destructive element, viz., the glowing hot current, the more especially as no pyrometer has been made as yet with which temperatures up to 600 degrees Celsius and more can be continuously determined.

By the application of this pyrometer it will be possible to control at any time the apparatus and the stokers, and in cases where Siemens' or other similar apparatus are employed, to determine the exact time when the same must be reversed. The pyrometer, likewise, indicates any disturbing influences occurring in the flues, variations in the fuel, in the atmosphere, &c., and thus offers the best guarantee that the heating of the apparatus is not forced to such an extremity as to cause the destruction of the pipes, stop-valve, &c.

1071. Copper Pyrometer, for determining the temperature of blast when highly heated. *E. A. Cowper.*

The temperature of the blast is readily ascertained by heating a small piece of copper in it, and then dropping the copper into a pint of water in a copper vessel surrounded by a non-conductor, and where a thermometer shows one degree for every fifty degrees the copper had been heated. The thermometer is provided with two scales, one fixed, showing the temperature of the water, the other sliding, showing the temperature of the blast.

1074a. Water Pyrometer. *C. W. Siemens.*

The pyrometer consists of a copper vessel, capable of holding rather more than a pint of water, and well protected against radiation by having its sides and bottom composed of a double casing, the inner compartment of which is filled with felt. A good mercury thermometer is fixed in it, having, in addition to the ordinary scale, a small sliding scale, graduated and figured with 50 degrees to 1 degree of the thermometer scale ; there are also some cylindrical pieces of copper provided with the pyrometer, each accurately adjusted in size, so that its total capacity for absorbing heat shall be 1–50th that of a pint of water.

In using the pyrometer, a pint (0·568 litre, or 34·66 cubic inches) of water is measured into the copper vessel, and the sliding pyrometer scale is set with its zero at the temperature of the water as indicated by the mercury thermometer ; a copper cylinder is then put into the furnace or hot blast current the temperature of which it is wished to ascertain, and is allowed to become heated for a time varying from 2 to 10 minutes, according to the intensity of the heat to be measured.

It is then to be withdrawn and quickly dropped into the water in the copper vessel, where it raises the temperature of the water in the proportion of 1 degree for each 50 degrees of the temperature of the copper. The rise of the temperature may then be read off at once on the pyrometer scale, and, if to this is added the temperature of the water as indicated on the mercury thermometer before the experiment, the exact temperature required is obtained.

For very high temperatures platinum cylinders may be employed instead of copper.

1074b. Electrical Pyrometer.   *C. W. Siemens.*

The electrical resistance of metal conductors depends upon their dimensions, material, and temperature; an increase of the latter causing a corresponding increase of resistance. The law of this increase is known.

Thus, the resistance of a conductor being ascertained at zero centigrade, it can be calculated for any temperature, and *vice versâ;* if the resistance can be found by measurement, the temperature can be calculated. This is the principle upon which Siemens' electrical pyrometer is based.

A platinum coil of a known resistance at zero centigrade· is coiled on a cylinder of fire-clay, protected by a platinum shield, which is placed in an iron or platinum tube, and then exposed to the temperature to be determined. Leading wires are arranged to connect this coil with an instrument suitable for measuring its resistance, and from this resistance the temperature can be calculated.

The instrument supplied for this purpose is a differential voltameter.

The differential voltameter consists of two separate glass tubes, in each of which a mixture of sulphuric acid and water is decomposed by an electrical current passing between two platinum electrodes. The gas which is generated is collected in the long cylindrical and carefully-calibrated top of the tube, and its quantity is read off by means of a graduated scale fixed behind the tubes.

Movable reservoirs are provided communicating with the tubes to regulate the level of the liquid.

The current of the battery is divided (by passing a commutator) into two circuits, one of which consists of standard resistance in the instrument and the platinum electrodes in one tube; the other, of the resistance to be measured and the electrodes in the other tube. The quantities of gas developed in the two tubes are in inverse proportion to the resistances of their respective circuits, therefore one of the resistances, viz., that in the instrument, being known, the other can be calculated.

Directions for use :—Fill the battery glasses with pure water, or, in case of the power of the battery decreasing, with a solution of sal-ammoniac in water. Connect the poles to B and B' on the commutator. Expose the small end of the pyrometer-tube, as far as the cone, to the heat to be measured, and connect the terminals x, x', c to the ends of the leading cable bearing corresponding letters. Connect the other end of the leading cable to the terminals x, x', c on the voltameter.

The differential voltameter is to be filled with the diluted sulphuric acid through the reservoirs, the india-rubber cushions being lifted from the top of the tubes. The commutator is to be turned so that the contact springs on both sides rest on the ebonite. The liquid in both tubes is to be regulated to the same level (zero of scale), and the india-rubber cushions to be let down again. Give the commutator a quarter of a turn, and the development of gas will commence almost immediately. Turn the commutator half round every ten seconds to reverse the current. Keep the current passing until the liquid has fallen in the tubes to at least 50 degrees of the scale, then put the commutator in its first position, so that the contact springs rest on the ebonite; read off the level of the liquids on the scale marked V, and the scale marked V̇; find these numbers in the table under V and V, and the intersecting point of the lines starting from these figures gives the temperature. For a new experiment adjust the levels as before.

1077. Bailey's Patent Civil Engineers' Pyrometer, for ascertaining the temperature of flues, &c., with sheath and box-wood handle to enable managers of works and others to carry it about with them for use when necessary. *W. H. Bailey & Co.*

1080. Bailey's Patent Flue Pyrometer, for testing the temperature of boiler flues, hot-air chambers, stoves, galvanisers, &c. *W. H. Bailey & Co.*

1080a. Hobson's Patent Hot Blast Pyrometer.
Joseph Casartelli.

In this instrument the aim is to tone down the temperature of the blast by an admixture of a constant proportion of cold atmospheric air, so that the highest temperature likely to have to be recorded is brought within the range of a good mercurial thermometer. The hot blast is introduced in the form of a jet, which by suitable arrangement is made to induce a stream of atmospheric air ; the mixed stream then passes on, and impinges on the bulb of the thermometer. The scale has been laid down by experiment, and the instrument gives the same reading as Siemens' copper ball pyrometer. It is found that pressure does not affect the result ; and, as all the instruments are made exactly alike, the same result is invariably obtained. By the use of this instrument much time is saved, and the result is more trustworthy than with any other instrument in use.

1080b. Casartelli's Improved Pyrometer, for ascertaining the temperature of flues, stoves, &c. *Joseph Casartelli.*

This instrument consists of two different metals of different ratios of expansion, and any permanent set which may take place in the metals is compensated by the fact that the set will take place in opposite directions. The scale is laid down by experiment. It is so constructed that it is only necessary to expose one half of the stem to the action of the heat.

1081. Bailey's Patent Bakers' Pyrometer, "Bakers' Guide," for ascertaining the temperature of bakers' ovens, and enabling them to prevent the possibility of the bread becoming burnt, by keeping the oven at one uniform temperature.
W. H. Bailey & Co.

1082. Wood and Bailey's Patent Blast Furnace Pyrometer, for ascertaining the temperature of hot blasts.
W. H. Bailey & Co.

IV.—CALORIMETRY.

1063a. Original Apparatus of **M. Regnault** for ascertaining **Specific Heat** by observing **Refrigeration.**
College of France, Paris.

1097a. Diagrams (3), representing the **Apparatus of M. Regnault,** by M. d'Obelliane.   *Polytechnic School, Paris.*

1064. Lavoisier's original Calorimeter.
Conservatoire des Arts et Métiers, Paris.

Fabre and Silbermann's original Calorimeter, for measuring the heat disengaged in combustion.
Conservatoire des Arts et Métiers, Paris.

1064a. Original Vessel, by **Dulong,** for measuring **Specific Heat** by **Refrigeration.**   *Polytechnic School, Paris.*

1064b. Original Apparatus, by **M. Regnault,** for ascertaining the **latent Heat** of **Steam** at different pressures.
College of France, Paris.

1057. Apparatus employed by Dr. Andrews in his experiments on the amount of heat disengaged in the combination of hydrogen and other combustible gases with oxygen.   *Dr. Andrews, F.R.S.*

The gases contained in a cylindrical vessel of thin copper are exploded by the ignition of a fine platinum wire, and the heat is measured by the rise of temperature of the water in a calorimeter, capable of being rotated gently round its horizontal axis.

1058. Apparatus for determining the amount of heat produced in the combination of liquids and solids with oxygen.
Dr. Andrews, F.R.S.

1058a. Fabre and Silbermann's Calorimeter.
L. Golaz, Paris.

This apparatus consists of :—1. A large brass vessel closed by a cover in two parts and open in the centre. 2. A thin copper vessel coated with silver in the interior. 3. A vessel like the preceding, fitted with a cover. The calorimeter is supported in the centre of the second enclosure. The space between it and the second enclosure being lined with the skin of a swan. In the interior of this vessel is placed an agitator and a thermometer, made with great precision. All these different enclosures are placed in a large and solid triangular support, having two plates united by three copper columns.

The combustion chamber is a thin copper vessel, having at its upper extremity an opening closed by a plug flush with the side of the chamber, and provided with a tube by which the gases and the products of the combustion are discharged. A helix formed by a thin tube is fixed to this opening, and so arranged that the gases pass through it from end to end, and are discharged at the upper extremity, whence they are collected in vessels placed for that purpose. A second tube is fixed to the combustion chamber reaching to the lower part, and serves for introducing the oxygen into the chamber. The stopper is formed of a piece of metal exactly fitting the upper part of the combustion chamber, and carrying two tubes, by one of which can be introduced the pipes for hydrogen or oxygen according to the nature of the combustion desired. The other is used for watching the combustion. For that purpose it is closed at its lower end by a threefold plate, of glass, quartz, and alum ; a small looking-glass is placed at the upper end of this tube, and by this means all that is going on in the interior can be seen.

1058b. Regnault's Apparatus for determining Specific Heat of Gas. *L. Golaz, Paris.*

This apparatus, constructed for M. Regnault by the exhibitor, consists of three parts:—

1. This affords the means of obtaining a current of gas, the velocity of which is constant and can be regulated as desired.

2. The bath, which imparts a determinate initial temperature to the gas.

3. The calorimeter in which the gas parts with its excess of heat.

The first part of the apparatus is composed of a copper reservoir, holding about 35 litres, and provided with two stop-cocks. It is placed in a vessel filled with water, which keeps it at a temperature sensibly constant, and always accurately known. The central reservoir is full of gas under a pressure determined by a syphon pressure gauge.

The flow of the gas is regulated by a delicate valve formed of a vertical tube, in which a close fitting spindle works by means of a micrometer screw. The end of the spindle is conical, and fits into a corresponding seating at the bottom of the tube. The diameter of the spindle is slightly diminished about one centimeter above the cone. The gas passes into the valve through a capillary opening at the apex of the conical seating, and passes out round the diminished part of the spindle to the outlet pipe, which is about a centimeter above the inlet. In connexion with the outlet pipe is a syphon pressure gauge. The amount of clearance given to the valve is indicated by a graduated disc fitted to the top of the spindle.

The oil-bath which heats the current of gas consists of a cylindrical vessel, in which is a helical tube ten metres in length. Its extremity is connected with the calorimeter by means of a nozzle, in which is fitted a small tapering glass tube by means of a cork, through which it passes. The bath has a cover at the top, through the middle of which passes a square rod carrying an agitator ; a short tube fixed to the cover, near the exit of the gas, allows of the introduction of a thermometer, so that the temperature of the gas before entering the calorimeter can be ascertained.

The boiler is preserved from immediate contact with the surrounding air by means of a metallic jacket, which serves at the same time as a support ; the base of this support is furnished with levelling screws ; an iron column placed behind supports a rod with a clamp holding a horizontal bar, at the end of which is a pulley over which passes the cord sustaining the agitator. The bath is warmed by a gas lamp.

On leaving the bath, the gas goes into the calorimeter, which consists of a cylindrical vessel of thin brass, provided with a nozzle through which the gas enters, and 4 shallow cylindrical boxes, through all of which the gas passes successively. These boxes communicate with one another and with the lower vessel by means of tubes suitably arranged ; on leaving the uppermost box, the gas discharges itself into the atmosphere by a short tube. A very thin strip of brass bent spirally is in the interior of each box ; so that the gas on leaving the principal vessel traverses successively the superposed boxes, circulating the while for a long time in contact with their cold sides and in the ducts which offer sections sufficiently large for resistance to be but very feeble. All these boxes are contained in a cylindrical vessel made of brass, into which a determined volume of water, at a known temperature, is poured. In this first envelope is a thermometer and an agitator. To preserve the calorimeter from irregular currents of air it is enclosed in a second vessel, the bottom of which is provided with three small cones made of wood or cork, on which is placed the calorimeter. This vessel has a hole in the side to allow the tube of the calorimeter to pass through, and a cover in which are two holes, one of which affords a passage for the discharge tube, and the other for the stem of the agitator. It has, moreover, a third opening through which the thermometer is introduced. Finally the calorimeter and its envelopes are supported and fixed at their proper height by means of an adjustable foot.

The details of the experiments made, and the latest information published by M. Regnault, will be found in the "Mémoires de l'Academie des Sciences," vol. 26, p. 41.

1058bb. Regnault's Apparatus for determining the Specific Heat of Solids.　　　　　　　*L. Golaz, Paris.*

This apparatus, made by the exhibitor for M. Regnault on a new pattern, enables experiments to be made with great ease and rapidity. It is composed of three distinct parts:—1. The stove for heating the body which is to be investigated. 2. The apparatus for the production of steam. 3. The calorimeter.

1. This consists of three concentric tubes closed at their upper ends by a copper plate attached by bolts to the circumference of the outside tube, at the centre there is an opening of the size of the central tube and closed by a metallic plug with double sides, in the centre of which there is another opening in which a thermometer is fixed. At the lower end of the tubes there is bolted a metallic plate similar to the one above, and terminating in a cone; the central tube also has an outlet at the base of this piece, the second enclosure descends near the sides of the cone, the third enclosure is attached to this piece. The vapour issuing from an apparatus to be described further on reaches the lower part or base of the cone, between the first and second enclosure, circulates all round and warms the centre, leaves this chamber by holes made at the top of this tube and passes once more into the third enclosure, whence it goes into the neighbouring condenser.

2. The apparatus for producing the steam is a copper boiler furnished with three tubes, the first of which, placed in the centre, receives an apparatus of copper in which the steam on its way back is condensed; above the condenser is a funnel fed by a current of water which accelerates condensation; a return tube reaches to the bottom of the boiler and serves as a passage for the condensed vapour. The second tube, which is the largest, is used as a conductor for the vapour to the stove. The third one, which is small in diameter, brings the vapour which can be condensed in the stove directly into the boiler. The condenser is bound to the enclosures by tubes provided with junctions.

3. The calorimeter is composed of two concentric vessels supported on a wooden stand with square columns, and on one of them is placed a copper block, which serves to fix the thermometer. Under this support there is screwed a copper carrier, which slides on a guide of the same metal fixed to the table of the apparatus. A sliding wooden screen intercepts the radiation of the rest of the apparatus from the calorimeter. A chair suitably placed supports the stoves, and a support connected with the table serves to fix at a convenient height the apparatus for producing the vapour.

The body to be studied is placed in a small metallic basket suspended in the middle of the central tube of the stove ; a thermometer indicates the temperature of this part of the apparatus. When the moment has come for bringing down the body into the calorimeter the screen must be lifted up, and the carrier which supports the thermometer made to slide on the copper guide until it reaches a certain point fixed on the guide. With one hand the basket is let slip, and with the other the detent of a rapidly acting spring appliance is released. The body being thrust into the calorimeter it must be brought back to its former position and the screen then lowered. This apparatus, which is easily worked, enables M. Regnault's experiments on this subject to be repeated most accurately.

1058c. Jamin's Calorimeter. *L. Golaz, Paris.*

This instrument, made for the designer, is constructed on the same principle as M. Favre's, but being only used for demonstrations its construction is simpler. It consists of a cylindrical vessel of glass, provided with an iron cover to which are fixed two iron tubes reaching down into the cylinder, the stem of the thermometer being placed vertically in the cover. A small glass funnel is placed over a stop-cock terminating in a capillary glass tube, which serves for filling the apparatus.

1058d. Bunsen's Absorptiometer. *L. Golaz, Paris.*

This apparatus consists of a wooden stand, to which is fixed an iron ring carrying two vertical supports, united at the top by a circle to which is fixed a plate on hinges. Along one of the columns is placed an iron tube, having at its upper end an iron funnel ; at the lower it enters the vertical opening of a three-way cock. One of the horizontal openings communicates with a chamber made in the centre of the stand, and the other communicates with an outlet ; thus it is possible either to make the mercury run into the chamber, or, by turning the cock another way, to permit it to escape.

On the upper surface of the wooden stand has been made a deep groove, at the bottom of which is fixed a washer of vulcanized india-rubber. A glass cylinder, the ends of which are perfectly true, rests in this groove, and fits the top of the stem ; the plate, which is used as a cover, is provided with an india-rubber disc, and rests on the top of the glass cylinder.

The absorption-tube is open at its lower end, and cemented into a socket made of iron, having a screw thread which fits into an iron block cut away at the sides. A rotary motion about its axis is sufficient to cause the opening or closing of this tube, an india-rubber washer being placed at the junction. On the part of the block which is not cut away are fixed two springs of steel, which retain it in the hollow of the stand and in two diametrically opposite grooves which are placed there ; a slight friction is thus produced, permitting however the opening and shutting of the absorption tube. At the centre of the higher plate of the cover is an iron socket on which is stretched, by means of a ring screwed to it, a piece of india-rubber ; in lowering the cover the point of the tube enters into the socket, and the tube is found perfectly fixed at the centre.

1058e. Regnault's Air Thermometer. *L. Golaz, Paris.*

This apparatus is composed of a glass cylinder terminating in a capillary stem bent at right angles, and supported by a metal stand, having in its centre a tube on which is placed the stem of the thermometer. The screw adjustment, with two pointers, is used to take, by means of the cathetometer, the height of the mercury in the small glass vessel which is placed underneath the apparatus. Above is arranged a glass cylinder which serves to hold the ice in the experiment. A bent tube lets the water which runs off escape.

1058f. Regnault's Apparatus for determining Specific Heat of Liquids. *L. Golaz, Paris.*

This apparatus consists of a cylindrical copper vessel, having at its lower extremity a lateral conical appendage. In the axis of the vessel there is a cylinder of brass which receives the liquid to be experimented upon, and which is furnished at its upper part with a vertical tube, into which is inserted a thermometer, giving the temperature of the liquid. The cylinder terminates in a stop-cock inclined downwards, and having at its lower extremity a tube which, after having traversed the conical space, is soldered to the cylinder, enters the tube of the calorimeter at the other side of the stop-cock, and terminates in a tube which goes through the wall of the cylinder. It is by this tube that the key of this stop-cock enters, it being worked from the exterior. A hole made in the hand of the key establishes a current of air which drives the liquid into the calorimeter, at the same time that the stop-cock is turned for that purpose.

A double wooden screen, fixed perpendicularly, cuts off all radiation from the vessel which is being heated or from the calorimeter. Two iron arms fixed to the upright of the screen support the vessel at such a height that a gas burner may be placed underneath.

The calorimeter consists of a brass cylinder, into which passes the liquid experimented on, and of a flat vessel containing a thin screw-shaped blade for the purpose of condensing the vapour which might be generated by the liquid in the vessel. A brass receptacle surrounds this first calorimeter, a second one, resting upon feet made of wood or cork, surrounds the first. The height and the position which must be given to the calorimeter are regulated by a stand with a movable column. The description of the experiments with this instrument is to be found in "Mémoire de l'Académie," tome 26, page 262.

This apparatus is constructed of glass, copper, or platinum, according to the nature of the liquids upon which it is desired to operate. The one exhibited is made of platinum.

1058g. Favre's Mercury Calorimeter.
L. Golaz, Paris.

This apparatus consists of a large spherical reservoir of cast-iron, resting on a support of iron, surrounded with a non-conducting substance. Underneath is made a vertical opening, through which a steel piston passes, being set in motion by a screw provided with a winch-handle. Near the centre of the sphere there is a cast-iron tube on which is screwed a steel stop-cock, terminating in a small glass ball, to which is fitted a carefully calibrated horizontal thermometer. The graduations on the stem of this thermometer correspond with its interior capacity; it is supported by two wooden columns, which rest upon a solid bench opposite the stem, and on the corresponding side are two iron tubes, placed obliquely with regard to the vertical diameter of the sphere. These two tubes, called "mouffles," are immovably fixed to the iron sphere, and extend to its interior. They are thus completely surrounded with mercury when the reservoir is full, an operation which is very easily accomplished by means of two stop-cocks placed opposite to the plunger. A small glass pipe of a particular shape is used for bringing liquids into the "mouffles." The divisions on the stem of the thermometer are read by means of a micrometer. To protect the large bulb of the thermometer from the influence of the temperature of the exterior, it is surrounded by a wooden box, in which is placed flannel and swan's down.

1060. Thermometric Tube for determining the calorific capacities of different liquids. *Elie Wartmann, Geneva.*

A thermometric tube, being part of the exhibitor's apparatus for the determination of calorific capacities in liquids. A full description of the method is printed in the number for May 1870 of the "Archives des Sciences physiques et naturelles." An electric chronoscope, such as Sir Charles Wheatstone's, expresses in thousandth parts of a second the time necessary for the cooling between two constant temperatures of the same body (the thermometric tube) when immersed in equal volumes of different liquids, at the same initial degree of heat.

1061. Apparatus made by De la Rive and Marcet for measuring the specific heat of Gases. A small copper calorimeter, containing a very thin serpentine gold pipe.
Lucien de la Rive, Geneva.

1063. Drawing of an Apparatus for determining the calorific capacity of liquids.
Dr. Leopold Pfaundler, Professor of Physics at Innsbruck.

In a box protected against draught there are placed two calorimeters—one filled with water, the other with the liquid to be tested. An electric current is passed through the two spiral wires, of equal resistance, which are inserted in the fluids.

Two paddles stir the liquids, and two thermometers measure the temperature.

The respective capacities for heat are calculated from the proportion of the increment of heat.

V.—RADIATION.

1056. Hargreaves's Thermo-radiometer, for measuring loss of heat by radiation from walls of furnaces, sides of steam boilers, &c. *James Hargreaves.*

The silver-plated copper vessel is filled with water and enclosed in the case, the blackened face then being exposed for a given time (say five minutes) to the radiating surface, a thermometer inserted in the neck of the vessel shows the elevation of temperature due to radiation. The heat is calculated as follows, either in calorics or British thermal units.

$$\frac{WS\,(T-t)}{a\,m} = x.$$ Where WS = weight and average specific heat of vessel

and its contents ; t, temperature of the same before exposure ; T, temperature of the same after exposure ; a, area of blackened face of vessel ; and m, time of exposure, whence may be calculated the amount of fuel necessary to replace the heat lost by radiation.

40075.   S

1062. Apparatus used in researches on the **Absorption** of **Radiant Heat** by gases and vapours.

Phil. Trans., 1861.   *Prof. Tyndall, F.R.S.*

1065. Pair of concave Reflectors of Prima German silver, 500 mm. in diameter, on brass stands, with supports for carbon and tinder, for experiments on radiant heat.

Warmbrunn, Quilitz, and Co., Berlin.

1066. Melloni's Apparatus for the investigation of the laws of radiant heat.   *H. Lloyd, Trinity College, Dublin.*

1072. Pouillet's Actinometer, for sidereal radiation.

Conservatoire des Arts et Métiers, Paris.

1072a. Actinometer, for measuring the intensity of **Solar Radiation.**   *J. Louis Soret, Geneva.*

It is composed of a tube, of the diameter of about 35 millimetres, closed at one end, and blackened inside. The end, which can be opened by removing the stopper, is furnished with a diaphragm. The central tube is encircled with a concentric brass wrapper, which has to be filled up with pounded ice, or with snow. The apparatus is upheld by a horizontal axle upon a wooden support. This axle is formed of a tube which, on one side, lets out the rod of the thermometer (lacquered and blackened), and on the other may be adjusted to an air-exhausting pump. The apparatus is directed towards the sun ; the orientation is obtained by means of the exterior appendiculæ.

See "Recherches sur l'intensité calorifique de la radiation solaire.—Comptes Rendus de l'Association Française pour l'Avancement des Sciences," 1st Session, Bordeaux, p. 282.

1072b. Actinometer.

M. Desains, Member of the Institute, Paris.

1073. Pouillet's Pyrheliometer, for observations on solar heat.   *Conservatoire des Arts et Métiers, Paris.*

VI.--ABSORPTION.

1084. Ice-making Machine ; system of Raoul Pictet & Co. *Raoul Pictet & Co., Geneva. Geneva Association for the Construction of Scientific Instruments.*

This machine manufactures ice by means of anhydrous sulphurous acid. This substance has the following advantages :—

1. It is without action upon metals and fatty substances.
2. It gives but slight pressures, never exceeding 4 atmospheres in a temperature of 30° centigrade.
3. It is free from all danger of ignition or explosion.
4. It is the least expensive volatile liquid.

This machine turns out 12 kilogrammes of ice per kilogramme of coal consumed.

1085. Apparatus for Freezing Water, constructed by Mr. Bieberich. Compare the adjoined instruction for use.

University of Munich (Prof. v. Jolly).

1086. Small Ammonia Ice Machine.

Vaast and Littmann, Halle.

VII.—CONDUCTION.

1091. Despretz's Apparatus for showing the **Conduction of Heat** in metals with 9 thermometers.

Warmbrunn, Quilitz, and Co., Berlin.

1090a. Three Original Bars, by **Despretz,** used in his experiments upon the **Laws** of **Conductivity.**

The Faculty of Sciences, Paris.

1088. Forbes' Iron Bar for **Thermal Conductivity,** with its crucible. *Prof. Tait.*

1089. German Silver Bar, of same size, cast for same purpose. *Prof. Tait.*

1059. Diacalorimeter. To measure the resistance which liquids offer to the passage of heat. *Frederick Guthrie.*

Two conical platinum vessels, having their bases accurately plane, are supported so that their bases are parallel, horizontal, and nearly in contact. The lower cone is fixed, and, being provided with a vertical tube fitting air-tight in which water stands at a known height, serves as an air thermometer and calorimeter. The upper cone can be adjusted by a micrometer screw at any distance from the lower one. Through the upper cone a current of warm water or steam is passed. Between the bases of the cones is introduced the liquid whose thermal resistance is to be measured.

1092. Ingenhousz's Apparatus for demonstrating the **Conduction of Heat,** with nine rods of different metals.

Warmbrunn, Quilitz, and Co., Berlin.

1093. Apparatus intended to produce the **Curves** of **Thermic Conductivity** on the surface of bodies.

M. Jannetaz, Paris.

On the glass support are placed plates coated with grease, which may be coloured. The current of a battery is made to pass through the small platinum ball, which is at the end of wires of the same metal, and which are put into contact with the plate.

1093b. Four Boxes, containing plates subjected to the action of the above apparatus, and showing the isothermic curves produced by the fusion of the grease upon—

    A. Bluish schist, from the valley of Sulvan, near Vernagaz.

    B. Coal schist, from neighbourhood of Motivon, near the Col de Voza.

    C. Another schist from the Col de Voza.

    D. Septypite from Tholy (Vosges).

    E. Clayey schist.

    F. Hyaline quartz, with faces cut parallel with the axis.

    G. Trooshte, silicate of zinc, with faces cut parallel with the axis.

    H. Gneiss, fine grained, from the Val Anzasca, Monte Rosa.

M. Jannetaz, Paris.

VIII.—POLARISATION.

1095. Forbes' Mica Plates for the polarisation of heat by refraction.   *University of Edinburgh.*

The mica is split into numerous thin films by careful application of heat, and fixed in the wooden tubes at the proper polarising angle.

IX.—MISCELLANEOUS.

Lens of Benedetto Bregans, 42 cm. in diameter, given to the Grand Duke Cosimo III. after 1690. With this lens Avesani and Yargioni, 13 years after the extinction of the Accademia del Cimento, made the well-known experiments on the combustion of the diamond and other precious stones. Afterwards it served the celebrated Sir Humphry Davy for his researches on the chemical constitution of the diamond.

The Royal Institute of " Studii Superiori," Florence.

Plaster Model of Thermodynamic Surface.

Prof. J. Clerk Maxwell.

1097. Apparatus employed by **Professors Stewart** and **Tait** for producing rapid **Rotation** in a **Vacuum.**

Prof. Balfour Stewart.

The driving shaft is of iron, and passes into the receiver through an iron tube containing mercury, which acts as a barometer.

This instrument was constructed and in part devised by R. Beckley, of the Kew Observatory, for an experiment conducted by B. Stewart and P. G. Tait. The mechanical principle of its construction is as follows :—The receiver in which the rotation takes place is in fact the extended vacuum of a barometer. Through the mercury of this barometer a shaft communicates a slow motion to some machinery in the receiver. The velocity of this motion is there greatly increased by means of a suitable train of wheels, until at length a vertical disc is made to rotate with great rapidity on a horizontal axis. If this disc be made to rotate in an ordinary obtainable vacuum it will become heated, and the question is to determine whether this heating of the disc is entirely due to the friction of the residual air, or to something else.

Two wires, covered with gutta percha, carried through the bed-plate of the apparatus, convey into the receiver a thermo-pile arrangement, by means of which the temperature of the disc may be measured. This thermo-pile may either measure the increased temperature of the disc after rotation by radiation—in which case it is fitted with a reflecting cone ; or it may be made to tap the surface of the disc after rotation—in which case there is an arrangement working through a barometric tube, by means of which the pile may be brought up to the disc in vacuo after rotation. Finally, there is an arrangement, working also through a barometric tube, by means of which a vessel containing some peculiar chemical substance may be uncovered in vacuo at will. The object of this is to reduce the pressure of the vacuum by chemical means. For instance, if we have a carbonic acid vacuum made as low as it can be made by ordinary means, a vessel containing moist potash would be uncovered, and a great part of the residual gas absorbed. The experimenters by this means have obtained a vacuum as low as 0·025 in. The conclusion from these experiments appears to be, that there is a certain heating of the disc which does not depend on the residual air.

1098. Effects of Heat on Nature. 9 Sénarmont crystals for heat. *Laurent.*

1100. Automatic Fire Extinguisher and Alarm. For ships, factories, and all places where steam apparatus is employed. *Sanderson & Proctor.*

The action of the apparatus is as follows :—

Thermometers are fixed on the ceiling or elsewhere in the room, and are set to complete an electric circuit at a given degree of heat, each thermometer being in connexion with a galvanic battery and electro-magnet. A steam valve is also fixed in the room, communicating with the valve at the boiler. As soon as a fire breaks out, and raises the temperature of the nearest thermometer to the given degree, the electric current is complete, and the electromagnet, by a very simple contrivance, opens the steam valve in the room and the valve at the boiler, the steam rushes into the room at the existing pressure of the boiler and effectually extinguishes the fire. Simultaneously with the action of the electro-magnet, an electric alarm and indicator are set in motion, thus giving prompt notice of the fire, and indicating its position.

1101. Prism of Rock Salt, 50 × 60 mm.
W. Steeg, Homburg vor der Höhe.

1102. Lens of Rock Salt, 75 mm. in diameter, and 300 mm. in radius. *W. Steeg, Homburg vor der Höhe.*

1103. Plate of Rock Salt, 60 × 60 mm.
W. Steeg, Homburg vor der Höhe.

1104. Diagram of the **Thermodynamic Relations of Aqueous Vapour** for pressure up to 30 atmospheres.
Oberbergrath E. Althaus, Breslau.

These diagrams have appeared on a smaller scale, and accompanied by a memoir entitled : The Boiler Systems for High Pressure, and their Application to Mining Engines, Part III., in the " Zeitschr. für das Berg-, Hütten-, und

" Salinenwesen, im Preussischen Staate. Herausg. im Minist. für Handel,
" Gewerbe, und öffentliche Arbeiten. XXIII. Bd., 5 Lief."

1104a. Two Photographs, representing the **Thermo-ventilator** when open and closed.   *M. Wellesen, Christiania.*

This instrument, which is useful as a means of promoting health, consists
of five parts, comprising a regulating balance for heat, and a valve for fresh
air, which is self-closing, and can keep out the cold. The smaller part of
the iron and brass tube (to which a foot or stand is fixed for exhibiting purposes) gives a clear idea of the system by which the thermo-ventilator is
applied to the internal orifice of the atmospheric tube.

1105. **Apparatus** of sheet metal, constructed by the exhibitor,
for demonstrating the **Causes of Disturbances in the Draught
in Chimneys,** as well as the influence of wind and weather.
It is at the same time a means of studying the general motion of
air produced by heat (ventilation). Compare the adjoined description.   *Prof. Dr. Meidinger, Carlsruhe.*

1105a. Apparatus for quickly communicating a **Given
Temperature** to a **Liquid,** and to maintain it at this temperature.
          W. Gloukhoff, Warden of Russian Standard Measures.

It is used in the determination of the density of the liquids; for the
verification of alcoholometers, hydrometers, &c. ; at normal, or any other temperature. The principal part of this apparatus is a cylindrical metallic
stirrer, immersed in the liquid to be experimented upon, contained in a very
solid glass cylinder. The warm or cold water, poured in small quantities, in
the interior hollow space of the stirrer, by means of a glass funnel introduced
in one of the two metallic tubes communicating with the hollow space, and a
few movements up and down of a stirrer, will very quickly warm or cool the
liquid, and thus bring it to a determined temperature. The syphon, introduced
in another tube, serves to empty the stirrer. In the glass disc, covering the
glass cylinder, are two apertures: the large one, for weighing solid bodies in
liquids, immersing in it the hydrometers, &c.; and the small one for a thermometer to determine the temperature of the liquid.

1105b. Apparatus for keeping a **Constant Temperature**
in an air or water bath.
          The Secondary Government School, Assen (Netherlands).

This gas regulator, constructed after the design of Dr. A. Van Hasselt,
teacher at the school for middle-class education at Assen (Netherlands), was
made principally by J. Van Rossum, servant in the laboratory.

The air thermometer, connected with the regulator and with a large Bunsen
cell, is partially filled with mercury, so that only the upper bulb and a
great part of the tube contain air.

When the desired temperature has been reached, the wire of platinum is
screwed down, so as to reach the mercury, at which moment a current from
the cell passes. This current passes through the wire, which is coiled round
the iron tube in the apparatus, through which the gas is flowing, and which is
used as an electro-magnet.

The little iron valve on the side of the tube, covered with a thin plate of
caoutchouc, which now shuts the tube, has a small aperture, which lets pass
gas enough to prevent the lamp from being extinguished.

SECTION 9.—MAGNETISM.

West Gallery, Ground Floor, Room F.

I. NATURAL MAGNETS.

Natural Magnet, mounted by Galileo, weighing six ounces. He presented it to the Grand Duke Ferdinand II. del Medici. It is made in the shape of an urn and holds a weight of ten pounds. *The Royal Institute of " Studii Superiori," Florence.*

Galileo was the first to arm natural magnets, and thus succeeded in making them bear very much greater weights; he proved, at the same time, that the same loadstone can bear a greater weight when in several pieces than when it is in one block; and that moreover its attractive power can be increased by loading it by degrees with an increasing weight.

Natural Magnet, armed by the Accademia del Cimento.
Accademia del Cimento.

The members of the Accademia del Cimento found that the power of this magnet was not lost by passing through very many substances; and that its attractive force varied according to its position with regard to the poles of the earth.

1106. Great Natural Magnet; one of the largest known. *See* Lamont, " Handbuch des Magnetismus," 1867, p. 107.
Teyler Foundation, Haarlem.

Weight, with the armature - - 152 kilograms.
Force - - - - - 114 „

1107. Natural Magnet, mounted in brass case, with steel poles, and soft iron keeper. *Elliott Brothers.*

1107a. Natural Loadstones (two), Russian, in perforated and painted metal cases. *Bennet Woodcroft, F.R.S.*

II. PERMANENT ARTIFICIAL MAGNETS.

1108. Collection of Artificial Magnets, lately forged by M. Van Wetteren, and magnetised at Teyler's Museum.
Teyler Foundation, Haarlem.

A. Single magnet, weight 2·17 kilogr.; greatest original force, 51·3 kilogr.; permanent force, 35·9 kilogr.

B. Compound magnet. No. 3,053. Weight, 1·52 kilogr.; greatest original force, 36·4 kilogr.; permanent force, 27·7 kilogr.
C. Two single magnets. No. 3,003. Weight, 0·51 kilogr.; greatest primitive force, 19·5 kilogr.; permanent force, 12·4 kilogr. No. 3,005. Weight, 0·52 kilogr.; greatest original force, 18·6 kilogr.; permanent force, 13·7 kilogr.

A third open keeper with movable side plates.

The box, C, forms a school apparatus, to demonstrate the effect of mutual contact of the magnets, and of the constitution of the keeper, on the distribution of free magnetism, on the weight that can be suspended, and on the deflections of the needle produced by the magnet when open, and when more or less closed.

See the memoir of Prof. Van der Willigen, which will soon appear in the Archives du Musée Teyler.

1109. Great Artificial Magnet, forged and magnetised in 1850 by Messrs. Van Wetteren and Logemann, according to the directions of Dr. Elias, whose property it is.

Teyler Foundation, Haarlem.

Newly magnetised at Teyler's Museum.
Weight -   -   -   -   - 28 kilograms.
Original force   -   -   -   - 260   „
Permanent force -   -   -   - 200   „

1109a. Large Artificial Magnet made of thin Plates.
M. Jamin, Paris.

1110. Photograph of a Horse-shoe Magnet, made by Johann Dietrich, of Basle, in 1755.
Prof. Hagenbach-Bischoff, Director, The Physical Institute in the Bernoullianum, Basle.

1111. Permanent Bar Magnets (pair of), in case.
James How & Co.

1112. Compound Horse-shoe Magnet. *James How & Co.*

1112a. Series of Permanent Magnets, bar form, in wooden cases for students. *Harvey, Reynolds, and Co.*

1113. " Logemann's " Magnet, being a powerful battery of steel-plate magnets. *Frederick Guthrie.*

1113a. Magnets by M. Jamin. *M. Bréguet, Paris.*

1113b. Magnet, Jamin's, with plates 0·05 m. in width.
M. Breguet, Paris.

1113c. Large Artificial Magnet, and fittings.
M. Jamin, Paris.

III. ELECTRO-MAGNETS AND ELECTRO-MAGNETIC ENGINES.

1115. Large Electro-Magnet, for showing magnetism and diamagnetism. *Elliott Brothers.*

1115a. Small Electro-Magnet by M. d'Obelliane, bearing thirty times its own weight. *Polytechnic School, Paris.*

1116. Three Electro-Magnets.
The Committee, Royal Museum, Peel Park, Salford.

One of which is a circular plate, $4\frac{1}{2}$ inches diameter; another, a plate $4\frac{1}{4}$ inches square; and the third, 4 inches, with one armature; the 4-inch magnet will sustain a load of 1,400 lbs. to 1,500 lbs. Made by the late R. Roberts, C.E., of Manchester.

1117. Powerful Electro-Magnet.
The Committee, Royal Museum, Peel Park, Salford.

Of the horse-shoe form, with an elliptical section. Made by the late R. Roberts, C.E., of Manchester.

1118. Surface Electro-Magnet made in 1840. When fully excited, the armature is retained with a force of upwards of a ton.
J. P. Joule, D.C.L., F.R.S.

1119. Electro-Magnet, on Joule's Construction, mounted so as to serve for supporting weights, or for experiments on Diamagnetism. *Dr. Stone.*

1119a. Electro-Magnet, by Repmann.
M. Breguet, Paris.

1119b. Electro-Magnet, by Jablochkoff.
M. Breguet, Paris.

1119a. Small Electro-magnet. *H. L. Rutter.*

1120. Tubular Electro-Magnets. *John Faulkner.*
In this system iron cases are used, of such size and thickness as ensures the utilisation of the maximum magnetic effect.
The object is to accumulate and utilise all or a portion of the electric power, and thereby when desirable prevent the loss that takes place in electro-magnets usually employed in electrical appliances.

1121. Diagrams (30), illustrating the **Great Waste** of **Power** in **Electro-Magnets** as heretofore made, and the economy of tubular electro-magnets. *John Faulkner.*

These diagrams are produced by scattering iron filings upon paper, prepared with paraffin, placed above ordinary electro-magnets and improved electro-magnets respectively.

1122. Objects illustrating the applications of tubular **Electro-Magnets.** *John Faulkner.*

1. A. Electro-magnets; various.
2. B. Electric bells; various.
3. C.   „   indicators; various.
4. D.   „   semaphore actuators; various.
5. E.   „   telegraph sounders; various.
6. „   „   „   key and sounder on one base.
7. „   „   „   „   „   separate.
8. „   „   „   sounder with movable cover.
9. „   „   „   „   with movable core.
10. „   „   „   „   with movable coil.
11. F.   „   brass and iron separators.
12. G.   „   pottery glaze iron extractors.

In the sounders a movable cover of wire is made partially or totally to cover the coil by means of a regulating screw. In case of partial failure of the battery, the instrument can be increased in power by a suitable alteration of the position of the wire cover.

1123. Froment's Engine. An electro-magnetic machine depending upon the successive attraction by fixed electro-magnets of bars of soft iron fastened on a wheel and parallel to its axis.
Frederick Guthrie.

The successive magnetisation and demagnetisation of the magnets is effected by the action of cams on the axis of the wheel, which lift ivory rollers, and so displace springs to which they are fastened.

1124. Helmholtz's Electro-magnetic Engine.
F. Rob. Voss, Berlin.

The advantage of this machine is that it is set in motion by a very small galvanic force; with two Bunsen elements it will drive one of Professor Helmholtz's double-syrens or one of his centrifugal commutators.

Professor Helmholtz has applied a contact-arrangement to the commutator of the Siemens' bobbin, which surpasses all former ones in that it avoids great surfaces of friction, thus attaining greater power and speed.

1124a. Electro-magnetic Machine, with velocity regulator. Helmholtz.
Physical Institution of Berlin, Prof. Helmholtz.

The current which drives the electro-magnetic machine is interrupted by the centrifugal regulator whenever the velocity exceeds a certain limit, whereupon the driving force ceases. Using a current which is only a little stronger than what exactly suffices for the normal velocity, exceedingly constant velocities of rotation are obtained. (Described by M. Exner in the Sitzungsberichte of the Vienna Academy, Math. Natura. section, vol. LVIII., part II., page 602.)

1125. Electro-magnetic Engine.   *F. Stöhrer, Leipzig.*

1125a. Electric Motive Power, acting on a pump.
M. Loiseau, junior, Paris.

1125b. Electric Motive Power, acting on a jet of water.
M. Loiseau, junior, Paris.

1126a. Apparatus by Th. Petrouchevsky, Professor at the University of St. Petersburg, for **measuring** the distance of the **Magnetic Poles** in straight magnets from their ends.

Imperial University of St. Petersburg.

A straight magnet is suspended horizontally at a point where one of the poles is approximately supposed to be. Another straight magnet, much shorter than the first, is placed in the same horizontal plane, perpendicularly to the suspended needle. A slow movement can be given to this second magnet along a divided scale, horizontal and parallel to the suspended needle. The mutual repulsive action of the two magnets is greatest when the direction of the axis of the small magnet passes through the magnetic pole sought, provided that the suspended magnet remains perpendicular to the second magnet. To fulfil this condition an auxiliary magnet is used, which is placed on the other side of the suspended magnet, and which is to produce an equal, but contrary, effect to the second magnet. This method of finding the poles has been described in Russian in the "Messenger of Mathematical Sciences" (Wiestin), and since, in Poggendorff's Annalen, Bd. 152, s. 42. It requires the use of three distinct appliances: 1st. A bifilar apparatus for suspending cylindrical magnets. 2nd. A measuring apparatus, having a small magnet that can be moved parallel to itself while remaining perpendicular to the direction of said movement. The apparatus has two divided scales, and is supplied with a microscope, two levels, and two small telescopes of short focus. These two appliances are placed alongside, and parallel to one another, and, approximately, in the magnetic meridian. The third (No. 3) apparatus consists of a divided scale placed perpendicularly to the suspended magnet. Along this may be moved a small magnet, of which the function has been explained above. The second apparatus enumerated was constructed by Mr. Brauer, of St. Petersburg from the designs of Professor Petrouchevsky.

1126b. Apparatus for so-called **"Normal" Magnetisation,** by Th. Petrouchevsky, Professor of Physics at the University of St. Petersburg. *Imperial University of St. Petersburg.*

An iron cylinder for magnetising, placed symmetrically in a magnetising coil, may receive very different distributions of the free magnetism, according to the length of the coil, but the distribution does not depend on the strength of the current. The inventor has discovered that the distribution of the remanent magnetism, or rather the distance between the poles, remains always the same, independently of the length of the coil or the strength of the current. At a fixed length of the coil, the magnetic poles of the electro-magnet and the poles of the remanent magnetism coincide, so that the electro-magnet, placed horizontally and perpendicularly to the magnetic meridian before a sensitive magnetic needle, acts upon it in the fixed direction (say, the direction of the meridian itself) during the circulation of, and after the cessation of, the current; in the latter case, the poles of the permanent magnetism act. The poles are then almost at the ends of the magnetising spiral, the length of which is about 0·8 of the iron cylinder, independently of the length and diameter of the cylinder. This arrangement of the poles of the electro-magnets is called "normal." The apparatus used in these researches are composed of two very distinct parts. The first apparatus contains two brass cylinders, surrounded with wire spirals; by turning one of these cylinders, the number of spirals can be lessened on the one, and increased on the other. Inside one of these cylinders is placed the iron to be experimented upon. This apparatus is set upon a table covered with marble plates moving in two directions, perpendicular to one another. The other apparatus is a needle, called "unipolar," suspended by means of a cocoon thread, with two microscopes.

1127. Ring of Elias for magnetising artificial magnets of large size. *Teyler Foundation, Haarlem.*

A magnet of 28 kilogrammes weight has been recently magnetised by this ring in Teyler's Museum.

Dr. Elias proposed more than 25 years ago his ring-coil for the production of artificial magnets of all dimensions, by an intense galvanic current. His artificial magnet here exhibited has lately been re-magnetised, with a slight modification of his method by Prof. Van der Willigen in the Teyler Museum, by this ring with the current of forty Bunsen elements of the usual large size.

1128. Apparatus for showing to an audience the effects of the superficial tension of liquids (Tomlinson's Cohesion Figures); magnetic curves or the movement of liquid films, &c.
Prof. W. F. Barrett.

1129. Bar of Metallic Nickel, 20 inches long and $\frac{1}{2}$ inch diameter. *George Gore, F.R.S.*

1130. Three Plates of Metallic Nickel, 6 inches long and $1\frac{1}{2}$ inch wide. *George Gore, F.R.S.*

1131. Small Nickel Horseshoe, for making a Nickel Magnet. *George Gore, F.R.S.*

1132. Plate of Metallic Cobalt, 6 inches long and $1\frac{1}{2}$ inch wide. *George Gore, F.R.S.*

1135. Horseshoe Magnet of **Nickel,** used by Sir William Thomson in his experiment on the effect of magnetism on the thermo-electric quality of nickel. Result published in the Transactions of the Royal Society for the year 1856, Bakerian Lecture.
Edinburgh Museum of Science and Art.

1136. Apparatus for showing a series of molecular and magnetic changes in a red-hot iron bar; and designed also to show the influence of traction, compression, and torsion upon such changes. The latter experiments have not yet been made with it.
George Gore, F.R.S.

See Philosophical Magazine, Sept. 1870.

1136a. Apparatus for exhibiting molecular changes occurring during the heating and cooling of iron wire.
Prof. W. F. Barrett.

When an iron or better a steel wire is raised to a white heat, a sudden contraction occurs at a dull red heat, the apparatus exhibits the phenomenon first noticed by Mr. Gore; when the wire is allowed to cool after being raised to a white heat a corresponding expansion at a red heat was found by Prof. Barrett, who also discovered that at this moment a crepitating sound was emitted, the wire resumed its magnetic state at this period, and further-more, a sudden reglowing of the wire takes place. These phenomena at this critical temperature are also associated with changes in the conductivity and thermo-electric property of the wire, as discovered by Prof. Tait. *See* Phil. Mag. Dec. 1873, Jan. 1874, and Report Brit. Ass., 1875, p. 259.

1136b. Apparatus designed by G. Gore, F.R.S., for exhibiting the effects of stress upon the magnetisation of iron.
Prof. W. F. Barrett.

1136c. Gore's Apparatus for exhibiting the torsion of an iron wire produced by axial and transverse magnetisation.
Prof. W. F. Barrett.

1136d. Apparatus and Diagrams, to illustrate the elongation of unstrained iron produced by magnetisation.
Prof. W. F. Barrett.

1137. Coulomb's Torsion Balance for magnetic and electric observations.   *Warmbrunn, Quilitz, and Co., Berlin.*

1137a. Electro-Magnetic Helix, on wooden tube. Used in electro-torsion experiments.   *George Gore, F.R.S.*

1137b. Two Wooden Tubes, upon which to form electro-magnetic helices for electro-torsion experiments.
George Gore, F.R.S.

1138. Apparatus for the **demonstration** of **Magnetic Friction,** constructed by the late M. Kleemann in Halle.
Prof. Dr. Dove, Berlin.

V.—APPARATUS FOR INVESTIGATIONS CONNECTED WITH DIAMAGNETISM.

1139. Glass Tubes prepared by **Faraday** for testing the magnetic and diamagnetic character of **Gases.**
The Royal Institution of Great Britain.

The tubes containing the gas to be examined were suspended in the magnetic field of a powerful magnet, the result being either attraction or repulsion of the tubes as the gases they contained were either magnetic or dia-magnetic.—Phil. Trans., 1850.

1140. Bars of Borate of Lead Glass, made and used by **Faraday,** for the action of magnets on polarized light.
The Royal Institution of Great Britain.
Phil. Trans., 1845.

1141. **The Diamagnetic Box of Michael Faraday,** containing spheres, cubes, and bars of diamagnetic metals; tubes of various liquids, bars of borate of lead, glass, various crystals, cradles, supports, &c., used by Faraday in his researches, on diamagnetism. *Prof. Tyndall, F.R.S.*

1142. Instrument used in researches on the **Polarity** of the **Diamagnetic Force.** *Prof. Tyndall, F.R.S.*

Phil. Trans., 1856.

1143. Specimen of " Faraday's Heavy-glass."
George Gore, F.R.S.

1144. Electro-Magnet for Induction and Diamagnetic Experiments, made in 1850, of a broad plate of annealed iron, so as to obtain a large induced power from a small voltaic source.
J. P. Joule, D.C.L., F.R.S.

The coil is composed of a bundle of copper wires, and has a resistance about equal to that of a Daniell's cell, exposing a surface of one foot square.

1144a. Large Diamagnetic Apparatus, with glass case, rods, &c. *Warmbrunn, Quilitz, & Co., Berlin.*

1143a. Magnetic Bench, for showing the **Rotation of the Polarized Ray.** *Dr. W. H. Stone.*

1144a. Electro-Magnet, for showing the rotation of a polarised ray. *Dr. Stone.*

1144b. Vacuum Tubes (2), to show the rotation of the spark round the magnetic poles. *Dr. Stone.*

VI.—APPARATUS FOR THE OBSERVATION AND REGISTRATION OF THE TERRESTRIAL MAGNETIC ELEMENTS.

DIP AND INTENSITY INSTRUMENTS USED IN MAGNETIC SURVEYS ON SHORE AND AT SEA.

EXHIBITED BY THE ADMIRALTY—HYDROGRAPHIC DEPARTMENT.

1145. Dip Circle for observations at sea, fitted with special arrangements for finding the magnetic meridian. By Nairn and Blunt ; date, 1772–1834.

This instrument may be considered as intermediate in construction between that made by Nairn for Captain Phipps, in his voyage towards the North Pole in 1773, and the Fox circle introduced by Mr. R. W. Fox in 1834, hereafter described.

It is suspended by an universal joint, from a wooden stand carrying one adjusting screw. The needle, 9 inches long, with steel axles, vibrates within a circle graduated to 20', and the ends of the axis are fitted to work in the agate holes of two adjustable screws in the vertical bars supporting the circle, and otherwise strengthening the instrument. The sliding pointers on the graduated circle are intended to be adjusted to the mean position of the needle when the motion of the vessel causes it to vibrate on either side of the dip. The screw on the under side of the circle works the metal supports on which the needle is placed until adjusted in the agate holes. A thermometer graduated to $-38°$ is placed inside the instrument.

The peculiar arrangement for ascertaining the magnetic meridian consists of a small compass gimballed at the end of a wooden arm. The other end of this arm has a brass fitting to fix on pins in the graduated circle on the top of the frame. The motion of the arm in azimuth causes the whole apparatus to move in the wooden stand until the Dip Circle is in the magnetic meridian, as indicated by the compass.

1146. Dip Circle and Intensity Apparatus. Fitted with arrangements for ascertaining the magnetic meridian by three methods. By Dollond ; probable date, 1776–1834.

The Dip Circle is made after the pattern described by the Hon. Henry Cavendish in the Phil. Trans., vol. xlvi., in which the dipping needle rolls upon horizontal agate planes, and a contrivance is applied for lifting it off and on to the agates at pleasure. A milled headed screw works this lift, and an adjacent butterfly screw, an arrangement for causing the needle to vibrate. The vertical circle is graduated to 20' ; the outer circle of the base plate is also graduated to every 45°.

The direction of the magnetic meridian may be ascertained by two methods other than that usually adopted:—1. An edge bar horizontal needle fitted with an agate cap may be placed on the steel point fixed to a balanced axis provided for placing on the agates like the dipping needles. The coincidence of this needle with the plane of the vertical circle shows the latter to be in the magnetic meridian. 2. The same edge-bar needle can be placed on a pivot screwed in the centre of the graduated circle at the bottom of the travelling box.

Of the three dipping needles, two are flat and one cylindrical and sharply-pointed. The axes are made of gun metal, and one of the flat needles is fitted with a brass cone on Mayer's principle.

Intensity Observations.—For this purpose the box which carries the dip circle is fitted with two apertures filled with glass, and a torsion circle on the top. The two flat needles, one of gun-metal for eliminating torsion, and the other for horizontal vibrations, have metal pins screwed into the centres by means of which they are attached to the stirrup suspended by silk fibres from the torsion circle. The vibrations are observed through the glass sides, and the magnetic meridian by the edge-bar horizontal needle before described.

This apparatus closely resembles that used by David Douglas on the north-west coast of America and the Sandwich Islands in 1829–34.

1147. Dip Circle. By Robinson ; date, 1830–75.

In this circle the needles are 6 inches long, flat and pointed. They move on agate planes in the centre of a graduated circle, and observations are read off by means of lenses fixed in the ends of a moveable arm centred on one of the glass sides of the instrument.

The advantages of this form of dip circle are:—1. That both the needles can be read off for nearly every angle of dip. 2. Portability, from compactness of stowage in the box, as the vertical circle is fitted so as to be readily detached from the horizontal.

An instrument of this kind was used by Major Estcourt during the survey of the river Euphrates in 1836.

1148. Dip and Intensity Circle invented by R. W. Fox, F.R.S. By Mr. George, of Falmouth; date, 1834–75.

The principal object of this instrument is the observation of Dip and Intensity at sea, and when placed on a properly constructed gimballed table this can be accomplished, except in very bad weather.

The needles are flat, tapering from the axis to a point, and 6·9 inches long. The axles are finely pointed, and work in the jewelled holes fitted to the bracket and centre of a concentric disc in the back of the instrument, which also carries the bracket. The grooved wheel on the axis is used for carrying the hooks and deflecting weights in intensity observations. In the holes in the cross arms of the verniers at the back of the circle, the deflectors, N and S, are screwed for dip and intensity observations, and are set at any required angle by means of the graduated circle. Of the two large thumb screws in the back of the moveable disc, one works the bracket when mounting the needle and vice versâ; the other works the clamp. At other times they are used in conjunction for the purpose of moving the disc when altering the bearings of the needles in the jewels. The pointed projection between these screws, when rubbed by the ivory disc, opposes the effect of friction in the needle and jewels.

The needles are packed in metal cases with screw ends, and may thus be used as deflectors.

Instruments of this construction have been largely and successfully used in the various magnetic surveys made at sea in H.M.'s ships.

1149. Hansteen's Intensity Apparatus ; date, 1819–50.

This form of intensity apparatus is that first adopted by M. Hansteen in his magnetic survey of Norway and the Baltic shores in 1819–24, and since largely used by various observers. The vibrating needle is cylindrical, pointed at the ends, 2·65 inches long and 0·15 inches in thickness. It is suspended from the moveable pulley at the end of the brass tube by a fibre of silk secured to a brass strap and loop in its centre. By means of the pulley the needle can be adjusted to the required height in the vibrating box.

The value of the observations depends on the permanency of the magnetic condition of the needle.

1150. Portable Magnetic Dip Circle, 3½ in. needle, made for, and used by, the late Sir John Shuckburgh. The dividing is very fine, and believed to be Ramsden's.   *G. J. Symons.*

GENERAL.

1176. Photograph of an Inclinatorium (Dip Circle), by Daniel Bernoulli, completed by Johann Dietrich of Basle, in 1751.
Prof. Hagenbach-Bischoff, Director, The Physical Institute in the Bernoullianum, Basle.

This instrument gained the Prize of the Academy of Paris in 1743.

40075.   T

1177. Dip Circle, for determining the magnetic inclination, adapted to needles of various lengths. (Barrow, London.)

H. Lloyd, Trinity College, Dublin.

1178. Theodolite Magnetometer, 9-inch circle, and collimator magnets. (Jones, London.)

H. Lloyd, Trinity College, Dublin.

1179. Dip Circle or inclination compass. *James How & Co.*

1180. Magnetometer, Kew pattern, constructed to determine the magnetic axis and the magnetic moment of a magnet, and the direction and intensity of the magnetic force of a given place.

Elliott Brothers.

The instrument consists of two distinct parts. For the observations of the deflection magnet, the copper box screwed to the centre of the azimuth circle is used; underneath this passes, through the centre, a divided metal bar with a vernier carrying a magnet; at right angles to this bar is the observing telescope. The hollow vibration magnet, with a scale on glass at one end and a collimating lens at the other, is observed through another telescope. The latter magnet is suspended in the mahogany box above the copper box.

1184. Declinometer for sea and land observations.

Carl Bamberg, Berlin.

The instrument is furnished with gimbals for use at sea, but may be fixed for observations on land. The magnetic system, which is provided with a mirror, oscillates upon a point, and is constructed for reversal; the adjustment is effected by means of a collimator telescope, and orientation from terrestrial and astronomical objects.

1185. Deviation Magnetometer, for determining the magnetic relations on iron vessels. *Carl Bamberg, Berlin.*

The deviation magnetometer enables determinations of deviation (horizontal and vertical) to be read off on points, and also determinations of horizontal and vertical intensity by oscillations and deviations. A small telescope serves for determining orientation from-terrestrial and astronomical objects. The instrument may be mounted on the same stand as the compasses.

1190. Drawing of a Dip Circle. *J. P. Joule, F.R.S.*

The needle, constructed of a thin ribbon of annealed steel, weighing 20 grains, is furnished with an axis made of a wire of standard gold. This axis is supported by threads of the Diadema Spider attached to the arms of a balance suspended by a fine stretched wire. The whole is hung by a wire which can be twisted at the head through 180°. At the bottom is attached a paddle immersed in castor oil, which brings the instrument speedily to rest in a fresh position. The deflections are read off by a short-focus telescope, placed on an arm revolving on an axis in the centre of the circle. With this instrument the dip can be determined within the fraction of a minute of a degree in less than a quarter of an hour.

With this drawing is exhibited a specimen of the THREAD of the DIADEMA SPIDER, also THREAD of the DIADEMA SPIDER COCOON.

1191. Portable Unifilar Magnetometer. An instrument for determining the horizontal intensity of terrestrial magnetic force; and also the declination.

Kew Committee of the Royal Society.

It consists of two parts : one for determining the time of vibration of a suspended magnet; the other for determining the amount of deflection it pro duces when caused to act upon a second needle.

In addition there is a third magnet, which is subsequently suspended, and its position referred to the astronomical meridian, by means of a mirror, which serves to allow of an observation of the sun's azimuth being made.

Used by the Rev. S. J. Perry, F.R.S., during the late Transit of Venus Expedition to Kerguelen.

1192. 12-inch Dipping Needle.   *Royal Society.*

1192a. 12-inch Variation Needle.   *Royal Society.*

1193. Kew Pattern Dip Circle.
Kew Committee of the Royal Society.

Dip circle of the pattern adopted by the Kew Observatory, having needles 3½ ins. long, which are read by microscopes, carried by a circle in front of the needle frame. It is also provided with accessory needles, for determining total force, after the method of Dr. Lloyd.

1196. Portable Theodolite for the observation of the **Magnetism of the Earth,** constructed by Dr. Meyerstein, Göttingen.
Prof. Dr. A. Kundt, Strasburg.

1197. Edelmann's Telescope with graduated Scale, for reading reflecting instruments (small).
M. Th. Edelmann, Munich.

1198. Edelmann's Telescope with graduated Scale, for reading reflecting instruments (large).
M. Th. Edelmann, Munich.

1199. Edelmann's Telescope with graduated Scale, for two observers.   *M. Th. Edelmann, Munich.*

1200. Edelmann's Rider, for observing two objects at the same time by means of a telescope with graduated scale.
M. Th. Edelmann, Munich.

1201. Edelmann's Declination Magnetometer.
M. Th. Edelmann, Munich.

1202. Edelmann's Variation Instrument for **Declination.**   *M. Th. Edelmann, Munich.*

1203. Edelmann's Variation Instrument for **Horizontal Intensity.**   *M. Th. Edelmann, Munich.*

T 2

1204. Edelmann's Variation Instrument for Vertical Intensity.   *M. Th. Edelmann, Munich.*

1205. Edelmann's Magnetometer for declination, vertical and horizontal intensity.   *M. Th. Edelmann, Munich.*

1206. Portable Magnet Theodolite.
          M. Th. Edelmann, Munich.

1207. Weber's Earth Inductor, new construction.
          M. Th. Edelmann, Munich.

1207a. Book, containing special treatises on some of the above instruments.   *M. Th. Edelmann.*

1207b. Book, containing the description of the above-named apparatus.   *M. Th. Edelmann.*

1209. Photograph of **Gauss's Bifilar Magnetometer.**
Magnetic Department of the Observatory, Göttingen, Prof. Dr. Schering, Director.

1210. Photograph of the **observing Telescope,** belonging to the above.
Magnetic Department of the Observatory, Göttingen, Prof. Dr. Schering, Director.

1211. Photograph of **Gauss's Earth Inductor.**
Magnetic Department of the Observatory, Göttingen, Prof. Dr. Schering, Director.

1212. Photograph of the **Multiplier and Needle.** For the determination of **Inclination** and **Absolute Intensity.**
Magnetic Department of the Observatory, Göttingen, Prof. Dr. Schering, Director.

1213. Photograph of the **Auxiliary or Deviation Needle.**
Magnetic Department of the Observatory, Göttingen, Prof. Dr. Schering, Director.

1214. Photograph of the **Theodolite** for the determination of the **Declination.**
Magnetic Department of the Observatory, Göttingen, Prof. Dr. Schering, Director.

1215. Description of the above-named instruments ; Gauss's Works, vols. 1–7.
Magnetic Department of the Observatory, Göttingen, Prof. Dr. Schering, Director.

1216. Dipping Needle with microscopes for the observation of the needle points, constructed by Dr. Meyerstein in 1843.
Magnetic Department of the Observatory, Göttingen, Prof. Dr. Schering, Director.

1217. Microscopic Apparatus for the determination of the **Collimation** of **Dipping Needles,** constructed by Dr. Meyerstein in 1843.
Magnetic Department of the Observatory, Göttingen, Prof. Dr. Schering, Director.

1218. Reflecting Dip Circle, after Dr. Meyerstein.
Physical Institute of the University of Göttingen, Prof. Dr. Riecke.

The instrument is so constructed that the magnetisation can be effected either by touching with a steel magnet, or by means of electric coil. In order to carry out the latter, the case is carefully taken off and the coil pushed over pillar and needle.

1219. Model for the illustration of the **Deviation** in **Iron Ships,** after Neumayer, constructed by the Joint-stock Company for the Manufacture of Meteorological Instruments, formerly Greiner and Geissler.
Hydrographical Department of the Imperial Admiralty, Berlin.

This model represents an important apparatus of instruction, which illustrates all the phenomena of deviation and compensation of the compasses. The apparatus is in use in the institutions of the Imperial Navy and schools of navigation.

1220. Gauss's Magnetometer, constructed by Breithaupt and Son, Cassel, with apparatus for suspension.
Polytechnic School in Cassel, Dr. Gerland.

The instrument has been constructed, according to the instructions and under the supervision of Professor *W. Weber* in Goettingen, by Messrs. *Breithaupt and Son.* Since the magnetometer with which *Gauss* and *Weber* had carried out their magnetic labours, and which is identical with the one here exhibited, has not been sent to London, this instrument may be considered as one of the oldest of its kind in the present exhibition.

1221. Photographic self-registering Horizontal Force, or **Bifilar Magnetometer,** constructed in 1847, at the Kew Observatory, by Mr. Francis Ronalds.
Kew Committee of the Royal Society.

Described in the British Association Report for 1849.
The magnet, 15 inches long, is suspended in a loop of fine wire, by means of a pulley, forming a bifilar arrangement. It carries, attached to its lower side, a light brass bar, which moves a little shutter in front of an oil lamp, allowing a pencil of rays to pass through a hole in it. The light is then thrown, by means of a lens, upon a daguerreotype plate, which is steadily drawn upwards by means of a clock.

The curves upon the daguerreotype plates were sometimes etched in, and engravings subsequently worked off, or tracings were made upon sheets of gelatine, which, being preserved, allowed the silvered plates to be repeatedly used.

This instrument was superseded by the improved magnetographs erected at Kew by Mr. Welsh in 1857, which have since remained in almost continuous action.

The suspension frame originally fitted has been replaced by one not belonging to the instrument when in use.

1222. St. Helena Magnetometers.

(1.) Declinometer instrument and telescope, used at St. Helena, 1840–1849.

(2.) Bifilar magnetometer and telescope, used at St. Helena, 1840–1849.

(3.) Vertical force magnetometer, used at St. Helena, 1840–1849. *Kew Committee of the Royal Society.*

The three magnetometers, the declination, horizontal force, and vertical force instruments, respectively, were made by Grubb, of Dublin, and formed one set of those used in the Colonial Magnetic Observatories, founded by the Government in 1840. The instruments were described in the Report of the Royal Society Committee of Physics, &c.

These instruments were erected at St. Helena in 1840, and constantly observed from that date until 1849.

The declinometer consists of a magnet bar, suspended by fibres of untwisted silk, and carrying a collimator arrangement of lens and scale the whole being enclosed in a cylindrical casing, perforated with windows, through which the scale is viewed by means of a telescope.

The bifilar is a somewhat similar arrangement, but the support of the magnet is formed of two parallel wires, which are twisted so as to bring the magnet into a position at right angles to the meridian.

The vertical force magnetometer is a light magnet 12 inches long, carrying a brass frame with cross wires at each end ; it is supported by a steel knife edge, bearing on agate planes, and its movements are observed by microscopes, fitted with micrometers, by which the position of the cross wires on the magnets is read.

1223. Declination Compass, used by Sir J. Richardson and Capt. Pullen.
Kew Committee of the Royal Society.

It consists of a square glazed box, containing a compass card, which is formed of a light metal divided circle, and two spring needles, connected to an agate cap in the centre. This is mounted so that it can either be suspended by a silk thread, or rest upon a point in the ordinary manner.

Two microscopes are fixed vertically above it, so that the divisions on the circle may be read by them, concave metallic reflectors being fitted to them for the purpose of illuminating the scale at the time of reading off.

1224. Portable Apparatus for vibrating a Magnet, used by Capt. Barnett, in H.M.S. "Thunder," in 1841.
Kew Committee of the Royal Society.

It is a glazed box 4 inches square, standing on a levelling stand, and carrying a brass suspension tube 6 inches in length. It also has an ivory circle fixed to its bottom.

There are two magnets 2·7 inches long, each of which when not in use is kept in a separate little copper box, where, fitted to an armature, it is embedded in iron filings.

1225. Dip Circle used by **Sir James C. Ross.**
Kew Committee of the Royal Society.

A Robinson dip circle, with four 6-inch needles, supported on agate planes, and read off direct on the circle of the instrument.

1225a. Apparatus for **discovering the Magnetic Poles** in Magnets and Electro-Magnets, by Professor Th. Petrouchevsky.
Imperial University of St. Petersburg.

A fine metal wire is stretched within the magnetic meridian, above a sensitive marine compass. Within a distance of 0·6ᵐ from the compass, the pole of which is to be determined, and below the same wire, a magnet is placed, so that its two arms are horizontal, and perpendicular both to the meridian and to the wire. The loaded needle of the compass usually deviates through the effect of the magnet, but in a special case it may be made to remain within the plane of the meridian. To effect this the line passing through the two curved magnet poles must pass also within the plane of the meridian, a thing easily accomplished. In the case of electro-magnets, besides the two bobbins forming part of the electro-magnet, a third bobbin is used, intended to compensate the effect of the first two upon the loaded needle. The description of this method, which presumes that both poles are equidistant from the extremities of the magnet, as also the results of experiments, are published in the Russian work "Cours de Physique Expérimentale," by M. Petrouchevsky, and since in the "Annalen der Physik," by Poggendorf.

1226. Forbes' Hemispheres for illustration of Lectures on the Earth's Magnetism and Temperature.
University of Edinburgh.

1227. Gambey's Declination Compass.
Conservatoire des Arts et Métiers, Paris.

1228. Self-registering Bifilar Magnetometer, with apparatus for determining the temperature-correction of the magnets employed in several automatic instruments, from the displacements of the photographic trace due to observed changes of temperature.
Chas. Brooke, F.R.S.

1229. Self-recording Magnetometer.
Chas. Brooke, F.R.S.

Rough home made apparatus, by which the first automatic records of magnetic variation by reflected light were obtained. The cylindrical lenses are water-lenses.

1229a. Self-registering Balanced Magnetometer, with compensation for changes of temperature, and warm water envelope for testing the same. The compensation is effected by the weight of the column of mercury in a thermometer tube.
Chas. Brooke, F.R.S.

1229b. Self-registering Barometer.

Chas. Brooke, F.R.S.

1229c. Photographic Apparatus, for registering simultaneously the variations of both the above instruments.

Chas. Brooke, F.R.S.

1229d. Self-registering Bifilar Magnetometer, with compensation for changes of temperature, and warm-water envelope for testing the same.   *Chas. Brooke, F.R.S.*

The compensation is effected by diminishing the lower interval of the double suspension, by means of the differential expansion of glass and zinc, in proportion to the diminished magnetic energy of the bar, due to elevation of temperature.

Photographic apparatus for registering the variation of the above, by means of a reflected pencil of light.

1230. Photographic self-registering Declination Magnet, constructed in 1846, at the Kew Observatory, by Mr. Francis Ronalds.

Kew Committee of the Royal Society.

Described in the Philosophical Transactions for 1847, vol. I.

The magnet, 2 feet long, when in use, was suspended by a silken skein, 9 feet long, on its under side; it carries a brass bar, from one end of which hangs a perforated metal plate, which, moving in front of a lamp, permits a pencil of light to fall upon a daguerreotype plate, carried slowly upwards by a clock suitably arranged.

The magnet is surrounded by a damper, made by electrotyping a frame of mahogany with copper. Both are enclosed in double wooden cases, having both surfaces covered with gold paper.

This instrument was superseded by the improved Kew magnetographs, which have been in almost continuous action since 1858.

1231. Instrument for the **determination** of the **position** of the **point** of **convergence** of the rays of the **Aurora Borealis,** both when it is below the horizon and also when it is above the horizon at the appearance of the Corona.

Prof. Heis, Münster.

The ball, resting in the pan, can after a few trials be brought into such position that several diverging pencils of the aurora borealis on the northern or the southern sky are, when properly viewed, covered by the rod which passes through the centre of the ball. The point of this rod, which can be moved up and down in the ball, shows, when the instrument is set according to the astronomical meridian, the azimuth and altitude of the converging point of the aurora pencils. This point of convergence does not exactly coincide, as the exhibitor has shown at the time of the great display of aurora borealis, Feb. 4th, 1872, with the point towards which the inclination needle directs. From the deviation of the two points, the height of the aurora can be calculated.

The instrument, which is easily manipulated, is much recommended to arctic explorers.

Instrument for navigators in the Arctic Regions for ascertaining the connexion of the Northern Lights with terrestrial magnetism, and for determining

the altitude of the Northern Lights.   By means of an instrument designed by
the exhibitor, the point of convergence of the north light rays is to be
accurately determined, as well when at the appearance of the corona it is
situated above the horizon, as when it is below the same, and in regard to
altitude and azimuth.   By the deviation of the point of convergence from the
direction of the dipping needle, the height of the north light rays can be
calculated.

1231b. Compass with **Diamond Pin.**
Ernst Winter, Hamburg, Eimstustel.

SECTION 10.—ELECTRICITY.

WEST GALLERY, GROUND FLOOR, ROOM F.

I.—APPARATUS FOR PRODUCING AND MAINTAINING DIFFERENCES OF ELECTRICAL POTENTIAL.

a. FRICTION AND INDUCTION MACHINES.

1233. Electrical Machine, having ebonite plate 3 feet in diameter. *Frederick Guthrie.*

This machine gives sparks 13 inches long.

1233a. Priestley's Electrical Machine. *Royal Society.*

1267. Winter's Machine, with 18-inch ebonite plate and condenser attached, for the accumulation of electricity.
Elliott Brothers.

Large Double-Plate Vulcanite Friction Electric Machine. *Dr. W. H. Stone.*

1233a. Large Electrophorus, with ebonite plate, 360mm. diameter. *Warmbrunn, Quilitz, & Co., Berlin.*

1234. Carré's Electric Machine. *Prof. W. F. Barrett.*

This is an induction machine or continuous electrophorus, but the loss from the inductor is replaced by a small attached frictional machine.

1234a. Froment's Electric Pendulum.
Prof. W. F. Barrett.

The pendulum is kept in motion by a current passing round the electro-magnet which lifts a small weight that is released as the pendulum descends. A control clock is in connexion with the pendulum.

1234b. Crystal of Tourmaline, mounted to show Pyro-Electricity. *Prof. W. F. Barrett.*

This crystal during heating or cooling exhibits polarity at its extremities. It is pivoted on a diamond cap.

1234c. Electrical Machine, by Singer, used by Mr. Francis Ronalds in his early experiments in the discovery of the electric telegraph ; described in his work on the electric telegraph, dated 1823. *Kew Committee of the Royal Society.*

It is an ordinary cylinder machine of blue glass, standing on glass columns.

1235. Bertsch's Machine. *Frederick Guthrie.*

A negatively excited sheet of ebonite leans against a revolving disc of the same material. On the other side of the revolving disc, one above and one below, are electric rakes. The conductor in connexion with the lower rake becomes negatively charged, the other one positively.

1236. Holtz's Machine. *Frederick Guthrie.*

A good example of this machine in one of its original forms, with windows and armatures. It gives a current of sparks over an interval of 8 inches.

1237. Electrical Machine, based on Holtz's principle, with ebonite discs. *Dr. L. Bleekrode, The Hague, Holland.*

This machine is constructed for generating electricity on the principles of induction as first employed by Holtz. The form is very much simplified, and the only material used is ebonite (india-rubber combined with sulphur). Two forms are constructed by the exhibitor ; the single ebonite machine with one fixed disc and another rotating before it, and the double ebonite machine. The latter consists of one fixed disc with paper armatures placed in the ordinary way but on both sides, a double system of conductors, and two rotating discs. The construction is no more complicated than that of the single machine, yet the quantity of electricity is exactly doubled.

The advantages of the machines constructed in this way, supported by experience of more than two years, may thus be briefly stated :—

(1.) The ebonite machines, constructed on the system of the exhibitor, with ebonite of a good quality (which may be easily had but must be carefully chosen) are at least as powerful in their action as the machines with glass discs, but they surpass them in being less costly, not liable to be broken, and much less dependent on the condition of the atmosphere. This must be appreciated in England, where, as is the case in Holland, glass electrical machines (working by induction) often remain inoperative owing to atmospheric moisture.

(2.) Although of very simple construction, they are very useful and powerful machines.

(3.) From a theoretical point of view they present many interesting properties when compared with machines in which glass is employed, and this led to the conclusion that they differ in their mode of producing electricity. An experimental investigation of this machine, stating its peculiarities, has been published in Poggendorff's Annalen, 1875, No. 10, pp. 278, 279.

1241. Di-Electrical Machine. *M. Carré.*

4560. Photograph of the " Cecchi Electrical Machine," formed of two discs, which are placed partially one over the other, the one of caoutchouc, and the other of glass, with parallel axes. *Prof. Filippo Cecchi, Florence.*

The Cecchi electrical or dielectrical machine is composed of two discs with parallel axes. The upper disc is of india-rubber, and is supported by an axis of glass ; the one below is of glass on an axis of metal. The axis of the glass disc has on one side a large pulley, and the axis of the other disc a small pulley, and by means of a continuous cord, not crossed, there is transmitted to the caoutchouc disc a rotatory motion eight or ten times faster than that of the glass-disc ; both the discs turn in the same direction. The discs are partially placed one above another, and are very close but without touching. The upper part of the caoutchouc disc passes between two arms furnished with metallic points, and connected with a large sphere of brass insulated at the extremity of a long glass rod. To this sphere is attached the hook of a condenser or else a Leyden jar formed by a barometer-tube with very thick walls. The lower part of the same disc passes before a comb of metallic points, called a T-comb, which communicates with the external armature of the condenser, and with two friction cushions of the glass disc, and then with the ground, and also with an exciter formed by a tube of brass with a ball at the end. When the discs are revolving, the large sphere becomes charged with negative electricity. This machine with discs of 80 centimètres diameter has given sparks of the length of 42 centimètres free in the air.

1242. Holtz's Machine, with four plates.   *M. Ruhmkorff.*

1242a. New form of Holtz's Machine.
Francis Pizzorno, Bologna, Italy.

This machine has the property of working well whatever may be the hygrometric condition of the air, since the two glass plates are placed in a crystal case hermetically closed. The air of the case is maintained continually dry by means of drying substances.

The conductors issue from the case and unite at the two Leyden jars which appear on the front of the figure.

1242b. Fixed Disc for a Holtz's Machine.
Augustus Righi, Professor of Natural Philosophy, Bologna.

The greatest possible difference of potential between the conductors of a Holtz's machine depends on the difference of the potentials of the paper surfaces carried by the fixed disc. But this latter difference is limited by the discharges which continually occur along the fixed disc. An ebonite plate is joined perpendicularly on the fixed disc, separating it into two parts, so that the discharges must follow the two faces of the plate. The potentials of the paper surfaces are increased, and the sparks between the conductors become longer.

This machine possesses four rows of points, namely, the two rows of points of the conductors, and the two rows of points obliquely communicating.

1242c. Large Electrical Machine, with double ebonite plates and Waiter's ring, formerly the property of Lord Lindsay.
Dr. Stone.

1268. Replenisher. Designed by Sir W. Thomson for restoring electricity to the Leyden jar of his quadrant electrometer.
Elliott Brothers.

A small charge being given to the Leyden jar, the replenisher increases or decreases the difference of potentials between the two coatings of the jar by a constant per-centage per half turn.

1243. Old Electrical Machine, with glass cylinders, one of which is covered with sealing wax, so as to obtain both positive and negative electricity.

The Council of King's College, London.

1244. Nairne's Early Electrical Machine, with glass globe.   *The Council of King's College, London.*

1245. Cylinder Electrical Machine.

The Council of King's College, London.

1246. Plate Electrical Machine, with four rubbers.

The Council of King's College, London.

1247. Armstrong's Electric Boiler or **Hydro-Electric Machine.**   *The Council of King's College, London.*

1248. Volta's Electric Lamp, or apparatus for lighting gas by means of an electric spark.

The Council of King's College, London.

It contains a leaden bottle for the generation of hydrogen gas. In the orifice are two wires separated from each other, which are connected to the two plates of an electrophorus. One of the wires is connected with the tap, so that the upper plate of the electrophorus is raised at the same time that the hydrogen is allowed to escape at the orifice, and the spark from the electrophorus sets fire to the hydrogen and thus lights the lamp.

1729. Glass Globes for producing Electricity by rubbing with the hand.   *The Council of King's College, London.*

The globes are caused to revolve by means of multiplying wheels and a band of rope. The globes may be exhausted, and they then become luminous; the greatest amount of electricity or "fire" was obtained from them when they were exhausted. In the one with a large brass cap, a small wooden disc could be inserted with threads distributed round its edge; when the globe was excited the threads stood out from the edge of the disc. Constructed about A.D. 1720.

1249. Induction Electric Machine. *T. Rob. Voss, Berlin.*

As there is no glass in Germany which insulates perfectly, Professor Helmholtz has used Leyden jars made of ebonite or vulcanite, which can keep electric charges for 14 days, or 14 times longer than the glass jars of Kirchhoff.

The advantages of this instrument are:—(1.) That the quadrants with the needle and mirror can be easily removed, so that any change in the needle or misplacement of the mirror may be examined with certainty.

There are new arrangements in the Leyden jar for raising or turning the needle without shaking the entire instrument (a thing to be avoided).

1249b. Combined Holtz's and Bertch's Induction Machine, with arrangement for separating the same.

Harvey, Reynolds, and Co.

1249c. Induction Electric Machine. *J. Teller, München.*

The fixed disc has neither holes nor cuts, which were hitherto considered indispensable.

1249d. Toepler's Induction Machine.
Royal Institution of Great Britain.

The apertures usual in the fixed disc are here dispensed with as unnecessary, the disc is thus rendered less breakable, and a greater action is obtained. The apparatus is very simple in construction and can easily be taken asunder for cleaning. The driving disc is at the same time utilised as the exciter of electricity.

1250. Holtz's Electric Machine, with fixed induction surface.   *Borchardt, Hanover.*

1251. Holtz's Electric Machine, with movable induction surfaces.   *Borchardt, Hanover.*

1542d. Electric Machine, with large ebonite disc.
C. Etler.

1252. Machine for **exciting Positive** and **Negative Electricity.**   *E. Stöhrer, Leipzig.*

It has the form of a small disc electric machine. According as one or other brass ball at the end of the caoutchouc frame is taken hold of, a quantity of positive or negative electricity is obtained.

b. GALVANIC BATTERIES.

1285a. Apparatus for Volta's Fundamental Experiment, with arrangement for chloride of calcium, two brass, one copper, one zinc plate, and insulating handle.
Warmbrunn, Quilitz, & Co., Berlin.

1253. Water Battery.
The Council of King's College, London.

1254. Daniell's Battery, employed in researches by Professor Daniell.   *The Council of King's College, London.*

1255. Early Voltaic Batteries:—
Babington's battery.
Cruikshank's   „
Wollaston's   „
Sturgeon's   „
The Council of King's College, London.

1256. Hare's Calorimotor, or Deflagrator.
The Council of King's College, London.

1257. De La Rue's Powder Chloride of Silver Battery.
Tisley and Spiller.

1257a. Forty Cells of a Rod Chloride of Silver Battery,
being part of a battery of 8,040 cells.
Warren De La Rue, F.R.S., and Hugo W. Müller, F.R.S.

The elements consist of a flattened silver wire and a zinc (non-amalgamated) rod. The electrolytes are a solution of chloride of ammonium, 23 grammes to a litre of water, and fused chloride of silver cast on to the silver wire.

When the terminals are not connected, the battery is quite inactive ; one such battery has been in action since November 1874. In order to prevent contact between the chloride of silver and the zinc rod, the rod of chloride of silver is encased in a tube of vegetable parchment open at both ends. The cell is a glass tube closed with a paraffin stopper.

Such a battery will evolve three cubic centimetres of mixed gases per minute, if connected up with a voltameter containing one volume of sulphuric acid, and eight volumes of water.

This battery is particularly well suited to experiments with a large number of elements, on account of its constancy.

8040 cells give a spark in air between a point (positive) and a plate (negative) of 0·342 inch (8·68 mm.) ; the striking distance (*distance explosive*) is shorter when the point is negative and the length of the spark is materially affected by slight differences in the form of the point; the point used is parabolic in form and is made of copper wire 0·125 inch (3·175 mm.) in diameter, the plate 1·1 inch (27·94 mm.) in diameter. The length of the spark between a point and a plate appears to be in excess of the ratio of the square of the number of cells ; for example 8,000 will give a spark 64 times as long as the spark from 1,000 cells. Between two spherical surfaces of 3 inches (76·2 mm.) radius and 1·5 inch (38·1 mm.) in diameter ; the striking distance (*distance explosive*) is only 0·038 inch (2·1 mm.) ; between spherical surfaces the law of the striking distance being in the ratio of the square of the number of cells does not hold. When a resistance is introduced into the circuit of 6,000,000 Brit. Ass. units (ohms), a series of intermittent brilliant sparks is obtained like those from an electrical machine ; but without resistance when the spark jumps the ordinary voltaic arc is formed. The current of 8,040 cells passes through a residual hydrogen vacuum of 38 mm. tension in a tube 1 6 inch (40·6 mm.) diameter, and 27 inches (58·58 centimeters) between the terminals ; the current of 1,200 cells passes with a tension of 0·2 mm.

1258. Portable Medical Battery, with modified form of De La Rue's chloride of silver and zinc elements.
Tisley and Spiller.

1259. New Galvanic Battery for Domestic Purposes.
Aurel de Ratti.

This Zinc-Carbon Battery is charged with a saturated solution of sulphate of magnesia or Epsom salts, a very cheap material. The flask above the cell is filled with crystals of this salt, on which a saturated solution of the same is poured until the flask is quite full. The cork with the glass tube is then forced down till the solution rises and fills the tube. A small cork is loosely fixed in the open end of the glass tube. No air must be allowed to remain in the tube or flask. The latter is now inverted, and the tube introduced through the round hole in the lid. The flask will be held in position by the projecting cork fitting into the round hole. The end of the glass tube will thus be immersed in the solution in the jar. The carbon plate is finally pushed through the square aperture in the lid, and by a simple manipulation the cork is pushed from the open end of the glass tube.

1259a. Muirhead's new Manganese Battery.
Warden, Muirhead, and Clark.

The positive plate is of zinc of a hollow cylindrical form placed in a perforated vitreous chamber. The negative plate is of platinized carbon, surrounded with lumps of platinized carbon and di-oxide of manganese. The exciting liquid is a solution of muriate of ammonia. Its electro-motive force is 1·6 volts, internal resistance, 2 ohms. Electro-motive force of a Daniell cell, 1·1 volts.

1260. Gas Voltaic Battery devised by W. R. Grove, Esq.,
M.A., F.R.S., Professor of Experimental Philosophy in the London Institution (now The Hon. Sir W. R. Grove), and described by him in a communication read before the Royal Society, May 11th, 1843. Experiments with this battery are described in a postscript, dated July 7th (Phil. Trans., 1843, p. 91).
London Institution, Finsbury Circus, E.C.

It consists of a series of Woulfe's bottles, into the necks of which glass tubes closed at one end are fitted by grinding; each tube contains a slip of platinum foil, coated with finely divided platinum, the slip being connected with a wire sealed into the end of the tube, and terminating outside in a little cup; the cups being filled with mercury, the tubes may be connected by wires dipping into the mercury. When the Woulfe's bottle and its tubes are filled with dilute sulphuric acid, and one of the tubes is then charged with hydrogen and the other with oxygen, in quantities such as will allow the platinum to touch the acid, and the ends of a wire are dipped into the cups at the tops of the tubes, an electric current is produced. At the same time the gases in the tubes gradually diminish in volume, the volume of hydrogen which disappears being double that of the oxygen; the current being generated, in fact, by the formation of water.

1260a. The Original Nitric Acid Battery.
The Hon. Sir W. R. Grove, F.R.S.

1261. Grove's Gas Battery, made by Spencer and Son, Dublin.
Prof. W. F. Barrett.

The current in this battery is produced by the gradual union of the gases oxygen and hydrogen, which fill the alternate upright glass tubes. Strips of platinum passing down the tubes serve for making metallic connexion with the gases.

1262. Constant Gas Voltaic Battery devised by W. R. Grove, Esq., M.A., F.R.S., Professor of Experimental Philosophy in the London Institution (now The Hon. Sir W. R. Grove), and described by him in a communication to the Royal Society, dated May 30th, 1845.
London Institution, Finsbury Circus, E.C.

To charge the apparatus, the stopper is removed from the end of the tube, and the glasses are filled to the top of the narrow platinum plates with acidulated water; acid is also poured into the end vessel, so as to cover the lump of zinc. The hydrogen which is evolved by the action of the zinc on the acid gradually expels the air from the main channel, and, when this is judged to be the case, the stopper is inserted; the hydrogen will now rapidly descend in all the tubes until the zinc is laid bare, and then remain stationary. A gas battery is now obtained, the terminal wires of which will give the usual voltaic effects, the atmospheric air supplying an inexhaustible source of oxygen, and the hydrogen being renewed as required by the liquid rising to touch the zinc; by supplying a fresh piece of zinc when necessary it becomes a self-charging battery, which will give a continuous current; no new plates are ever needed; the electrolyte is never saturated, and requires no renewal except the trifling loss from evaporation, which indeed is lessened, if the battery be in action, by the newly composed water.

1263. Element of M. Becquerel's Sulphate of Copper Battery. *Conservatoire des Arts et Métiers, Paris.*

1264. Twelve Elements of Galvanic Batteries on different systems, by Ruhmkorff. *Conservatoire des Arts et Métiers, Paris.*

1266. Smee's Battery. Six cells, with arrangement for raising the plates out of the cells. *James How & Co.*

1266a. Set of Six Cell Smee's Batteries. *E. Cetti and Co.*

1269. Grove's Nitric Acid Battery. *Elliott Brothers.*

1270. Faure's Nitric Acid Battery. *Elliott Brothers.*

The advantages offered in this battery are, greater constancy; less inconvenience from fumes, the porous cell being a stoppered bottle; and it not being necessary to amalgamate the zincs, common salt being used in the outer cell.

1271. Glass Battery Cell, with two carbon and two zinc plates. *Keiser and Schmidt, Berlin.*

1272. Glass Battery Cell, with two carbon and one zinc plate. *Keiser and Schmidt, Berlin.*

1273. Glass Battery Cell, with one carbon and one zinc plate. *Keiser and Schmidt, Berlin.*

1274. Dipping-Battery, with 10 elements, of the exhibitors' own construction. *Keiser and Schmidt, Berlin.*

1275. Dipping-Battery, with 16 elements, with pachytrope of the exhibitors' own construction. *Keiser and Schmidt, Berlin.*

1276. Battery for Field Telegraph Service, constructed for the Prussian Railway Battalion according to the plan of Captain Witte. *Keiser and Schmidt, Berlin.*

1277. Leclanché Cell, for working house telegraphs, modified by the makers. *Keiser and Schmidt, Berlin.*

1277a. Yeates' Improved Leclanché Cell.
Prof. W. F. Barrett.

New form of Leclanché cell, the position of the zinc and carbon are reversed, thus rendering it less liable to polarisation, and hence far more constant than the ordinary form. Made by H. Yeates, King Street, Covent Garden.

1278. Drawing and Description of a Galvanic Battery, with arrangement for combining the elements ad libitum.
Dr. Tasché, Giessen.

1279. Portable Battery for Electro-therapeutic Purposes. 24 elements.
Prof. Beetz, Munich.

1279a. Round Immersion Battery, with automatic break for medical purposes.
J. Teller, München.

By this arrangement powerful action is obtained, and a very constant current, even with great resistance. The consumption of zinc is (in consequence of self-amalgamation in the acid chromate of mercury solution) very small, and this result is also favoured by the small immersion, which is limited by the slide on the upright bars. By the automatic interrupter, the battery can be used also with intermittent current, and such a battery current (because without alteration of poles) is to be preferred to the action of an induction current.

1280. Portable Battery, with Ebonite Insulations for the investigation of tension phenomena. 16 cells.
Prof. Beetz, Munich.

1281. Delicate Battery, with four platinum-zinc elements, two silk conducting strings, with eight reserve plates in a case.
Kgl. Chirurgische Klinik, Breslau, Prof. Dr. Fischer.

1282. Delicate Battery, with four carbon-zinc elements.
Kgl. Chirurgische Klinik, Breslau, Prof. Dr. Fischer.

1283. Delicate Battery, with two carbon-zinc elements, two conducting strings, and three reserve plates in case.
Kgl. Chirurgische Klinik, Breslau, Prof. Dr. Fischer.

1284. Small Battery, with two platinum-zinc elements.
Kgl. Chirurgische Klinik, Breslau, Prof. Dr. Fischer.

In galvanocaustics (the art of destroying diseased portions of tissue by means of the electric current) the batteries used generally consist of four very large Grove or Bunsen elements. Wires proceed from the battery to a piece of platinum, which is to be raised to a red heat. This collection shows Middeldorff's original arrangement, as used in Breslau, and also recent modifications.

Two Secondary Elements, by Planté.
M. Bréguet, Paris.

Battery of 20 Secondary Elements, by Planté.
M. Bréguet, Paris.

Battery of 20 Secondary Elements, by Planté.
M. Bréguet, Paris.

c. THERMO-ELECTRIC BATTERIES.

Thermo-electrical Pile of Nobili, composed of 12 elements disposed in rays.
The Royal Institute of " Studii Superiori" at Florence.

Thermo-electrical Pile of Nobili, divided into three small ones of 12 elements each, to be combined at will.
The Royal Institute of " Studii Superiori" at Florence.

53. Thermo-electrical Pile of Nobili for the experiments on radiant heat, composed of 37 elements, and furnished with a conical mirror.
The Royal Institute of " Studii Superiori" at Florence.

It is well known that Leopoldo Nobili, who was for several years professor at the Royal Museum of Physical Science and Natural History of Florence, was famous for the construction of thermo-electrical piles, of which he made much use in his important experiments on radiant heat, partly carried out in conjunction with the celebrated Melloni.

1298a. Nobili's First Thermo-Electric Battery.
Prof. Dove, Berlin.

1298b. Melloni's First Thermo-Electric Pile.
Prof. Dove, Berlin.

1298c. Antinori's First Apparatus for Induction Sparks.
Prof. Dove, Berlin.

This apparatus was bought at an auction in Florence, after Nobili's decease.

1286b. Thermo-Electric Apparatus by Seebeck.
1. Ring of copper and antimony.
2. Cylinder of copper and antimony, 46 mm. in diameter and 22 cm. in length.
3. Six circular discs ; diameter 10 cm. ; of copper, brass, and other alloys.
4. Square disc, 16 cm.
5. Two rods, Bi. Sb.
Prof. Dr. Dove, Berlin.
(Property of the Royal Academy of Sciences at Berlin.)

1265. Pouillet's Thermo-Electric Battery.
Conservatoire des Arts et Métiers, Paris.

1285. Thermo-electrical Battery, bismuth and antimony.
Geneva Association for the Construction of Scientific Instruments.

U 2

1286. Nobili's Thermo-Electric Pile, of 54 pairs of bismuth and antimony bars, soldered alternately together; the smallest difference of temperature between the two faces of the pile develops a current, readily indicated by a suitable galvanometer.

Elliott Brothers.

1235c. Apparatus intended for producing **Thermo-electric Currents** in a special manner.

Imperial University of St. Petersburg.

It consists of ten straight electro-magnets, with their poles joined alternately so as to form a zig-zag. The iron cores are not in direct contact but are connected by small brass cylinders to which they are soldered at their extremities. These small cylinders carry brass plates and rods placed alternately. When the apparatus is to be used, these plates are heated approximately up to 100° centigrade; the brass rods are cooled with crushed ice, and then the galvanic current of a six element battery (carbon, zinc, chromic liquid,) is passed through the bobbins of the electro-magnets, odd or even numbers. The thermo-electric current produced by the iron cores produces a strong deviation of the needle of a sensitive galvanometer of small resistance.

The fact of the heterogeneousness of magnetised metals and non-magnetise metals was discovered by Sir W. Thomson. The apparatus here described is constructed by Prof. T. Petrouchevsky, Professor at the University of St. Petersburg. The first experiments made with this apparatus, slightly altered however, have been briefly described in Russian in the "Journal of the Russian "Society of Chemistry," and in that of "The Physical Society of the Univer- "sity of St. Petersburg," Vol. 6, Section Phys., p. 107 (1874).

1287. Noë's Thermo-electric Battery of 96 pairs. Convenient for lecture experiments. *George Gore, F.R.S.*

Attains its maximum power in about one minute. May be heated to low redness. Decomposes water freely. Will excite an electro-magnet to sustain 2 cwt. It has an arrangement, or "current transformer," by means of which its entire power can be employed with three different combinations of its elements, viz., as 96 by 1, 48 by 2, or 24 by 4, and changed instantly from one combination to another. The connexions of the "transformer" require no cleaning. Made by W. J. Hauck, Vienna.

1288. Small Single-cell Apparatus, with platinum-plates, for showing the thermo-electric properties of liquids.

George Gore, F.R.S.

(*See* Philosophical Magazine, Jan. 1857.)

1289. Single-cell Apparatus, for examining the thermo-electric properties of liquids. *George Gore, F.R.S.*

(*See* Proceedings of the Royal Society, 1871.)

1290. Large Single-cell Apparatus, with platinum-plates, for showing the thermo-electric properties of liquids.

George Gore, F.R.S.

1291. Four-cell Apparatus, with copper plates, for showing the thermo-electric properties of liquids. *George Gore, F.R.S.*
(*See* Philosophical Magazine, 1857.)

1292. Twelve-cell Apparatus, with platinum-wire electrodes, for examining the thermo-electric properties of liquids.
George Gore, F.R.S.
(*See* Proceedings of the Royal Society, 1871.)

1293. Model of the most improved form of apparatus for investigating the thermo-electric properties of liquids. Used with ribbons of platinum, gold, palladium, and silver.
George Gore, F.R.S.

1297. Thermo-Battery. *Siemens and Halske, Berlin.*

1297a. Thermo-Electric Pile, small student's form, nickel-plated. *Harvey, Reynolds, and Co.*

1298. Thermo-Electric Pile (Noë's system), with 64 elements, heated by gas. The electro-motive power equal to six Bunsen elements. *P. Dörffel, Berlin.*

The elements, consisting of a round rod (positive) and thin wires (negative), are arranged in two opposite rows of 64 elements each, whose heating bars (cast of positive metal and protected against the flame by copper casing) project in a row into the open space between the elements, so that they are all alike heated by the stand of Bunsen burners below, and convey the currents to the elements. The cooling of the other junctions is effected by means of metallic cooling-plates attached to them, supported by the wooden frame under the elements. The electro-motive force is equal to 6 Bunsen or 120 Jacobi-Siemens' units. The resistance = 2·45 Siemens' units. The pile contains a Dove's Pachytrope, in order to arrange the elements in groups of 1, 2, or 4, by means of which the resistances and the electro-motive force may be changed.

1299. Thermo-pile (Noë's system), with 20 elements in radiating arrangement, heated by gas. The electro-motive power is equal to one Bunsen element. *P. Dörffel, Berlin.*

Here the elements are arranged radially, so that the heating bars all run to a middle point, where they can be heated by the single flame of a Bunsen burner. The cooling is done with metal plates which are rolled into a tubular form, and serve at the same time as stands for the battery. The electro-motive force is equal to 1 Bunsen or 20 Jacobi-Siemens units. This apparatus (as also the next, 1300) is recommended for small experiments in electrolysis, &c.

1300. Thermo-electric Pile (Noë's system), heated by a spirit lamp, with 20 smaller elements, and consequently of greater resistance. The electro-motive power is equal to one Bunsen element. Resistance equal to 0·52 Siemens' units.
P. Dörffel, Berlin.

1301. Thermo-electric Pile (Noë's system), heated by a spirit lamp, with 10 smaller elements. Its electro-motive power is equal to 0·5 Bunsen element.   *P. Dörffel, Berlin.*

Designed specially for medical use, in connexion with a small induction apparatus. Should long action be desired it is well to place the battery with lamp in a vessel with water, to avoid the great heating its small size involves, and to increase the action.

1294. Thermo-Electric Battery or Clamond Pile.
Thermo-Electric Company.

The poles or generators are constructed of zinc and antimony, both being metals bearing great electrical properties. The electricity is ' given out without any intermediate agency, except heat, which is generated as gas ; coke or charcoal is consumed. Economy in maintenance, and cleanliness in application, gives this arrangement an advantage over other batteries, and the current obtained is constant and free from polarisation or exhaustion.

1301a. Thermo-Electric   Generator   (Clamond's Patent). Constructed either for electrotyping, plating, gilding, or telegraphy. A pile of 100 bars, with a gas jet burning 4 feet per hour, will deposit an ounce of copper per hour.
Thermo-Electric Generator Company (Clamond's Patent).

The Thermo-Electric Piles or Generators are constructed of elements, one pole of which is tinned iron, the other being an alloy of two parts of antimony to one of zinc. The iron is cast into the alloy, and thus a perfect connexion is made. The pairs thus formed are then laid side by side, and being cemented together, form a ring or crown (the cement used is a mixture of asbestos and silicate of soda) ; one crown being complete another is laid above it, though insulated from it by the same cement, and so on, giving the pile a cylindrical form. The junctions are heated thus : Up the centre of the pile is placed a perforated earthen tube and gas issuing from a Bunsen's jet burns at the perforations, heating an iron core red hot, which radiates its heat to the junctions of the pairs, thus the flame never impinges on the metals, and all oxidization, &c. is obviated; the heated air passes over the top of the iron core, and curling down, escapes by a pipe from the bottom of the pile. The elements of each crown are connected in series, but the terminals of every crown are brought into a wooden support and can be connected at will for high tension or great quantity. As a standard of power the following may be used : —
A 100 bar pile consuming 4 feet of gas per hour has E.M.F. 5 volts., Int. Res. 1 ohm.
A 240 small tension bar pile, consuming 4 feet of gas per hour has E.M.F. 12 volts., Int. Res. 6 ohm.
Piles are also made to be heated by coke or charcoal, and a battery having an E.M.F. of 20 volts, and Int. Res. of 4 ohms burns 2 lbs. of coke per hour. Petroleum is also used for heating the piles.

1301b. Thermo-Electric Pile of Hydrogenium.
Prof. Dewar.

Consists of alternate layers of Palladium and Hydrogenium; electromotive force equal to that of iron and copper.

1096. Thermo-electric Diagram for teaching purposes.
(Trans. Roy. Soc. Edin., 1872–3.)   *Prof. Tait.*

d. INDUCTION COILS.

1303. Ruhmkorff's Coil. *Frederick Guthrie.*

The current from a galvanic battery passing through a spiral of copper wire magnetises the soft iron bars placed within it, which by their attraction so move a steel spring as to interrupt the current. The current being thus broken, the magnetisation ceases until the current is again restored. The result is a very rapid making and breaking of the current in the spiral or primary wire. Outside the primary are many miles of fine insulated copper wire called the secondary wire. Connected with the primary by wires, one on each side of the contact breaker, is a tin-foil condenser. This absorbs the extra-current when the primary is broken, and serves to augment the secondary when the primary is made. The interior magnetism acts in the same direction.

1304. Six-inch Induction Coil, and Browning's Spark Condenser, for obtaining spectra of metals by the induction spark. *John Browning.*

When using the spark condenser the amount of coated surface introduced may be varied at pleasure, and the density of the spark thus regulated.

1304a. Apps' Patent Induction Coil, giving sparks of 17 in. in air, with a battery of five Grove's cells, platinum, 5 × 3 in. immersed. *Alfred Apps.*

1304b. Henry's Induction Coils.
The Council of King's College, London.

1304aa. Large Induction Coil, with thick secondary wire (10½ miles in length), and improved form of contact-breaker by which a long interval of contact is obtained. *Horatio Yeates.*

This is wound on the plan proposed by Dr. Fergusson (in two divisions), the secondary wire which is No. 32, B. W. G. is 10½ miles long. The primary wire, No. 8, B. W. G. is wound in two laps.

The condenser is composed of 70 sheets of tinfoil, 26 × 16, insulated with paraffine paper.

The contact-breaker, which is so formed as to give a long interval of contact, is also furnished with an adjustment by means of which the coil can be worked with a very small battery, and maximum results obtained with the largest suitable battery.

1305. Large Induction Coil, with Foucault's break; will give 18-inch sparks. A cube of glass which was pierced by this coil. *M. Ruhmkorff.*

1305a. Electric Necessaire, containing Ruhmkorff bobbins.
M. Loiseau, jun., Paris.

1306. Induction Apparatus for medical purposes.
Keiser and Schmidt, Berlin.

1307. Induction Apparatus for medical purposes.
Keiser and Schmidt, Berlin.

1308. Induction Apparatus for medical use.
Keiser and Schmidt, Berlin.

1309. Spark Induction Machine, No. 1, with armature and Geissler's tubes.   *Keiser and Schmidt, Berlin.*

1310. Spark Induction Machine, No. 2, with armature and Geissler's tubes.   *Keiser and Schmidt, Berlin.*

1311. Spark Induction Machine, length of spark, six millimeters.   *Keiser and Schmidt, Berlin.*

1312. Spark Induction Apparatus, length of spark, one centimeter.   *Keiser and Schmidt, Berlin.*

1313. Spark Induction Apparatus, length of spark, 4·5 centimeters.   *Keiser and Schmidt, Berlin.*

1314. Spark Induction Apparatus, length of spark 8 centimeters.   *Keiser and Schmidt, Berlin.*

1315a. Great Induction Coil.

66 Leyden jars and fittings.
6 stands for jars.
The Royal Polytechnic Institution.

e. MAGNETO-ELECTRIC MACHINES.

1303a. First Induction Machine, called **" de Pixii,"** constructed under the direction of Ampère.
College of France, Paris.

1686. Galvanometer of Nobili.
The Royal Institute of " Studii Superiori" at Florence.

Rose of Metallic Colours obtained by means of electricity by Leopoldo Nobili.
The Royal Institute of " Studii Superiori" at Florence.

A very great number of liquids and metals were subjected by Nobili to experiments in order to obtain, by means of electricity, the coloured rings which bear his name; and exceedingly important are his observations on the complete scale of colours which he succeeded in forming and which yet exist at Florence.

1687. Original Magneto-electrical Machine of Leopoldo Nobili and **Vincenzo Antinori,** which gave, on the 30th of January 1832, the first spark, before Leopold II.
The Royal Institute of " Studii Superiori" at Florence.

This is the first machine with which the induction spark was obtained by means of an artificial magnet. The anchor (*ancora*) tied by a thread, and movable like a lever round a pivot, is detached from the loadstone by a blow, while at the same time, by means of a spring, the electrical circuit is interrupted, so that the spark appears at the point of interruption. Faraday, who was the first to observe the induction currents, obtained the spark by using an electro-magnet; the arrangements, however, of Nobili and Antinori were the first that constituted a true magneto-electrical machine. And we may also remark that, at the same time, Nobili and Antinori obtained the electric spark, likewise, by means of the natural magnet of the museum, an enormous parallelopiped of about 50 × 65 × 78 centimeters in dimension.

1114. Saxton's Magneto-Electric Machine. Copy of the original machine made by Mr. Saxton, and exhibited by him before the third meeting of the British Association, held at Cambridge in the year 1833.   *John O. N. Rutter.*

This machine was made specially for the contributor by Mr. Saxton, immediately after the meeting at Cambridge, and has been in his possession ever since. It is capable of producing sparks, shocks (through the tongue), and decomposes water. It also reproduces the ordinary phenomena of electro-magnetism.

The machine is described in Daniell's "Introduction to Chemical Philosophy," 1843, p. 585, sec. 873.

1316. Ladd's Dynamo-Magneto-Electric Machine.
                               William Ladd & Co.

Invented March 1867. (*See* Proceedings of the Royal Society, No. 91, 1867.)

This was the first machine with two armatures, one being employed to excite the electro-magnets and the other to produce an electric current, which may be used for any purpose to which a battery is applicable.

1316a. Ladd's Dynamo-Magneto-Electric Machine, with two wires on one armature. This machine will heat 15 inches of platinum wire.   *William Ladd & Co.*

1317. The first Magneto-Electric Machine, with circular magnets, 1866.   *William Ladd & Co.*

1318. Magneto-Electric Machine, with circular magnets, larger form, 1867.   *William Ladd & Co.*

1319. Magneto-Electric Machine (direct current).
                               James How & Co.

1320. Magneto-Electric Machine (Duchenne's form).
                               James How & Co.

1321. Magneto-Electric Machine (Clark's form). An early machine by Logemann, of Haarlem.   *James How & Co.*

1322. Electro-Magnetic Coil Machine, for medical application. Primary and secondary currents.   *James How & Co.*

1323. Magneto-Electric Machine, with alternate current for production of light.   *La Société l'Alliance.*

A magneto-electric machine, with four discs or 64 bobbins with alternate current for the production of light. This machine requires a driving power of 3 H.P. when revolving from 400–450 times per minute.

1324. Magneto-Electric Machine, to be worked by hand or steam.   *La Société l'Alliance.*

This machine with eight bobbins is for the purpose of demonstration with direct and alternate current, and can be worked by hand or steam.

1325. Experimental Magneto-Electric Machine, the first constructed in which electricity and magnetism, rendered active by the expenditure of mechanical force, were made to act and re-act on one another in such a way as to greatly increase the development of their forces.   *S. Alfred Varley.*

This machine was the first of its class, and acted on what was a new principle at the date of its construction. The new principle consisted in making electricity and magnetism, rendered active by the expenditure of mechanical force, act and re-act on one another in such a way as to greatly increase the development of their forces. In this machine iron bobbins wrapped with insulated wire are revolved between the poles of very feeble magnets made of soft iron. The electricity (small in amount when the machine is first put in motion) which is developed in the insulated wire of the bobbins passes, by means of a commutator, through convolutions of insulated wire surrounding the soft iron magnets, and renders them more highly magnetic. The magnetism of the soft iron magnets being thus increased, develops a correspondingly increased quantity of electricity in the revolving bobbins, which re-acts on the soft iron magnets, rendering them still more highly magnetic.

The expenditure of mechanical force giving motion to the machine is greater as the magnetism and electricity developed increase, the consumption of mechanical force having relation to the quantity of electricity rendered active.

1326. Gramme's Magneto-Electric Machine, for electrotyping.   *H. Fontaine.*

1327. Gramme's Magneto-Electric Machine, for electric light.   *H. Fontaine.*

1328. Gramme's Magneto-Electric Machine, for electric light of great power.   *H. Fontaine.*

1329. Gramme's Magneto-Electric Machine, for demonstration.   *H. Fontaine.*

1315. Magneto-induction Machine.
                    Gustav Baur, Stuttgart.

This apparatus, containing several electro-magnets and a current regulator, is furnished with double coils of wire, and may be used to set in action electric apparatus of very various resistance and with very quick interruption of current, e.g., Ruhmkorff coils. In general, any experiments may be made with it that are made with batteries of 1-6 Bunsen elements. It is suitable, for medical purposes, galvanocaustics, &c., and, if a part of the rotating electro-magnets be wound with fine wire, for production of a constant current up to 60 Meidinger elements.

1330. M. Le Roux's Electro-Magnetic Apparatus, for showing the effect of magnetism on copper discs.   *M. Ruhmkorff.*

1331. Model of a Magneto-Electric Machine, designed to illustrate the advantage gained by the use of an electro-magnet in place of the usual permanent magnet. *William Raynor.*

This model of a magneto-electric machine is one that has been constructed for the purpose of showing the great increase of the electric current by the use of an electro-magnet in place of the permanent magnet, when such magnet is excited or charged by a galvanic cell; and this plan may be applied with advantage to all magneto-electric machines using soft iron magnets.

1331a. Sketch and Description of an Improvement in Magneto-Electric Machinery. This improvement is obtained by using two armatures on one spindle, to be fixed one on each side of the magnet, and placing them at right angles to each other, so that one is in full action when the other is changing its polarity. *William Raynor.*

1332a. Portable Magneto-Electric Machine, with double coiled magnet. *Harvey, Reynolds, and Co.*

1249a. Volta Faradaic Machine, with arrangement in the pocket for taking shocks ; also giving interrupted or continuous current, the batteries being of the constant Leclanché form. *Harvey, Reynolds, and Co.*

1336. Dynamo-Electric Light Apparatus, making 480 revolutions per minute, with an expenditure of 6 horse-power gives a light of from 12,000 to 15,000 normal candles. *Siemens and Halske, Berlin.*

In these machines the inner iron core is fixed. Around this core revolves a German silver bobbin, upon which is wound in a peculiar manner eight double circuits of covered copper wire, these circuits terminating in the metallic segments, which are successively brought as two opposite poles into contact with the wire brushes. The magnetic field in which this bobbin revolves (on its own axis) is formed by electro-magnets, the continuation of the cores of which are curved iron bars, and these bars are so arranged as to be brought as near as possible to the revolving bobbin. The current given by these machines is continuous and in one direction.

1336a. Dynamo-Electric Light Apparatus. This machine gives a light of 4,000 normal candles with 850 revolutions of the armature per minute, with an expenditure of work equal to 3 horse-power. *Siemens and Halske, Berlin.*

1336b. Dynamo-Electric Light Machine producing a light equal to 1,000–1,300 normal candles with about 1,100 revolutions of the induction cylinder per minute and an expenditure of 1 to $1\frac{1}{2}$ H.P. The machine is 640 millimetres in length, 540 millimetres width, and 225 millimetres height. *Siemens and Halske, Berlin.*

1336c. Magneto-Electric Machine to give a constant current. The apparatus has 50 steel magnets, and the current produced is equal to that from 8–10 Bunsen's elements. The armature is rotated by hand. *Siemens and Halske, Berlin.*

1336a. Various Examples of Magneto-Electric Apparatus.

Electro-magnetic machine by Gramme, with a Jamin magnet of 0·08m.

Magneto-electric machine, by Gramme, for electric light of 150 burners.

Magneto-electric machine, by Gramme, with electro-magnet for laboratory.

Magneto-electric machine, by Gramme, with a Jamin magnet (small model).

Magneto-electric machine, by Gramme, with a Jamin magnet (large model with fly-wheel and treadle).

Exploder, with Jamin magnet (large model) with bobbin, cable, and key.

Exploder, with Jamin magnet (medium model).
M. Bréguet, Paris.

1336d. Magneto-Electric Machine. *H. L. Rutter.*

f. OTHER MODES OF PRODUCING ELECTRICITY OR ELECTRIC CURRENTS.

1337. Apparatus, designed to obtain electric currents by means of the combined action of gravity and motion. Preliminary experiments only have yet been made with it.
George Gore, F.R.S.

1338. Apparatus for investigating electric currents produced by the friction of different metals. (Not yet completed.)
George Gore, F.R.S.

1341. Delezenne's Circle. An instrument for developing electrical currents by the agency of terrestrial magnetism.
Elliott Brothers.

1342. Apparatus by which **Forbes** procured an **Induction Spark** from a **Natural Magnet.** *Trans. R. S. Edin.,* 1833.
University of Edinburgh.

II.—APPARATUS FOR REGULATING THE PLACE AND TIME AT WHICH THE EFFECTS OF AN ELECTRIC DISCHARGE OR CONTINUOUS ELECTRIC CURRENT ARE PRODUCED.

1343. Six Specimens of Tubular Binding Screws for making electrical connexions. *George Gore, F.R.S.*

1343a. Specimens of Wire for Electric Apparatus.
1. Copper wires, covered with gutta-percha (1, 2, 25).
2. Copper wires, covered with gutta-percha and cotton (3).
3. Copper wires, covered with cotton (4, 5, 6, 9, 11, 12, 13, 14, 26).
4. Copper wires, covered with silk (7, 8, 10, 15 to 23, 27).
5. Elastic poires, with their cordons (24).
6. Covered wires of various metals, with statement of their resistance. *Madame Bonis, Paris.*

1235a. Apparatus for uniting several **Galvanic Elements** into one of large surface, so as to preserve the entire strength of the current by lessening the resistance.
Imperial University of St. Petersburg.

To effect this, two metal cylinders, each bored with seven holes parallel to the axle, and fitted with screws, are used. In one cylinder, the six ends of conductors connected with the positive electrodes (anodes) of the elements are inserted, and in the other, the six conductors connected with the negative electrodes (cathodes). The seventh hole is reserved for a conductor, the section of which is equal to the total section of the six other conductors. This last conductor issues from the opposite side of the cylinder. The apparatus here described is used by Prof. Petrouchevsky in St. Petersburg.

1345. Single Plug Key, to close or break contact for long or short durations. *Elliott Brothers.*

1346. Fall-hammer, to obtain perfectly equable closing of a circuit.
Prof. Engelmann, Physiological Laboratory and Ophthalmological School, Utrecht.

On a brass prismatic lever, movable round a horizontal axis, slides the bridge, a copper cover having underneath two amalgamated copper points. On depressing a spring the bridge falls from a nearly vertical position, and plunges the bridge into two mercury vessels, movable on a horizontal slide, and connected with the battery. A spring prevents the bridge from rebounding. Velocity of fall to be regulated by moving the bridge on the lever with corresponding displacement of the mercury vessels on the horizontal slide.
The bridge being in the primary circuit of an induction apparatus, the breaking is every time to be effected at another place of the circuit, before lifting the bridge from the mercury, in order to prevent oxidization of the mercury by the spark.
The instrument can easily be managed with one hand.

1347. Firing Key, for torpedoes, &c. A simple contact key, with a movable piece of vulcanite, which can be brought between the two platinum contacts to prevent fatal results by accidentally closing the circuit. *Elliott Brothers.*

1348. Apparatus for reversing the direction of an Electric Current. Used with an electro-magnetic torsion apparatus. *George Gore, F.R.S.*

(*See* Philosophical Transactions of the Royal Society, Vol. 164, p. 529.)

1348a. Current-reversing Electrode.
J. Teller, München.

This gives a more convenient change of current than the commutators so far as electro-medical apparatus is concerned.

1349. Double Reversing Key, used for cable testing.
Elliott Brothers.

1350. Thomson's Reversing Key, used in connexion with the electrometer, for facilitating the measurement of the electrostatic capacity of a cable or condenser.   *Elliott Brothers.*

1351. Lambert's Key, constructed for charging or discharging cables and condensers.   *Elliott Brothers.*

1352. Spottiswoode's Rapid Break, for use with Intensity Coils.   *Tisley and Spiller.*

By means of this break the discharge in vacuum tubes can be regulated, and the motion of the stratifications diminished or rendered stationary, as required. See Proceedings Royal Society, June 10, 1875.

1354. Forms of discharge on making and closing an induction current.   *Prof. Donders, Utrecht.*

The trial with the noematachograph to have the instant of stimulation registered on the chronoscopic line by the current itself, led to the discovery :
1. That the discharge can form a long series of sparks.
2. That the electricity disappears more slowly when the spring rests on metal, more rapidly when it rests on a plate of mica, than in the form of sparks making holes in the paper. (Compare Onderzoekingen gedaan in het phys. labor. Ser. 2, T. III. 1870 ; and Wiedemann, Die Lehre vom Galvanismus und Electromagnetismus, 2ᵉ Auflage, 1874, B. II., s. 360.)

1355. Drawings showing the patent system of lightning conductors applied to buildings.   *J. W. Gray & Son.*

1356. Model of mid-section of a Ship, showing the patent system of lightning conductors applied to vessels in Her Majesty's service, &c.   *J. W. Gray & Son.*

1357. Indestructible Solid Copper Tape Lightning Conductors. Small and large sizes.   *Sanderson & Proctor.*

This form of lightning conductor possesses the greatest conducting surface. Hitherto it has been made in short lengths riveted together ; now it is made in any length without joints, thereby offering less resistance to the free passage of the electric current.

1358. Copper Rope Lightning Conductors, improved. The smallest and largest sizes.   *Sanderson & Proctor.*

1359. New Lightning Conductor Apparatus.
Prof. Carl Wenzel Zenger, Prague.

This apparatus consists of lightning conductors arranged symmetrically, balls being used instead of conical points. A plan shows its application to the I. R. Real School, and to the National Theatre at Prague.

1360. Top of Lightning Conductor. The lower part is made of gun metal, the upper of copper, and the extreme point of gold or silver. Constructed according to the instructions of Professor Ed. Hagenbach-Bischoff, in Basle. *G. Linder, Basle.*

The electricity escapes easily through good conductors from points and edges; the point does not oxidize in the atmosphere, and being a good conductor is not liable to be melted by electricity.

1361. Needle of Lightning Conductor, brass gilt.
Geneva Association for the Construction of Scientific Instruments.

1362. Lightning Conductors (various kinds).
John Faulkner.

Two photographs of expedients for applying lightning conductors to high spires and factory chimneys, and for the repair of high spires and chimneys.

1363. Models of Lightning-Conductors of the latest construction. *Mittelstrass Brothers, Magdeburg.*

1364. Apparatus serving **for the separation and collection of induced currents,** constructed by Dr. Th. Tasché, manufactured by Staudinger & Co., in Giessen.
Dr. Tasché, Giessen.

1364a. Current Analyser, with glass axis, made by Jung, of Giessen. *Physical Institute (Univ. of Giessen), Dr. Buff.*

The "current analyser" can be occasionally used for experimental research in voltaic induction, to separate the two induced currents, and to study the proportion of their intensities or electro-motive forces. *See* Poggendorff's Annalen, Vol. 127, p. 57.

1365. Binding Screws for **Galvanic Work.**
M. Th. Edelmann, Munich.

1366. Current-key for **Beetz's Compensation Method.**
M. Th. Edelmann, Munich.

III.—APPARATUS FOR ACCUMULATING ELECTRICITY.

1367. Leyden Jar of five and a half square feet coated surface. *Teyler Foundation, Haarlem.*

This jar is one of the 100 jars arranged in four cases, by which Van Marum constructed a battery of 550 square feet coated surface. The coatings of tinfoil have been renewed recently; but all is restored in the form in which it was used by Van Marum.

See Van Marum, " Machine Électrique," II., p. 195.

1368. Leyden Battery of 15 jars.
Teyler Foundation, Haarlem.

This battery is one of 16 used by Van Marum for his famous experiments, giving a total coated surface of 225 square feet.

The coatings of tinfoil have been renewed recently; the bottom of tea-lead in the case is also restored; and the outer coating of the case bottom, which Van Marum also made of tea-lead, has been replaced by zinc.

See Van Marum " Machine Électrique," I. p. 155, and II. p. 3.

1369. Battery of 10 one-gallon Leyden Jars.
Frederick Guthrie.

This battery stands in a mahogany frame. The jars stand upon perforated zinc. There is an arrangement for drying them by a current of hot air. The spark from this battery deflagrates a platinum wire a foot long.

1369a. Series of Leyden Jars, with connectors, and ebonite covers, 10 pieces from 90 to 100mm. high.
Warmbrunn, Quilitz, & Co., Berlin.

1369b. Battery of Leyden Jars, consisting of six jars 312mm. high, in mahogany case.
Warmbrunn, Quilitz, & Co., Berlin.

1369c. Cylinder, on insulating support.
Warmbrunn, Quilitz, & Co., Berlin.

1369d. Cylinder, with elder-pith balls.
Warmbrunn, Quilitz, & Co., Berlin.

1369e. Sphere, on insulating support, with two movable hemi spheres on ebonite rods.
Warmbrunn, Quilitz, & Co., Berlin.

1369f. Large dissected Leyden Jar.
Warmbrunn, Quilitz, & Co., Berlin.

1370. Spiral Leyden Jar.
Frederick Guthrie.

Two sheets of ebonite, alternating with two sheets of tinfoil, are rolled up together. The central knob is connected with the inner edge of one of the foils; the brass girdle is connected with the other sheet. A Leyden jar is thus formed which is compact with a large surface.

1371. Mica-plates for insulating electrical apparatus.
Max. Raphael, Breslau.

Mica can be rendered electrical by the least friction, hence its frequent employment as an excellent insulating material, especially on account of the facility with which it can be worked.

1372. Two Large Condensers, consisting of Leyden jars, each 400 millimeters high and 200 millimeters in diameter.
Borchardt, Hanover.

1372a. Adjustable Disc Condenser, which has also been used as a spark micrometer. *Sir William Thomson.*

It was in this instrument that the sound produced in an air condenser by a sudden change of potential was first heard. The lower part of the cell is arranged to hold cups of pumice-stone impregnated with sulphuric acid.

1372b. Cylindrical Condenser, for measuring capacity in absolute electrostatic units, described in Messrs. Gibson and Barclay's paper in the Transactions of the Royal Society for 1871.
Sir William Thomson.

1372c. Condenser for the Holtz-Bertsch Electrical Machine. *Mottershead & Co., Manchester.*

IV.—APPARATUS FOR PRODUCING AND OBSERVING EFFECTS OF ACCUMULATED ELECTRICITY.

1373. Apparatus for demonstrating the fundamental laws of electrical and magnetical attraction and repulsion, made according to the instructions of Professor Ed. Hagenbach-Bischoff, in Basle.
G. Linder, Basle.

The ebonite rods are negatively electrified when rubbed with fur, and positively electrified when rubbed with gun-cotton. (*See* Carl, Repertorium der Experimental-Physik, VIII., p. 75.)

1374. Insulated Pith-Ball Stand, with mahogany arm ; the arm is itself a box in which the pith balls may be placed when it is not in use. *The Council of King's College, London.*

1648. Series of Elder Pith Figures, Butterflies, &c.
Warmbrunn, Quilitz, and Co., Berlin.

V.—APPARATUS FOR PRODUCING AND OBSERVING EFFECTS OF THE DISCHARGE OF ACCUMULATED ELECTRICITY, WITH SPECIMENS OF PERMANENT RESULTS PRODUCED.

1375. Photographs of **Sparks** from a **Holtz Machine,** in cold and in heated air. (See Trans. R. S. Edin., 1874-5.) Taken by an instantaneous process, a quartz lens being employed.
Prof. Tait.

1375b. Spark Micrometer after Riess.
Warmbrunn, Quilitz, & Co., Berlin.

1376. Apparatus used by M. Rijke to measure the distances at which the spark of the Leyden jar passes.
Prof. Dr. P. L. Rijke, Leyden.

1377. Vacuum Tube for electric discharge. 1856.
Teyler Foundation, Haarlem.

Masson in Paris used a Torricellian vacuum sealed by the lamp, in his extensive researches on the electric spectrum. Some time afterward, Dr. Geissler, in Amsterdam, made this Torricellian vacuum at the instigation of Prof. Van der Willigen, now director of the Teyler Museum, whose property it is at present. The experiments with this tube are described in Poggendorff's Annalen, vol. xcviii. p. 487, 1856. Subsequently Dr. Geissler in Bonn constructed his various well-known and beautiful tubes. This tube contains a little mercury and carbonic oxide gas.

1377a. Four Geissler's Vacuum Tubes. *E, Cetti and Co.*

1377b. Collection of Geissler's Tubes.
Dr. H. Geissler, Bonn.

1377c. Tube, by Geissler, for two gases.

Tube, by Geissler, forming a diadem.

Tube, by Geissler, forming a diadem.

Tube, by Geissler, with inner spiral.

Tube, for liquids, with six spirals.
Alvergniat Frères, Paris.

1377d. Three Vacuum Tubes, to show the connexion between the resistance of rarefied air and the phenomenon at the cathode, the so-called negative glow   *Prof. Hittorf, Münster.*

Apparatus (A) made by Dr. Geissler, of Bonn, consists of two balls which communicate together by two tubes of equal width, one short and one of spiral form 3¾ mètres long. The electrodes of aluminium wire pass through the balls and end in the short tube so that there is a free interval of only 2 mm. between them. The opening current of the Rhumkorff coil passes, in consequence of the great rarefaction of the air in the tube, not by this short path, but prefers the longer one. If the latter be stopped by closing the glass cock, the passage is effected, but only at much greater tension, by the short path. The tube (B) has the same arrangement, but without the glass cock. It is used in place of (A) where the air is able to penetrate and the required vacuum ceases. The glass vessel (C) consists of a wide reservoir and a cylindrical tube, each of which holds one of the two equally long wires as electrodes. The tension of electricity with which passage occurs is much greater when the wire in the narrow part serves as cathode than when it is anode. This may be shown if a spark micrometer be introduced in the induction current near the tube, and for each of the two directions the interval of the balls be determined with which the current takes the path through the tube. If the wire in the wide reservoir, being cathode, be placed in conductive connexion with the third aluminium wire, which is in the beginning of the cylindrical tube, the current can no longer pass over in the latter to the former This, therefore, loses its negative light, and only with the greater tension, such as occurs with the other direction, is the passage of electricity effected (Cf. Pogg. Ann., Bd. 136, p. 197.)

1377e. Three Tubes of Glass, with rarefied air to show the magnetic behaviour of the negative light.

Prof. Hittorf, Münster.

The aluminium wire, which is quite sheathed with glass, with the exception of the end, is taken as cathode of the opening induction-current. The straight discharge from the cross section, when the tubes are brought over and between the poles of an electro-magnet, behaves like a flexible conductor which is fixed at one end and at the other freely movable, and follows the Laplace-Biot laws. (Cf. Pogg. Ann., B. 136, p. 213.)

1377f. Three Glass Tubes of Rarefied Air and Sulphide of Calcium, to show the phosphorescence of the negative electric light. *Prof. Hittorf, Münster.*

The negative electric discharge which, with great rarefaction, occurs at a cathode with small surface, raises the conducting particles of gas to a very high temperature. When strong induction currents are used, these, notwithstanding their small mass, are capable of raising the surface of badly-conducting solid bodies with which they come into contact to a red heat. This heating, which the negative discharge gives in much greater degree than the positive, produces with the best light-givers, like sulphide of calcium, a light of dazzling intensity.

1378. Gassiot's Star. *Frederick Guthrie.*

This exhibits (1) the varieties of the electric discharge through various rarefied gases in tubes of different shapes, and (2) by being rotated shows by the retention of images the intermittent nature of the discharge.

1379. Block Specimen of Glass, $2\frac{1}{2}$ inches high, penetrated vertically by an electric discharge. (By Ruhmkorff, of Paris.)

George Gore, F.R.S.

1380. Effect of Lightning. Portion of a half-sovereign and a fragment of sheet iron fused together by a discharge of lightning in the colony of Natal. This and other coins were in a tin box, of which this fragment alone remained after the passage of the discharge. *Robert James Mann, M.D.*

1381. Metals fused into Glass by Lightning.
Alfred B. Harding.

Frame No. 1 consists of strips of zinc, tin, and lead, fused into glass by an actual flash of lightning, collected by means of "exploring wires" stretched over the grounds of the late Andrew Crosse, and conveyed into his electrical room, as shown in the stereograph. It was here accumulated in the great Leyden battery of 50 jars, and passed thence by dischargers through the metals, which were burnt into the glass on which the strips were laid.

Frame No. 2 contains composite strips of copper and iron, gold and tin, and gold, silver, and copper, fused in like manner.

A photograph of the Leyden battery, with which the experiments were performed, accompanies the frames.

1382. "Thunder House," or model to illustrate the identity of lightning and electricity, and the use of lightning conductors in protecting buildings—said to be the first model of the kind, and to have been made by Dr. PRIESTLEY with his own hands.

Conrad Wm. Cooke, M. Soc. T.E.

1383. Old Electric Egg (beginning of last century). The property of Prince Pless.   *The Breslau Committee.*

The great age of the instrument appears both from tradition and from the style of the wooden frame and the nature of the brass work. It is certainly one of the oldest instruments of the kind.

1384. Apparatus employed by Sir Charles Wheatstone to determine the **Velocity** and **Duration** of the **Electric Discharge.**

Rotating mirror. Spark disc. Early rotating disc with balls and sliding rod.   *The Council of King's College, London.*

1385. Riess' Spark Micrometer.   *F. Rob. Voss, Berlin.*

1386. Apparatus for testing Lightning Conductors.
M. Th. Edelmann, Munich.

1387. Electrograph, for the production of electric sand-figures, constructed, from the plan of the exhibitor, by M. Th. Edelmann.   *Prof. W. von Bezold, Munich.*

This serves for the study of the nature of the electric discharge in simple or branched circuits, with the aid of sand-figures. The figures exhibited have partly been produced with this apparatus, partly under the air pump, and by means of a caoutchouc solution transferred from the ebonite plate to black tissue-paper. They are accordingly not copies, but true originals, produced by the discharge.

1388. Framed Table, with electric sand-figures.
Prof. W. von Bezold, Munich.

VI.—APPARATUS FOR PRODUCING AND OBSERVING EFFECTS OF CONTINUOUS ELECTRIC CURRENTS.

a. HEATING AND LUMINOUS EFFECTS.

1389. Diagram showing the **Amounts** of the **Electro-motive Force,** and the **Peltier** and **Thomson Effects** in a **Thermo-electric Circuit** of **Iron-Copper,** both junctions being at temperatures under the neutral point. For teaching purposes.
Prof. Tait.

1390. Peltier's Apparatus, for studying the thermal effects of currents in circuits composed of two or more metals.
Conservatoire des Arts et Métiers, Paris.

b. CHEMICAL EFFECTS.

1391. Apparatus for the polar **Decomposition** of **Water** by means of atmospheric electricity or the currents of the ordinary electrical machine. The gases are collected in fine thermometer tubes, by which means their absorption by the electrolyte is avoided.

Dr. Andrews, F.R.S.

1391a. Warmbrunn, Quilitz, and Co.'s Apparatus for Decomposition of Water, peculiar construction, with graduated tubes for the separated gases and for detonating.

Warmbrunn, Quilitz, & Co., Berlin.

1392. Bottle, containing fragments of pure **Electrodeposited Metallic Antimony.** *George Gore, F.R.S.*

(*See* Philosophical Transactions of the Royal Society, 1857, 1858, and 1862.)

1393. Two Specimens of **Electro-deposited Antimony;** one of the explosive, and one of the pure variety.

George Gore, F.R.S.

1394. Rare Specimen of pure Carbon, deposited by means of an electric current upon a rod of platinum.

George Gore, F.R.S.

c. ELECTRIC DIFFUSION AND CHANGE OF SURFACE-TENSION.

1395. Apparatus for producing Vibrations and Sounds, and an intermittent electric current by means of the electrolysis of a solution of cyanide of potassium and mercury with electrodes of mercury. *George Gore, F.R.S.*

The effects are produced by the alternate rapid formation and destruction of films upon the positive electrodes. (*See* Proceedings of the Royal Society, Vol. 12, p. 217.)

1395a. Electro Capillary Force Machine, after Lippmann.

R. Jung, Heidelberg.

To set this machine in action, the two wide glass vessels are first filled to a height of 1 to 3 cm. with mercury, placed in position in the glass trough, and then two thirds filled with pure dilute sulphuric acid. Then the two bundles of thin glass tubes are pushed repeatedly down into the mercury, so that the air is driven out, and the interstices are quite filled with mercury and acid. The bundles are then fixed by screws to their frames, so as to be about half immersed in the mercury, and to stand in equilibrium in the middle of their respective vessels. If the cups of the key be now filled with mercury, and the crank, which works it, so placed that the current is reversed a little before the opposite crank comes to its dead point, the machine (having been connected with the poles of a Daniell battery) will commence working, and may make as many as 100 revolutions in a minute. A Meidinger element keeps the machine in action for months.

1395b. Apparatus for electric osmose.

Prof. Hittorf, Münster.

In each of the three divisions formed in the glass cylinder by the clay plates the electric endosmose (when the vessel is quite filled with the solution of an electrolyte) is produced or prevented according as, on passage of the current, the three openings are free or are closed. With the arrangement it is proved that the transference of the ions is quite independent of the electric endosmose. (Pogg Ann., Bd. 96.)

d. Effects due to the Force between Currents and Magnets.

1396. Apparatus for showing the Rotation of a Bar-magnet on its axis by the passage through it of an electric current.   *George Gore, F.R.S.*

(*See* Proceedings of the Royal Society, Vol. 24, p. 121.)

1397. Apparatus for showing the Rotation of a Copper Wire upon its axis between the poles of two magnets by passing through it au electric current.   *George Gore, F.R.S.*

(*See* Proceedings of the Royal Society, Vol. 24, p. 121.)

e. Effects due to the Force between Currents and Currents.

1398. Apparatus for demonstrating the Laws of Ampère.
Geneva Association for the Construction of Scientific Instruments.

The mode of suspension used in this apparatus allows the conductor to make a complete revolution.

The current passes from the movable conductor into an annular cup, concentric with the axis of motion and filled with a conducting liquid.

All the conductors are made of aluminium so as to lessen their weight as much as possible.

The apparatus may be used for a great number of experiments; it is specially adapted for the following demonstrations:—

1. Parallel currents in the same direction attract one another, and those in contrary directions repel one another.
2. Inclined currents in the same direction attract one another, and those in contrary directions repel one another.
3. The attraction and repulsion of the same current are equal.
4. A sinuous current acts like a rectilinear current of the same general direction and having the same extremities.
5. A closed current takes a direction perpendicular to the magnetic meridian.
6. A solenoid has the essential properties of a magnet.
7. The elements of the same current repel one another.

The mutual action of magnets and currents is demonstrated by means of the same apparatus, by replacing one of the currents by one or more magnets.

1398a. General Table, by Ampère, with apparatus used by him in the discovery of the action of currents.

College of France, Paris.

1399. Apparatus for demonstrating the action of Metallic Discs in movement upon a metallic wire used as a voltaic conductor. *Prof. Daniel Colladon, Geneva.*

Experiment performed on 4th September 1826, in presence of the Paris Academy of Sciences, by Messrs. Ampère and Colladon.
Bulletin de Sciences Mathématiques, by De Férussac, vol. 6, p. 212.

1400. Model of a Circular Railway, for showing the rotation of a metal ball upon it by the passage of an electric current. *George Gore, F.R.S.*
(*See* Philosophical Magazine, Feb. 1859.)

1400a. Small Model of a Circular Railway, for showing the rotation of a metal ball upon it by the passage of an electric current. (*See* Philosophical Magazine, February 1859.)
George Gore, F.R.S.

VII.—APPARATUS FOR REGULATING THE STRENGTH OF ELECTRIC CURRENTS.

1402. Wheatstone's Rheostat, or changeable resistance, for quickly adding or removing a low resistance. *Elliott Brothers.*

1403. Voltastat and Voltameter combined.
Frederick Guthrie.

Air-tight through the stopper of a cylindrical vessel containing dilute sulphuric acid pass the following: (1.) Two platinum wires coated with glass. (2.) A long and wide tube open at both ends, the lower end reaching to the bottom of the cylinder. (3.) A tube opening freely beneath the stopper and above it by a very fine capillary aperture. The platinum wires are enlarged into platinum plates of triangular form, with their apices downwards, and further apart than their bases. Increase in the current passing by means of the wires between the electrodes causes the liquid to rise higher in the manometer tube, and also, by laying bare the electrodes, increases the resistance.

1404. Voltaic Compensator, an apparatus for maintaining the electric current derived from any sort of voltaic battery at constant intensity.
Elie Wartmann, Professor of Natural Philosophy at the University of Geneva.

A full description of this instrument is printed in the "Archives des Sciences physiques et naturelles," January 1858. In addition to the principal current, which, if constant, would do the work required, there is an auxiliary one, the strength of which is kept down by inserting an additional resistance. This resistance diminishes with the weakening of the principal current, and the consequent increase of the auxiliary current compensates that weakening.

1404a. Regulator, by Foucault.   *M. J. Duboscq, Paris.*

1405. Apparatus to make the **Electric Light,** derived from a **Voltaic Battery,** constant in its position and intensity.
Elie Wartmann, Professor of Natural Philosophy in the University of Geneva.

This apparatus, called *fixateur* for the electric light, was used in the years 1856 and 1857 for lighting the harbour of Geneva. A full description is to be found in the " Archives des Sciences physiques et naturelles," December 1857. By means of an electro-magnet and of gravity, two points of carbon are placed and kept at such a distance that the light produced by the current of an electric battery may be as bright as possible.

1406b. Regulator of Electric Light, by Serrin, with glass globe.   *M. Bréguet, Paris.*

1406. Regulator of Electric Currents, after the plan of M. Mascart.   *M. Redier.*

1407. Regulator for the Electric Light.   *M. Carré.*

1408. Artificial Charcoal Sticks for the Electric Light.
M. Carré.

1409. Electric Lamps, 1 small and 1 large. These lamps are automatic in their motion ; in them the carbon points are caused to approach or recede from each other.
Siemens and Halske, Berlin.

1409. Electric Lamps. These lamps are automatic in their action, in them the carbon points are caused to approach or recede from each other as required, without the aid of clockwork.
Siemens and Halske, Berlin.

VIII.—APPARATUS FOR DETECTING AND MEASURING DIFFERENCES OF ELECTRIC POTENTIAL AND CURRENTS OF ELECTRICITY.

a. ELECTROSCOPES AND ELECTROMETERS.

1410. Two Repulsion Electrometers constructed and used by Van Marum.   *Teyler Foundation, Haarlem.*

1411. Small Pocket Electroscope used by **H. B. de Saussure** during his excursions in the Alps.
M. H. Henri de Saussure, Geneva.

1412. Insulating Stand, with **Air Chamber,** artificially dried by sulphuric acid, used in connexion with first portable atmospheric electrometer. *Sir William Thomson.*

This stand was ordinarily attached to the top of the electrometer, as figured in Nichol's Cyclopædia, Art. Electricity (atmospheric), and in Thomson's Reprint of Papers on Electrostatics and Magnetism, XVI., 263. Sometimes—as in observations to determine in absolute measure the electric force in the atmosphere, on the sea beach, and in boats in Brodick Bay, Isle of Arran (reprint XVI., 281), and, with the assistance of Dr. Joule, on the Links of Aberdeen (British Association meeting, 1851)—the stand was detached from the electrometer and laid on the ground at a distance from it with connexion by fine wire to the insulated part of the electrometer, which also was placed on the ground, and was read by observer lying as close to the ground as possible.

1413. Atmospheric Portable Electrometer, No. 2, altered for first trial of divided ring principle for a quadrant marine electrometer, and used successfully on board the "Great Eastern," though not in connexion with the cable, in 1865.
Sir William Thomson.

This instrument has not been repeated, nor described in print, but it may yet do good service at sea. Made by James White, Glasgow.

1413a. Sir Wm. Thomson's Quadrant Electrometer, with most complete adjustments and of most perfect construction.
James White.

A descriptive pamphlet accompanies the instrument.

1413b. Sir William Thomson's Portable Electrometer, with most complete adjustments, and of most perfect construction. *James White.*

1413c. Electric Sensitizer. *Sir William Thomson.*

This instrument is an induction electric machine, and is used with the portable electrometer or other electrometer for testing the potential of a conductor without removing any part of its charge.

1414. Atmospheric Portable Electrometer, No. 4, altered to a plan for marine electrometer, which was discarded soon after trial. *Sir William Thomson.*

1415. Atmospheric Portable Electrometer, No. 5. Perfected portable electrometer, on same general plan as No. 1, described fully in Friday Evening Lecture to the Royal Institution, May 18th, 1860 (Thomson's Reprint of Papers on Electrostatics and Magnetism, XVI., 277). Made by James White, Glasgow. *Sir William Thomson.*

1416. Atmospheric Portable Electrometer, No. 10. First of new plan described in report on electrometers and electrical measurements (British Association Report for 1867, Committee on Standards of Electrical Resistance, and Thomson's Reprint of Papers on Electrostatics and Magnetism, XX., 368).

Sir William Thomson.

In this first instrument the attracting disc turns with a micrometer screw instead of moving in a geometrical slide, as in the portable electrometers now made. The receptacle for pumice and sulphuric acid was dangerously placed in the roof. This instrument was designed and first tried in the Island of Arran in 1862. Made by James White, Glasgow.

1417. Attracted Disc Heterostatic Station-Electrometer on same electric principle as latest portable electrometers, but with mechanism inverted.　　　*Sir William Thomson.*

This instrument is of convenient dimensions and general plan for stationary observations of atmospheric electricity and various electrostatic measurements. Made by James White, Glasgow.

1418. Large Portable Electrometer of same general plan as No. 10, altered to measure distance between two metallic conductors giving sparks with electro-motive force measured by another electrometer, in continuation of Smith and Ferguson's measurements. (Proceedings of the Royal Society, 1860, and Thomson's Reprint of Papers on Electrostatics and Magnetism, XIX., 320.)

Sir William Thomson.

Numerous accurate experiments were made many years ago by this piece of apparatus, but the results have not hitherto been published. Made by James White, Glasgow.

1420. Station Electrometer.　　　*Sir William Thomson.*

This electrometer was used by Professor Everett in his two years series of observations on atmospheric electricity at Windsor, Nova Scotia, described in the Transactions of the Royal Society of London for 1868. Its electric principle is the same as that of No. 8 of the perfected form of portable electrometer of the first kind. (See Thomson's Reprint, xvi. 777.)

1421. First divided Ring (semi-circular) **Electrometer.**

Sir William Thomson.

This was used for several years in the University of Glasgow, and described in the Accademia Pontificia dei Nuovi Lincei, February 1857, and in Thomson's Reprint, xviii. 311.

The movable electrified body projects from one side of the bearing wire far enough to travel over the flat semi-circular rings and experience their electric force. It is kept electrified by a fine platinum wire dipping in sulphuric acid, which forms the outside coating of the Leyden jar below it.

1422. Attracted Disc Electrometer, with double micrometer screw, arranged, to give the same period of free oscillation with different forces at different distances. *Sir William Thomson.*

The lower disc is insulated, the upper connected with the metal work of the case of the instrument and of the micrometer screws by a spiral spring by which it is hung. By turning the torsion head, the upper end of the spring and the sight-marks with movable stops for the lower end of the spring, are moved through different distances, of which the former is $1\frac{1}{2}$ times the latter.

The instrument exhibited was made 15 or 20 years ago. The present condition of the spiral spring shows that it has become elongated through time, without stress, because the hook at its lower end, bearing the disc, rests firmly against the lower stop, with the stop in the lowest position that the micrometer screws allow.

1423. First Mirror divided Ring (semi-circular) **Electrometer,** used at Kew for recording atmospheric electricity.

Sir William Thomson.

A specimen of the curve by which it recorded the atmospheric potential is published in Thomson's reprint, xvi. 292. Specimen sheets of its actual work accompany the instrument.

1424. First Trial Apparatus, towards mirror quadrant electrometer. *Sir William Thomson.*

This instrument was first designed for marine use. The mirror and needle are supported on a stretched bundle of silk fibre, as are the needle and magnets of the marine galvanometer. The electric connexion between the needle and the inside coating of the Leyden jar is made by a spiral of fine platinum wire. These peculiarities were tested and found to work moderately well in the trial instrument now exhibited, but have never been repeated ; nor does it seem very desirable they should be repeated, as the balancing of the needle on this plan, with sufficient accuracy for good work at sea, would probably be more troublesome than the object would justify. The electric action of this instrument was found so promising that immediately instruments were constructed on the same electric plan for use on land. The shape and dimensions of the suspended needle and of the electrified surroundings of the mirror are precisely the same as those of the quadrant electrometers now made. The improvements upon this original working model consist of geometrical slides for the quadrants, mechanical details regarding the suspension, the substitution of a fine platinum wire hanging down into the liquid in the bottom of a tall Leyden jar for the platinum spiral, and the addition of a replenisher and gauge for the charge of the jar.

1425. Divided Ring (semi-circular) **Electrometer,** described in Nichol's Cyclopædia, article Electricity (Atmospheric).

Sir William Thomson.

1426. Improved Helmholtz's Quadrant Electrometer.

T. Rob. Voss, Berlin.

This is well suited for school use, as it is not very expensive, and its action is, in proportion, as good as that of larger machines.

1427. Electrometer. *E. Stöhrer, Leipzig.*

1428. Kohlrausch's Torsion-electrometer.

Prof. Wüllner, Aix-la-Chapelle.

1429. Kohlrausch's Sine-electrometer, with two needles of different magnetic moment. *Prof. Wüllner, Aix-la-Chapelle.*

1430. Kohlrausch's Condenser.

Prof. Wüllner, Aix-la-Chapelle.

All three pieces of apparatus were manufactured by Th. Schubart, of Ghent and Marburg.

The detailed description and theory of Kohlrausch's various apparatus may be found as under :—

Poggendorff's Annalen, Vols. 72 and 74 for the torsion-electrometer.

    ,,   ,, Vol. 88 for the sine-electrometer.

    ,,   ,, Vols. 75 and 88 for the condenser.

The apparatus which are exhibited show the forms which Th. Schubart (late of Marburg, and now of Ghent) makes at the present time. A good description of the present forms is given in Wüllner's "Lehrbuch der Experimental-Physik," Vol. 4, 3rd edition, p. 159, and p. 299.

1431. Edelmann's Quadrant-electrometer.
M. Th. Edelmann, Munich.

1431a. Electrometer for measuring potentials, and particularly the potential of an accumulator at the moment of the discharge.   *Prof. Augustus Righi, Bologna.*

1431b. Induction Electrometer.
Prof. Augustus Righi, Bologna.

A caoutchouc tube carrying several copper rings is wrapped round the non-insulated pulleys (2, 4). If the insulated inductor (1) is charged, the rings go from (2) to conductor (3), with charges of contrary sign, and these charges remain in the insulated conductor (3). For, as the rings touch the conductor (3) by means of a little pulley placed in its interior, the charge preserved by any single ring is slight. For a very small charge of the inductor, the conductor (3) acquires a charge great enough to be shown by a gold leaf electroscope. If the inductor is uncharged, and the pulleys and the rings are of the same metal and very clean, the conductor (3) remains uncharged.

In open places the conductor (3) is charged by the sole influence of atmospheric electricity.

1432. Ronalds' Electrical Apparatus, as employed by him at the Kew Observatory; consisting of a principal conductor, with its glass support, umbrella, and heating apparatus; its voltaic collecting lantern; Volta's electrometers and sights; a Henley electrometer; a Gourjon galvanometer; a discharger, or spark measurer; and a Bennet's gold-leaf electroscope.
Kew Committee of the Royal Society.

Apparatus erected in the equatorial room of the Kew Observatory, in 1843, by Mr. Francis Ronalds, for the purpose of observing atmospheric electricity, described in the British Association Report for 1844.

It consists of a principal conductor, which is a stout copper tube, passed through a large aperture lined with sealing-wax in the roof of the building in which the instrument was placed, and carrying an inverted copper tray, to exclude rain.

The tube is supported by a stout glass core, which is kept in a dry state by a copper funnel passing up its interior, kept constantly heated by a small lamp.

A second lamp is enclosed in a Volta's collecting lantern, fixed to the top of the collecting tube.

To the cross arms at the base of the tube are attached severally:—Volta's electrometers with ivory scales, and sights for accurately determining the angles of the deflection of the straws; a Henley, or quadrant electrometer;

a galvanometer, by Gourjon (the property of Sir C. Wheatstone); a discharger, or spark measurer; and a Bennet's gold-leaf electroscope.

Continued regular observations were made with these instruments for several years.

The wooden stand now exhibited was not the original table upon which they were placed; that, a fixture in the Observatory, having been destroyed.

The Henley electrometer has also been replaced by a less perfect instrument.

1433a. Box Electroscope, avoiding the 'faults of common gold leaf electroscopes. *Prof. Beetz, Munich.*

1433b. Bifilar Electroscope, with copper, zinc, and condenser-plates for showing Volta's fundamental experiments, and tourmaline for showing pyro-electricity. *Prof. Beetz, Munich.*

1434. Thomson's Divided Ring Electrometer and **Gauge,** formerly in use for recording atmospheric electricity, at the Kew Observatory.
Kew Committee of the Royal Society.

This instrument, which consists of two parts, the electrometer and the gauge, was erected at the Kew Observatory in 1861, in connexion with a photographic recording apparatus, and worked there for about four years, producing daily records of the fluctuations, &c. of atmospheric electricity, which were discussed by Professor Everett, and the results published, together with a description of the instrument, in the Philosophical Transactions for 1868, Pt. I.

It has since been replaced by an improved quadrant electrometer.

1435. Singer's Gold Leaf Electroscope, for lecture purposes. Large size. *Elliott Brothers.*

1436. Quadrant Electrometer, being a modification of Sir W. Thomson's delicate quadrant electrometer, used for measuring the difference of potential between two conductors.
Elliott Brothers.

1437. Peltier Electrometer, for measuring the electrical tension of a charge by the repulsion of a light aluminium needle, which receives a directive force from a very small magnet attached to it. *Elliott Brothers.*

1437a. Peltier Electrometer.
The British Telegraph Manufactory, Limited.

1438. Capillary Electrometer, after Lippmann.
R. Jung, Heidelberg.

A glass tube *a*, filled to a height of about 85 cent. with mercury, and ending below in a fine point, dips in a cylinder *b*, so that its point presses lightly against the side, where there is a microscope *c* placed horizontally. The bottom of the cylinder contains mercury, and above this there is dilute sulphuric acid which covers the point of the tube; a platinum wire, connected with one terminal of the apparatus and protected by a glass tube from the sulphuric acid, dips into the mercury. The long tube is connected above with a small glass inverted U-tube, in which a platinum wire is fused, which, reaching down into the mercury in the tube, forms the upper electrode. The other end of the U-tube is connected by caoutchouc tubing with an air press, which, on its other side, is connected with a manometer. A little mercury is first forced through the fine point by means of the press. The microscope is then so placed that the zero point of the eyepiece micrometer coincides with the image of the meniscus of mercury in the capillary tube; then the electric source to be measured is brought into the circuit of the apparatus, its negative pole being connected with the upper electrode. The mercury forthwith retires, and can only be brought back by a determinate pressure with the press.

1439. Coulomb's Torsion Balance, for measuring magnetic and electric attraction and repulsion.   *Elliott Brothers.*

1439a. School Form of Coulomb's Torsion Balance.
Harvey, Reynolds, and Co.

b. GALVANOMETERS.

61. Galvanometer for hydro-electrical currents, with which Matteucci discovered, in 1844, the muscular current.
The Royal Institute of " Studii Superiori " at Florence.

The Copley medal was awarded by the Royal Society to Matteucci for his electro-physiological labours, and above all for his discovery of the muscular current.

Galvanometer of Nobili, with astatic system, and bobbin composed of eight threads to be united at pleasure.
The Royal Institute of " Studii Superiori " at Florence.

Galvanometer of Nobili, with astatic system for the hydro-electrical current.
The Royal Institute of " Studii Superiori" at Florence,

56. Magnetoscope of Nobili, composed of an astatic system suspended in a glass bell furnished with a graduated circle.
The Royal Institute of " Studii Superiori " at Florence.

The first galvanometer upon the astatic system was made by Nobili, in 1825. He afterwards so improved galvanometers, that they could be adapted to every kind of current, and be perfectly similar to one another As to the magnetoscope, he made use of it to detect the slightest traces of magnetism.

1235b. Large-sized Galvanometer, for demonstrating the principal applications of Ohm's formula.
Imperial University of St. Petersburg.

It consists of a strong brass ring, below which are two long plates, fitted at their extremities with an adjustment for uniting the galvanic couples parallel-wise (in quantity). Another ring, with more than 400 turns of wire, serves to study the combination of the couples in a different way. The two rings are united together, and can be set different distances from the needle. The

needle is furnished with two index-arms crossing each other at right angles,
the four ends of which are bent downwards at right angles, so as to mark
the deflection upon a graduation placed round the outer vertical surface of a
cylindrical ring.

This apparatus, for demonstration, is constructed according to the directions
of Prof. Petrouchevsky, Professor at the University of St. Petersburg.

1440. Galvanometer with variable resistance.
*Prof. Dr. R. A. Mees, Director of the Physical Labora-
tory of the University of Groningen.*

This galvanometer can be used for hydro-electric as well as thermo-electric
currents. By its aid the dependence of the sensibility of a galvanometer on
its resistance can be demonstrated.

1441. Reflecting Astatic Galvanometer, with coils of low
resistance, and with telescope and scale, for measuring thermo-
electric currents of very low intensity. Focal distance of telescope
and scale about three metres. Made by Ruhmkorff, of Paris.
George Gore, F.R.S.

1442. Galvanometer, by **Colladon,** with wires insulated by
a special method. This instrument was used by the inventor for
measuring the intensity of currents produced by electric friction
machines, by electricity from the clouds in 1826, and by electric fish in 1831. *Prof. Daniel Colladon, Geneva.*

The inventor used this same galvanometer in 1831, for studying the distribution of the electric poles upon torpedoes, and the strength of currents produced by animal electricity.

Annales de Chimie et de Physique, 1826, vol. 33.
Péclet, Traité de Physique, 1832, vol. 2, p. 221 to 225.
Mémoires de l'Académie Royale des Sciences, Institut de France, vol. 10,
p. 74.

1443. Balance Galvanometer, giving indication of current
in grains or other weights in scale pans, which may be adjusted to
any standard. Invented by exhibitor in 1848.
William Sykes Ward.

1444. Galvanometer, designed by **Colladon** for **Currents**
produced by **Electro-Statical Charges.**
*Geneva Association for the Construction of Scientific In-
struments.*

Apparatus for demonstrating that the electricity drawn from friction
machines, from the clouds, &c., produces currents, of which the direction and
the intensity are measured by the deviation of the magnetic needle of a
galvanometer.

1445. Marine Galvanometers used on board H.M.S. "Agamemnon" and the U.S. Frigate "Niagara," in the Atlantic Cable
Expeditions of 1858. *Sir William Thomson.*

A light mirror, weighing 30 milligrammes and 9 millimetres in diameter,
with single needle cemented to its back, suspended by stretched platinum

wire in centre of field of coil composed of two bobbins of fine copper wire. Micrometer screws to adjust zero by torsion of upper and lower parts of platinum wire. The first words transmitted across the Atlantic were from the "Agamemnon approaching the Irish coast, and were read on one of these instruments on board the "Niagara" approaching Newfoundland. (Encyclo pædia Britannica, Art. Telegraph (Electric), VII. 6, and VIII. 4.) Made by White and Barr (now James White), Glasgow.

This instrument was sent out for use on the first Red Sea cable (1859), but did not arrive until after the failure of the cable. It returned dismantled, and was never set in action again. It has been superseded by the siphon recorder. Made by White and Barr (now James White), Glasgow.

1446a. One of the First Mirror Galvanometers, made for the reading of messages through submarine cables, and used for that purpose at Newfoundland in 1858. This galvanometer has an arrangement for altering the intensity and direction of the directing force by a double motion of the directing magnet,—one varying its distance from the suspended needle and mirror, the other giving the magnet a motion in azimuth. This arrangement is still used in the Astatic Mirror Galvanometer for testing submarine cables.

Sir William Thomson.

1447. Ironclad Marine Galvanometer, used on board the "Great Eastern" in the Atlantic Cable Expedition of 1866, and subsequently by Mr. Willoughby Smith in the Mediterranean and Red Sea. *Sir William Thomson.*

This instrument is the first ironclad marine galvanometer, and the first with suspension by stretched silk fibre instead of platinum wire. Made by James White, Glasgow.

1447a. Differential Galvanometer, constructed specially for testing the locality and nature of faults in submarine cables. It is the first instrument in which *shunts* were used for practical electrometric purposes. A shunt is applied to one of the wires of the coil so as to multiply the reading of a rheostat *ten* times. It also was used for telegraphic reading purposes and for measuring the discharge from cables. It was made in 1858, and was in constant use for many years. *W. H. Preece.*

1447b. Thomson's Mirror Galvanometer, with hinged coils, and **Shunt** for same. *Warden, Muirhead, and Clark.*

1447a. Sir William Thomson's Astatic Mirror Galvanometer, with hinged coils, for testing purposes.

Warden, Muirhead, and Clark.

1447b. Lamp for Mirror Galvanometer.

Warden, Muirhead, and Clark.

1447c. Portable Testing Galvanometer on gimbals, with resistance coils and shunts. *Warden, Muirhead, and Clark.*

1447d. Sir William Thomson's Dead-beat Mirror "Speaking" Galvanometer for receiving signals on long submarine cables, provided with scale, stand, and lens. The mirror carrying the magnets is confined in a small air-chamber which can be contracted at will. The compressed air acts like a cushion, and "damps" the motion of the mirror, thereby preventing oscillations.   *Warden, Muirhead, and Clark.*

1448. Absolute Galvanometer or Magnetic Dynamometer.   *Frederick Guthrie.*

A current traverses in succession four spirals embracing soft iron cores. Two of the so formed electro-magnets are fixed and two movable together in a horizontal plane by means of the suspending torsion thread. The spirals are such that the magnets repel one another. If, when no current is passing, a beam of light reflected from a mirror attached to the movable pair falls in a certain place, then, when a current passes, the torsion screw head must be turned so as to force the magnets up to the same distance as before. The repulsion or angular torsion is proportional to the square of the strength of the current.

1449. Galvanometer for measuring large currents in definite units. Graduated in Weber units for use with the electric light, &c., and in ounces of silver deposited per hour, for use in electroplating and other forms of actual work.   *John T. Sprague.*

1450. Patent Universal Galvanometer, indicating current and resistance in definite units of measurement.
John T. Sprague.

This galvanometer contains four circuits, having 1, 10, 100, and 1,000 fold degrees of action on the needle, enabling it to be used for large or small currents. The patent dial is graduated to indicate the current in actual units, either the British Association, Weber, or in chemical equivalents. It is also graduated to show the total resistance of the circuit in ohms without the aid of a resistance instrument when used with a Daniell cell. By using a fixed resistance it shows the electromotive force of the circuit in volts.

1450a. Galvanometer for Projections.
M. J. Duboscq, Paris.

1450b. Galvanometer, for thermo-electric currents.
Luzard, Paris.

1451. Rhé Electrometer of Marianini, for observing electric discharges between the atmosphere and the earth.
Robert James Mann, M.D.

This instrument was planned by Professor Melsens. It contains a coil of copper wire which is to be made continuous with the system of a lightning rod, or with the earth wire of a telegraph line. When an electric spark passes through the coil a soft iron bar in its interior is magnetised, and a traversing magnetic needle pivoted above the coil is then deflected by it out of the north and south line of the earth's magnetism towards either the east or west. When the interior iron has been magnetised it must be replaced by a neutral bar before another observation can be made.

1452. M. Becquerel's Electro-magnetic Balance.
Conservatoire des Arts et Métiers, Paris.

1453. Pouillet's First Compass for Sines and Tangents. *Conservatoire des Arts et Métiers, Paris.*

1454. Sine-Tangent Galvanometer, for use at will either as a sine or tangent galvanometer. *Siemens and Halske, Berlin.*

1455. Aperiodic Galvanometer, with telescope and scale.
Siemens and Halske, Berlin.

The needle of this galvanometer is suspended in a copper ball, which acts as a damper, preventing vibration in any new position given to the needle. The needle itself is in the shape of a thimble cut away longitudinally on opposite sides, and by this arrangement the magnetic intensity is considerably increased, and the inertia of the magnet reduced. Du Bois-Raymond has shown that the "damping" of an astatic needle can be carried so far that the needle does not vibrate, but directly takes up its position of deflection, and these he has termed "aperiodically vibrating needles." Dr. Werner Siemens has attained the same end with simple non-astatic magnets, by means of certain forms of the vibrating magnets, and of the damping copper mass. The vibrating magnet consists of a steel thimble, from which two opposite sides are cut away parallel to the axis. This horseshoe magnet vibrates in a cylindrical space in a copper ball, which forms the centre of the wire coils.

1456. Inclination Galvanometer. *Dr. Werner Siemens.*

Intended for use particularly with the selenium photometer (No. 895). The coil of the galvanometer is wound horizontally; the needle vibrates in a vertical plane and carries a mirror which reflects the image of a finely photographed scale (placed above) into the optical axis of a microscope.

1457. Galvanometer for testing Lightning Conductors, constructed by the exhibitors for the Prussian Royal Engineers.
Keiser and Schmidt, Berlin.

1458. Mirror Multiplier. *E. Stöhrer, Leipzig.*

1459. Galvanometer, showing both inclination and declination. *E. Stöhrer, Leipzig.*

The broad brass frame which carries the magnetic needle can be turned about its axis ; likewise the vertical support in its base. Thus the needle can be enabled to move in various planes, and the altered action of the force measured by observation of the vibrations.

1459a. Current Measurer, with arrangement for very strong electric currents. *Gustav Baur, Stuttgart.*

1460. Edelmann's Mirror-galvanometer for absolute measurements. *M. Th. Edelmann, Munich.*

1461. Large Wiedemann's Galvanometer.
M. Th. Edelmann, Munich.

1462. Small Wiedemann's Galvanometer.
M. Th. Edelmann, Munich.

1463. Edelmann's Compensation-galvanometer.
M. Th. Edelmann, Munich.

1464. Edelmann's Lecture-galvanometer.
M. Th. Edelmann, Munich.

1465. Edelmann's Small Mirror-galvanometer.
M. Th. Edelmann, Munich.

1466. Edelmann's Pocket Compasses, with absolute measurements for electro-therapeutics.
M. Th. Edelmann, Munich.

1467. Du Bois-Raymond's Astatizing Magnet.
M. Th. Edelmann, Munich.

1468. Ordinary Compensator.
M. Th. Edelmann, Munich.

1469. Edelmann's Induction Galvanoscope.
M. Th. Edelmann, Munich.

1470. Large Galvanometer for **Electro-therapeutic Purposes.**
M. Th. Edelmann, Munich.

1471. Differential Galvanometer, very delicate, with two coils. (Becker, London.) *H. Lloyd, Trinity College, Dublin.*

1474. Apparatus to illustrate some of the laws of electrical rotation of Faraday and Ampère. *Yeates & Sons.*

1475. Tangent Galvanometer. *James How & Co.*

1476. Thomson's Astatic Reflecting Galvanometer, round glass case, for lecture purposes. *Elliott Brothers.*

The most sensitive instrument yet constructed for detecting the presence of a current and measuring its magnitude.

1477. Thomson's Astatic Reflecting Galvanometer, in square brass case, made by Elliott Brothers. *Prof. W. F. Barrett.*

The upper coil of this galvanometer has a very high resistance, the lower a very low one; the coils are used independently, but require only one lamp, and hence can readily be changed without any disturbance by merely changing the position of the plug.

1478. Set of Shunts, for use with galvanometers, to reduce the angle of deflections of the needle 10, 100, or 1,000 times by shunting off the current, so that only $\frac{1}{10}$, $\frac{1}{100}$, or $\frac{1}{1000}$ of the current passes through the galvanometer, and the remainder through the shunts. *Elliott Brothers.*

1479. Simple Horizontal Galvanometer, coiled with stout wire of low resistance, intended especially to show thermo-electric currents. *Elliott Brothers.*

1480. Tangent Galvanometer, designed by Gaugain, for measuring electro-motive force, resistance, &c.   *Elliott Brothers.*

The instrument is composed of two coils, separated by a distance equal to one half their diameter; contains three sections of stout wire, which can be connected at pleasure. This instrument can be used with very powerful currents.

1481. Speaking Galvanometer, used for signalling through submarine cables. The mirror and magnet are doubly suspended; glass in front and back protect the mirror and magnet from draught and dust.   *Elliott Brothers.*

1482. Oil Vessel Galvanometer, as used in military schools, for testing and signalling.   *Elliott Brothers.*

The coils consist of two sections of high and low resistance, which can easily be brought into circuit by a switch in front of the instrument. To reduce the oscillations of the needle, the instrument is provided with a glass vessel for liquids, into which the vane of the needle dips.

c. ELECTRO-DYNAMOMETERS.

1472. Electro-Dynamometer. Apparatus for the determination of the strength of electric currents by measuring the action between different parts of the current itself.
        H. Lloyd, Trinity College, Dublin.

1483. Electro-dynamometer, for measuring electric currents which are constantly being reversed in direction. Made by Dr. Meyerstein, of Göttingen.   *George Gore, F.R.S.*

1484. Edelmann's Absolute Dynamometer,
        M. Th. Edelmann, Munich.

1485. Fine-wire Electro-dynamometer.
        M. Th. Edelmann, Munich.

d. VOLTAMETERS.

1486. Voltameter.
        Geneva Association for the Construction of Scientific Instruments.

The tube of the voltameter is divided on glass into sixteenths of cubic centimetres.

1486a. Voltameter for teaching purposes.
        Physical Institute (Univ. of Giessen), Dr. Buff.

This apparatus for the decomposition of water is described in Liebig's Annalen, Vol. 93, p. 256. It is intended to render the gases evolved available for analysis or otherwise.

The connecting piece, fitted above with caoutchouc tubing, serves for ordinary purposes. By means of the apparatus, not only may the reconstitution of water be shown, but the phenomena of heating and ignition, *e.g.*, of a steel spring. For such experiments 12 Bunsen pairs are necessary, connected in series.

1486b. Silver Voltameter, with platinum vessel.

Prof. Beetz, München.

A convenient modification of the Poggendorff voltameter.

1486c. Mercury Voltameter.

The Physical Science Laboratory of the Technological Institute, St. Petersburg (Russia).

This apparatus is intended for measuring the strength of currents by the reduction of mercury. It consists of two electrolytic glass bowls joined by a tube, and an apparatus for measuring the reduced mercury volumetrically. A more detailed description, with drawing and directions for use, accompany the instrument.

1732b. Electrotype Apparatus, with stand, and Poggendorff's silver voltameter. *Prof. Hittorf, Münster.*

The vessels serve in investigating the processes by which the electrolytes in aqueous, alcoholic, or other solutions afford passage to the electric current. They enable the changes occurring in the electrodes to be kept distinct and fully determined by quantitative chemical analysis. In soluble electrolytes the ions can be certainly determined, and so the primary decompositions distinguished from the secondary. Further, the apparatus shows the ratio of the velocities with which the two ions move in opposite directions. (*See* Poggendorf's Annalen, Bd. 89, 96, 103, and 106.)

IX.—APPARATUS FOR MEASURING ELECTRICAL RESISTANCE AND CAPACITY.

1487. Wheatstone's Bridge of the simplest construction, especially used for conductivity and low resistance tests. The bridge is provided with one pair of equal resistances and one standard resistance. *Elliott Brothers.*

1488. Thomson's Circular Sliding Resistance, of great importance in cable tests in connexion with the quadrant electrometer for measuring resistance, tension, or potential of cables and batteries, and for detecting faults in cables.

Elliott Brothers.

1488a. Megohm or 1 million Ohms resistance, valuable for measuring very high resistances, for determining the constants of galvanometers, &c. *Elliott Brothers.*

1487a. Revolving Coil used in the determination of the **" British Association Unit of Electrical Resistance,"** 1863-4. (The property of the British Association.)

Prof. Clerk Maxwell.

The coil forms a closed conducting circuit. As the coil revolves about a vertical axis, the horizontal component of the earth's magnetic intensity produces an alternating current in the coil. A very small magnet is suspended at the centre of the coil, and is deflected from the magnetic meridian by the current in the coil. When the coil revolves rapidly, the alternations of the current do not produce any sensible vibration of the magnet, and the tangent of the permanent deflection is a measure of the conductivity of the coil.

1487b. Resistance Coil in an annular brass case, and embedded in paraffin. The resistance of the revolving coil was compared with that of this coil during the experiments.

Prof. Clerk Maxwell.

1487c. Pridge Arrangement used in the above comparisons.

Prof. Clerk Maxwell.

1489. Plate, showing **Dr. Bosscha's Method** of determining the **Ratio of two resistances,** with explanatory note.

J. Bosscha, Professor, Royal Polytechnic School, Delft.

1490. Plate, showing **Dr. Bosscha's Method** of determining the **Ratio of two electro-motive forces,** with explanatory note.

J. Bosscha, Professor, Royal Polytechnic School, Delft.

1491. Cylindrical Condenser for measuring Capacity in absolute electrostatic units. *Sir W. Thomson.*

1492. Adjustable Disc Condenser, also used as a spark micrometer. *Sir W. Thomson.*

1492a. Early Rheostat, given by Faraday to Sir Charles Wheatstone. *The Council of King's College, London.*

1492b. Divided Condenser, as designed by Major Malcolm, R.E., for comparing the electrostatic capacities of cables, the electro-motive force of batteries, &c. *Elliott Brothers.*

This condenser is in 12 subdivisions, as follows, viz.:—·001, ·002, ·002, ·005, ·01, ·01, ·02, ·05, ·1, ·1, ·2, ·5 = 1 microfarad; so that any capacity from $\frac{1}{1000}$ to 1 microfarad can be obtained.

1493. Rheostat with large copper wire.

The Council of King's College, London.

1494. The Original Wheatstone's Bridge.

The Council of King's College, London.

1494a. Rheostat (Agometer), with platinum wire. Invented by M. H. Jacobi, in the year 1841.

Physical Science Cabinet of the Imperial Academy of Sciences, St. Petersburg.

1495. Universal Resistance-Box.
Siemens and Halske, Berlin.

This instrument is intended to measure by simple manipulation the resistance of wires, the electro-motive force of batteries, and their internal resistance. It has three branch resistances, and a complete series of resistances from 1 to 10,000 S. U.

1496. Complete Bridge. *Siemens and Halske, Berlin.*

This bridge, besides the branch resistances, and variable resistance, includes battery reverser, galvanometer key, and short circulating keys, so arranged that one of the branch resistances or the variable resistance can be cut out instantaneously.

1495a. Large Lane's Measuring Flask.
Warmbrunn, Quilitz, & Co., Berlin.

1497. Edelmann's Small Rheochord.
M. Th. Edelmann, Munich.

1498. Resistance Apparatus with Weber's absolute units.
M. Th. Edelmann, Munich.

1499. Universal Compensator, for measuring resistances and electro-motive forces. *Prof. Beetz, Munich.*

The apparatus exhibited by Prof. Beetz was manufactured from his plans by M. Th. Edelmann.

1499a. Compensator, by Fresnel, made by Froment.
Polytechnic School, Paris.

1500. Post Office Resistance Coil, as used by the Government telegraph authorities, for testing the resistance of telegraph lines, cables, batteries &c. *Elliott Brothers.*

These coils are constructed on the Wheatstone's bridge principle, and the arrangement allows of the measurement of resistances from $\frac{1}{100}$ to 1,000,000 ohms.

1501. Dial Resistance Coil, for testing the resistance of telegraph lines, cables, batteries, &c. *Elliott Brothers.*

These coils are capable of greater accuracy than other resistance coils. For each dial only one plug is used to bring the desired resistance in circuit, for every change, therefore, the same number of plugs is necessary.

1502. Preece's Balance, for comparing resistances in connexion with a set of resistance coils. *Elliott Brothers.*

This instrument is on the principle of the Wheatstone's bridge, and contains four pairs of equal resistances, a commutator, and two keys.

1503. Set of Resistance Coils, German silver, with Wheatstone's bridge, for testing the resistance of telegraph lines, cables, batteries, &c. These coils allow of measurements of resistance from $\frac{1}{100}$ to 1,000,000 ohms. *Elliott Brothers.*

1504. British Association Unit or Ohm, of German silver wire, equal to 10^7 absolute electro-magnetic units, the standard of electrical resistance. *Elliott Brothers.*

1504a. Large Set of Resistance Coils, with Bridge (a new form and arrangement). *Warden, Muirhead, and Clark.*

1504a. Large size set of Resistance Coils, containing a resistance of 11,110 ohms in the aggregate, with proportional coils of 10, 100, 1,000, and 10,000 ohms. The units, tens, hundreds, and thousands are in separate sets; the resistances are thrown into circuit by inserting the plugs.
Warden, Muirhead, and Clark.

1504aa. Medium size set of Resistance Coils, with proportional coils of 10, 100, 1,000 ohms.
Warden, Muirhead, and Clark.

1504cc. Small set of Resistance Coils.
Warden, Muirhead, and Clark.

1504cd. Set of Proportional Coils, containing three pairs of 10, 100, 1,000 ohms respectively. To be used with 1504cc.
Warden, Muirhead, and Clark.

1504ce. Set of Standard Resistance Coils, 10, 100, 1,000 ohms. *Warden, Muirhead, and Clark.*

1504cf. Set of Galvanometer Shunts, $\frac{1}{9}$, $\frac{1}{99}$, $\frac{1}{999}$, $\frac{1}{9999}$ of the resistance of the galvanometer.
Warden, Muirhead, and Clark.

1504b. New combined Set of Thomson and Varley's Resistance Slides (a new form and arrangement).
Warden, Muirhead, and Clark.

1504c. Set of Resistance Coils (small portable set), containing 10,000 B A units in the aggregate, and set of proportional coils for same. *Warden, Muirhead, and Clark.*

1504d. Condenser, 20 microfarads.
Warden, Muirhead, and Clark.

1504e. Small portable Standard Condenser, capacity 1 microfarad in aggregate, divided in : 1 : 2 : 3 : 4 microfarads.
Warden, Muirhead, and Clark.

1504f. Standard Condenser, $\frac{1}{3}$ microfarad.
Warden, Muirhead, and Clark.

1504d. Condenser or Accumulator of 20 microfarads capacity. This form is principally used in telegraphing through long submarine cables. It is placed between the cable and the receiving instrument, or between the latter and the earth in order to render the signals more distinct.

Warden, Muirhead, and Clark.

1504f. Standard Condenser of 5 microfarads capacity subdivided into tenths. *Warden, Muirhead, and Clark.*

1504g. Standard Condenser of 5 microfarads capacity.

Warden, Muirhead, and Clark.

1504g. Sine Inductor. Apparatus for producing an induction current by the rotation of a magnetic disc for determining resistance. *L. Waibler, Darmstadt.*

1507. Conductivity Apparatus, for comparing the conductivity of wires of copper, or any other metal or alloy, with 100 inches of pure copper weighing 100 grains, 200 inches weighing 100 grains, or 300 inches weighing 100 grains. The conductivity is determined by resistance, length, and weight.

Elliott Brothers.

X.—STANDARDS OF COMPARISON FOR MEASUREMENT OF ELECTRICAL MAGNITUDES.

1505. Weber's Resistance Coils, No. 3 and No. 5, certified to represent the following resistances :—

No. $3 = 60717 \times 10^5$ Weber's absolute units.

No. $5 = 59440 \times 10^5$ „ „

J. Bosscha, Professor, Royal Polytechnic School, Delft.

Comparisons of these coils were published by Dr. Schröder van der Kolk in Poggendorff's Annalen, vol. 110, p. 465.

The coil No. 3 was used in the first determination of the electro-motive force of a Daniell's cell in absolute measure, executed in 1857 by Dr. Bosscha. This electro-motive force was found to be 1026×10^8 Weber's absolute units. Applying Sir William Thomson's theorem (Phil. Mag., Series IV., vol. ii. p. 557), the mechanical equivalent of heat derived from this result is 432·1 as compared with MM. Favre and Silbermann's determination of the heat generated by the action of zinc on sulphate of copper ; 419·5 as compared with Mr. Joule's experiments on the heating power of a Daniell's cell (Phil. Mag., Series IV., vol. iii.), discussed in Poggendorff's Annalen, vol. 103 ; 437·3 as compared with similar experiments of Professor Lenz (Pogg. Ann., vol. 59, p. 203), discussed in Poggendorff's Annalen, vol. 108, p. 162 ; and 421·1 as compared with similar experiments of Mr. Joule, related in "Memoirs of the Lit. and Phil. Soc. at Manchester," vol. vii. p. 94, and discussed in Poggendorff's Annalen, vol. 108, p. 168.

1506. Siemens' Resistance-Unit.

Siemens and Halske, Berlin.

This is a wire resistance equivalent to the resistance offered by a prism of pure mercury one square millimetre in section and one metre long at 0° centigrade, which is the basis of the Siemens'-standard unit.

1506a. Normal Mercury Unit in glass spiral.
Siemens and Halske, Berlin.

1504g. Clark's Standard Element.
Warden, Muirhead, and Clark.

This is the most constant voltaic cell known as regards electro-motive force. It is used as a practical standard of electro-motive force. The poles are pure zinc and mercury. The exciting paste is composed of protosulphate of mercury boiled in a saturated solution of sulphate of zinc.

XI.—APPARATUS FOR THE APPLICATION OF ELECTRICAL PRINCIPLES TO PRACTICAL PURPOSES.

a. TELEGRAPHIC APPARATUS.

1652. Electric Telegraph, original apparatus as it was made under the direction of its original discoverer, Th. Sömmering, in München, 1809.   *K. Sömmering, Frankfort.*

1653. The Volta's Pile, then used for the above, together with the 10 original silver, and 10 original zinc plates.
K. Sömmering, Frankfort.

1654. The Original Conducting Wire, as it was made under the direction of the discoverer in 1809–1811, and tested in the Isar.   *K. Sömmering, Frankfort.*

1655. Alarum belonging to Sömmering's Telegraph.
K. Sömmering, Frankfort.

1508c. Portion of the **original Telegraph,** laid by Sir Francis Ronalds in 1816, in his garden at Hammersmith, and described in his book in 1823.   *Latimer Clark, Westminster.*

The original wooden model of the dial of the instrument used with the above telegraph.

1508aa. Electric Telegraph. A portion of the original copper wire and glass tube buried by Sir Francis Ronalds in 1816 for the purpose of his then new discovery of electric telegraphy; found after several months' search by the exhibitor in 1862.
Captain Henry Hill.

1508aa. Two Drawings illustrating plans by Sir Charles Wheatstone, in 1840, for the manufacture of submarine electric telegraph cables, and proposed plans for laying the same.
Sir Charles Wheatstone.

1680a. First Electro-magnetic Needle Telegraph, invented in the year 1830 by Baron P. L. Schilling (born in 1786 at Reval), consisting of :—

Two electro-magnetic alarum apparatus ; two ditto, and gyrotropes (key-board, indicator) ; one current indicator ; one multiplicator for the signs (sign receiver) of the magnetic needle to the right or to the left respectively, turning of the white or black side of a paper disc towards the spectator.

Physical Science Cabinet of the Imperial Academy of Sciences, St. Petersburg.

1508b. Portion of the **first line of Telegraph,** laid by Cooke and Wheatstone in 1837 between Euston and Camden stations. It consists of five wires and was worked with their earliest or "hatchment" dial instrument. *Latimer Clark.*

1419. Part of **Cooke** and **Wheatstone's First Working Telegraph.** *Edinburgh Museum of Science and Art.*

1680. Specimen of the First Telegraph Line, 1837.
R. S. Culley.

This specimen of the first Telegraph line was dug up on the railway incline between Euston and Camden. It was laid down in connexion with the first experiments made with Cooke and Wheatstone's earliest instrument in 1837.

1689. Cooke and Wheatstone's A B C Instrument, 1840. *Reid Brothers, London.*

The escapement wheel on the axle of which the pointer is fixed is controlled by electro-magnets.

The communicator is outside, and concentric with the indicator dial, and consists of a cog-wheel working into two smaller wheels. The cog-wheel is turned by a handle, and the battery contacts are made by small wooden cylinders inlaid with metal fixed on the smaller wheels.

The wheelwork is driven by a mainspring.

1690. Nott and Gamble's Step by Step Pointer Telegraph, 1846. *Reid Brothers, London.*

Electro-magnets act on a ratchet-wheel by means of clicks attached to their armatures. On the axle of the ratchet-wheel the pointer is fixed ; the latter is moved forward through a space equal to the distance between two letters for each making and breaking of the battery contact. A simple tapper or pedal key is used for sending the currents, and the pointer is allowed to rest for a short interval when it is opposite to the letter desired to be indicated. The instrument is furnished with an alarum, the bell being struck by a hammer attached to the armature of an electro-magnet provided for the purpose.

1681. Cooke and Wheatstone's Double-Needle Telegraph, with Alarum. Earliest form, with 6-inch coils and astatic needles. *Reid Brothers, London.*

These instruments were used on the line erected between Paddington and Slough for the purpose of exhibiting the invention in 1844, and were the identical instruments by the aid of which Tawell was caught.

1682. Double-Needle Instrument. As used by the Electric Telegraph Company in 1846.   *Reid Brothers, London.*

Used for all public business, and for the principal railway circuits, with 6-inch coils and astatic needles. The coils were afterwards reduced to one inch. The double needle was superseded for public purposes by the Bain recording instrument from about 1850. It is still used by some railway companies.

1683. Cooke and Wheatstone's Single-Needle Instrument. (Early form.)   *J. A. Warwick.*

Used by the Electric Telegraph Company in 1846 on unimportant railway circuits.

The specimen exhibited has a "crutch" handle commutator, which was the earliest form used ; later the "drop" handle was introduced.

1508d. Two Photographs of the **Telegraph Apparatus** used by **Gauss and Weber.**

Physical Institute of the University of Göttingen (Prof. Dr. Riecke).

One of the photographs represents the magnetic inductor which serves for giving the signals, the other the unifilar magnetic rod with multiplier serving for receiving the signals.

1508. Two Original Drawings of plans, sections, and elevations of machinery of Sir C. Wheatstone's proposed scheme for a sub-marine telegraph cable, 1840.   *Robert Sabine.*

1508a. Portion of the **First Submarine Telegraphic Cable.** This cable was laid by Mr. T. R. Crampton, C.E., in 1851, and established the practicability of submarine telegraphy.

Thomas Russell Crampton.

Submarine telegraphy in 1851 was deemed by most engineers and the public to be visionary if not impracticable.

Its extension over the whole world since its first practical introduction in 1851, has been immense, and its advantages incalculable. It was established in the following manner :—Various propositions were from time to time put forth to effect the object, but few people were prepared to take the risk until a company was formed having most influential men on the direction, who advertised in the usual manner for subscriptions. Such, however, was the want of confidence felt in the scheme, that only about two per cent. of the necessary capital was subscribed, and this money was consequently returned to the applicants. Notwithstanding this apathy of the public some of the directors and their friends did not cease to entertain a full conviction of its possibility, and they subsequently consulted Mr. T. R. Crampton, C.E., on the subject, and offered to assist towards providing the funds if he felt sufficiently confident of ultimate success. Mr. Crampton undertook the entire charge and responsibility of the form, construction, and laying of the cable, and also took upon himself rather more than one half the pecuniary risk, the other half of the money being found by Lord de Mauley, Sir James Carmichael, Bart., Messrs. Davies Son and Campbell (the solicitors of the Company), the Hon. F. W. Cadogan, and Mr. Haddon.

The cable was in the same year (1851) successfully laid between Dover and Calais by Mr. Crampton.

The great risk the parties ran can be better appreciated from the fact that three successive attempts by other parties to establish submarine cables between England and Ireland occurred soon afterwards, which all failed.

The above-named gentlemen were also instrumental in laying the next successful cable between Dover and Ostend.

This latter was constructed in a similar manner to the original one, and they are both still in operation.

No fundamental change has yet been effected in the form and mode of construction of heavy cables, thus proving satisfactorily that the first type of heavy submarine cable laid upwards of twenty-five years ago is practically right in principle.

1508f. Magneto-Electric Machine termed a **"Thunder Pump."** *Reid Bros.*

1508g. Cooke and Wheatstone's Coil, 1837, with front and back needle ; coil wound with cotton covered wire.
Reid Bros.

1508h. Reid's Patent Circular Coil, 1848, with front and back needle complete ; coil wound with cotton covered wire.
Reid Bros.

1446. First Instrument for **recording Signals** through **Long Submarine Cables,** by curve of perforations produced by sparks from a Ruhmkorff coil guided by a platinum wire moved by a needle under the influence of the varying current from the cable. *Sir William Thomson.*

SPECIAL COLLECTIONS ILLUSTRATING THE HISTORY OF ELECTRIC TELEGRAPHY, CONTRIBUTED BY H.M. POSTMASTER GENERAL.

1508e. Cooke and Wheatstone's Earliest Needle Telegraph, 1837.

The letters are indicated by the convergence of two needles. The five line-wires required for the instrument were inserted in grooves in a triangular piece of wood, and wire laid underground.

1509. Cooke and Wheatstone's Four-Needle Telegraph, 1838.

Some of the letters are indicated by the convergence of two needles, as in the five-needle instrument, the rest by one or more movements of the needles to the right or left.

1548. Cooke and Wheatstone's Revolving Disc Telegraph, 1840.

A step-by-step instrument. The letters of the alphabet are arranged round a paper disc fixed on the axle of an escapement wheel.

The letters are presented at an opening in the front of the case.

The escapement is similar to the " echappement-à-cheville," and is *controlled* by an electro-magnet.

There are as many teeth in the escapement wheel as there are letters on the revolving disc ; the latter moves from one letter to the following for each current sent.

The train of wheelwork is actuated by a mainspring.

1549. Cooke and Wheatstone's Pointer Telegraph Instrument, 1840.

This instrument is similar in all respects to the revolving disc telegraph, excepting that a pointer takes the place of the revolving disc.

1550. Cooke and Wheatstone's Magneto-Eléctric Communicator, used with their revolving disc or pointer alphabetical telegraph, 1840.

This communicator is so arranged that a current is sent when its spoked wheel is turned through a distance equal to that dividing the letters engraved upon it.

The commutator fixed on the axle of the revolving electro-magnet is so constructed that the magneto-electric currents are all in the same direction.

1516. Old Form of Double-Needle Instrument, with six-inch coils.

1512. Double-Needle Instrument, newer form.

Used by the Electric Telegraph Company.

1514. Portable Double-Needle Instrument, which can also be used for testing wires.

Each of the needles worked by two finger keys behind the case.

1515. Model Double-Needle Instrument with a Key Commutator, 1849.

1511. Single-Needle Instrument, modern form, with handle commutator.

Used by the Electric Telegraph Company and by railways.

1517. Series, showing the several Forms of Coil and Needle used by the Electric Telegraph Company.

a. The original form.　6-inch coils.
b. Holmes' diamond needle.　1-inch, 1848.
c. Clark's Needle.
d. S. A. Varley's coil with soft iron needle magnetised by induction.
e. Spagnoletti's do.
f. Brittan's do.

1518. Early Train Signalling Instruments.

a. Cooke's first " block " instrument used on the Norfolk Railway, about 1845.
b. Step by step train indicator used on the London and South-western Railway.
c. Do.　　do.　South-eastern Railway.
d. Signalling instrument used for starting and stopping the endless rope by which the Blackwall Railway was first worked, about 1840.

1510. Alarum with Centrifugal Hammer. Used in connexion with Cooke and Wheatstone's first needle instruments.

Moved by wheelwork and mainspring; released by an electro-magnet. Used in connexion with Cooke and Wheatstone's first needle instruments.

1513. "Thunder Pump."

Henley's Magneto-Electric Machine, used for ringing alarums, etc., commonly known as the "Thunder Pump." Used by the Electric Telegraph Company.

1554. Early form of Wheatstone's Resistance Coils.

1552. Bain's Chemical Telegraph. (Incomplete.) **First form of recording Instrument used in England,** 1846.

The chemically-prepared paper is wrapped round the cylinder. An iron style presses on the paper and writes dots and dashes in a spiral line as the cylinder moves endwise while revolving.

The signals are formed manually by pressing a key, or automatically by a spring, making contact through perforations previously punched in a paper ribbon.

1551. Bain's Perforator for double dot or Steinheil Alphabet, 1846.

The perforations are made by circular punches on a band of paper. The third key on the right is for advancing the paper without punching, for spacing between letters and words.

1532. Bain's Chemical Telegraph as used by the Electric Telegraph Company in 1850 in place of the double needle.

The paper ribbon was prepared with yellow prussiate of potash and nitrate of ammonia ; the style is of iron. The Steinheil code, dots in two parallel lines, was occasionally used, but was entirely superseded by the Morse code of dots and dashes.

1531. Morse Embosser, about 1853. Used by the Electric Telegraph Company.

This superseded the Bain instrument and was in its turn superseded by the inkwriter.

The dots and dashes of the Morse alphabet are made by a rounded steel point fixed at one end of a lever, the other end being furnished with an armature and attracted by an electro-magnet worked by a relay and local battery.

1533. Various forms of Ink-writers. (Old.)

a. Breguet's ink-writers.
b. Ink-writer with inking pad and reservoir.
c. Do.   do. ink bottle.

1529. Keys used with the Bain and Morse Telegraphs, by the Electric Telegraph Company.

a. Simple spring key.
b. Key for sending a short reversal after each signal, two sets of batteries being required. When the key is up, the line wire is connected to the receiving apparatus.

c. Wheel key. A constant current is maintained on the line, and signals are made by depressing the key and thus reversing the current. A switch is used for making the necessary alterations to the connexions for sending and receiving.

1528. Plunger Signalling Key, used by the London District Telegraph Company.

This key is used instead of the drop handle key of Cooke and Wheatstone's instrument, or the two-pedal key in Highton's instrument. An ordinary single-needle indicator is used in connexion with it.

1533. Whitehouse's Relay, 1854.

A small permanent horseshoe magnet oscillates between the pole pieces of an electro-magnet. The adjustment is effected by the attraction of another small permanent magnet.

1534. Varley's Mill, 1855. Used by Electric Telegraph Company.

This contrivance was used in connexion with relays and translators. When a current passes through the coils the armature is attracted rapidly in the usual way, but when the current ceases the armature returns slowly to its normal position, by the intervention of a wheel, pinion, and fly, thus lengthening the contact between the line lever and the spacing battery. This is now effected by utilising the extra current from the magnet. (*See* 1566.)

1535. Andrews' " Pump Relay " or " Spacer," used by the United Kingdom Telegraph Company.

This is a contrivance to produce a similar result to Varley's mill. A loosely fitting piston with a ball valve moving in a cylinder filled with oil is used instead of the fly.

1536. Zinc Sender, used by Electric Telegraph Company.

Used for sending a short reversal after each signal in the Morse system, to assist in discharging the line wire, in cases where the double-current key (1562) could not be used.

The coils of this apparatus are wound with fine wire of high resistance, and are placed as a " leak " or derived circuit on the line wire at the sending end. A single current Morse key is used. When this is depressed a portion of the current passes through the coils and moves the tongue of the relay over in contact with the screw stop which is connected to the reversing battery, the other pole of this battery being in connexion with the earth.

The back stop of the Morse key is in permanent electrical connexion with the tongue of the relay, and thus, when the key is raised and its lever comes into contact with the back stop, a reverse current will pass out to the line; but a portion of this reverse current also passing through the coils of the zinc sender will immediately move the tongue to the opposite side, the stop against which it rests being in connexion with the relay of the receiving apparatus. Thus after each reversal following a marking current the instrument is in a position for receiving, and the receiving station can stop the sending station during transmission. A smaller battery than the sending one is used for the reversal. A spring is fixed on the tongue on the battery side to lengthen the contact.

1539. Simple Electro-Magnetic Relay, used by the United Kingdom Telegraph Company. First form.

The simplest form of relay, consisting of an electro-magnet; its armature is attached to one end of a lever, the other end playing between two limiting stops, the local battery circuit being closed when the armature is attracted.

1540. Earliest Form of Relay with an inducing Magnet, used by the Electric Telegraph Company.

The coil is wound on a reel of soft iron, upon each end of which a hollow "casing" or cap of the same material is fitted, almost completely encasing the coil in soft iron. The armature is shaped thus, and is magnetised by induction from a compound bar magnet placed behind. The crescent-shaped portion plays between the inner ends of the casings, which for that purpose do not quite meet, but leave the central portion of the coil exposed.

An ordinary magnetic needle pivoted below the coil is acted upon by the latter, and serves as an indicator to call attention.

The armature is held up against knife-edge bearings by two helical springs, and the adjustment is effected by varying the tension of one of them.

1541. Later form of Relay, used by the Electric Telegraph Company, 1856. A few are still in use by the Post Office.

A horizontal bar of soft iron is pivoted vertically, and is free to move in the interior of two cylindrical bobbins. The ends of the bar which project beyond the bobbins play between the poles of horsehoe permanent magnets fixed at each end. The relay is adjusted by moving the stops, and consequently the soft iron bar, to one side or the other.

1519. Henley's Magneto-electric Double-Needle Instrument, 1848. Used by the British and Irish Magnetic Telegraph Company.

The needles only move on one side of their vertical position, and the signals are made up of the single and combined movements of the two needles.

This instrument requires two line wires, and is worked by the magneto-electric current generated by moving the handle or handles.

The interior needles are small straight bar magnets, playing between the semicircular pole pieces of an electro-magnet. The needle remains on the side on which it is left by the last current which passes through the coils, and does not return to its vertical position by gravity, as in Cooke and Wheatstone's needle instrument.

1520. Henley's Magneto-electric Single-Needle Instrument, 1848. Used by the British and Irish Magnetic Telegraph Company.

The dots and dashes of a modification of the Morse alphabet are represented by the duration of deflection of the needle on *one* side only of its normal position. It is worked by the magneto-electric current generated by moving the handle. Its construction is precisely similar in principle to Henley's magneto-electric double-needle instrument. It was used on less important lines than those on which the latter instrument was adopted.

1521. Highton's Needle Telegraph, 1848. Used by the British and Irish Magnetic Telegraph Company.

A horseshoe or circular magnet within a circular coil worked by a reversing key.

The signals are similar to those of Cooke and Wheatstone's single needle, but the alphabet is different.

Needle and coil used with this instrument.

1522. Highton's Needle Telegraph, smaller form. Used by the British and Irish Magnetic Telegraph Company.

Identical in construction with No. 1521, which it superseded.

1523. Highton's Needle Telegraph, last form. Used by the British and Irish Magnetic Telegraph Company.

With key detached. Identical in construction with the two former instruments, but the signalling key is detached from the indicating portion, and placed near the edge of the desk before the operator for greater convenience of working.

1524. Key, for Highton's single Needle Telegraph. Also used with Bright's Bell Instrument.

When the right-hand key is depressed a current in one direction is sent out to the line, and the pointer of the indicating instrument moves to the right. When the left-hand key is depressed a current in the opposite direction is sent, and the pointer moves to the left.

1525. Bright's Bell Telegraph, 1855, with Relay. Used by the British and Irish Magnetic Telegraph Company.

The single needle alphabet is produced by striking two bells of different tones, the hammers being actuated by electro-magnets worked by a relay and local battery. The relay consists of two electro-magnetic bobbins placed side by side, their ends being furnished with pole pieces turning inwards. Between these pole pieces at each end of the bobbins the ends of permanently magnetised needles pivoted on vertical axes play; these needles are so placed as regards their polarity, that a current in one direction moves the needle which closes the local circuit of the right-hand bell, and a current in the opposite direction moves the other needle which closes the local circuit of the left-hand bell. The signalling key used with this instrument is similar to that used with Highton's single needle.

1526. Bright's Direct Bell Instrument, 1870. (Model.)

Model of a Bright's bell instrument, in which the bells are struck by hammers attached to the magnetic needles of the relay.

1542. Relay for Bright's Bell Instrument.

A relay or repeater for relaying the signals of the Bright's Bell instrument used on the long lines of the British and Irish Magnetic Telegraph Company, when owing to leakage at the supports, from bad weather or other causes, the direct currents were too weak to work the whole length   In construction it is similar to the relay fixed on the bell instrument itself, but it consists of duplicate bobbins and magnetic needles.

1527. Highton's Gold Leaf Telegraph, 1846.

A movable conductor formed by a strip of gold leaf is placed in proximity to one pole of a permanent bar magnet, and moves to the right or left of its normal position according to the direction of the current passing through it   It was designed as a substitute for the needle and coil of Cooke and Wheatstone's instrument, but was never brought into practical use.

[This apparatus was described by Cumming in 1827, in his "Electrodynamics"; it is also mentioned in the *Encyclopædia Britannica*, 7th ed., Art "Voltaic Electricity," and in the treatise on "Electromagnetism" in the *Library of Useful Knowledge*, 1832.]

1543. Wheatstone's Type Printing Instrument, 1841.

Steel type punches are fixed at the extremities of separate radiating springs placed round the circumference of a horizontal wheel, which is fixed on the quickest wheel of the train of wheel work, and governed by a dead beat escapement actuated by electro-magnets. The paper band passes under the type, and the printing is performed by an electro-magnet which causes a hammer to strike the proper punch when it is opposite the paper. Alternate layers of white and blackened paper are employed to receive the impressions of the punch.

1544. Theiler's Synchronous Type Printer, 1854.

Two currents are required for each letter, one to start the instrument and another to print. The type-wheel returns to zero after the printing of each letter.

1546. Theiler's "Step by Step" Type Printer, 1863–64.

The type-wheel fixed on the axle of the escapement is controlled by reversals, and the printing is performed by the wheelwork, which is brought into action by an electro-magnet, the local battery circuit of the latter being closed by a vibratory arrangement, which does not act until the type-wheel is stopped at the letter required.

1547. Dujardin's "Step by Step" Type Printer, 1865.

A step-by-step instrument in which the escapement of the type-wheel is controlled by reversals, the electro-magnets acting on the anchor of the escapement being worked by a polarised relay in the line circuit, and a local battery.

The operation of printing is performed by an electro-magnet, the local battery circuit of which is closed by the anchor of the escapement at the end of each oscillation.

When the apparatus is running, and no key is depressed, the currents passing through the printing magnet are too short to cause it to attract its armature; but when a key is depressed and the type-wheel stopped at any particular letter the duration of the current is lengthened, the printing magnet attracts its armature, and an impression is made.

An electro-magnetic "cut off" arrangement is used in connexion with the printing magnet, which causes the printing current to be equal in duration whether the key be held down for a longer or shorter time.

This instrument was used for a short period on a wire between London and Edinburgh in 1865, by the Electric Telegraph Company.

1545. Hughes' Type Printing Instrument.

This is a purely synchronous instrument. The instruments at each end of the line run at the same speed.

There are as many keys on the key-board as there are letters on the type-wheel, and the keys are so arranged that by pressing any one of them a current is sent to line at the moment the letter it represents on the type-wheel

Z 2

is opposite the printing hammer. The current actuates the printing hammer by releasing a detent in the wheel-work which drives the hammer.

There is a peculiarity in the electro-magnet which releases the detent. Its cores are polarised by a permanent magnet sufficiently strongly just to retain the armature in contact with its poles. The current overcomes this polarity and releases the armature, which is drawn away by a spring. It is replaced by the wheel-work connected with the hammer.

1537. Andrews' Relay, for Hughes' Type Printing Instrument, 1868, used by the United Kingdom Telegraph Company.

This relay is adapted for relaying the short currents required for working the Hughes' instrument ; its peculiarity consists in the relayed currents being equal and independent of the line current.

Series of Apparatus for Recording the Steinheil or Single Needle Code.

The first recording telegraph was Steinheil's, completed in 1837. The code was formed of dots recorded in two lines, those in one of the lines being formed by the +, those in the other line by the − current. The code resembles that of Cooke and Wheatstone's single needle.

The system, if ever brought into practical use, was superseded by that of Morse, in which the code is comprised of marks of varying lengths.

But, as the dash occupies thrice the space and thrice the time of a dot, attempts have been made from time to time to introduce apparatus registering the Steinheil or single needle alphabet.

1564b. Unpolarised Recorder, for registering the single needle code.

1564c. Unpolarised Recorder, for registering the single needle code.

1564d. Polarised Recorder, for registering the single needle code.

1564e. Polarised Recorder, for registering the single needle code.

The negative current prints a *dot* (■), the positive current *two dots* $\left(\begin{smallmatrix}■\\■\end{smallmatrix}\right)$ transversely. The first represents the dot, the second the dash of the Morse alphabet. The letter A is written thus . . . . ■ .

1564f. Non-polarised Recorder, for registering the single needle code. Richard Herring's system.

The negative current prints a dot (▪), the positive current a transverse dash (|), so that the printing resembles the Morse alphabet. The letter A is written thus ▪ | .

1564g. Modification of Wheatstone's Automatic Transmitter, used in connexion with Richard Herring's system.

1564h. Modification of Wheatstone's Perforator, for his automatic system, arranged to punch the single needle code, used in connexion with Richard Herring's system.

Apparatus used by the Post Office in 1876.

1561. Wheatstone's A B C Instrument, Communicator, and Indicator.

This instrument is used on wires leased to private persons, and for the least important postal lines. It is well adapted for these purposes on account of not requiring skilled labour for its manipulation. The instrument is worked by a magneto-electric machine within the instrument.

Switches used with this instrument.

Alarum ,, ,,

1560. Single Needle Instrument.

Used on lines of a class superior to those on which Wheatstone's A B C is employed. It is but little liable to derangement, and requiring no adjustment is preferred to recording or acoustic apparatus for circuits of secondary importance.

1558. Siemens' Direct Writing Morse Inker, with Morse Signalling Key and Galvanometer.

This instrument is well known, being used in almost every country; its advantage over the Morse embosser consists in the greater legibility of the signals, which can be read in any light, and it requires much less power to work it. This form is used on lines of moderate length, where the direct line current is sufficient to actuate the electro-magnet without the assistance of a relay and local battery.

1559. Siemens' Local Ink-Writer.

Used in connexion with a relay and local battery on lines where the direct current is not sufficiently strong to actuate the direct ink-writer.

1565. Sounder.

This instrument is of American origin; it is used instead of the ink-writer on some of the lines of the Postal Telegraph Department, the dots and dashes of the Morse alphabet being read by sound from the clicking of the armature lever instead of from the paper band of the recording instrument; it possesses the advantage of not fatiguing the eye of the operator, and allows him perfect freedom for writing the message.

1562. Double Current Morse Key.

It can be arranged either to be worked as a single current Morse key, or to send double currents, as desired.

1564a. Double-current Key, with levers instead of the ordinary springs.

These keys send double currents or reversals. By the *marking* current the tongue of the distant relay is moved to the battery contact, by the reversed or *spacing* current it is brought back to its rest stop. This arrangement increases the range of adjustment, and as a current passes to line between the signals the effect of leakage from other wires is in great measure neutralised.

A switch, or a *zinc-sender* (1536), is required to alter the connexions when changing from sending to receiving.

1572. Siemens' Relay used with the ink-writer.

1563. Stroh's Polarised Relay, with Inducing Magnet.

This relay is similar in construction to the electrical portion of the Wheatstone " Receiver," No. 1557.

1564. New form of Non-polarised Relay.

This consists of two vertical electro-magnetic bobbins the cores of which are prolonged. Two soft iron needles are fixed on a vertical axle passing between the coils, one at the top and one at the bottom end of the axle. These needles are placed crosswise, so that when the cores are rendered magnetic the four poles of the bobbins tend to turn the needles on their vertical axis in the same direction. Two adjustments are provided, one the ordinary spring adjustment, the other for moving the needles nearer to or further from the poles.

This arrangement is much more sensitive than the old form of non-polarised relay with horseshoe electro-magnet and armature.

1566. Double Current Translator, used for relaying double current Morse signals.

The signals are relayed from the lever of the sounder, the latter being worked by a polarised relay in the line circuit. The necessary alterations to the connexions for restoring the apparatus to its normal condition when transmission in either direction has ceased is accomplished by an automatic switch called a " spacer," which consists of an electro-magnet, armature, lever, and stops. The local circuit of this magnet is closed by the relay which works the sounder when a marking current is sent, and by another relay placed in the line circuit when a spacing current is sent, so that the armature is continually held down while a station is sending.

The armature is prevented from rising during the short interval between the currents by a " shunt " placed across the terminals of the electro-magnet coil, which forms a path for the extra current direct when the local circuit is broken, and delays the demagnetisation of the iron core.

1567. Duplex Translator for relaying the signals on long circuits, worked on the duplex system.

This apparatus consists of two sets of duplex instruments combined.

1569. Testing-Box Galvanometer.

Used for testing wires ; the requisite changes of connexions for receiving and sending currents are made by a peg switch in front of the instrument.

1530. Switches or Commutators for making Changes of Wires in Circuits.

a. Earliest switch for double needle, fixed at Normanton about 1857.
b. Tumbler switch.
c. Umschalter or Universal switch.

1555. Wheatstone's Automatic Telegraph, 1867. The Perforator for preparing the paper ribbon.

There are three punches. The centre punch perforates small holes in the centre of the ribbon, which form a rack by which the paper is carried forward in the *transmitter*. The centre lever acts on this punch only ; the other two act upon the centre punch, but in addition perforate large holes for the Morse dot and dash.

They are struck either by a small mallet or by pistons actuated by compressed air.

1556. The Transmitter.

In Bain's and other automatic systems electrical contacts are made through the holes in the paper. In Wheatstone's the holes act like the *cards* of a jacquard loom, and *control* the movements of the mechanism by which the contacts are made.

The apparatus can be adjusted to produce signals in three ways.

(*a*.) By persistent currents, the *marking* current lasting during the whole time of the dot or dash ; the *spacing* or reverse current filling up the intervals between them. It acts precisely in the same way as the double-current key. (1562.)

(*b*.) By *intermittent* currents, where a signal is commenced by a short *marking* current, and ended by an equally short *spacing* current ; no current whatever being sent in the interval, the line being disconnected.

(*c*.) By *compensated* currents, where signals are commenced and ended as in (*b*.), but instead of the line being disconnected in the interval a large resistance is inserted, so that the last current sent continues to flow, but is weakened.

1557. Receiver, 1867. This is a very sensitive ink-writer.

The electrical portion consists of two vertical electro-magnetic bobbins, the iron cores of which are furnished with pole pieces. Two soft iron pieces or tongues are fixed on a vertical axle, and are magnetised by induction from a horseshoe magnet placed near them ; they play between the pole pieces of the electro-magnet. The marking disc is in connexion with an arm attached to the vertical axle, and is pressed against or removed from the paper ribbon according as the current passes in one direction or the other through the coils. The tongues of the electro-magnets are set neutral, so that they remain on the one side or the other after the current which has moved them has ceased to act.

The speed of the wheelwork can be regulated at pleasure to suit the rapidity of the signals.

1568. Translator for relaying the signals of Wheatstone's transmitter. (Post Office, 1876.)

It is somewhat similar to the double-current translator (1566). It differs in that the currents are sent forward by the tongues of the line relays themselves, instead of by sounders in a local circuit. These tongues are set so as to have no bias on either side.

The automatic switch is similar in principle to 1566,

1570. Regulator Clock and Apparatus for the distribution of Greenwich time current at provincial stations.

The current is sent on the wires used for message traffic, and it is therefore necessary to shift them from the telegraph apparatus to the time apparatus. To ensure regularity, this must be done automatically by means of a clock.

The wire by which the signal is to be received from the central station, as well as the wires on which the time signal is to be distributed, are connected to the levers of an electro-magnetic switch (No. 1571).

About 20 seconds before 1.0 p.m. a local current sent from the clock actuates this switch, causing the levers to be shifted from the stops which are in connexion with the ordinary instruments to those connected with the relay which fires the gun. When the time current passes through the relay it closes a local circuit in which the gun fuze and the battery are included.

About 20 seconds after the time current has passed, the local current actuating the electro-magnetic switch is interrupted by the clock, and the levers of the switch to which the line wires are attached are again placed in connexion with their respective instruments.

The clock itself is corrected daily in the following way :—

It is kept at a slightly gaining rate, and is stopped automatically by a detent which acts on a pin projecting from the escape-wheel when the hands point to 1.0 p.m. The pendulum, however, continues swinging.

When the time-current is received the outgoing current from the time-relay passes through an electro-magnet inside the clock which liberates the detent.

The clock is so arranged that a second electro-magnetic switch may be actuated, and the time current distributed in a similar way at 10.0 a.m.

1571. Automatic Time-Switch for switching line-wires from their instruments and connecting them to the time-relay.

1573. Galvanometer for showing incoming time-current.

1574. Galvanometer for showing outgoing time-current.

1553. Early form of " Chronofer," or time-current sending apparatus, used in London about 1852. (Incomplete.)

1575. Lightning Protectors or Dischargers (various).

a. Varley's Lightning Protector (original form).
b. „   „
c. „   „
d. „   „
e. „   „
f. „   „
g. „   „
h. „   „
i. „   „
k. „   „   Varley's vacuum (modern form). Tube
l. „   „   protector, vacuum tubes in water-tight
m. „   „   compartment for outdoor use.
n. „   „   Latest form of fine wire protector, Post
Office pattern, consisting of a metallic bobbin, electro-nickel plated, in connexion with the earth. It is wound with a fine wire covered with silk and passed through melted paraffin wax, the wire being placed in the line circuit. The metallic bobbin is carefully lacquered to secure insulation in damp situations.

1576. Insulators. Various.

Original ring insulator.
Cone insulator, original form (1845).
„   „   later   „
„   „   form for replacing broken insulators.
Walker's double-cone insulator (brown).
„   „   „   (white).
Insulator (1847), in which the bolt by which the wire is suspended is insulated by mastic cement in the cavity at the top of the insulator.

Brown earthenware insulator, with zinc cap (1850).
Do., with groove for wire.
Do., glass instead of earthenware.
(1851) brown earthenware.
     ,,   glass.
Clark's invert insulator, corrugated.
Varley's double-cup insulator, fitted.
Ditto, in separate parts, 1862.
Varley's No. 11 wire double-cup insulator.
Brown earthenware insulator called (Z).
Double shed invert compressed ware.
Single   ,,   ,,   ,,
South Devon invert insulator (brown).
   ,,   ,,   ,,   (white).
Andrews' invert with ebonite inner shed.
   ,,   ,,   bolt tube   ,,
   ,,   ,,   bolt tube and shed.
Ebonite insulator, single shed.
Old form of shackle.
Bright's shackle.
Modern terminal insulator.

TELEGRAPHIC APPARATUS CONTRIBUTED BY THE IMPERIAL GERMAN TELEGRAPH DEPARTMENT.

1608. Copy of the Electro-chemical Telegraphic Apparatus of S. T. Sömmering, the first German telegraphic apparatus, constructed in 1809.

1609. Drawings of the Russian Councillor Schilling's (of Kannstadt) **Apparatus,** being the first needle telegraph (two sheets).

1610. Copy of the Electro-magnetic Telegraphic Apparatus of Gauss and Weber, of Göttingen, made and used from 1833 to 1838. It consists of—

(a.) The signal sender (by means of induced current).

(b.) The signal receiver (a magnetic rod with multiplying coil and mirror).

(c.) Telescope for reading off the deflexions of the magnet; with stand.

1611. Copy of the Electro-magnetic Telegraphic Apparatus of Steinheil, in München, 1837.

1612. Electro-Magnetic Indicator, combined with type-printing apparatus of Siemens, of Berlin, constructed in 1846. Patented in Prussia.

1613. Bell Machinery for Railways (Siemens), 1847.

1614. First Gutta-percha Press for covering conductors with insulating material without a seam. After Siemens' model, and constructed by Fonrobert and Bruckner. By machines after this model numerous subterranean lines in Germany and Russia were constructed from 1847 to 1851, and submarine lines are now made in the same manner.

1615. Plate Lightning Conductor, employed first by Siemens in 1848 between Eisenach and Frankfort

1616. Double Style Apparatus, with writer in relief (Siemens).

Constructed by Siemens in 1853. The lines made by Siemens and Halske, the Warsaw-Petersburg, and other telegraph lines, were at first worked with apparatus of this system.

1617. Polarised Relay for the above. Oldest form without a steel magnet. The magnetic induction of the core and armature is effected by a branch current of the local battery. (Siemens, 1851.)

1618. Key for the Polarised Relay.

1619. Magneto-electric Current Generator, with 28 pairs of magnets.

1620. Dynamic Mine Exploder (Siemens, 1850).

1621. Polarised Relay for Double-style Apparatus, with steel magnets (Siemens, 1852).

1622. Hand-puncher for Self-acting Sender.

1623. Transmitter for Self-acting Sender.

1624. Receiver for Self-acting Sender (Morse with suspended magnets).

1625. Relay for Self-acting Sender (with suspended magnets).

Employed by Siemens in 1855 in the first experiments made to employ short alternating currents of equal duration for production of Morse writing with aid of a polarised relay.

1626. Mile Resistance, with movable steel lever; Siemens, 1854.

1626a. Mile Resistance.

1627. Original Colour-writer of John, Prague, 1854.

1628. Relay with suspended magnets and double coils for conversation. Frischen and Siemens, 1854. Patented in Prussia, with resistance scale divided according to miles of copper wire, 1 diameter. As made by Siemens since 1848.

1629. Polarised Relay, with a horseshoe electro-magnet and steel armature. Siemens, 1855. This form is still in use.

1630. Induction Coil.

1631. Induction Key.

1632. Apparatus System for submarine conduction. Siemens. Constructed for the Red Sea Telegraph Cable.

1633. Back Current Discharger for submarine conduction. Siemens, 1857. Used first in the Red Sea.

1634. Colour-writer by Siemens. First construction. A small colour arm, fastened on a universal joint, dips into an open reservoir of adjustable level. Patented in England in 1862.

1635. Hair Needle Galvanoscope of Siemens, 1869.

GENERAL TELEGRAPH APPARATUS.

1576aa. Telegraph Apparatus, &c.
General Direction of Russian Telegraphs.

1. Atlas, containing details of the construction of the Central Telegraphic Office in St. Petersburg.
2. Lightning conductor, for cable.
3. Lightning conductor, for station.
4. Insulator, with double cup (large model), "Korniloff's."
5. Insulator, with double cup (small model), same make.
6. Insulator, with double cup (large model), "Bélotine's."
7. Insulator, with double cup (small model), same make.
8. An element, "Meidinger's."

1750a. Telegraphic Instruments.
Messrs. Siemens Brothers.

1. A B C telegraph instrument with finger keys (under glass cover to show working parts).
A magneto-electric instrument not requiring battery; used for house, private, and railway telegraphs.
2. A similar instrument in wooden case.
3. A B C telegraph instrument with handle movement, in round case, not requiring battery; used for house, private, and railway telegraphs.
4. A similar instrument in square case.
5. Standard ink recording telegraph instrument with self-starting and self-stopping arrangement; for terminal stations.
6. A similar instrument arranged for translation.
7 Portable telegraph sounder for military purposes (Major Malcolm's pattern), non-polarised.
8. Polarised circular relay with double contact.
9. Bi-polar relay with double contact under glass cover.
10. Portable condenser of ½ microfarad capacity, with 2 terminals and short circuit peg.
11. Sub-divided condenser of 1, 2, 2, 5, 10, 10, microfarad capacity, with 4 spring pegs and one short circuit peg.
12. Sub-divided condenser of 10, 10, microfarad capacity, with 3 terminals and short circuit pegs.

13. Subdivided condenser of $\frac{1}{4}$, $\frac{1}{2}$, 1, 2, microfarad capacity, with spring pegs and short circuit peg.

14. Sliding rheostat of 10,000 units resistance.

15. Sliding rheostat of 8,400 units resistance.

16. Sliding rheostat of 1,300 units resistance.

17. Resistance coil of 1,000 ohms.

18. Portable resistance bridge for electrical testing, with 3 pairs of proportional coils, viz., 10, 100, 1,000, 10, 100, 1,000, and a set of 16 coils of the aggregate resistance of 10,000 units. All the coils are of platinum-silver alloy. The apparatus also includes battery key, galvanometer key, and reversing commutator.

19. Plate lightning protector for telegraph stations. The line plate and earth plate are of brass, transversely grooved and mounted on marble slab.

20. Lightning protector for telegraph stations. A fine silk covered wire forms part of the line circuit, and is wound round a grooved copper cylinder which is to be placed in connexion with earth.

21. Tubular lightning protector for telegraph stations. The line tube and earth tube are transversely grooved.

22. Universal galvanometer (portable), which serves for the following purposes : —

1st. Measuring electrical resistances, the instrument being arranged as a bridge :

2nd. For comparing electromotive forces. Professor E. du Bois Reymond's modification of Poggendorff's compensation method being used :

3rd. For measuring the intensity of a current, the instrument being simply used as a sine galvanometer.

The instrument consists of a sensitive galvanometer which can be turned in a horizontal plane, combined with a resistance bridge (the wire of which bridge instead of being straight is stretched round part of a circle). The galvanometer has an astatic needle suspended by a cocoon fibre, and a flat bobbin frame wound with fine wire. The needle swings above a cardboard dial divided in degrees; as, however, when using the instrument the deflection of the needle is never read off, but the needle instead always brought to zero, two ivory limiting pins are placed at about 20 degrees on each side of zero. The galvanometer is fixed on a graduated slate disc, round which the platinum wire is stretched. Underneath the slate disc three resistance coils of the value of 10, 100, and 1,000 Siemens' units are wound on a hollow wooden block which protrudes at one side ; this projection carries terminals for the reception of the leading wires from the battery and unknown resistance. The adoption of three different resistance coils enables the measuring of large, as well as small resistances, with sufficient accuracy.

The whole instrument is mounted on a wooden disc which is supported by three levelling screws, so that it may be turned round its axle. On the same axle a lever is placed which bears at its end an upright arm, carrying a contact roller. This roller is pressed against the platinum wire round the edge of the slate disc by means of a spring acting on the upright arm, and forms the junction between the A and B resistance of a Wheatstone's bridge, which resistances are formed by the platinum wire on either side of the contact roller, one of the three resistance coils forming the third resistance of the bridge.

23. Sine galvanometer (portable). The magnetic needle moves on a steel point within the wire coil, and graduated segment inside a circular brass case which turns in a graduated circle.

24. Detector galvanometer (portable) with 3 coils.

25. Detector galvanometer with oscillating magnet

26. Differential galvanometer, constructed on the double shunt principle.

27. Sounder telegraph instrument with split armature.

28. Set of ink recording instruments for duplex telegraphy, according to the differential system.

29. Double current Morse key with automatic switch for sending and receiving position, to be used with a single battery.

30. Morse key for testing purposes.

31. Double current Morse key for testing purposes.

32. Double current Morse key for signalling.

33. Discharging key.

34. Magneto induction bell with inductor of 6 magnets, for signalling on railways, mines, inclines, &c. without the use of batteries.

35. Magneto induction bell, similar to the above.

36.   Ditto   ditto.

37. Magneto inductor of 6 magnets, without bell.

38. Dynamo-electric exploder for low tension (quantity) fuses, used for blasting and torpedo firing purposes.

This apparatus consists of an ordinary Siemens' armature, which is caused, by the turning of a handle, to revolve between the poles of an electro-magnet. The coils of the electro-magnet are in circuit with the wire of the armature, and the residual magnetism of the electro-magnet cores excites at first weak currents, which pass into the electro-magnet coils, increasing the magnetism of the core and inducing still stronger currents in the armature wire, to the limit of magnetic saturation of the iron cores of the electro-magnets.

The current thus generated in the machine is sent into the wire or cable leading to the fuse by the opening of a shunt key. The fuse offers a great increase in its resistance at some point, by the interposition of a badly-conducting substance ; the consequent action is that the piece of bad conductor is highly heated, causing ignition of the explosive substance contained in the fuse.

The coils, whether of armature or electro-magnets, of the quantity exploder are wound with wire of large diameter to a total resistance of 8 to 10 units in about 2,000 windings ; electric currents possessing great heating power, but of low intensity, are therefore generated by this machine.

39. Dynamo-electric exploder for high tension fuses, also called spark fuses (intensity).

The construction is similar to the above, but this exploder has its coils, both of the armature and the electro-magnets, wound with fine wire to a total resistance of 2,000 to 2,500 Siemens' units, in about 17,000 windings. Upon causing the armature to revolve, currents are generated of great intensity, and passed by the automatic action of the machine into the wire or cable leading to the fuse, when an electrical spark passing between the separated conductors in the fuse inflames the explosive priming.

40. Case of electric cable samples.

41.   Ditto   ditto.

42.   Ditto   ditto.

1576ab. Wheatstone's Improved Automatic Telegraph System. Receiver and transmitter.

The British Telegraph Manufactory, Limited.

Sounder.

The British Telegraph Manufactory, Limited.

Wheatstone's Cryptograph.

The British Telegraph Manufactory, Limited.

Reversing Switch.
The British Telegraph Manufactory, Limited.

Double Plate Lightning Guard.
The British Telegraph Manufactory, Limited.

S. A. Varley's Lightning Guards (two on board).
The British Telegraph Manufactory, Limited.

1576c. Wheatstone's large Dial Indicator, for use with *A B C* instruments.
The British Telegraph Manufactory, Limited.

1576h. Double Current Key, for Wheatstone's Automatic Telegraph. *The British Telegraph Manufactory, Limited.*

1576i. Double Current Key, for Wheatstone's Automatic Telegraph. *The British Telegraph Manufactory, Limited.*

1576j. Rheostat for Duplex Telegraphing.
The British Telegraph Manufactory, Limited.

1576k. Testing Key.
The British Telegraph Manufactory, Limited.

1576l. Testing Key.
The British Telegraph Manufactory, Limited.

1576m. Testing Key for Discharges.
The British Telegraph Manufactory, Limited.

1576n. Lightning Guard.
The British Telegraph Manufactory, Limited.

1576o. Polarised Relay with Wheatstone's Patent Chain Adjustment.
The British Telegraph Manufactory, Limited.

1576p. Wheatstone Magnetic Counter with Magnet and Wheel showing Mode of Application.
The British Telegraph Manufactory, Limited.

1576q. Magneto-Electric Clock with Steam Engine Movement. *The British Telegraph Manufactory, Limited.*

1576r. Wheatstone's Roman Type Printer for use with the A B C Telegraph.
The British Telegraph Manufactory, Limited.

1576s. Wheatstone's Portable A B C Telegraph, specially arranged for Military Purposes.
The British Telegraph Manufactory, Limited.

1576t. Wheatstone's Alphabetical Magneto-Electric Telegraph. *The British Telegraph Manufactory, Limited.*

1576u. Polarised Morse with Wheatstone's System of Adjustment and Inking.
The British Telegraph Manufactory, Limited.

1576v. Translating Morse with Wheatstone's System of Inking. *The British Telegraph Manufactory, Limited.*

1576w. Resistance Box.
The British Telegraph Manufactory, Limited.

1576x. Resistance Box.
The British Telegraph Manufactory, Limited.

1576y. Large Standard Resistance Box.
The British Telegraph Manufactory, Limited.

1576z. Commutator.
The British Telegraph Manufactory, Limited.

1582. Meidinger's Galvanic Element.
Prof. Dr. Meidinger, Carlsruhe.

1583. Meidinger's Galvanic Element (larger size).
Prof. Dr. Meidinger, Carlsruhe.

For working telegraphs, electric clocks, and bell apparatus, also for electroplating in silver or gold.

1584. Box, containing a battery of 21 elements for producing constant currents for medical use. Entirely new construction of the exhibitor. *Prof. Meidinger, Carlsruhe.*

The glasses are filled with a solution of bichromate of potash and sulphuric acid in water. The gutta-percha covered rods are introduced into the wide glass, and the wires into the mercury tube, with which is connected a platinum wire, leading to the platinum sheets. After working the metal connexions are taken out of the glasses. The liquid is allowed to remain in the glass till it is exhausted; it may serve for 100 one-hour operations. In transport, the liquid does not escape, even in violent shaking. The stoppers with inserted glasses are never taken out. Filling is accomplished by pouring the liquid through the wide tube; emptying by inverting the glass, when the liquid flows out through the fine air passages. The gutta-percha under the zinc is gradually worn away in proportion as the (amalgamated) zinc is dissolved; only about a millimètre of free surface of zinc is necessary. The narrow trough is filled with very dilute zinc vitriol, and serves for regulating the strength of current by displacement of the zinc pole. On drawing out a peg, the liquid escapes below into a central vessel.

1585. Colour-writer (Morse), North German pattern.
L. E. Schwerd, Carlsruhe.

This is furnished with contact-arrangement for translation, and with commutator for sending the current through both electro-magnet coils either successively or simultaneously; in the latter case the resistance of the coils should be only $\frac{1}{4}$.

1586. Box Relay, N. German pattern.
L. E. Schwerd, Carlsruhe.

1587. Key, N. German pattern.   *L. E. Schwerd, Carlsruhe.*

1588. Colour-writer (Morse), S. German pattern.
                                    L. E. Schwerd, Carlsruhe.
This apparatus (much used in Baden and Bavaria) is adapted for either constant or working current.

1589. Pendulum Relay, S. German pattern.
                                    L. E. Schwerd, Carlsruhe.

1590. Galvanometer, S. German pattern.
                                    L. E. Schwerd, Carlsruhe.
The deflections of the needle correspond closely between 15° and 35° to the strength of current.   The instrument is sensitive and easily portable.

1591. Key, S. German pattern.   *L. E. Schwerd, Carlsruhe.*

1591a. Andrew's Guillotine Relay.   *W. Andrews.*
The armature is a permanent horseshoe magnet, placed horizontally, and playing between two electro-magnets, the one above the other below the armature.
The upper electro-magnet is of the ordinary form, the lower one has the cross-piece removed and the poles of a horseshoe permanent magnet connected to its cores instead.   The current passes through both coils, which are so connected that the electro-magnets act oppositely upon the permanent horseshoe armature, one attracting while the other repels.
The upper electro-magnet is movable in a vertical direction, and can be shifted up and down by an adjusting screw.   This forms the chief adjustment, but a spiral spring is also provided for the purpose.

1591b. Morse Ink Writer, with **Relay** for translating connexions.   *Warden, Muirhead, and Clark.*

1607. Uno-electric Pile with Galvanometer.
                              Landsberg & Wolpers (Hanover).

1635a. "Edison's Electric Pen," and **"Autographic Press."**   *Thos. D. Clare.*
This pen consists of a small electro-magnetic engine on the top of a holder, which is used as a pen, and works a needle which pierces the paper, making 5,000 to 6,000 fine holes per minute, so that in writing such is the rapidity of the motion of the needle that the point does not drag or tear the paper.
The pierced paper or " stencil " is placed in a frame, and an inked roller is passed over, which fills the fine perforations with ink.   A sheet of paper is then placed below the written paper or stencil, and the roller is again passed over once or twice, when a perfect fac-simile is obtained.
These fac-similes can be produced at the rate of four to six per minute, and one writing or stencil will suffice to print 1,000 copies.

1657. Original Five-Needle Telegraph Dial.
                        The Council of King's College, London.

1658. Two-Needle. Telegraph.
                        The Council of King's College, London.

1659. Three A B C Telegraph Sending Instruments, showing gradual improvements.
The Council of King's College, London.

1660. Horizontal Sending and Receiving A B C Telegraph Instrument.
The Council of King's College, London.

1661. Two A B C Telegraph Receiving Instruments.
The Council of King's College, London.

1662. Punching Instrument, for preparing the paper for the transmitter. *The Council of King's College, London.*

1663. Transmitter.
The Council of King's College, London.

1664. First Electric Key, constructed by Sir Charles Wheatstone. *The Council of King's College, London.*

1665. Wheatstone's First Relay Instrument.
The Council of King's College, London.

1666. Printing Telegraph.
The Council of King's College, London.

1666a. Charge and Discharge Key.
Warden, Muirhead, and Clark.

1666b. Reversing Key. *Warden, Muirhead, and Clark.*

1666c. Short Circuit Key. *Warden, Muirhead, and Clark.*

1666d. Testing Key, for testing copper resistance. The battery circuit is completed first, then the galvanometer circuit by one movement of the finger. *Warden, Muirhead, and Clark.*

1666e. Perforated vitreous chamber, for interior of cell No. 1666d. *Warden, Muirhead, and Clark.*

1666f. Pair of Brequet Crossley A B C Telegraph Instruments. *Louis John Crossley.*

1558a. Model of a Morse Printing Telegraph (blue writer), with key, galvanometer, and paper stands.
A. Herbst, Berlin.

1667. Apparatus specially intended **to illustrate the Morse System of Telegraphy** in primary and secondary schools. *J. Cauderay, Lausanne.*

For the clockwork is substituted a small crank, which can be turned by hand. This apparatus works by means of a single Bunsen cell, of small size. Its low price, 40 frs., places it within the reach of every school.

40075. A a

1667a. Arm of Insulators, for land wires,
Warden, Muirhead, and Clark.

No. 1. Andrews white porcelain, ordinary, No. 8.
No. 2. Andrews white slot and hole at top.
No. 3. Andrews white slot.
No. 4.   Do.   brown, No. 8.
No. 5.   Do.   do.   No. 11.
No. 6.   Do.   stoneware, No. 8.
No. 7. Brown porcelain, Post Office pattern.
No. 8. Clark's invert insulator.
No. 9. Spanish.
No. 10. New South Wales.

No. 11. Spanish larger hook.
No. 12. Varley's, No. 9.
No. 13. Indian Pattern.
No. 14. Warden's.
No. 15. Relays, No. 11.
No. 16. **U** shackle.
No. 17. Wardens perforated iron hook for iron poles.
No. 18. **Z** insulator.
No. 19. Bell hood (small).
No. 20.   Do.   (large).
No. 21. Bright's shackle.

1667b. Arm, with eleven **Insulators.**
Warden, Muirhead, and Clark.

1667c. Arm, with fourteen **Insulators.**
Warden, Muirhead, and Clark.

1669. Submarine Telegraphic Cables. *Geminiano Zanni.*

In the above specimens the conductor (consisting of one or more copper wires) is enclosed within a series of soft iron wires, which unite to form a strand or core by passing the combined series of wires through a bath of molten tin, or any other comparatively soft metal. The core thus combined into a solid body is protected from corrosion. It is then coated with gutta percha, or any insulating material, over which tin foil is wrapped to exclude moisture. A band of hemp (immersed in tar) is wound round the core to protect the metal foil from injury or corrosion.

1670. Ward's Dead Beat Telegraph.
William Sykes Ward.

Two delicate coils of fine wire are suspended on points around the poles of powerful permanent magnets; the motions are limited so as to give distinct indications without tremulous vibrations. Patented in 1847.

1671. Signalling Key for Telegraphy, to send into a line positive or negative currents.   *Elliott Brothers.*

1671a. Small registering Electro-Magnet, capable of supplying 500 signals per second.
Marcel Deprez, 16, Rue Cassine, Paris.

1675. Apparatus for sending **simultaneously** in opposite directions **Telegraphic Despatches,** between two stations with one single wire in the line.
Elie Wartmann, Professor of Natural Philosophy at the University of Geneva.

A full description is to be found in the number for March 1856 of the " Bibliothèque universelle de Genève." The keys, when worked, close two different circuits, which are to be of such directions and intensities that they neutralise each other in the two sets of coils of the branches of the electro-magnets.

1678. Apparatus for **transmission** of two **simultaneous Telegrams** in the **same Telegraphic Wire,** from one station to another.

Elie Wartmann, Professor of Natural Philosophy at the University of Geneva.

A description of the method and apparatus is printed in the " Archives des Sciences physiques et naturelles," Geneva, November 1860.

1679. Apparatus for permitting any two **Stations** on the same **Telegraphic Wire** to communicate immediately without the aid of the intermediate ones, so that the despatch remains secret between them.

Elie Wartmann, Professor of Natural Philosophy at the University of Geneva.

For complete description see the " Bibliothèque universelle de Genève," May 1853. It consists of the following parts :—
a. Sender ; b. Regulator ; c. Indicator ; d. Interruptor. The principles of the instrument are: 1. To break the communication of every intermediate station with the earth ; 2. To maintain the contact of their electro-magnets with the poles in order to facilitate the transmission of direct currents between the two stations to be united ; 3 To let the officer in the station from which the despatch originates know that it is really received in the proper one ; 4. To cut off immediately the communication after the receipt of an answer ; 5. To reinforce the current of the battery in the first station by means of the batteries in all intermediate stations.

1238. The First Instrument used to **Electrify** the Ink-Bottle of the Siphon Recorder. *Sir W. Thomson.*

This was the first instrument used for producing the electricity required to electrify the ink-bottle of the siphon recorder. What is now known as the mouse-mill, referred to in Clerk Maxwell's " Electricity and Magnetism," is a modification of this instrument, driven by intermittent electro-magnetic force. Described in Thomson's reprint of " Papers on Electro-statics and Magnetism," xxiii. 416–419.

1239. Modified form of 1238. *Sir W. Thomson.*

1240. Further developed form of **1238.**

Sir W. Thomson.

One of the applications of this is to multiply indefinitely the electro-static indications obtainable from a feebly electrified body on the same principle as Nicholson's Revolving Doubler, and as the rotating induction instrument exhibited by Mr. C. F. Varley at the International Exhibition of 1861.

1683a. Automatic Lightning Guard.

Warden, Muirhead, and Clark.

1683a. Eggington Automatic Lightning Guard, for protecting submarine cables from the effects of lightning and powerful earth currents. The land wire is automatically put to earth by the currents themselves and the cable left safely insulated.

Warden, Muirhead, and Clark.

1690a. Double Indexed Telegraphic Post, with alphabetical receiver, indicator, and printer; manipulator on Chambrier's system.   *M. Deschiens, Paris.*

The whole of this apparatus is enclosed in a case for protection.

1691. Stock Exchange or **Bourse Telegraph Instrument.**   *Siemens and Halske, Berlin.*

A battery is employed only at the central station, from which the messages are sent by depressing lettered finger keys on the transmitter, which are so arranged that the letters most commonly occurring are nearest at hand. Only one line wire is required. The receiving instrument prints in Roman type, and has a double type wheel carrying figures and letters, so that by depressing a changing key in the transmitting apparatus letters or figures may be printed at will. The type wheel automatically returns to its zero position after each printed signal, and is therefore always in adjustment.

1692. Magnetic Telegraph Apparatus.
                                    Siemens and Halske, Berlin.

The transmitter may be any dial instruments in which the currents are generated by magnets. The receiving instrument prints upon a paper riband from a double type wheel, at the same time indicating on a dial the letters sent. The type wheel is of vulcanised caoutchouc instead of metal.

1693. Two Automatic Cylinder Transmitters and printing instruments working **Duplex,** *i.e.,* so arranged that messages may be transmitted from both stations simultaneously.
                                    Siemens and Halske, Berlin.

The transmitter is an automatic finger key instrument, in which the letters to be transmitted are formed in Morse code upon a cylinder of wires, these wires being pushed forward by levers connected with the finger keys, so that by depressing the keys corresponding to the letter to be sent, the complete Morse signal is formed in the wires, which afterwards are brought by revolution of the cylinder into contact with levers connected to the battery and line.

1694. Automatic Type-printer, including transmitter, receiving instrument, and relays.   *Siemens and Halske, Berlin.*

The transmitter is an automatic finger key instrument, in which the currents to be transmitted are controlled by a cylinder of wires, these wires being pushed forward by levers connected with the finger keys, so that by depressing the key corresponding to the letter to be sent the necessary combination is formed in the wires, which afterwards are brought by revolution of the cylinder into contact with levers connected to the battery and line, and by this means the numbers of currents requisite to bring the type wheel of the receiving instrument to the desired letter are sent into the line. The relay consists of a double armature of aluminium bobbins wound with silk covered wire, which are drawn into or expelled from an intense magnetic field, according to the direction of the current received in the coils from the transmitting instrument, the bobbins causing contact to be made by their movement. The receiving instrument has a double type wheel worked by an escapement, which is actuated by two electro-magnets brought into play by the currents received from the relay.

1694a. Printing Telegraph, invented by M. H. Jacobi, academician, in the year 1850. Two uniform apparatus for two stations.

Physical Science Cabinet of the Imperial Academy of Sciences, St. Petersburg.

1694e. Charge and Discharge Keys (new form).

Warden, Muirhead, and Clark.

1701. Magneto-electric Morse Ink Printing Telegraphic Apparatus. *Geminiano Zanni.*

The above instrument, consisting of the mechanical clockwork as in an ordinary electric Morse apparatus, has in addition a compound magnet arranged with an induction coil. The motion is given to the coil for the production of the magnetic current by the same clockwork as drives the paper forward; thus dispensing with the use of the voltaic battery.

1703. Pair of Undemagnetisable Coils, designed in 1866. *S. Alfred Varley.*

The magnetic needles inside these coils are made of soft iron rendered magnetic by induction, instead of being made of tempered steel magnetised.

As the needles are magnetic only by virtue of the permanent magnets in their neighbourhood, the influence of powerful currents, induced by lightning, upon them, can only be momentary; consequently the telegraphic circuit is not liable to be interrupted by the demagnetisation of the needles, which so frequently occurs in needle telegraphs of the ordinary construction.

Extract from the Handbook of Telegraphy, by R. E. Culley, Engineer-in-chief of Telegraphs to the Post Office, 5th edition, pages 199, 200 :—

" The greatest improvement which has been made in the needle instrument is the introduction of an induced magnet of soft iron for the needle in place of the permanent steel magnet."

1704. Instrument for sending double-curb signals into submarine cables. *Sir William Thomson.*

This instrument was constructed in 1858, but not completed in time for trial on the transatlantic cable before its failure in the September of that year.

1705. Electro-magnet Relay for reaction produced by the magnetism emanating without regulation. Extremely rapid, and produces a novel double effect, something like the cut of a whip.

M. Guyot d'Arlincourt.

1706. Relay for Rapid Transmission.

M. Guyot d'Arlincourt.

Application of this relay for the transmission into the rapid apparatus. In this same apparatus two other relays called whipcuts discharge the line with each emission of the current.

1707. Autographic Apparatus. *M. Guyot d'Arlincourt.*

Employment of the circular vibration of the branches of a diapason for regulating the working of the apparatus. New correcting system for ascertaining at once the isochronous action of two apparatus that are not regulated perfectly one upon the other.

1708. Dial Printing Telegraph.   *M. Guyot d'Arlincourt.*

Printing apparatus with electro-magnet of two planes and two effects; one of the planes at the open ends of the electro-magnet governs the receiver, the other at the heel produces the impression of the letter. When the positive and negative currents are thrown alternately and without interruption, the two planes act together, and the local circuit of impression is open. On the interruption of any current, the plane at the heel acts alone, the two planes are no longer united, and the printing takes place.

1709. Morse Sounder.   *Dumoulin Froment.*

1709a. Morse's " Sender and Receiver."
Dumoulin Froment, Paris.

1709d. Photograph of Telegraph Punching Apparatus.   *C. H. G. Olsen, Christiania, Norway.*

1709e. Photograph of Telegraphic Printing Apparatus.   *C. H. G. Olsen, Christiania, Norway.*

1709f. Photograph of Drawings of Telegraphic Apparatus.   *C. H. G. Olsen, Christiania, Norway.*

1709g. MS. detailed Description of Automatic Printing Apparatus.   *C. H. G. Olsen, Christiania, Norway.*

1709h. Drawing of Telegraphic Apparatus.
C. H. G. Olsen, Christiania, Norway.

1709i. Two Copies of Pamphlet on Electric Telegraphy.   *Senor F. Riano.*

1708a. Automatic Printing Telegraph, constructed by C. H. G. Olsen, Christiania.
C. H. G. Olsen, Optician, Christiania.

This apparatus consists of two principal parts.

1. *The punching apparatus,* on which the messages are prepared thus:— By depressing one of the keys of a keyboard in the apparatus, holes will be punched in a paper ribbon, corresponding to letters and signs marked on the keys. Then this ribbon is rolled up and applied in—

2. *The printing apparatus,* where it is run over a metallic roller. Whenever a hole appears, a steel pin will make contact with the roller and send a current to the line. The current will liberate the armature of a magnet, by which a wheel with engraved signs and letters will revolve, and another paper ribbon, applied under this wheel, is lifted into contact with it. If the wheel has its right position to the writing on the perforated paper, the letter desired will appear printed on the paper. (2 photographs, 7 drawings, and a more detailed description follow.)

1709c. Contact Breaker, with two discs, which can be displaced, and sliding springs on glass inlets.
Prof. Dr. Dove, Berlin.

(*See* Transactions of the Royal Academy of Sciences at Berlin, 1841, p. 296.)

1695a. Model of Aerial Telegraph, by Chappe.
Telegraphs Department, Paris.

1695b. Model of Aerial Telegraph, by Monge.
Telegraphs Department, Paris.

1695c. Model of Aerial Telegraph, by Bréguet and Bettancourt. (This model belongs to the Conservatoire des Arts et Métiers.) *Telegraphs Department, Paris.*

1695d. Model of the First French Electric Telegraph, with two needles. *Telegraphs Department, Paris.*

1695e. Electric Telegraph, by Hughes, with latest improvements. *Telegraphs Department, Paris.*

1695f. Autographic Apparatus, by Meyer, constructed by Mr. Hardy. *Telegraphs Department, Paris.*

1695g. Froment's Dial Telegraph.
Telegraphs Department, Paris.

SPECIAL APPLICATIONS.

1576d. Electric Fire Alarum, for warehouses, &c.
The British Telegraph Manufactory, Limited.

1696. Electric Alarum for indicating the heating of Axle Bearings. *C. & E. Fein, Stuttgart.*

This is intended to show the heating of an axle in its bearings, and for this purpose is placed in an opening in the cap piece, so that its lower end touches the axle. As soon as the temperature rises, contact is made and a bell rings. The metal of the bearings, plummer block, &c. serve instead of return wires, so that only one insulated conductor is required.

1697. Group of **Apparatus,** comprising an electric control clock, with six locking signal studs. *C. & E. Fein, Stuttgart.*

This is for use in manufactories, public buildings, &c. The dial is a paper disc, which is changed daily, and revolves once in 12 hours; it contains as many rings as there are stations. By pressing down a knob at any of the stations an electro-magnet inside the clock draws its anchor forward, and a mark is made on the dial. As the dial is very large, minutes can be read off easily. When a new dial is put on, the knob in the centre is to be turned so as to set the clock to the right time. The clock will go for several days, but it is better to wind it up daily. The signalling knobs can be kept locked up, and may be opened by the watchman with a key which is common to the whole apparatus.

1698. Battery Case, containing six large Meidinger's cells for working the above apparatus. *C. & E. Fein, Stuttgart.*

1699. Lamp with Double Screen, slide and adjustable lens, scale stand, scale, &c. To be used with reflecting galvanometers generally. *Elliott Brothers.*

1700. Magneto-electric Railway Passenger Signalling Apparatus.   *Geminiano Zanni.*

This is for enabling the passenger in any part of a railway train to communicate with the guard or driver by simply touching a button, without the use of a voltaic battery.

1702. Magnetic Bells and Signals.   *Geminiano Zanni.*

In the above the mechanical electric bell moved by clockwork, at present in use, is dispensed with. The motion of the coils is caused by moving a bell-pull lever half a revolution. Self-acting apparatus may, by this invention, be arranged for giving alarm in case of fire or burglary without the use of a voltaic battery; by adopting clockwork to set the coil in motion, the bell or signal would act on opening a door or window, or by the heat resulting from fire.

1673. Electric Domestic Bells (three).   *J. Round.*

1674. Electric Detective Bell.   *J. Round.*

1728a. Electrical Communicator.   *Garnham & Co.*

This invention enables a perfect communication to be maintained between guards and drivers, and passengers and guards, by means of electricity. The want of a thorough system of communication between guards and drivers on trains has long been felt, as it is well known that the guard in the after part of a train cannot at all times hear the driver's whistle. By means of this electrical communication a perfect code of signals can be maintained. In the case of danger, passengers can readily give an alarm to the guard, and this passenger signal continues ringing until the guard replaces it, and it therefore indicates the compartment of a train in which the alarm bell was sounded. The apparatus consists of a simple battery with the necessary communicators; it is simple, inexpensive, and not liable to get out of order, and is also readily applied to all existing trains.

1576g. Magnetic Bell, with tell-tale.
The British Telegraph Manufactory, Limited.

1702a. Electrical Alarum with Leclanché's Battery.
Mariais, Paris.

This apparatus consists of a watch placed on a box which contains an electrical bell worked by two elements of a chloride of ammonium (called Leclanché's battery). The wire from the zinc is fixed in the binding screw No. 2, the wire from the carbon in the binding screw No. 1. When the small hand of the watch touches the piece which turns in the circular groove, the bell rings until the hand is released by turning the small handle which is at the side of the piece. The binding screw No. 3 which is above is to be used for the wire from the carbon when the apparatus is used as an ordinary electric bell. The polished stem which carries No. 4 binding screw can only be used for six hours, when the rod which carries the platinum point is clamped in the hole.

1702b. Lamp with Self-lighting Electrical Apparatus and Bichromate of Potash Battery.   *Mariais, Paris.*

The battery inside is charged with bichromate of potash mixed with about one-tenth water. The lamp, the small tube of which can be seen, is an ordinary benzolin lamp with sponge; when once filled it is put in its place. By pressing on the central button the lamp is made by the mechanism of the apparatus to approach the platinum lighter (inflamateur), which is in the centre. On letting the button go, the lamp is lighted. The cover of the lamp should be replaced by the hand in order to prevent evaporation.

The box contains eight spare platinum plates.

1709b. Detector of Disorder in Pneumatic Telegraph Tubes. *M. Bontemps, Telegraph Inspector, Paris.*

   a. With bell.

   b. With differential manometer.

1703. Magnetic Ship Signal. *Geminiano Zanni.*

The above signal is composed of two parts, the sender and the receiver. For actuating the sender, merely turn the handle over to the order required to be sent, the dial of the receiver will indicate this order, and also repeat it back to the sender.

The magnetic current is caused by the motion of turning the handle or crank to the order; the use of the voltaic battery is dispensed with.

1709c. Elisha Gray's Telephone, an instrument for transmitting musical notes by means of electricity.

Warden, Muirhead, and Clark.

A. Board for transmitting the four notes of the common chord, and two receiving instruments, B, C.

A reed vibrating the note to be transmitted is caused to interrupt the current entering the line. At each vibration the interrupted current reproduces the note at the distant station by there setting a similar reed in vibration as in the electro-magnetic receiver B, or by the friction of the finger (through which the current is made to pass) on the zinc face of the revolving sounding board. By means of the telephone several messages may be sent along a wire at once, a different note being employed for each.

APPLICATION TO NAVAL AND MILITARY PURPOSES.

1711a. Military Telegraph.

M. Trouvé, 6, Rue Thérèse, Paris.

This apparatus is characterised by the junction into one single object of the three parts which constitute a telegraphic post, viz., the monitor, the manipulator, and the reflector; and also by the smallness of its size, which enables a man to carry it on his back, with its cable, like a soldier's knapsack. To this apparatus are joined four time-pieces; the first two are tellers transmitting the despatch by sound. The two others have dials, movable and printed on both sides; the one has letters and figures, the other letters only, with blanks for tracing words, which allows of orders being sent as rapidly as by word of mouth.

FUZES.

(*a.*) WIRE OR LOW TENSION ELECTRICITY FUZES.

1712. Earliest form of Fuze used by the Royal Engineers for the explosion of mines by electricity (first used in removing the wreck of the "Royal George," Spithead, 1839).

F. A. Abel, F.R.S.

1713. Mining Fuze, platinum wire, improved construction by Chemical Department, Woolwich. Wire, ·003 inch in diameter (Abel's electric fuze, No. 1).   *F. A. Abel, F.R.S.*

1714. Detonators, platinum wire, Chemical Department, Woolwich. Wire, ·003 inch diameter.   *F. A. Abel, F.R.S.*

1715. Detonator (low tension), latest construction, with bridges of platinum-silver or indio-platinum wire (·0014 inch in diameter).   *F. A. Abel, F.R.S.*

1716. Statham and Brunton's Fuze, 1854. Composition : sulphide of copper and fulminate of mercury.   *F. A. Abel, F.R.S.*

1717. Von Ebner's Fuze, 1867 (earliest fuzes constructed by Von Ebner about 1855). Composition : coke, sulphide of antimony, and chlorate of potash ; primed with gunpowder.
F. A. Abel, F.R.S.

1718. Abel's High Tension Fuze, 1858. Composition : subphosphide of copper ; primed with gunpowder.   *F. A. Abel, F.R.S.*

1719. Abel's Fuze for Royal Engineer Service, 1862. Composition : subphosphide of copper ; primed with gunpowder.
F. A. Abel, F.R.S.

1720. Submarine Fuze, for instructional purposes only. Composition : blacklead and fulminate ; priming, sulphide of antimony and chlorate of potash.   *F. A. Abel, F.R.S.*

1721. Beardslee Fuze, 1864. Streak of blacklead between the poles ; primed with gunpowder.   *F. A. Abel, F.R.S.*

1722. Abel's Land Service Detonator. Composition : phosphide of copper ; primed with fulminate of mercury.
F. A. Abel, F.R.S.

1723. Submarine Detonator. Composition : graphite and fulminate of mercury ; priming, fulminate.   *F. A. Abel, F.R.S.*

1724. Gun Tubes (low tension). Original form, McKinlay. Gunpowder priming.   *F. A. Abel, F.R.S.*

1725. Abel's Gun Tube (high tension). Original form, Gunpowder kept damp by chloride of calcium (1857).
F. A. Abel, F.R.S.

1726. Gun Tube (high tension). Service, Abel's. Composition : subphosphide of copper, with tube of gunpowder.
F. A. Abel, F.R.S.

1727. Gun Tube, naval, platinum, silver-wire, ·0014 inch in diameter. Composition : gunpowder and gun-cotton, with tube of gunpowder.   *F. A. Abel, F.R.S.*

1728. Series of Specimens illustrating the development of the applications of Electricity to the explosions of mines, guns, &c.

1. Original form of wire fuze.
2. Statham fuze.
3. Platinum wire gun tube and section.
4. Experimental high tension fuze.
5. Abel's electric fuze and section. No. 1.
6. Abel's electric gun tube and section. No. 4.
7. Abel's platinum wire fuze and section. No. 5.
8. Austrian electric fuze (Von Ebner's) and section.
9. Abel's submarine fuze and section. No. 2.
10. Abel's electric detonator (land service) and section. No. 5.
11. Abel's electric detonator (submarine) and section. No. 7.
12. Naval electric gun tube and section.
13. Low tension detonator and section. No. 9.

Glass case and printed labels. *War Office.*

1576e. Magnetic Exploder and Fuzes.
The British Telegraph Manufactory, Limited.

c. ELECTRO-CHEMICAL APPLICATIONS.

1730. Application of the Electrotype Process to the Reproduction of Works of Art.
Messrs. Elkington and Co.

Reproduction of a medal. This is an object composed of two parts only, viz., the obverse and the reverse.

a and *b.* Moulds in gutta percha, &c. of the obverse and the reverse.
c and *d.* Moulds with the copper deposited in them.
e and *f.* Copper deposits removed from the moulds.
g. The two sides or portions of the medal as deposited, soldered together ready for finishing.
h. Electro-deposited copy or reproduction of the medal, gilt or silvered.
Reproduction of a small candlestick :
a. The moulds of the base and two sides of the nozzle in gutta percha, &c.
b. Moulds with the copper deposited in them.
c, d, and *f.* Copper deposits of the two sides of the nozzle of the candlestick, and the base removed from the moulds.
g. The same soldered together and fitted.
h. The candlestick silvered and completed.

1730a. Reproduction of the " Strauss " Tankard.
South Kensington Museum.

a. A copy or reproduction complete with fictile ivory body or drum.
b. Mould of the " drum."
c. Cast in fictile ivory.
d. Moulds with the copper deposited in them.

e. Copper deposits removed.
f. The same soldered together to form a metal "drum."
g. Moulds of the handle.
h. Moulds with the copper deposited in them.
i. Copper deposits removed from the moulds.
j. The same soldered together as part of the tankard.
k. A group of the other details of the tankard deposited in copper and removed from the moulds.
l. The tankard as soldered together and fitted with copper "drum," ready for gilding, &c.
m. Completed metal copy of the tankard, silvered and parcel gilt.

1730b. Group of Ferns, &c., in a basket, as an illustration of the method of coating natural objects, however delicate, with copper, and afterwards silvering or gilding the same.

The objects are first prepared with a metallic surface, then immersed in a solution of sulphate of copper, and afterwards electro-plated with gold or silver.

1730c. Electro Jewellery.   *Gustave Trouvé, Paris.*

1731. Single Cell Apparatus, used in the Electrotype process.   *J. How and Co.*

This consists of a glass outer cell furnished with a perforated shelf, with an upright porous vessel containing a zinc rod well amalgamated. The porous cell is charged with dilute sulphuric acid, and the object to be coated attached to the binding screw by a copper wire, and suspended in the outer glass vessel, which is filled with a saturated solution of sulphate of copper, a supply of crystals of this sulphate being placed upon the perforated shelf for the purpose of keeping up the strength of the solution.

Smee's Battery, used in the Electrotype process.
          J. How and Co.

This consists of a central platinised silver plate for the negative element between two zinc plates, connected together by a clamp. It is charged with dilute sulphuric acid.

1731a. Daniell's Constant Battery, used in the Electrotype process.   *J. How and Co.*

The outer copper cell forming the negative element; the positive element consists of a rod of zinc placed in a porous cylinder. To charge the battery the porous cell is filled with dilute sulphuric acid and the outer cell with a saturated solution of sulphate of copper, crystals of the sulphate being placed upon the perforated shelf of the outer cell for the purpose of keeping up the strength of the solution.

1731b. Bunsen's Battery, used in the Electrotype process.   *J. How and Co.*

This consists of a negative element of carbon contained in a vessel of porous earthenware, which is surrounded by a cylindrical zinc positive element in an outer glazed earthenware cell. It is charged by placing concentrated nitric acid in contact with the carbon, and dilute sulphuric acid in contact with the zinc. The zinc is required to be kept well amalgamated.

1731c. Glass Decomposing Cell, used in the Electro-type process. *J. How and Co.*

Consists of a glass vessel fitted with two insulated brass bars, with attached binding screws. A piece of sheet copper is hung upon one of the bars, and the object to be coated on the other. The vessel is nearly filled with a strong solution of sulphate of copper. The zinc pole of a battery (say Daniell's) is connected with the bar to which the mould or object to be coated is attached, and the copper pole to that upon which the sheet copper is suspended. This apparatus will also answer for depositing silver and gold.

1732. First Electrotype Reproduction obtained by M. Jacobi. *Conservatoire des Arts et Métiers.*

1732a. Ten Copper Plates, engraved by the electro-chemical process, the proofs from which are in the accompanying album.
Erhard, Paris.

A proof, fresh printed from a lithographic stone, or an autograph, or a proof from a copper plate wanting to be renewed, is by this process transferred to a copper plate, and produces in a few moments an engraving en creux as clear as that from the original plate, which is in no wise damaged by the operation. By this process it is possible : 1. To avoid having to preserve stones, at once brittle and cumbersome. 2. To reproduce a worn out plate, and so secure unlimited copies. 3. To effect, upon the copper plate, corrections not feasible upon the original plate worn out by previous printings.

1732c. Electric Multiplier. *Prof. George Fuller, Belfast.*

XII.—SPECIAL COLLECTIONS OF APPARATUS.

4566. Photographs of collection of **Apparatus formerly belonging to Volta.**
Liceo Volta of Como, Cabinet of Physics and Chemistry, Prof. Giovanni Gambara.

1. Small voltaic pile.
2. Small voltaic pile with circle of cups.
3. Electrophorus.
4. Two wooden discs covered with silk.
5. Electroscope.
6. Apparatus for igniting hydrogen gas by the electric spark.
7. Glass electrical pistol of Volta.
8. Eudiometer of the same.
9. Small squares of zinc and copper for generating electricity.
10. Glass tube containing mercury, for determining the coefficients of the expansion of air.
11. Small case made expressly for the protection of the above instruments

1733. Original Apparatus with which **Faraday** obtained the Magneto-Electric Spark.
The Royal Institution of Great Britain.

A welded ring of soft iron six inches in diameter, $\frac{7}{8}$ of an inch thick, one part covered by a helix A containing about 70 feet of insulated copper wire occupying about nine inches in length upon the ring. The other part covered by a second helix B containing about 60 feet of insulated copper wire. The helices are separated from each other at their extremities by half an inch of the uncovered iron.

The iron ring was converted into a magnet by passing a voltaic current through the helix A. This induced an electric current in the helix B, and a small spark was for a moment seen at the carbon terminals.—Phil. Trans. 1831.

1733a. Siberian Loadstone and Spark Apparatus. This was the loadstone employed by Faraday in his experiments on magneto-electric induction, from which he first obtained the induction spark. (*See* Exp. Researches, vol. II.).

The Council of King's College, London.

1734. Faraday's original Apparatus for Magneto-Electric Induction by a permanent magnet.

The Royal Institution of Great Britain.

A paste-board tube is surrounded by a helix C of insulated copper wire. The diameter of the tube allows a cylindrical bar magnet to pass freely into it. The terminal wires of the helix are connected with a galvanometer. On the introduction of a permanent bar magnet into the helix, and on its withdrawal from it, currents of electricity were induced in the helix which caused a deflection of the galvanometer needle.—Phil. Trans. 1831.

1735. Faraday's Rotating Rectangle for illustrating the inductive action of the earth.

The Royal Institution of Great Britain.

The wire rectangle provided with a commutator for collecting the currents was attached to a galvanometer, and rotated across the line of the magnetic meridian, the electric current induced in the rectangle deflecting the galvanometer needle.—Phil. Trans. 1852.

1736. Various Helices, Spirals, &c. used by **Faraday** in his researches on Magneto-Electric Induction, &c.

The Royal Institution of Great Britain.

Phil. Trans., 1831.

1737. Magnet made by **Static Electricity,** with note by Faraday.   *The Royal Institution of Great Britain.*

" A magnet made at the London Institution by an electric discharge from 70 square feet of charged surface. Present, Sir H. Davy, Pepys Jordan, Bostock, and Faraday."—Note by Faraday.

2434. Metallo-Chrome. A specimen prepared by Nobili, and presented by him to the Royal Institution.   *Mrs. Faraday.*

1739. Diagrams of magnetic Curves, prepared by Faraday.

Mrs. Faraday.

1740. Coils and Helices, used by Faraday in his magneto-electric researches.   *Mrs. Faraday.*

XII.—SPECIAL COLLECTIONS.

1741. Model frequently **used by Faraday** during his researches on the rotation of a ray of polarised light by electricity and magnetism.   *Mrs. Faraday.*

1742. Block of Glass pierced by sparks from an induction coil. Presented to Faraday by M. Ruhmkorff, 1861.
Mrs. Faraday.

1232a. Daguerreotype Portrait of Faraday.
Mrs. Charles Cowper.

1232b. Daguerreotype Portrait of Faraday and Brand.
Mrs. Charles Cowper.

1742b. Original Apparatus, by Arago and Matteuci.
Polytechnic School, Paris.

1742a. Apparatus constructed by **A. De la Rive** for demonstrating the rotatory motion which an electric discharge in rarefied gas performs around a magnet.
De la Rive Collection.   The property of Messrs. Soret, Perrot, & Sarasin, Geneva.

This apparatus consists of an electric shell, perforated by a soft iron cylinder, magnetised by placing it on one of the poles of an electro-magnet. The electric discharge is produced between the extremity of the soft iron cylinder, which is insulated by a glass tube, and a metallic ring which encircles the soft iron. As soon as the soft iron is. magnetised, the electric discharge begins to revolve around it. Previous to Ruhmkorff's improvements in induction-coils De la Rive made this experiment with the current of the "Armstrong" machine.

1742aa. Diagrams and MS. by Henry Cavendish (40 sheets), the property of his Grace the Duke of Devonshire.
The Cavendish Laboratory, Cambridge.

These form part of the hitherto unpublished papers of Henry Cavendish. The experiments described were made in the years 1771-2-3, and relate to electrical measurements.

He first establishes the law of the inverse square by proving that if a globe is insulated inside a hollow sphere, and then put in connexion with the hollow sphere, and the whole highly electrified, and if the globe is then insulated and the hollow sphere removed, the globe is found to be without charge.

He then describes a method of measuring the electric capacities of bodies, and applies it to a globe, to discs of different sizes, to squares of different sizes and substances, and to long cylinders and wires, comparing his results in each case with his calculations.

1742c. Apparatus devised by **De la Rive & Sarasin** for demonstrating that the electric discharge in a rarefied gas, turning under the power of a magnet, draws with it in its rotatory motion the gas which transmits it, and all bodies, sufficiently light, that it meets with in its course.
De la Rive Collection.   The property of Messrs. Soret, Perrot, & Sarasin, Geneva.

A bell-glass stands upon a platten, which itself must be placed upon the pole of an electro-magnet. This glass being filled with rarefied gas, the electric discharge is completely effected between the central brass ball, and a ring of the same metal constituting the other electrode. A sail-wheel is adjusted inside this ring, so that its vertical paddles may be upon the direct line from the ball to the ring, whereby the discharge strikes it in revolving under the action of the magnet. It then gives it a rotatory direction as soon as the direction of the magnet is changed.

(*See* Archives des Sciences, vol. 45, p. 387; Philosophical Magazine, vol. 44, p. 149.)

1743. Induction Coil by Bonijol, an old Genevese maker. *De la Rive Collection. The property of Messrs. Soret, Perrot, & Sarasin, Geneva.*

Bonijol constructed a great number of electrical apparatus under the direction and with the advice of G. and A. De la Rive.

This coil, in which by means of a single medium sized element of constant power, induced currents of considerable force may be produced, was frequently used by A. de la Rive in his researches.

See "Archives de l'Electricité" by De la Rive, 1841, Vol. 1, p. 280.

1744. " Bréguet " Thermometer, used and referred to by **A. De la Rive** in his works upon the causes of voltaic electricity, and upon the properties of magneto-electric currents. *De la Rive Collection. The property of Messrs. Soret, Perrot, & Sarasin, Geneva.*

1746. Galvanic Battery, by **A. De la Rive.** Modification of the **Grove Battery,** with nitric acid on the exterior. Constructed by the Geneva Association for the Construction of Scientific Instruments. *De la Rive Collection. The property of Messrs. Soret, Perrot, & Sarasin, Geneva.*

The nitric acid is placed in a large glass phial, and in sufficient quantity to serve for a long time without being changed.

The diaphragm containing the acidulated water and the zinc closes perfectly the orifice of the phial ; when the battery is taken to pieces, this diaphragm is replaced by a glass stopper. By this arrangement, the disengagement of nitrous vapour is avoided. This battery may thus safely remain in the experimenting room, close to the apparatus, and is especially suited for working a Ruhmkorff coil. Two elements suffice for a medium sized coil. De la Rive constantly used this apparatus for his researches upon induced currents, and always left it in his laboratory.

1747. Floats, constructed by **Gaspard de la Rive.** *De la Rive Collection. The property of Messrs. Soret, Perrot, & Sarasin, Geneva.*

Apparatus intended for the demonstration, in a simple manner, of the "Laws of Ampère" upon the reciprocal action of currents. The conductors can be adapted directly to a small floating battery.

1748. Apparatus, by **A. De la Rive,** for the derivation and the relative measurement of **Induced Currents,** and used by him in his studies upon rarefied gases.

De la Rive Collection. The property of Messrs. Soret, Perrot, & Sarasin, Geneva.

This was used for diverting into a galvanometer a very small portion of an induction current, which passes through a glass trough filled with distilled water, in which are dipped two platinum wires, joined to the galvanometer. The current passing through the thin liquid thread placed between the wires is partially diverted into the galvanometer, which thus measures a quantity proportioned to the total intensity of the induced current that may go through the trough.

The deviation of the needle of the galvanometer increases proportionally with the distance between the wires.

This apparatus was often used by De la Rive in his researches on the passage of the induced current through rarefied gases.

1748a. Apparatus used by **De la Rive and Sarasin** to demonstrate that **Rarefied Gases,** crossed by inductive discharge, become condensed under the action of magnetism.

De la Rive Collection. The property of Messrs. Soret, Perrot, & Sarasin, Geneva.

The induced current passes through the tube with two compartments, one of which is placed between the two poles of a powerful electro-magnet; the glass cock is rapidly turned round, thus interrupting the current, and separating the two compartments. These are afterwards brought into communication with a very sensitive manometer, and it is then found that the pressure is a little greater in the one which has been between the two poles of the magnet, and which will also contain the negative electrode.

(*See* Archives des Sciences Physiques et Naturelles; new period, vol. 41, p. 5; and Philosophical Magazine, 4th series, vol. 42, p. 211.)

1748b. Apparatus used by **A. De la Rive** in his Studies upon the **Magnetic Rotatory Polarisation** of **Liquids.** (Made by the Geneva Association for the Construction of Scientific Instruments.)

De la Rive Collection. The property of Messrs. Soret, Perrot, & Sarasin, Geneva.

1. Nicol polariser.

2. Tube for holding liquids, with double wrapper for the heating, required for studying the influence of the temperature upon the phenomenon.

3. Analyser, with graduated circle and special register, invented by M. Thury. This consists of two tangent discs in ivory, the one supported by the analysing Nicol, the other by the pinion that helps to turn it, and of a horizontal ruler. A pencil mark made by this ruler on both discs shows the position of the Analyser. Instead of having to read each observation separately, which is inconvenient in experiments made in the dark, the corresponding mark to every observation is simply noted, and afterwards read altogether. (*See* Archives des Sciences Physiques et Naturelles, vol. 38, p. 209.)

1749. Apparatus designed by **Auguste De la Rive,** for the demonstration of the **Electric Theory** of the **Aurora Borealis.**

Made by the Geneva Association for the Construction of Scientific Instruments.

De la Rive Collection. The property of Messrs. Soret, Perrot, & Sarasin, Geneva.

A large sphere, made of wood, represents the earth. Two iron cylinders represent the two extremities of the terrestrial magnetic axis. They penetrate into two globes filled with rarefied air, which simulate the higher regions of the polar atmosphere. The electric discharge, which takes place in this rarefied air, following rays all around a point situated in the prolongation of the terrestrial axis, turns about this point, and so turns in a different direction at either pole, when the two cylinders are charged by means of a horse-shoe electro-magnet, in accordance with the observations upon the rotation of the rays of the Aurora Borealis. See "Archives des Sciences physiques et naturelles," 1862, vol. 14, p. 121. Philosophical Magazine, 4th series, vol. 23, p. 346.

1750. Metallic Plates, for Watch Case, used by A. De la Rive in his first experiments in galvanic gilding.

Lucien De la Rive, Geneva.

1745. Photographs of a Special Collection of Instruments used by Volta.

Royal Lombardian Institution of Science and Letters.

1ST PLATE :
1. Electrophorus, with mastic cake, designed by Volta.
2. Condensing electrometer, the same which Volta made use of to demonstrate metallic electricity.
3. Columnar pocket pile, adopted by Volta to demonstrate his theory at the Institute of Paris in the presence of Buonaparte.
4. Original letter of Volta.
5. Lamp for hydrogen gas, which is ignited by the electrophorus. It has the form of the lamps which Volta introduced so largely into Germany.
6. Apparatus employed in the first researches of Volta, for collecting and rendering appreciable the smallest quantities of electricity.

2ND PLATE :
The same instruments on smaller scales.

3RD PLATE :
Fac-simile of part of a letter of Alexander Volta to Professor Bartletti, dated Como, 15th April 1777.

1473. Apparatus for covering wire with silk for electric purposes.　　　*H. Lloyd, Trinity College, Dublin.*

1751. Polar-light Apparatus.

Prof. Lemström, Helsingfors, Finland.

Report of a speech of Dr. Lemström on his Polar-light apparatus, and the theory of the Polar-light.

This apparatus serves to prove that the polar light or aurora borealis is an electric current flowing from the higher regions of the atmosphere down to the earth.

A sphere of brass is fixed on a bar of india-rubber or ebonite 0·6 meter long, which is screwed to the board of the cross-shaped stand. A cylinder of india-rubber, 3 meters long, is fixed to the same board at about 0·7 meter from the sphere. From the cylinder proceeds a branch with an arm, both of india-rubber. On the arm are fixed 16 Geissler's tubes, the air in which has a pressure of about 0·5 millimeter. The lower ends of the tubes are pierced by platinum wires, which are directed towards the sphere, whilst at the upper end the platinum wires are, by means of their copper wires, in a metallic union ·with a button, and also in metallic union with the earth. From underneath the sphere a copper wire, well insulated with india-rubber, leads to the negative pole of a Holtz's electric machine (a machine by Carré of Paris was employed with great advantage), of which the positive pole is in metallic connexion with the earth. As soon as the machine is put in movement, the sphere, being charged, becomes negatively electric, and at the same time there goes through all the tubes a current of reddish-lilac light, so that they altogether form a shining bow-shaped belt. With an ordinary machine this phenomenon may still be observed when the lower ends of the tubes are at a distance of *two meters* from the sphere. This proves evidently that the electricity flowing out from (or into) the sphere not only traverses the layer of air that is between, but goes also with such power through the tubes that the gas therein becomes incandescent from the heat that the electric current produces, as is well known. In order that the electricity might more easily flow out in the air from the sphere, this latter is furnished with points. These points, as well as the metallic union between the upper end of the tubes and the earth, are of no absolute necessity, for the phenomenon may be produced without them, but in that case the distance between the sphere and the tubes must be considerably reduced.

The light produced by the apparatus proves clearly that a current of electricity may go through a layer of air of ordinary pressure 760^{mm} without producing the phenomenon, but if it meets in its way a space of rarefied air of low pressure (from 0 to 30^{mm} to 40^{mm}) the light immediately arises, the current causing the molecules of gas to become incandescent.

On the Theory of Polar-Light.

The knowledge we have acquired of the electric state of the earth proves that it is a conducting body, charged with a small quantity of negative electricity, and surrounded by the atmosphere, in general charged with positive electricity. Though this latter might be produced by an influence from the earth, it is still very probable that it proceeds from the process of evaporation, either directly or by the friction of vapour against particles of air. The atmospheric air possesses a very small conducting power for electricity when dry and at ordinary pressure, but the conducting power increases considerably as soon as the air becomes moist and rarefied. It has been proved by experiments that the conducting power is highest at a pressure between 5^{mm} and 10^{mm}, and amounts to 10,000 times that at a pressure of 760^{mm}. If the rarefaction of the air is carried further than 5^{mm}, the conducting power diminishes again, but very slowly. It is known that in proportion to the elevation above the surface of the earth the air becomes more and more rarefied according to the law expressed by the formula given by Laplace, and that consequently, at a certain elevation the earth is surrounded by a layer of air that has a pressure of only 5^{mm}; the conducting power for electricity in this layer is sufficiently great to allow of its being regarded as a conductor in comparison to the air in the lower regions, and even in the highest. The negatively electric earth is thus surrounded by a conductor of electricity concentric with it. All the positive electricity that attains the region of rarefied air at about 5^{mm}; or, as it might be called, this air conductor, submits almost to the same laws as if it were in a real conductor, and must thus be distributed in a manner depending on the influence of the electronegative earth. Part of the electricity, conducted by the vapours, remains on the clouds in the atmosphere and discharges in form of lightning; another part attains the region of rarefied air, owing to the fact that the vapour itself, submitting to well-known physical laws, rises to this elevation, and also because electricity tends to distribute itself on a conductor.

The manner in which the electricity distributes itself between the two conductors depends on their relative position as well as on their form. The earth may, with sufficient approximation, be considered as spherical, as well as the air conductor, but in their position relatively to each other it appears that the region of rarefied air at 5^{mm} approaches much nearer to the earth at the poles than at the equator, principally in consequence of the difference of the temperature of the air in the two places. If we assume the mean temperature of the air round the equator to be 25°, at the poles −12°, and everywhere on the air conductor − 60°, and we suppose at the same time the air everywhere half saturated with moisture, and that the temperature is reduced in proportion to the elevation, we find, if the above-mentioned formula of Laplace (1*) is applied, that the air conductor must at the equator be at an elevation of 37·47 kilometers, and at the poles 34·25 kilometers.

In consequence of this relative position, if the two conductors are regarded as conducting surfaces, the electric density on them both becomes about 9 per cent. greater at the poles than at the equator, and the power, by which the two electricities endeavour to unite, at least 20 per cent., but probably, if all the circumstances are considered, 30 or 40 per cent. greater at the poles than at the equator. In these facts then must be sought the principal cause of the accumulation of electricity at the poles of the earth and of the phenomenon occurring there, called polar-light or aurora borealis.

It is a remarkable fact that thunderstorms diminish in number as well as in intensity in moving from the equator towards the poles, and that at the 70th degree of latitude they cease completely, exhibiting once more in the highest north vestiges of their primitive intensity. In Finnish Lapland, for instance, thunderstorms are very uncommon, but when they occur they are extremely intense, and are almost always accompanied by thunderbolts. This peculiarity has probably its cause in the fact that the region of thunderclouds descends towards the earth in accordance with the same rule as the before-mentioned air conductor. The reduced number of thunderstorms is caused by the fact that the very source of electricity in the atmosphere, that is to say, the evaporation, is very much reduced; however, another important cause is here active, namely, the heightened conductive power that the air possesses in consequence of its greater saturation with moisture, whereby the electricity becomes unable beyond a certain latitude to remain upon the clouds, until it has attained a greater tension, but is conducted down to the earth in form of a slow current, visible in the polar-light.

It results from experience, with a high degree of probability, that the polar-light is an electric phenomenon, for its effects are of the same nature as those of the electric currents. Thus the polar-light causes disturbances in terrestrial magnetism, induces currents in telegraphic wires, and furnishes a spectrum of nine bands, which, with one exception, coincide with those produced by an electric current passing through rarefied air. Thus there is no doubt that the polar-light is caused by an electric current passing from

---

(1*) X = 18·393 metres (1 + 0·002837 Cos. 2 ϕ) $\dfrac{(1 + 0·004T + t)}{2}$ $\dfrac{l\,H}{h}$ where X signifies the elevation, ϕ the latitude, T the temperature at the surface of the earth, t at the upper point H, and h the reduced height of the barometer for the same points.

the upper rarefied layers of air to the earth; this current, during its passage through the rarefied air, produces light phenomena that cannot arise in denser layers of air.

The polar-light apparatus now exhibited shows that an electric current passing from an insulated body does not produce any light in air of normal pressure, but as soon as it rises to the rarefied air in the Geissler's tubes, there is directly produced a phenomenon very like the real polar-light. In the apparatus the upper end of the tubes is in union with the earth; this is by no means necessary, for the light phenomenon is also produced if this union be removed, provided that in such case the tubes be brought a little nearer to the insulated sphere. For the rest, the earth represents here the wide space of rarefied air that we find beyond the limits of the air conductor, and which serves here as an electric reservoir.

To consider now how the polar-light on a large scale is formed in nature. As before said, the earth and the air conductor hold to each other the position above mentioned, and the two electricities, the negative electricity of the earth and the positive electricity of the air conductor, endeavour with a certain force to unite in a belt around the north pole. The insulating power of the denser air prevents this reunion; but if we assume that the equilibrium is attained, the union will instantly take place as soon as this insulating power is diminished or the electricity on the conductor augmented. The first case, which probably is the most ordinary, happens if a southerly wind carrying a quantity of vapour attains the polar regions; for instance, the belt, where the vapour, in consequence of the cold, is condensed into a fluid form, reduces considerably the insulating power of the air and enables the electric current to flow through it. The same thing would occur if a layer of clouds happened to enter into this belt; the upper end of the cloud would become negatively electric, the lower one positive, and thus the distance between the two conductors would in fact be diminished. The electric current would go from the air conductor to the cloud, and through this latter to the earth. Similar phenomena are observed in the polar regions, for the upper edges of the clouds are not unfrequently seen shining with a yellowish light stretching considerably upwards, whilst no light is discernible under the cloud because of the air there having attained a density sufficient to prevent the current from producing light.

For further details of the polar-light and its theory, see Archives des Sciences Phys. et Natur. de Genève, 1875 (Sept. and Oct.), and in January 1876, as well as to two essays published in the years 1869 and 1873 in the same scientific journal, all which articles are more or less the result of observations made in the arctic regions. Besides these may be mentioned the works upon polar-light of the American natural philosopher Loomis, Rep. of Smithsonian Ins., 1865, &c.

4558. Photograph of a Dielectrical Machine, system of Prof. Cecchi.   *Prof. Filippo Cecchi, Florence.*

4559. Photograph of an Electric Motor, on the plan of Prof. Cecchi, of Florence. Two rectilineal electro-magnets with polar masses of iron are alternately magnetised, and by means of levers and eccentrics produce the motion.

Prof. Filippo Cecchi, Florence.

The electro-magnetic motor of Prof. Cecchi, of Florence, is set in movement by two rectilineal and parallel electro-magnets placed horizontally upon a wooden base.   Each of these electro-magnets bears at one end a mass of soft iron, while from the other projects a metal point, which by means of a small band moves a long lever of the third order, and this last moves by means of another very long band an eccentric, situated on the axis of revolution.   On this same axis there is also a commutator which serves to send a current of electricity alternately into one and the other of the electro-magnets.   This motor, on equal conditions, has a force superior to that of many other electro-magnetic motors.

SECTION 11.—ASTRONOMY.

<small>West Gallery, Ground Floor, Room L., and the Terrace overlooking the Horticultural Gardens.</small>

I.—INSTRUMENTS FOR DETERMINING THE PLACES AND MOTIONS OF THE HEAVENLY BODIES.

a. Astrolabes.

1752. Suspension Astrolabium. A very old astronomical instrument made in 1525. *Prof. Buys-Ballot, Utrecht.*

1752a. Astrolabe, date 1374. This instrument affords one of the oldest illustrations of the use of Arabic numbers.
A. C. Baldwin.

1752a. Ancient Astrolabe, by Petrus Raimondi, 1375.
A. C. Baldwin.

This instrument is 4 inches in diameter, with five interchangeable discs, astronomical circles, and two pointers. On the back, a zodiacal calendar. Round the edge is the inscription, " + hoc astralabium : fuit factum : et cum armillis verifficatum Barghinone : anno Xri 1375. per Petrum Raimodi de domo regis Aragonum : lati Barghinone : 41 longi 39."

1753. Astrolabe, constructed for Sir Francis Drake, prior to his first expedition to the West Indies in 1570.
Royal Naval Museum, Greenwich.

This instrument is said to have been preserved in the Stanhope family till 1783. It was subsequently presented to King William IV., who in 1833 deposited it in Greenwich Hospital.

1754. Astrolabe. Ivory, mounted with gilt ormolu. A figure of the Creator is engraved outside. It still retains the original compass and needle. Nuremburg. Dated 1585.
Rev. J. C. Jackson.

1754a. Ancient Astrolabe, supposed to have belonged to the Spanish Armada. *Robert J. Lecky, F.R.A.S.*

This instrument was found under a rock in the island of Valencia, Ireland, 1845, within view of the place where three vessels of the Spanish armada were wrecked, and from the style of its finish and workmanship is supposed to have belonged to one of them.

1755. Persian Astrolabe.
The Royal United Service Institution.

Presented to the United Service Institution, May 1842, by Major-General Sir John May, K.C.B., K.C.H.

1756. Ptolemy's Planisphere or Astrolabe, made in 1601, by Michael Coignet, at Antwerp. (See the works of Gemma Frisius, Metius, Lansberghen, &c.)
H. G. Van de Sande Bakhuyzen, Director of the Observatory, Leyden.

1757. Ptolemy's Planisphere or Astrolabe, made in the beginning of the 16th century. (See the works of Gemma Frisius, Metius, Lansberghen, &c.)
H. G. Van de Sande Bakhuyzen, Director of the Observatory, Leyden.

1757a. Large Bronze Astrolabe, which belonged to King Philip II., of Spain.   *Archæological Museum, Madrid.*

Near the handle are the following inscriptions, giving the date and name of the maker:—
Philippo Rege.   Gualterus Arsenius Frisi Nepos Lonanii fecit an 1555.
The diameter of this instrument is $0 \cdot 59^m$.

1757b. Bronze Astrolabe, made in the 16th century.
Diameter, $0 \cdot 32^m$.   *Archæological Museum, Madrid.*

1757c. Arabian Astrolabe. Made at Toledo, A.D. 1067.
Diameter, $0 \cdot 24^m$.   *Archæological Museum, Madrid.*

Astrolabe made at Toledo as stated in the Arabic inscription engraved at the back : "In the month of Shawan.   One of the works superintended by "Ibrahim Ibn Said, the Muazini, Assohli, at Toledo.   In the 459th year of "the Hegira."
This year began on the 21st of November of 1066, and ended on the 10th of the same month of 1067.   The famous astronomer Ararquiel, called by Guillermo Anglicus, *Pater Isaac*, lived at Toledo at this period, and this instrument may have been made under his directions.   It contains five plates giving the latitudes.   The translations of the inscriptions will be found in the interesting study on astrolabes, published by Dr. Eduardo Saavedra in the "Museo Español de Antiguedades," V. VI., pp. 402-414.   It is worthy of notice that European numbers are to be met with on several of the plates, which would make it appear probable that the instrument had been used by some Christian astronomer not knowing Arabic.   The number of towns mentioned, ranging from Persia to the coast of Movono, and several Spanish towns, is much more extensive than on similar instruments of the kind.

1757d. Bronze Arabian Astrolabe. Diameter, $0 \cdot 21^m$.
Archæological Museum, Madrid.

Astrolabe without name of maker, date, or locality ; but it may be surmised from the Arabic inscriptions on it that this instrument was made at Morocco, in 1774.

1757e. Memoir written by Don Eduardo Saavedra upon several Arabic astrolabes existing in Spain, published in the "Museo Español de Antiguedades."

Archæological Museum, Madrid.

1757f. Bronze Astrolabe, without date or maker's name, and was probably made in Italy in the 16th century.

Ministry of Marine, Madrid.

1758. Two Astrolabes, with four double sights; belonging to H.H. the Prince of Pless. *The Breslau Committee.*

1759. Astrolabe, with movable sun-dial; belonging to H.H. the Prince of Pless. *The Breslau Committee.*

32. Astrolabe of Lord Dudley.

The Royal Institute of " Studii Superiori," Florence.

Robert Dudley, Duke of Northumberland, spent many years of his life at the Florentine Court. He died on the 6th of September 1649, at Florence. Although not an Italian instrument, his astrolabe is exhibited as a mark of courtesy to England.

Astrolabe of a Florentine author, with seven change discs, one of them with a geographic planisphere.

The Royal Institute of " Studii Superiori," Florence.

Astrolabe of an Italian author, with a planisphere of Rojas.

The Royal Institute of " Studii Superiori," Florence.

Astronomical Ring of Ignazio Danti, a celebrated astronomer and mathematician of Perugia.

The Royal Institute of " Studii Superiori," Florence.

Fra Ignazio Danti, a Dominican, erected, under Cosimo I. (1537-1564), an armillary circle and a sun-dial, which are yet to be seen, on the façade of the Church of Santa Maria Novella in Florence.

1760. Arabian Planisphere. *Royal Museum, Cassel.*

A small apparatus, of Arabian origin, remarkable for its great age and comparatively good workmanship. It is in the form of a plate made of brass, 16 c.m. in diameter, provided with means for hanging up, and has a movable alhidada. The front side shows a planisphere, the back various divisions and tables covered everywhere with Arabian characters. On nine transposable and one fixed plate can be shown 19 different planisphere drawings for the same number of polar altitudes.

1761. Combined Planisphere and Astrolabe.

Royal Museum, Cassel.

A very strong plate of brass, 37·5 cm. in diameter, with an arrangement for free suspension. On the one side is a planisphere with web, on the other an astrolabe with double sight. The instrument dates from the 15th or 16th century.

1762. Astronomical Circle, after Gemma Frisius, made in 1572, by Gualtherus Arssenius, grandson of Gemma Frisius.

H. G. Van de Sande Bakhuyzen, Director of the Observatory, Leyden.

b. ALTAZIMUTH INSTRUMENTS.

1763. 12-inch Altazimuth. *Troughton & Simms.*

Circles, 12 inches in diameter, divided into spaces of 5' arc; by means of the attached microscopes these spaces are further subdivided, one division upon the micrometer = 1″ arc.

1763a. Pocket Altazimuth on Small Stand for altitudes, compass bearings, &c., improved by Francis Galton, F.R.S., and specially adapted for travellers. *L. Casella.*

1763b. Altazimuth Instrument, made by Dollond with double altitude circles. *W. Watson and Son.*

The instrument stands on three adjusting foot screws, above which is the azimuth circle with a telescope attached. The azimuth circle is an arrangement of one circle turning within another, so made that their upper surfaces are in the same plane, and the inner edge of the larger in contact with the outer edge of the smaller. The outer edge of the smaller circle carries the divisions, and the inner edge of the larger circle three verniers; the graduation is to 10 minutes of arc; above these is the large conical axis of the azimuth circle, from the top of which spring rectangular arms carrying the altitude circles, and telescope. The telescope is mounted between two 12-inch circles, the peripheries of which are divided on silver to 10 minutes of arc, each circle has two verniers and is furnished with tangent screw movements. Lamp with graduating aperture.

1764. Comet Seeker, on a stand, with horizontal and vertical motion, constructed by Professor Kaiser.

H. G. von Bakhuyzen, Director of the Observatory, Leyden.

The stand possesses the advantage that the eyepiece, which is fitted with a total reflecting prism, remains at the same height in all the positions of the telescope, whilst the axis of the eyepiece remains horizontal.

4554. Photograph of the Vertical or Zenith Telescope, for the observation of stars culminating near the zenith.

Prof. Lorenzo Respighi, Director of the Royal Observatory of the Campidoglio, Rome.

The vertical or zenith telescope is intended for the measurement of the zenith distance of stars culminating near the zenith, with the use only of the wire micrometer and mercurial horizon, without need either of inversion or level. The telescope being directed towards the nadir, on the reflecting horizon placed at a great distance below the object-glass, it is possible simultaneously to determine the nadir or the prime vertical with a fixed equatorial thread, and to collimate, with the movable thread, the reflected image of the stars that cross the meridian in the field of the telescope, and then to measure the zenith meridian distance with the micrometric screw. In order that the observation may be completely normal, it is necessary that the reflection of the stellar rays extend over the whole object-glass, which will be effected when the distance D of the reflecting horizon from the object-glass of the telescope satisfies the condition $D = \dfrac{A}{2 \tan Z}$, where A is the aperture of the object-glass, and Z the zenith distance of the star.

In Table VI. is photographically represented the zenith telescope of the Royal Observatory of the Campidoglio, constructed by Signor Ertel of Monaco, from a design given by Professor Respighi. The telescope is mounted as in transit instruments, and may also be used as a meridian telescope: the object-glass, made by Signor Merz, of Monaco, has a clear aperture of $0 \cdot 108^m$, with a focal distance $1 \cdot 582^m$ The micrometer is composed of 11 fixed equatorial threads and two movable threads, with a fixed meridian thread. The eye-piece can be moved with the greatest ease, whether in the meridian or vertically. The micrometer can be turned 90° to render the threads parallel to the meridian, and then the instrument serves as a transit-instrument.

1765. Four-inch Achromatic Telescope, on altazimuth stand, with quick and slow motions, in altitude and azimuth.

John Browning.

1766. Instrument for Easy Determination of Time by Equal Altitudes of Different Stars.

Colonel Zinger, Pulkowa.

This instrument is constructed for easy application of the method developed lately by Colonel Zinger for exact determination of time by equal altitudes of different stars (*see* Vierteljahrsschrift des Astronomischen Gesellschaft, 1875). The principal condition is unaltered relation of level to telescope in going over from a star in the east to another in the west by having found its vertical axis. The divided circles serve only for setting the telescope. For using only bright and well determined stars down to the (3 or 4) magnitude, 9 minutes of time will be in the average sufficient to give time with the accuracy of $0 \cdot 1^s$. The eyepiece of the telescope is provided with a micrometer, to enable the observer to get with the same instrument, by approximately equal altitudes of two stars near the meridian, exact determination of latitude.

c. Transit Circles and Quadrants.

1730. Model of the Greenwich Transit.

South Kensington Museum.

1767. Photograph of Meridian Circle, with object glass of 6″ aperture and 6′ focal length, circle of 2′ diameter; under construction for the Observatory of Strassburg.

A. Repsold and Sons, Hamburg.

In this instrument the microscopes are attached to the heads of the cast-iron columns, which also carry the bearings of the axis. One of the pivots contains an objective (2″ aperture), and the other a plate with a hole; this arrangement made at the suggestion of Prof. Winnecke, serves as a collimator for controlling the position of the axis. Perfectly central illumination of the field of view is effected by a small mirror at the back of the objective. The cells of the objective and eyepiece can be interchanged; there is an arrangement for Nadir observations, also for observations of the reflected image of a star, with a Nadir distance of 8°–60° by means of a movable mercurial horizon.

1768. Photograph of Transit Instrument, with azimuth circle and a broken telescope of 3″ aperture.

A. Repsold and Sons, Hamburg.

When the instrument is reversed, which can be very quickly effected, the level and lamp remain hanging, and the clamping-screw need not be loosened. There is a micrometer at the eyepiece.

1769. Photograph of Transit Instrument, with straight telescope of 3·3″ aperture.   *A. Repsold and Sons, Hamburg.*

In reversing it is not necessary to loosen the clamping screw, and the level remains in position. There is a micrometer at the eyepiece.

1770. Transit Instrument.
> *Prof. Dr..C. Bruhns, Leipzig, and August Lingke and Co., Freiberg.*

The transit instrument is to be used in determination of time and of latitudes in the prime vertical.

The iron stand is on three feet, and can be set horizontal by means of two levels at right angles to each other. On the stand is a cradle with two supports to carry the telescope, which is always horizontal. There is an arrangement by which this cradle, together with the telescope, can be inverted. In order that the inversion may be as easy as possible, a spring is fixed below the stand, which comes into play when the telescope is inverted, and carries nearly the whole weight of the cradle and telescope. On one side of the cradle is a revolving pivot, which can be brought down between two screws, and according to the position of these screws can be turned through small angles. This movement in Azimuth enables observations to be made with the instrument, not only in the meridian, but also in a vertical arc through the polestar, and the index on the cradle which points to a division on the arc gives the position of the cradle in Azimuth. The graduation is so arranged that the interval between two lines gives about 10′, and can be read accurately up to whole minutes.

On the lower stand there is a boss with screws, and when the pivot of the cradle, which rests between the screws, is raised, and the screws at the same time loosened, the telescope can with ease be set in the prime vertical, and by the graduated scale the accurate position in the prime vertical can be read.

The telescope has an aperture of 73 mm., a focal distance of 80 cm., and two achromatic eyepieces, with magnifying powers of 60 times and 90 times, also two sun-glasses. The telescope is clamped at half its length, and the clamping apparatus is so arranged that the telescope suffers no pressure. Besides, there is at the middle point of the telescope a support with two screws, in order that the telescope may rest firmly when it is clamped : by this means it has a larger base. In front of the objective is a prism (made by Schröder, of Hamburg, who also supplied the objective) giving total reflection, and giving the same aperture as the objective. The prism is fastened by six screws on the hypothenuse surface against a strong spring; this spring is so strong that no turning of the prism can occur while alterations owing to change of temperature may take place. On one side there are also parts against which the prism is held by two strong screws on the opposite side. In the hypothenuse surface of the prism there is a dull spot through which the central illumination of the threads is effected. By a revolving arrangement the illumination can be regulated ; this arrangement is particularly simple.

The telescope carries the head of the eye-piece, and as it can have only a horizontal direction, the eye-piece must remain always in the same position. A micrometer is attached to the eye-piece which can turn through 90° ; this can be used to determine micrometrically the difference of the zenith distance of stars which have nearly the same north and south zenith distance.

There are garnets in the bed plates of the supports so as to lessen the friction. For turning the telescope there is a ring of gutta-percha. The telescope carries a level, which can remain permanently on it; one division of which = $1''\cdot43$. It has also an altitude arc graduated to $10'$ and reading to $1'$, by means of a vernier. The Nadir distance can be found by an artificial horizon. One division of the micrometer head = $\frac{1}{100}$th of a revolution, and the pitch of the screw is about $\frac{1}{100}$th of a Paris inch.

By this instrument can be determined—

1. The time on the meridian.
2. The time in the vertical of the Polestar.
3. When it is turned through $90°$, the latitude, according to the method of Bessel.
4. The latitude, according to Talcott's method, where the difference between the north and south zenith distance is measured by the micrometer screw.

1771. Portable Catoptric Transit Instrument, with a telescope horizontally resting in collars, and revolving only on its own optic axis. Invented and constructed by C. A. Steinheil, sen[r].

Conservatorium of the Math. and Phys. Collections of Bavaria (Prof. Dr. Seidel).

The construction of this instrument is described by its inventor in *Schumachers Jahrbuch für* 1844 (Stuttgart and Tübingen, published by Cotta), p. 3, *et seq.*

1771a. Reflecting Transit Instrument.
W. Watson and Son.

In this the image of the star is received on a mirror, and then viewed by looking down through a small telescope placed in such a position as to be most convenient for observation; the telescope is stationary, but the mirror moving in the plane of the meridian, can be directed to any point in it, and observations taken without those inconvenient positions of the observer, so often necessary with transits of ordinary construction.

1775a. Arabic Quadrant in bronze, made by Ahmed Ibn Abd el Rahman. *Archæological Museum, Madrid.*

Its radius is $0\cdot21^{m}$. It has an inscription in Ensic letters of the latter end of the middle ages giving the name of the maker.

" Quadrant Orario " of Cosimo I. dei Medici, to the latitude of $43° 45'$, made in Florence in 1566, probably by Ignazio Danti.
The Royal Institute of " Studii Superiori," Florence.

Interesting likewise, because it contains calculations relative to the reform of the calendar, afterwards carried out by Pope Gregory XIII.

" Quadrant Orario " to the latitude of $43° 45'$, made in Florence in 1565 by Giovan. Battista Giusta.
The Royal Institute of " Studii Superiori," Florence.

" Instrumento del Primo Mobile " of Apiano, executed in Florence in 1568 by Ignazio Danti.
The Royal Institute of " Studii Superiori," Florence.

1772. Azimuth Quadrant, constructed by Metz in 1700.
H. G. Van de Sande Bakhuyzen, Director of the Observatory, Leyden.

1722a. Davis' Quadrant or Back Staff. The first marine instrument for taking altitudes of the sun with reference to the visible horizon. This instrument was the prelude to Hadley's quadrant and the reflecting sextants and circles of the present day. The property of the Royal Astronomical Society, presented by R. J. Lecky.   *Robert J. Lecky, F.R.A.S.*

1773. Astronomical Quadrant, said to have been the property of Napier of Merchiston, the inventor of logarithms.
University of Edinburgh.

The telescopes attached are evidently of much more recent and clumsy workmanship than the instrument itself. They are reported to have been added by a "college bailie" (in the days when the university was under the government of the town council), who fancied that he was thereby enhancing the value of his gift to the university.

1774. Quadrant, by Butterfield, of Paris.
Kew Committee of the Royal Society.

A brass quadrant, on a wrought-iron pedestal, carrying a telescope, with object glass ½ in. in diameter, and 2 ft. 3 ins. focal length. The quadrant is divided with a diagonal scale, and is provided with a case for hanging a plumb-line.

1775. Quadrant, formerly belonging to Tycho Brahe.
Royal Museum, Cassel (Director, Dr. Pinder)

This instrument is the astronomical quadrant of Tycho Brahe. The altitude quadrant, as well as the azimuth divided circle, are made of brass; the first is divided into sixths of a degree, the second into whole degrees, which can be read by a simple pointer, but without verniers. The radius of both circles is 40 cm., and the stand is constructed of cast iron.

1776. Two Quadrants, with double sights, old; the property of H.H. Prince Pless, Fürstenstein.   *The Breslau Committee.*

1777. Quadrant, for the observation of the height of the sun, old; the property of H.H. Prince Pless, Fürstenstein.
The Breslau Committee.

1779. Small Semicircle, with double sights, for observing the heights of the sun, old; the property of H.H. Prince Pless, Fürstenstein.   *The Breslau Committee.*

1777a. Quadrant, by Langlois.   *Paris Observatory.*

1778. Pillar Sextant, on staid, with artificial horizon.
John Browning.

These instruments are intended for use in an observatory, or otherwise on land, for the purpose of obtaining accurate time.

1781. 12-inch Astronomical Quadrant, by Bird, employed in the observations of the transit of Venus.

Royal Society.

1782. Prismatic Circle, constructed in 1843, from the design of Professor Kaiser.

H. G. Van de Sande Bakhuyzen, Director of the Observatory, Leyden.

By means of two observations the measurement of the angle can be obtained without any instrumental error, excepting those of division. Angles can be measured from 0° to 170°.

1783. Kaiser's Prismatic Circle, constructed on the same principles as the preceding. The construction of the stand, and of a few details, have been improved in this second model.

H. G. Van de Sande Bakhuyzen, Director of the Observatory, Leyden.

d. EQUATORIALS.

1784. Equatorial Telescope by Abraham Sharp.
The Council of the Yorkshire Philosophical Society.

Abraham Sharp, eminent mathematician, mechanist, and astronomer descended from an ancient family at Little Horton, near Bradford in York shire, was born about 1651. He was apprenticed to a merchant at Manchester, but his genius led him strongly to the study of mathematics, both theoretical and practical. By the consent of his master, he quitted business and removed to Liverpool, where he studied mathematics, astronomy, &c., and where for a subsistence he opened a school, and taught writing and accounts. He had not been long at Liverpool when he fell in with a merchant from London, in whose house the astronomer, Flamsteed, then lodged. Sharp contracted an intimate friendship with Flamsteed, by whose interest and recommendation he obtained a more profitable employment in the dockyard at Chatham, where he continued till his friend and patron, knowing his great merit in astronomy and mechanics, called him to his assistance in contriving, adapting, and fitting up the astronomical apparatus in the Royal Observatory at Greenwich, which had been recently built, about 1676. He was principally employed in the construction of the mural arch, which in 14 months he finished, greatly to the satisfaction of Flamsteed. According to Smeaton this was the first good instrument of the kind, and Sharp the first artist who cut accurate divisions upon astronomical instruments. When it was constructed Flamsteed was 30, and Sharp 25 years of age. Sharp assisted Flamsteed also in making a catalogue of nearly 3,000 fixed stars, with their longitudes and magnitudes, their right ascensions and polar distances, with the variations of the same while they change their longitude by one degree. Among other indications of great genius, it was stated that Sharp made most of the tools used by joiners, clockmakers, opticians, and mathematical instrument makers. The telescopes he made use of were all of his own making, and the lenses were ground, figured, and adjusted with his own hands. He died July 18th, 1742, aged 91.

1784b. Gregorian Telescope, made by the celebrated John Hadley, A.D. 1726, and handed down in his family as being the first telescope of the kind ever made.

Cambridge Observatory, J. C. Adams, Director.

1787. Small Universal Equatorial, formerly belonging to the late Dr. W. H. Wollaston. *H. Wollaston Blake, F.R.S.*

1785. Eight and a half inch Reflecting Telescope, with parabolised silvered glass mirror. *John Browning.*

Equatorially mounted, with powerful driving clock, battery of improved achromatic eye-pieces, double prism solar eye-piece for observations of the Sun. Position micrometer, and new double image micrometer, with rotating hour circle, to facilitate finding objects without calculations.

1786. Four and a half inch Reflecting Telescope, with parabolised silvered glass mirror, on parallactic stand, for following the heavenly bodies with a single motion. *John Browning.*

This instrument was contrived for educational purposes; the mirrors are warranted to be of such quality as to bear well a power of 500 diameters.

1788. Equatorial, small, capable of carrying a telescope of 3 to 3½ inch aperture with perfect steadiness. *Yeates & Sons.*

1789. Photograph of Equatorial Mounting of a Refractor, of 9″ aperture and 13′ focal length, with micrometer and position-circle at the eyepiece. *A. Repsold and Sons, Hamburg.*

The declination circle can be read close to the eyepiece of the telescope; it is illuminated by the lamp which illuminates the cross-wires in the field of view of the telescope. By means of one perpendicular roller beneath the centre of gravity, the pressure of the hour axis on the bearings is removed. The clockwork is arranged in the head of the cast-iron column of which the foot is below the floor.

1790. Orbit-sweeper. Equatorial arrangement of a telescope of 6″ aperture, 8′ focal length, with third axis, 1874. Constructed for the Observatory of Strassburg. (Photograph.)

A. Repsold and Sons, Hamburg.

In this equatorial arrangement the head of the declination axis carries at right angles the socket of a third axis, about which the telescope can be revolved, whilst its optic axis is inclined to the third axis at an angle of 90° ± 2° By means of this arrangement a quickly moving celestial body can be easier found, as the third axis is directed to the pole of the projection of its orbit. (See G. B. Airy's observations on an "Orbit-sweeper," in the monthly notices of the R. Astron. Soc.) The telescope can also be placed parallel to the third axis at the head of the declination axis, and has then only an equatorial movement. This instrument is provided with clockwork, and has a changeable polar distance from 0° to 66°.

1791. Model of the Great Melbourne Reflector, completed by Messrs. Grubb & Son in 1868. Scale, ¾ inch to a foot (1/16). *Howard Grubb, F.R.A.S.*

Diameter of great mirror, 48 inches.
Focus, 30 feet 6 inches.
Form, Cassegrainian.
The ventilated tube formed of steel lattice bars.
Quick motion in declination ⎫
Slow motion in declination ⎪
Slow motion in RA - - ⎬ available from eye end of telescope.
Clamping in declination - ⎭

1792. Model of the Great Refracting Telescope, of 27 inches aperture, for the new Imperial Observatory at Vienna, now in course of construction at Mr. Howard Grubb's new Astronomical Works, Rathmines, Dublin. Scale, 1 inch to a foot ($\frac{1}{12}$th).

Howard Grubb, F.R.A.S.

In this instrument the reading of all circles, right ascension as well as declination, is accomplished from eye end of great telescope.

Also quick motion in right ascension ⎫
„ quick motion in declination - ⎪
„ slow motion in right ascension - ⎬ All available from eye end of tele-
„ slow motion in declination - ⎪ scope.
„ clamping in right ascension - ⎪
„ clamping in declination - - ⎭

The one lamp hanging in end of declination axis illuminates—
Upper right ascension circle.
Declination circle on two opposite sides.
Bright and dark fields of micrometer.
Position circle of micrometer.
Field of 4-inch finder.

A second right ascension circle is available for reading from ground floor (south end), where also is a handle for quick setting, right ascension, and a sidereal clock face. The base of the instrument forms a chamber about 12 feet by $4\frac{1}{2}$ feet, in which is contained the clock.

1792a. Model of Equatorial Mounting for the 3-ft. Reflector at Parsonstown, recently erected for the Earl of Rosse. *Earl of Rosse, F.R.S.*

1793. Photograph of a **Heliometer,** with object glass of 4″ aperture and 5′ focal length. This instrument was used by the Russian expedition for the observation of the transit of Venus, 1874. *A. Repsold and Sons, Hamburg.*

The telescope revolves in the head of the declination axis. The scales on both halves of the objective can be read by one microscope, of which the micrometer is close to the eyepiece of the telescope, and the same microscope serves to read the metal thermometer on the head of the objective. The slides of the objective move simultaneously on cylindrical surfaces in opposite directions. The position circle can be read and all the movements made close to the eyepiece. The instrument is mounted equatorially with changeable polar distance from 0° to 66°, and is moved by clockwork.

1794. Photograph of an Equatorial Refractor constructed for the Observatory in Düsseldorf. *Carl Bamberg, Berlin.*

e. EYE-PIECES.

1795. Eye-piece Shutter for Telescopes. Allowing the aperture to be opened and closed by turning the head of the eye-piece.　　　　　　　　　　　　*Captain J. E. Davis, R.N.*

This is effected by fitting the kidney-piece with a fulcrum pin and a lever, the latter passing through the side, which is acted on by the head being turned. It obviates the necessity of the slide or kidney-piece being fitted with a protruding pin, which frequently is the cause of breaking the nail, or (with gloves on) not being felt ; the pin also often loosens, and drops out.

1796. Double-Image Micrometer or Eye-piece Helio-meter.　　　　　　　　　　*C. A. Steinheil Sons, Munich.*

In this instrument the images are formed by means of two rectangular prisms, each of which revolves on an axis giving measurements by a micrometer screw. The prisms reflect at an angle of 45°, and are placed in a pencil of parallel rays of light. Thus when the reflecting surfaces form an angle with each other, the pencils of rays do not issue mutually distorted, as in other heliometers, but remain central at all angles ; also the varying distance of the mirror from the plane of the image has no longer an influence on the definition of the images, which appear without parallax. A small telescope with its objective parallel to the axis of the telescope serves as eye-piece ; it is placed parallel to the telescope axis. The relative illumination of the images changes with the distance of the greatest diameter from the field of view, in which the images respectively move. Any illumination which is taken away from one image is added to the other ; the position-circle gives single minutes.

1796a. Zöllner Eye-piece, to be used as a star spectroscope for very small stars, and giving especially good definition.

　　　　　　　　　　　　　　　　　　A. Hilger.

f. MICROMETERS.

1797. Collection of the finest **Micrometer Screws,** of almost perfect accuracy, and a small instrument for observation.

　　　　　　　　　　　　　　　Hugo Schröder, Hamburg.

This apparatus serves for the examination of micrometer screws ; those exhibited are shown as examples of the great accuracy and delicacy which can be attained by cutting the screws according to the method invented by Hugo Schröder. A table showing the results of the examination of one of the screws by Dr. Vogel is subjoined.

1797a. Collection of Micrometer Screws, cut with Hilger's special apparatus, for astronomical purposes.

　　　　　　40 threads to the inch (1).
　　　　　100　　　,,　　　　,,　　(2).
　　　　　200　　　,,　　　　,,　　(1).
　　　1,000　　　,,　　　　,,　　(1).

　　　　　　　　　　　　　　　　　A. Hilger.

1798. Position Micrometer, constructed for the refractor of the Royal Observatory at Berlin.　　*Carl Bamberg, Berlin.*

1799. Four Micrometers, for Astronomical Telescopes.
F. W. Breithaupt and Son, Cassel.

Micrometer divisions on glass, for different kinds of astronomical observation. The one with circular divisions was used by Prof. Spörer, of Anclam, for the observation of the solar prominences at Aden during the total solar eclipse.

1800. Electro-magnetic Registering Apparatus.
M. Th. Edelmann, Physico-Mechanical Institute, Munich.

1800a. Stereo-Micrometer. Apparatus used with binocular telescopes, and measuring both angles and distances, for geodetic and astronomical measurements. Both eyes being employed, one measuring, the other observing, position and distance are thus given simultaneously. With stereo-micrometrical photographs of landscapes. *Prof. Carl Wenzel Zenger, Prague.*

1800b. Positive Eye-piece, with a bright line micrometer, invented by the exhibitor for the purpose of making measurements in the dark. *A. Hilger.*

II.—INSTRUMENTS FOR DETERMINING THE MO-LECULAR STRUCTURE OF THE HEAVENLY BODIES.

a. Spectroscopes.

1801. Spectroscope of Donati, which served him in the observations of the eclipse of the sun, made in Agosta (Sicily) the 20th of December 1870.
The Royal Institute of " Studii Superiori," Florence.

With the aid of this spectroscope Donati was able, during the eclipse, to observe the luminous lines of hydrogen. He afterwards constructed a second one, composed of 25 prisms of very great dispersive power. His premature death, however, robbed science of the important observations which would certainly have resulted from so perfect an instrument.

1802. Amateurs' Star Spectroscope. *John Browning.*

This instrument will show the lines in the spectra of stars of the second magnitude, when used with an object-glass only 3 in. in diameter. By detaching the cylindrical lens, the instrument may be used as a small direct vision spectroscope.

1803. Spectrum Apparatus, for the observation of the spectra of the fixed stars, planets, and nebulæ ; arranged after the spectrum apparatus of Boshkamper (belonging to the Observatory of Hamburg). *Hugo Schröder, Hamburg.*

The spectrum apparatus is constructed on the simple principle which has proved so successful at the Observatory of Bothcamp, with the difference, however, that this apparatus is arranged for absolute measurement, and that the one at the Observatory of Hamburg is attached at right angles to the principal axis of the refractor.

1804. Spectrum Apparatus, for observing the **Solar Prominences,** to be attached to the collimator of the spectroscope.

Hugo Schröder, Hamburg.

This spectrum apparatus, which is in reality a supplement of the first, can be fastened to the collimator of the other one with great readiness. The object of this apparatus is the observation and measurement of the solar spectrum as well as of the solar prominences. The principle on which it is constructed differs from that of the former in that the rays after once passing through the system of prisms do not issue from it in the same direction as they entered, but are refracted and dispersed by the heavy prisms of flint glass. By means of a rectangular prism of crown glass the rays are compelled to pass through the system a second time, and leave it in a direction parallel to that of their first entrance. By a second prism of crown glass the rays are reflected into the observing tube which is attached to the prism holder. The movement of the spectrum across the field of view, as well as the absolute measurement, is effected by turning the first prism of crown glass by means of the micrometer screw. This apparatus is, on account of its convenient and highly stable construction, particularly to be recommended for observers who have scanty room at their disposition, and yet wish to undertake accurate measurements.

1805. Star Spectroscope, after Dr. H. C. Vogel (described in the Berichte der königlichen sächsischen Gesellschaft der Wissenschaften, December 1873). *H. Heustreu, Kiel.*

This apparatus recommends itself for its simple construction, its varied application to all kinds of observations, and its reasonable price.

1806. Spectroscope made by Merz in Munich.

Prof. Dr. Winnecke, Strassburg.

1806a. Parts of a Solar Spectroscope, made by Elliott Brothers in 1869. *J. Norman Lockyer, F.R.S.*

In this instrument the prisms are brought to minimum deviation by means of a spring, suggested by Mr. G. W. Hemming, and the light is brought back through the prisms by a total reflection prism at the end of the train, on the plan first employed, it is believed, in this instrument, and suggested by the exhibitor.

1806b. Solar Spectroscope, with diffraction grating on speculum metal, presented to the exhibitor by Mr. Rutherfurd, of New York. *J. Norman Lockyer, F.R.S.*

1806c. Solar Spectroscope, used since 1868 in observing solar phenomena, made by Browning.

J. Norman Lockyer, F.R.S.

1806d. Slit arrangements for Spectroscopes.

J. Norman Lockyer.

1806e. Photographs of Hilger's Stellar and Solar Spectroscope. *A. Hilger.*

4552. Photographs representing various Scientific Instruments in the Observatory of the Collegio Romano, Rome.

Padre Secchi, Director, Rome.

No. 1. Spectroscope applied to the refractor of Merz, provided with a network by Rutherford.

No. 2. Spectroscope with network, applied to the refractor of Merz, with the cover adapted to the plate of the network.

No. 3. Spectroscope with five prisms, with double passage of the ray by reflection, and with automatic movement to bring within the eyepiece the various dispersed rays.

No. 4. Spectroscope with three angular prisms, and one with direct vision, enclosed in the collimator.

No. 5. Large objective prism of Merz, of 6 inches, for the spectrum analysis of the stars.

No. 6. Meteorograph of Padre Secchi.

No. 7. Curves traced with the meteorograph.

4555. Photographs representing the Daily Drawings of the Solar Chromosphere, made by the spectroscope of Professor Lorenzo Respighi.

Prof. Lorenzo Respighi, Director of the Royal Observatory of the Campidoglio, Rome.

On the 26th October 1869, at the Observatory of the Campidoglio, the spectroscopic observation and the daily representation of the chromosphere on the solar horizon were for the first time undertaken, and this work has been regularly continued up to the present time. The instrument used in these observations is an equatorial by Merz, with telescope of 4½ inches aperture, to which is applied a spectroscope with direct vision, by Hoffmann, with five prisms, with circle of position to fix the place of the various parts of the chromosphere and of the protuberances. Notwithstanding the small dimensions of the instrument and the moderate dispersion of the spectroscope, the chromosphere and the protuberances are clearly exhibited even in their smallest details.

The Photographs I., II., III., IV., and V. represent 140 drawings of the chromosphere made with this apparatus by Professor L. Respighi.

d. STELLAR PHOTOMETRY.

1807. Astronomical Photometer. Designed by Professor Thury.

Geneva Association for the Construction of Scientific Instruments.

The apparent brightness of a heavenly body seen in the telescope is gradually reduced by the movable diaphragm placed before the objective, and if necessary by the interposition of one or two dark mirrors placed behind the eyepiece in the square box, which is exposed with the diaphragm. The light is gradually reduced until the object is no longer visible. The aperture of the diaphragm is then shown upon a dial placed under the eye of the observer.

The full description of this apparatus is to be found in the "Archives des Sciences physiques et naturelles de Genève," 1874.

1808. Zöllner's Astrophotometer, for measuring the light of the heavenly bodies by comparison with that emitted by the brightest portion of the flame of a paraffin lamp.

Earl of Rosse, F.R.S.

The accuracy of the work done with this instrument depends on the fact that, though the total light emitted by the flame varies with its size, the *intensity* of the brightest part is appreciably constant. Two artificial stars are formed by means of a pin hole, a double concave lens, and a double convex lens, which appear in the field by reflexion from front and back faces of a plate of glass alongside of the image of the real star whose light passes through the plate. The intensity of the artificial star is varied, first by changing the pin hole, and finally by two Nicol's prisms, the colour being first matched with that of the star by means of a third Nicol, with a quartz plate between it and the first of the other two Nicols. The instrument is provided with object glasses of various sizes (and diaphragms) up to 2¾ inches, and, if fainter stars are to be examined, can be screwed on to the eyepiece of an equatorial instrument. A second arrangement, like the first, but without the quartz plate arrangement, forms an artificial star from moonlight, for comparison of the light of that body with the artificial star.

896. Photometer, constructed by Schwerd for the Observatory of Pulkowa.   *The Imperial Observatory, Pulkowa.*

In agreement with Prof. Argelander and M. Otto Struve, the late Prof. Schwerd of Speyer constructed, in 1863, four photometers of the same size, two for Russia (Pulkowa and Wilna), the third for the Observatory, Bonn, the fourth for his own use. The principle of the construction is that of comparing the light of different stars exhibited in the same field by telescopes of different aperture. The diameter of the diaphragms to be applied to the two object-glasses, and corresponding systems of lenses, for purpose of producing equal light and colour, gives the measure of the relative brightness. The two telescopes, one of 2·3 in. aperture and 4 ft. focal length, the other of 1·2 in. aperture and 2 ft. focal length, are parallactically mounted and moved together by the same clockwork (which is not exhibited), so that the images of the two stars keep constantly the same place in the field during observation. Being worked out in all parts with greatest care and on sound optical principles, it can hardly be doubted that this instrument perfectly answers its purpose ; but on account of the great number of constants to be determined for it, its use is rather difficult. Until now only two of these instruments have been practically applied, that of Schwerd himself, and the one constructed for Wilna. In both cases the first problem has been the determination of the co-efficient of extinction of light by the atmosphere of the earth.

1809. Astrometer for Reflecting Telescopes, invented by the contributor.   *E. B. Knobel.*

This instrument has been invented for determining the magnitudes of stars on the principle of limiting apertures. It consists of an equilateral triangular aperture, constructed of two plates, one forming the base and the other the opposite angle of the triangle, connected by a screw shaft of peculiar construction. The upper portion carrying the angle plate, being a *right handed*, and the lower connected to the base plate, a *left handed* screw. The pitch of the upper screw is *twice* that of the lower. By simply turning the milled head at the end of the shaft, the aperture is made smaller or larger within the limits of the triangle inscribed in the telescope tube, and zero. The instrument depending on the mathematical principle sin $30° = \frac{1}{2}$ the aperture is thus always accurately equilateral, and concentric with the mirror or object glass. The graduated base and the micrometer head give the side of the triangle, whence the aperture is readily obtained.

1810. Astro-Photometer, according to Glan's system.
Schmidt and Haensch, Berlin.

III.—OBJECTS ILLUSTRATING THE HISTORY OF THE TELESCOPE AND ASTRONOMICAL OBSERVATION.

1811. Galileo's Telescope. Object-glass, 50 mm. in diameter ; eye-glass, plano-concave, 23 mm. in diameter. It served for the most important discoveries and experiments of Galileo, and was constructed by himself.
The Royal Institute of " Studii Superiori," Florence.

1814. Galileo's Telescope. Object-glass, 38 mm. in diameter ; eye-glass, double concave, 19 mm. in diameter; made by Galileo.
The Royal Institute of " Studii Superiori," Florence.

1815. Object Glass of the diameter of 40 mm., **made by Galileo,** and with which he discovered the satellites of Jupiter on the 9th of January 1610, and first saw the spots on the sun. Having been broken, it was presented by Viviani to Prince Leopoldo dei Medici, who placed it in a frame finely wrought in ivory, with inscriptions.
The Royal Institute of " Studii Superiori," Florence.

Galileo's part in the invention of the telescope, and the merit due to him, clearly appears from the following extract of the answer given by him, in the " Saggiatore," to Padre Orazio Grassi, a Jesuit. " The share of credit that " may be due to me in the invention of this instrument (the telescope) and " whether I can reasonably claim it as my own offspring, I have expressed in " my ' Avviso Siderco,' which I wrote in Venice. I happened to be there when " the news reached me that a Dutchman had presented Count Maurice with a " glass, by means of which things far away appeared just as clearly as if they " were quite close at hand, but without any detail being mentioned. Upon " hearing this, I returned to Padua, where I was then living, and pondered over " this problem, and the first night after my return I found it out. The following " day I made the instrument. After that I immediately set to work to make " a more perfect one, which being completed six days afterwards, I took to " Venice, where so great a marvel attracted the attention of almost all the " principal gentlemen of that republic. Finally, by the advice of one of my " dearest patrons, I presented it to the prince in presence of the college. The " gratitude with which it was received, and the esteem in which it was held, " are proved by the ducal letters which I have still preserved, since they bear " witness to the generosity of His Serene Highness in confirming me for life in " my lectureship of the studio of Padua, with double the payment of that which " I had had beforehand, which in its turn was more than three times what

" any of my predecessors had enjoyed. These facts, Sig. Sarsi, did not take
" place in a forest or in a desert, they occurred at Venice, and if you had
" been there then you would not have put me down simply as a foster-parent
" of the invention. But perhaps some one may tell me that it is no small
" help towards the discovery or solution of any problem to be first of all
" apprised in one way or another of the truth of its conclusion, and to know
" for certain that it is not an impossibility that is being sought after; and
" that, therefore, the information and the certainty that the telescope had
" already been made, were of such use to me, that in all probability I should
" never have made the discovery without them. To this I answer that the
" help given me by the information I received, undoubtedly awoke in me
" the determination to apply my mind to this subject, and that without it
" I should very likely never have turned my thoughts in that direction; but
" besides this, that I cannot believe that the notice I had had could in any
" way render the invention easier; and I say moreover, that to find the
" solution of a problem already thought out and expressed requires far greater
" genius than to discover one not previously thought of; for in the latter
" case chance can play a great part, while the former is entirely the work of
" reasoning. We know that the Dutchman, the first inventor of telescopes,
" was simply a common spectacle-maker, who, handling by chance glasses of
" various kinds, happened at the same moment to look through two, the one
" concave, the other convex, placed at different distances from his eyes, and
" in this manner he observed the effect which was produced, and thus invented
" the instrument; but I, warned by the aforesaid notice, came to the same
" conclusion by dint of reasoning; and since it is by no means difficult to
" 'follow, I should like to lay it before you. This, then, was my reasoning;
" this instrument must either consist of one glass only, or of more than one
" glass, it cannot be of one alone, because its figure must either be convex, or
" concave, or comprised between two parallel superficies, but neither of
" these shapes alters in the least the objects seen whilst increasing or
" diminishing them, for it is true that the concave glass diminishes, and that
" the convex one increases them; but both show them very indistinctly, hence
" one glass is not sufficient to produce the required effect. Passing on to
" two glasses, and knowing that the glass of parallel superficies has no
" effect at all, I concluded therefore, that the desired result could not possibly
" follow by adding this one to either of the other two. Hence, I restricted
" my experiments to combinations of the other two glasses, and I saw how
" this brought me to the result. Such was the progress of my discovery, in
" which you see of how much avail was the knowledge of the truth of the
" conclusion. But Sig. Sarsi, or others, believe that the certainty of the
" result affords great help in producing it and carrying it into effect. Let
" them read history and they will find that Archites made a dove that could
" fly, and that Archimedes made a mirror that burned at great distances, and
" other admirable machines; now by reasoning on these, they will be able
" with very little trouble, and very great honour and advantage, to discover their
" construction, but even if they do not succeed, they will derive the benefit of
" being able to certify themselves that the ease of fabrication which they
" promised themselves from the pre-knowledge of the true result was much
" less than they had imagined."

We must add that many of the principal persons in Europe were anxious
to have Galileo's telescopes, and that he sent presents of them to the Grand
Duke of Tuscany, to Prince D. Antonio de' Medici, the Elector of Bavaria,
the Emperor Mathias, Cardinal Borghese, the Queen of France, the King of
Spain, the King of Poland, the Landgrave of Hesse, Giuliano de' Medici,
Cardinal Del Monte, Cardinal Montalto, to the Dukes of Acerenza and of
Acquaivra, &c., &c.; he also sent some to Holland, whence Constantine

Huygens writes in 1637, that it was impossible to find artificers in that country having skill enough to make telescopes of sufficient power to observe the satellites of Jupiter.

The principal discoveries made by Galileo by means of his telescope in chronological order, are as follows:—

He first of all observed the cause of the lunar spots, recognising in them the effect of shade produced by the heights which must exist upon the Moon, and thus proving that it was covered with mountains and table-lands; and he pointed out the way for determining their heights. He likewise perceived that the secondary light transmitted to us from the Moon proceeds from the reflection of the solar rays from the suface of the earth to that of the Moon.

Directing his telescope towards the Milky Way he was the first to certify the fact that it was composed of myriads of stars.

On the 3rd of January 1610, he discovered that three secondary planets revolve round Jupiter, and on the 13th of the same month he found the fourth. He named these satellites the "Stelle Medicee." He showed that they, in revolving round Jupiter, underwent eclipses, just as our moon does. He determined the length of their rotations, remarking that by their means it is possible to observe 1,000 eclipses a year, which would be of great utility for finding the longitude of any place.

In the month of August 1610, whilst observing Saturn, he noticed that it appeared to be continually accompanied by two planets which touched it at the extremities of the same diameter, so that he called it *tricorporeal*.

In the same month he first saw the spots on the Sun, determined the direction of their motion, and, with happy intuition, compared them to terrestrial clouds.

On the 30th September 1610 he found that Venus changed figure like the Moon, so he called it *horned* (*la falcata*).

He likewise observed Mars and its phases, and reasoned admirably upon the comets.

Taking his stand upon these astronomical observations, Galileo embraced, maintained, and defended the system of Copernicus, so that he had to undergo the cruel persecutions of Rome, which compelled him to utter that recantation, at the end of which he broke out into the now celebrated exclamation, "*Eppur si muove !*"

1816. Telescope of Torricelli, with an object-glass of 50mm. in diameter; eye-glass, plano-convex, 22mm. in diameter.

The Royal Institute of " Studii Superiori," Florence.

Torricelli was the first who set himself the task of solving the optical problem of finding the proper shape to be given to the surfaces of the glasses used for telescopes. After many months of study and hard work, as he himself relates, he succeeded in his object and the result was fully confirmed by experience. He did not make his lenses on metallic forms or moulds, but fixed them on pieces of slate with a cold mixture, in order to obviate the defects caused by the change of shape which the glasses undergo when warmed. To him we also owe the microscope formed of one lens, or rather of a little glass ball, which worked with a lamp.

1818. Day Telescope of Divini, from S. Severino. The diameter of the object-glass is 70mm.; the diameter of the first eye-glass and of the second lens, double-convex, 33mm. The third lens is wanting.

The Royal Institute of " Studii Superiori," Florence.

Divini constructed telescopes between the years 1646 and 1668, and even made some of the length of 72 palmi Romani.

1819. Day Telescope of Ippolito Mariani, a Florentine, surnamed *il Tordo;* first practical optician; taught by Galileo. Object-glass, 103 mm. in diameter. The eye-glass is composed of two lenses, plano-convex, of mm. 48 in diameter. The second and third lenses are similar to the eye-glass.

The Royal Institute of " Studii Superiori," Florence.

1820. Day Telescope of Campani, romano, object-glass 47 mm. in diameter; eye-glass, double convex, 27 mm. in diameter; second and third lenses the same.

The Royal Institute of " Studii Superiori," Florence.

Campani's well known telescopes are to be found in many observatories; some of them are 70, 100, 150, and even 210 palmi Romani in length.

1821. Telescope by Amici.

The Royal Institute of " Studii Superiori," Florence.

1830. Volume, Experiments in Natural Science Accademia del Cimento.

The Royal Institute of " Studii Superiori," Florence.

1831. Telescope, by Chr. Huygens. The objective ground and polished by him, and bearing his signature.

Prof. Dr. P. L. Rijke, Leyden.

Its focal distance is 3·906 m., and aperture 0·0616 m. The eyepiece is composed of three convex lenses. The lens *a*, the nearest to the objective, and the lens *b*, following it, have a focal distance of 0·105 m., and an aperture of 0·04 ; the lens *c* has a focal distance of 0·079 and an aperture of 0·038 m. The distance between *a* and *b* is 0·212 m. ; that between *b* and *c* 0·182 m. The eye should be placed at a distance of 0·058 from the lens *c*. The lenses *b* and *c* serve only to rectify the images.

1831b. Telescope by Campani.

Royal Museum, Cassel (Director, Dr. Pinder).

The tube is wood; it measures, when drawn out, 16 feet. This was bought in Rome by Landgrave *Charles.*

1831g. Two Telescopes (Achromatic), made by Dollond, about 1765, for the Russian expeditions for the observation of the transit of Venus in 1769.

The Imperial Academy of Sciences, St. Petersburg.

Object-glasses of 3·6″ and 2·8″ aperture, focal length 11·2′ and 8·5′. There are several of each size in possession of the Academy. As they are not designated by numbers or other distinct marks, it cannot exactly be made out which of them has been used by the different observers.

1832. Terrestrial Refractor, made by Van Deyl, at Amsterdam, in the year 1781. *Teyler Foundation, Haarlem.*

1831a. Huygens' Aërial Telescope. *Royal Society.*

1831b. Object-Glass by Huygens, 170 feet focal length.

Royal Society.

1831c. Object-Glass by Huygens. *Royal Society.*

1831d. Object-Glass, Venetian, of 90 feet focal length.
Royal Society.

1831e. Original Reflecting Telescope, made by Sir Isaac Newton. *Royal Society.*

1833. The Herschel 7-foot Telescope. The original instrument constructed by Sir W. Herschel.
Royal Astronomical Society.

The tube is 7 inches in diameter and 7 feet long. Both mirrors were finished by Sir W. Herschel's own hands; they are sound and whole, but are much tarnished, and the large mirror was damaged in a fire some years since. The framework of the stand is entire, but the moving screws, cords, &c. are useless in their present condition.

1784a. Newtonian Telescope, belonging to Sir W. Herschel, and used by him while living in Bath. He is said to have discovered the planet Uranus by its means. Focus, 7 feet; diameter of speculum, 6⅓ inch. *Edwin Smith.*

This telescope was purchased at Sir W. Watson's sale, Pulteney Street, Bath, about 1860. It had apparently remained after Sir W. Watson's death for some time in a lumber room of the house, and when purchased by Mr. E. Smith a paper of directions for the use of the different eyepieces was discovered in the drawer of the stand.

A portrait in oil of Sir W. Herschel, in one of the rooms of the same house, was sold at the same time. Dr. Brabant, of Marlborough Buildings in this city, who was a great friend of Sir W. Herschel, has often called on Mr. Smith to see the telescope, and repeatedly declared to him that this was the same instrument by which the planet Uranus had been discovered in 1781. It is supposed to have been made by Sir W. Herschel, while organist in the Octagon Chapel, Bath.

1834. 10-ft. Newtonian Reflecting Telescope by Sir William Herschel, with 8½ inch large mirror, small plane reflecting mirror, and several eye-pieces of various powers.
Rev. Robert Main, Director of the Radcliffe Observatory, Oxford.

This telescope was made by Sir William Herschel for the Radcliffe Observatory in the year 1812, and was received at the Observatory in April 1813; Sir William himself having come to Oxford to superintend the mounting and the adjustments of the mirror.

The correspondence with Dr. Robertson, who was then Radcliffe Observer, is preserved at the Observatory.

1834a. Eight plans of the Telescope, made in London at the end of the last century, under the direction of Sir William Herschel, for the Royal Observatory at Madrid.
Astronomical Observatory, Madrid.

These plans give an exact idea of all the details of the instrument and mounting.

The speculum was of 2 feet aperture and 25 feet focal length.

This instrument was sent from London in 1801, and set up at Madrid in 1804. Four years afterwards the French converted the observatory into a fort, the telescope was destroyed, the only part remaining being the speculum.

1836a. Small Hand Speculum Polishing Machine, constructed and used by Sir William Herschel to polish specula of

7 feet focus (6 in. aperture).   Smaller machines for smaller specula
were made of this construction, and are in part preserved.   The
polishing machines used to figure and polish large specula (18 in.
and 49 in. diameter) for the 20 foot and 40 foot telescopes used at
Slough and at the Cape of Good Hope appear to have been
of the same construction as this instrument and of proportionally
larger size.   *Prof. A. S. Herschel.*

1836b. Brass foot used in place of the lead block to carry
the furrowed pitch rubber or polisher revolving with the bed plate
of the machine, to whose centre it is fastened down by screws.
For the larger sized mirrors these brass plates were used as well
as for smaller sizes to replace the lead foundation or bearer of the
pitch, and some 18 in. ones besides this small one are preserved.

Prof. A. S. Herschel.

1836. Compound Speculum, of 2 feet aperture.

Earl of Rosse, F.R.S.

This is one of the earlier attempts of the late Earl of Rosse to construct
specula of considerable dimensions of the hardest and most reflective quality
of speculum metal.

To avoid the difficulty of casting the mirror in one, a cubical block of
speculum metal was sawed into laminæ, and these were laid side by side on a
ribbed backing of a zinc-copper alloy of the same co-efficient of expansion,
whose surface had been previously tinned; the whole was carefully brought up
to the melting point of tin, and melted tin applied to unite the whole. Though
superior in rigidity to the solid metal speculum afterwards successfully constructed, it was discarded in favour of the latter, owing to the injury to
definition through diffraction at the junctions of the laminæ.*

**1836aa. Speculum (Experimental), with Annulus se-
parate from Central Portion.**   *Earl of Rosse, F.R.S.*

Constructed for the purpose of attempting to correct spherical aberration
by advancing the annulus before it had been shown to be possible to produce
a paraboloid figure. Given up in favour of the *solid* speculum for same
reason as the last (diffraction).

1835. Discs of Optical Glass for Refracting Equatorial:
      1 Hard crown.
      1 Dense flint.   *Chance Brothers & Co.*

1835a. Series of seven Glass Parabolic Mirrors,
from 3½ in. to 15 in. in diameter, from 2 ft. to 10 ft. focus, silvered on the surfaces by Liebig's process.   *John Browning.*

**1836c. Achromatic Telescope with Rochon's Prys-
matic Micrometer** (in its main tube) with vernier reading to
1″, formerly belonging to the Rev. Dr. Peacock, F.R.S., Treasurer
to the Royal Astronomical Society, and probably the identical
instrument used by the late Mr. Francis Baily, F.R.S., &c., in his
observations of the Solar Eclipse of September 20, 1820.

William Lawton.

---

* N.B.—Another compound speculum of 3-foot aperture is still preserved,
but the smaller one is sent, as the weight of the other is considerable.

1836d. **Small Gregorian Telescope,** maker unknown, formerly the property of the late Dowager Viscountess Midleton, date about 1750.   *Hon. Miss Brodrick.*

1853. **Complete Transit of Venus Astronomical Equipment,** as used by the English expeditions.
                          The Astronomer Royal.

1853a. **One of the Telescopes** used in the "Transit of Venus" Expedition.
       The French Commission for Observing the Transit of Venus in 1874.

EYE-PIECES AND OBJECTIVES.

1837. **Objectives and Eye-pieces** of the 17th and 18th centuries, the greater part of which were ground and polished by Christian and Constantine Huygens.
                  Prof. Dr. P. L. Rijke, Leyden.

No. 1. Objective of 120 ft.
No. 2.   „   84 „
No. 3.   „   85 „
No. 4.   „   43 „ 7 in. } Bearing the signature of Const. Huygens.
No. 5.   „   43 „
Nos. 6 & 7 „   34 „ each
No. 8.   „   34 „ Bearing the name of Chr. Huygens.
No. 9.   „   34 „ Bearing the signature of Chr. Huygens.
No. 10.   „   10 ft. 8 in. Bearing the name of Chr. Huygens.
No. 11. Bearing the name of Hartsoeker.
No. 12. Objective of 32 ft. Bearing the signature of Marcell.
Nos. 1a and b. Eye-pieces of 7½ and 8 inch, to use with objective No. 1.
No. 2a.   „   6   „   „   „   „   2.
No. 5a.   „   4½   „   „   „   „   5.
No. 7a.   „   3⅝   „   „   „   „   7.
No. 8a.   „   3   „   „   „   „   8.
No. 10a.   „   2   „   „   „   „   10.

1838. **Photograph** of the **Lens** by which **Huygens** discovered **Saturn's Ring.**   *Prof. Buys-Ballot, Utrecht.*

This lens is stated to be the same by which Christian Huygens made out Saturn to be surrounded by a ring. It bears the inscription "X. 3 FEBR. " ꟙDCLV (Febr. 1655), Admovere oculis distantia sidera nostris."

1838a. **Two large Objective Glasses** mounted on brass.
                          M. Evrard, Paris.

1840. **Metal for a Newtonian Reflector,** with several wooden eyepieces, but without tube or mounting, by Hadley.
                          Royal Society.

IV.—APPARATUS FOR DETERMINING THE EARTH'S MOTION AND DENSITY, &c.

1841. **Apparatus used by Baily in repeating the Cavendish Experiment.**   *Royal Astronomical Society.*

Of this apparatus several parts were missing, but have been lately restored. The original portions are the long mahogany box with glazed ends for the torsion balance, and upright column in the middle for the suspension wires, and a box containing three small leaden, one brass, and two ivory balls, two brass cylinders, and one leaden lenticular weight. A full description of the apparatus will be found in the Memoirs of the Royal Astronomical Society, Vol. XIV.

1842. Gauss's Pendulum for demonstrating the rotation of the earth, executed in the year 1853 by Dr. Meyerstein.

Geodetic Institute of the Observatory at Göttingen (Prof. Dr. Schering, Director).

1878. Sphere bearing traces of M. Foucault's observations on the Rotatory motion of the Earth.

Conservatoire des Arts et Métiers, Paris.

V.—ASTRONOMICAL CLOCKS AND SUNDIALS.

1843. Astronomical Clock, with Sir G. B. Airy's Barometric Compensation. *E. Dent and Co.*

This clock has been fitted up with a Graham escapement; it is in other respects almost a counterpart of the new standard clock of the Royal Observatory, Greenwich.

It is found that the tendency of a clock is to lose with a high barometer, and gain with a low one. Compensation is effected in this way : there is a lever, one arm of which carries a float resting on the surface of the mercury in the cistern of a barometer tube ; the other arm carries a horseshoe magnet which faces the opposite poles of two bar magnets fastened to the pendulum bob. When the barometer rises the mercury in the cistern is depressed, so that the arm of the lever carrying the float falls whilst the other arm rises, thus bringing the horseshoe magnet closer to the bar magnets ; when the barometer falls the same action takes place in the opposite direction, thus increasing or diminishing a force acting in the same direction as gravity.

1843a. Three different Forms of Dipleidoscope. *E. Dent and Co.*

1. The simple form consists of two mirrors placed at an angle of about 60° and in front of them a plain unsilvered glass, the whole combination being mounted, for the sake of conveniently taking observations, in a small cast metal pyramid. The optical arrangement operates in this way. Rays from the sun fall upon the front glass, and part are reflected from it and form an image ; but the remaining part pass on, and meeting first one and then the other mirror, are reflected back through the front glass, and form a second image. The instrument is to be placed so that these images shall appear together in the field of view, a minute or two before apparent noon. Then what is seen is this : as the sun advances to the meridian the two images will approach, they will touch, and gradually cover one another (this observation gives the instant of apparent noon) ; they will continue to move on, and will finally leave one another ; and each of the observations of contact, superimposition, and parting contact will be each separately and together available for determining true time. The base-plate which accompanies the instrument, it is intended should be fastened out of doors, in such a position that its guide bar shall give the right direction to the small metal pyramid.

2. The mural form. This instrument only differs from the preceding in its method of mounting, which secures greater accuracy ; and a small telescope is added, for the purpose of obtaining a finer reading of all the observations.

(One of these instruments is mounted upon the balcony outside. Upon looking into the telescope (which must be shifted up or down until the images are seen) the two apparent suns will be visible in the field of view ; these will gradually approach as above described, and will be superimposed at the instant of apparent noon.)

3. The universal form. This is mounted upon an axis, which by means of a divided "latitude" scale can be placed parallel to the earth's axis. It has also another scale, divided from three hours before, to three hours after noon, and true time may be obtained at any quarter of an hour within this interval. The instrument is furnished with a compass, but this should not be used where more accurate methods of setting it can be employed.

1843b. Astronomical Signaller for the purpose of giving notice of the approach of stars, or any required point in R.A. to the meridian. *E. Dent and Co.*

1844. System of Patent Electro-Sympathetic Clocks.
James Ritchie and Son.

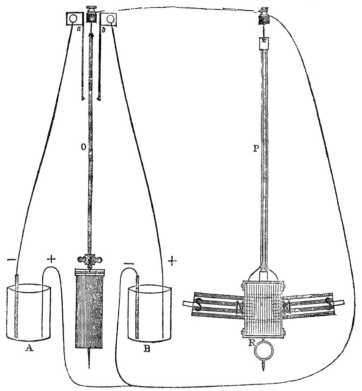

Fig. A.

These consist of—

1st. An ordinary clock, requiring periodical winding, to serve as a normal or motor clock for the system.

Wires from the reverse poles of two galvanic batteries are connected with slender insulated springs, so placed that the pendulum in vibrating touches each alternately, and transmits through the pendulum rod reverse currents to the line wire and subsidiary clocks.

This arrangement may be seen through the glass sides of the clock-case, and in Fig. A.

2nd. Electro-sympathetic clock. The pendulum consists of a coil of insulated copper wire. Within the coil (which forms the ball or bob) a double cluster of magnets, having their similar poles slightly separated, is fixed to the casing of the clock. The currents from the motor clock passing through the wire coil cause it to be alternately attracted and repelled by the magnets,

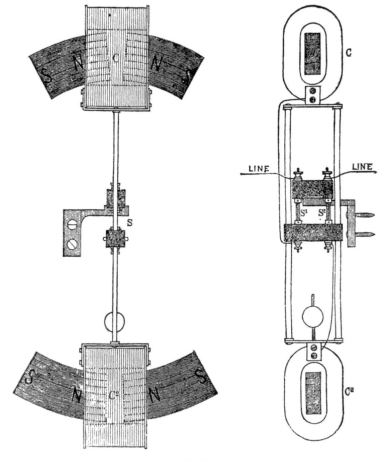

Fig. B.

thereby giving motion to the pendulum, and maintaining it in vibration. The connexions between the batteries and motor and sympathetic clock pendulums with the position of magnets are shown in Fig. A.

3rd. The sympathetic pendulum is also made with an ordinary ball or bob, carrying a single bar magnet passing into small coils of wire fixed to the case and placed in circuit with motor clock, as shown in the square case.

4th. To admit of a pendulum vibrating seconds being introduced within the limits of the usual office round clock-case, two small coils, one below and the other above the point of suspension, are used, having double magnets placed for each coil, through both of which the currents from the motor clock are passed.

This form of pendulum requires a more powerful battery to sustain its motion, and has less momentum than the long pendulums. Its construction is shown in the circular case, and in Fig. B.

5th. An arrangement to enable a pendulum vibrating half seconds to beat seconds is shown in Fig. C.

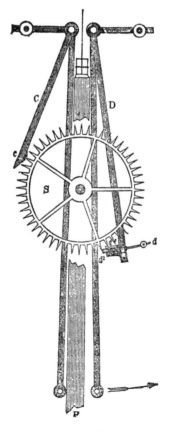

Fig. C.

D d

6th. The wheel-work and hands are propelled by two gravity arms, one on either side of the 'scape wheel, which are alternately raised by the pendulum in its vibration.

The impulse pallets and stops are so adjusted that the wheel is impelled forward half a tooth, and locked upon the opposite arm until released by the return of the pendulum. The action of the propelment is shown in the large-sized movement in the square case, and in Fig. D.

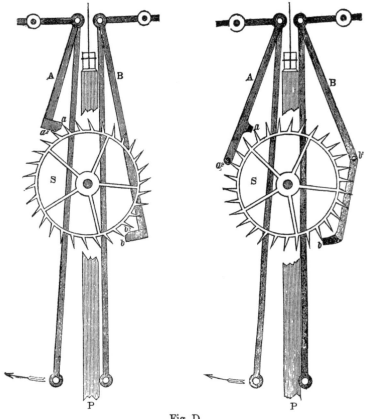

Fig. D.

In this system the standard or motor clock only requires winding, while the sympathetic pendulums, being controlled as well as driven by the electric currents transmitted by the motor clock, are caused to vibrate in unison, and so produce perfect coincidence of time on any number of clocks in connexion though miles distant from it. The electro-motive force being applied at the lower end of the pendulum, a long leverage is obtained requiring only a weak battery, which is less liable to derangement, maintains its constancy for a longer time, and requires less attention than a stronger one. Each subsidiary clock, being dependent upon its own pendulum for the time shown, is not affected by any trip or irregularity in the current, as the momentum acquired by the pendulum will carry forward the wheel-work for some time without any current. In systems where the electro-motive force is applied directly to move the wheel-work, a single miss of the current destroys the coincidence of time shown.

This electro-sympathetic system has for several years been in practical operation in the Royal Observatory and other public buildings in Edinburgh; the Royal Exchange, Manchester; the General Post Office, Glasgow: the Royal Dublin Society, Dublin ; Leith Harbour and Docks, and Aberdeen. The clocks at the two ports last named are about two miles distant from the motor clock. The system is likewise about to be introduced into Birmingham.

1845. Ancient Sundial, for showing the time in any latitude, scale for setting the sun's declination, and equation table.
Elliott Brothers.

1845a. Ancient Sun Dial with correction for latitude and variation, dated 1579. *A. C. Baldwin.*

1848a. Sundial, date 1575. *A. C. Baldwin.*

1846. Sissons' Universal Ring Dial, for finding the sun's declination and place in the ecliptic, the latitude of any place, and the hour of the day. *Adam Dixon.*

1847. Two Ring-shaped Equatorial Sundials.
The Breslau Committee.

1848. Two Sundials, with calendars.
The Breslau Committee.

1849. Compass, with Sundial, of the year 1597, with various movable discs for adjusting the zodiacal circle, &c.
Berggewerkschaftskasse, Bochum, Westphalia.

1849a. Plaster Cast of a Sundial. The original is made of stone and is preserved at the Archæological Museum at Madrid.
Archæological Museum, Madrid.

This was found with a number of other objects at Yecla, in the province of Alicante, Spain. At the upper part there probably was formerly an iron limb to mark the hours, on the Roman system of dividing them in couples. In order to use it the dial was placed in the shade, facing the north. A small spherical concave mirror was placed at a short distance, which reflected the light of the sun upon the dial, and by that means projected the shadow of the needle marking the hours.

At the base of the instrument are inscriptions in Greek characters, but the language in which they are written appears to belong to one of the Semitic class. A description of this dial was written by Sr. Saavedra, and published in "Discurso leido en la. Academia de la Historia de Madrid, porel Sr. Rada."

1849a. Poke or Pocket Dial. *John Ayling Blagdin.*

This was found about 25 years ago in a stream in Tillington parish, three miles from Petworth, whilst digging for the foundation of a bridge, and may be described as a ring of copper about 1½ inch in diameter, on the outer side of which are engraved letters indicating the names of the months with graduated divisions, and on the inner the hours of the day. In the inner slide there is a small hole which must be made to correspond to the month in which it is used. ' The circle is held up to the sun, the inner surface is then in shade and the sunbeam shining through the little orifice forming a point of light upon the hour marked on the inner side. In Knight's pictorial edition of Shakspere, " As you like it," p. 231, is a representation of a similar instrument.

The slide with the hole was wanting when it was brought to the exhibitor and was added by him, and the drawings of the whole were made by Mr. William Knight of Petworth.

1849b. Dial of a later date and more complicated construction.
John Ayling Blagdin.

The same rules as above apply to this instrument, which was given to the exhibitor's father about 45 years ago by the Rev. Deival of Duncton.

1849c. Brass Instrument, apparently used as a **Sun Dial.** Maker, Sauter, Petersburg. *W. Clinton Baker.*

VI.—SIDEROSTATS.

1851. 'sGravesande's Heliostat, with his Equatorial Clock. *Conservatoire des Arts et Métiers, Paris.*

1851a. Siderostat, by Foucault. *Paris Observatory.*

1851b. Heliostat, by Gamberg. *Paris Observatory.*

1851c. Heliostat, by Silbermann. *M. J. Duboscq, Paris.*

1851d. Heliostat, by Foucault. *M. J. Duboscq, Paris.*

1850. Universal Heliostat, designed by Col. Campbell, and executed by Adam Hilger, 192, Tottenham Court Road.
Lieut.-Colonel Archibald C. Campbell.
This instrument, after the polar axis has been set due north and south, and adjusted for latitude, will throw the light of a star or the sun in any required direction, and will keep it there by means of the clock ; all the slow motions in altitude and azimuth can be manipulated by the observer without stopping the instrument. The connexions are placed ready to the observer's hand, wherever he may be.
A small telescope is also attached to the polar axis, so that it may act as a finder for any object, which, when seen in this telescope, will be reflected by the mirror to the required spot. The mirror is 1 foot by 8 inches, and is a perfect plane. This was also constructed by Mr. Hilger.
Dimensions.—Iron stands, 30 ins. diameter ; height, 3 ft. 2 in. to centre of mirror. Mirror, 1 ft. by 8 in.

1851e. Photograph of a Siderostat, constructed by Messrs. Cooke and Sons of York, for the Royal Society.
J. Norman Lockyer, F.R.S.

VII.—CELESTIAL PHOTOGRAPHY.

a. INSTRUMENTS.

1852. The Kew Photo-heliograph, or Telescope, employed at the Kew Observatory for taking photographs of the sun's disc.　　　　　　　　　　　　*Kew Committee of the Royal Society.*

It was constructed, in 1857, by Ross, on the design and under the superintendence of W. De La Rue, Esq., at the cost of the Royal Society; and erected at the Kew Observatory, where occasional sun pictures were taken by its means until 1860, when it was dismounted, and taken to Spain, for the purpose of photographing the solar eclipse of that year. This it accomplished most satisfactorily, and a full account of its work was published in the Philosophical Transactions.

On its return to England, Mr. De La Rue established it at Cranford, where during the year 1861, almost daily, solar photographs were taken with it.

In 1862 it was again removed to Kew, and there maintained in constant operation until 1872. In 1873 it was transferred to the Royal Observatory, Greenwich, where it is now superseded by an instrument of more recent construction.

The diameter of the object glass is $3\frac{4}{10}$ ins., and its focal length 50 ins. An Huygenian eyepiece is employed for magnifying the image, and the instantaneous exposure of the plate is effected by causing a sliding plate, containing an aperture variable at will, to be rapidly drawn across the focus by a strong spring, which is released from the top by cutting a thread.

1852a. De La Rue's Model of Tower for the proposed employment of Huygens' Long Focus Object-Glasses in Solar Photography.　　　　　　　　　　　*Royal Society.*

1852a. Photo-Heliograph, constructed by Dallmeyer, and used for taking **Photographs of the Sun.** This consists of a telescopic camera equatorially mounted, and driven by clockwork.
　　　　　　　　　　　　　　　　　　　The Astronomer Royal.

The telescopic camera, total length about 8 feet, is made of brass tubing 5 inches in diameter, parallel for a length of 6 feet, when it opens out into a cone of about 2 feet in length, and sufficiently large at its extremity to receive the camera-screen, or sensitized plate, 6 inches square.

The object glass, of 4 inches aperture and 60 inches focus, corrected for coincidence of chemical and visual foci, occupies the other end of the tube furnished with the means of adjustment for focussing. The sun's image, produced by the object-glass at its focus, measures about half an inch in diameter, when it is enlarged, by a system of lenses termed a secondary magnifier (composed of two achromatised meniscus lenses turned concavity to concavity with an intervening space), to 4 inches on the camera screen. The secondary magnifier has all the necessary appliances for adjustment of focus. The difficulty to be surmounted in this arrangement is "optical distortion" in the enlarged image, which is, happily, almost entirely overcome.

Coincident in position with the small sun's image, formed by the object-glass, are perforations in the tube for the admission of sliders, containing apertures with cross-wires, glass reticules, &c. respectively; each capable of being placed concentric with the small image. At the same place also is the instantaneous shutter arrangement for effecting the exposures. This consists of a metal slide, perforated by a slit-opening. The shutter is actuated at one end by a spring, while at the other end a string, passing over a pulley and attached to a hook, can be made to hold the spring in a state of tension. This done, on the thread being cut or burnt, the spring is allowed to act, the shutter flashes across the image forming a cone of rays, exposes the sensitized plate, and the picture is produced. There are provisions for regulating the exposure by an alteration of the width of opening in the shutter, or by increasing or diminishing the tension of the spring.

The telescopic camera, attached to a bracket by means of two ring-clips or couplings, accurately turned, providing a motion in arc for the camera tube, read off by suitable scale and vernier, is bolted to the end of the declination axis.

Janssen's Apparatus. It is complete in itself and fits the same groove provided for the reception of the ordinary 6-inch collodion slide; the position planes of the sensitized plates also are identical in both, requiring no alteration of focus when using either description of slide. The apparatus is supported upon a flat piece of wood, or "carriage," the lower portion fits the camera groove, the upper part projects above the body of the camera and affords the means of attachment of a pinion wheel and arbor subsequently referred to. The carriage is perforated; the opening admits of a picture ½″ wide by 1″ high.

The sensitized plate is held in position by a circular cell of wood surrounded by a brass ring the periphery of which is racked with 300 teeth. It is free to revolve between three friction rollers, a pinion wheel originally cut with 12 teeth, of which 8 consecutive ones are removed, gears the racked edge of the cell, and when turned uniformly by a winch (situated near the centre of the telescope tube to prevent vibration) once per second during fifty seconds or for fifty revolutions, the plate-holding cell will have revolved *once* by an intermittent motion occupying *one-third* of a second and ceasing or *two-thirds*.

In front of the cell, *i.e.*, between the sensitive plate and the carriage opening, revolves a thin circular disc of brass, concentric with the plate but of smaller diameter. It is racked on its edge with 240 teeth, and is caused to revolve by a wheel having 60 teeth, which latter is mounted upon the pinion wheel arbor, and is therefore moved by the same winch that turns the plate-holding cell. The brass disc contains four apertures, equi-distant and otherwise corresponding in shape, set to the one opening in the carriage. (To regulate the exposure a mechanism is provided for simultaneously increasing or diminishing all four openings.)

The proportion of driver to follower being as 1:4 for one entire revolution of the winch the disc turns through one fourth of its circumference, so that when the winch is turned once in a second the cell containing the sensitized cell moves on for one third of that time, the brass disc or shutter moving also; after one third of motion the plate stops, but the shutter continues moving, and one of its four openings flashes across that portion of the plate, at that instant opposite the perforation in the carriage exposes and the photograph is taken. This is repeated for 49 times in succession. Thus 49 photographic images of a limited space are taken in 49 seconds of time.

Equatorial Mounting. This in outline resembles instruments constructed on the German plan, but is "universal," *i.e.*, admits of adjustment for any latitude up to 80°, either North or South of the Equator. A novel contrivance has been introduced for retaining clock gearing for great variations of latitude. Briefly the instrument combines all the most recent appliances for convenience of manipulation, and, though massive as regards construction, it may fairly be called portable. Eleven of these instruments have been sent to various parts of the world, without, it is believed, one single mishap.

1853b. One of the Photographic Apparatus used in the "Transit of Venus" Expedition. *The French Commission for Observing the Transit of Venus in 1874.*

1853c. Photographic Revolver, used in observing the transit of Venus. *M. Janssen, Member of the Institute, Paris.*

1853d. Photographic Impressions, obtained with the revolver. *M. Janssen, Member of the Institute, Paris.*

1854. Short Focus Mirror, spherical, for telescopes, corrected by two lenses of homogenous media, for reflecting telescopes and astro-photography. With pamphlet.
Prof. Carl Wenzel Zenger, Prague.

1856. Apparatus for the production of **Photographs of the Sun,** after Dr. Oswald Lohse. *A. Fuess, Berlin.*

1857. Stand with **Equatorial Motion** about a vertical and horizontal axis for a photographic camera of 6″ aperture, used by the German expedition to Kerguelen's Island, for the observation of the transit of Venus, 1874. (Photograph.)
A. Repsold and Sons, Hamburg.

The point of intersection of the horizontal and vertical axis produced is at the same time the centre of movement of the equatorial system, which consists of an hour axis and a declination arc. This arc is suspended from a double arm fastened to one end of the horizontal axis and parallel to the telescope, it turns round this arm remaining concentric with the centre of motion. When clamped in the head of the hour axis, it turns with it, and compels the double arm, and at the same time the telescope, to move equatorially about the horizontal and vertical axes. By this arrangement the position of a thread in the focus of the object-glass can always be controlled by the level attached to the telescope. (See P. A. Hansen, "Beschreibung eines Fernrohrstatins," &c., in the Berichten der Kgl. sächsisch. Ges. d. Wiss. Mathem. Phys. A., 1 Jul. 1870.) To enable the telescope to follow the diurnal movement there is a screw moved by hand in time with the ticks of a clock.

1858. Small Spectrograph. Simple apparatus for taking the sun's spectrum, consisting of a camera (without objective), and a Browning's pocket spectroscope.
Prof. H. W. Vogel, Berlin.

The small spectrograph serves for studying the chemical effect of the different parts of the solar spectrum upon substances sensitive to light and for ascertaining the varying intensities of the various parts of the solar light at different places and times. The slit is wedge-shaped, in order to have more light at one side of the spectrum than at the other. The more intense the chemical effect of a colour, the further it reaches towards the dark end of the spectrum. The apparatus is held in the hand, and so directed upon the sun that the spectroscope may throw no shade. No heliostat is required. The exhibitor has been able to use the instrument on board ship, whilst sailing from Brindisi to Ceylon. (Pogg. Ann., Bd. 156, p. 321.) The exhibited apparatus has been made by *Schmidt* and *Haensch*, Berlin.

1859. Specimen Impression made with the small spectrograph.　　*Prof. H. W. Vogel, Berlin.*

b. PHOTOGRAPHS.

1860a. Photographs of the Arrangement for Obtaining Solar Photographs, by means of Huyghen's lens of 123 feet focal length.　　*J. Norman Lockyer, F.R.S.*

1861. Photographs of the least refrangible end of the spectrum, by iron and other processes.　　*Capt. Abney, R.E., F.R.S.*

1862. Daguerreotype of the **Total Eclipse of the Sun** of the 28th of July 1851, taken at the Observatory of Königsberg.
　　Dr. Schur, Strassburg.

During the eclipse four photographs were taken. This one was formerly in the possession of Prof. A. C. Petersen, late Director of the Observatory in Altona, and after his death it became the property of his grandson, the exhibitor.

1863. Photographs of the **Sun,** taken with the Kew heliograph, and one of a scale put up for determining the amount of distortion produced by the instrument.
　　Kew Committee of the Royal Society.

The Kew Observatory possesses a set of these negatives, extending from 1858 to 1872, and it is now employed in accurately determining from them the positions and areas of the spots observed during the 10 years 1862–1872, during which they were uninterruptedly obtained.

They are photographed on collodion fibrine, and developed by pyrogallic acid.

The sixth picture in the frame is one of a series of views taken, of a standard scale, suspended to one of the galleries of the Pagoda in the Kew Gardens, distant 1,500 yards, for the purpose of determining the optical distortion of the heliograph.

930a. Photo-engraving of a Group of Sun Spots.
　　W. De La Rue, D.C.L., F.R.S.

Made from an original solar negative obtained with an equatorial reflector of 13 inches aperture, 10 feet focus, and a secondary magnifier attached to the eye end of the telescope. The negative being on a scale of 4 feet to the Sun's diameter.

1914d. Collection of Photographs, illustrating various expeditions for observing Total Eclipses of the Sun.
　　J. Norman Lockyer, F.R.S.

1914e. Three enlarged Photographs of the Moon, by Mr. Rutherfurd, of New York.　　*J. Norman Lockyer, F.R.S.*

1917a. Photograph of the Sun, by Mr. Rutherfurd, taken with his triple combination.　　*J. Norman Lockyer, F.R.S.*

1917b. Enlarged Photographs of the Sun, taken by M. Janssen.　　*J. Norman Lockyer, F.R.S.*

1917c. Photograph of the Sun, taken by Professor Win-lock, by a simple lens of 40 feet focal length.

J. Norman Lockyer, F.R.S.

1863a. Photographic Normal Spectrum of the Sun. Collection of enlarged comparison photographs, used in the research. *J. Norman Lockyer, F.R.S.*

1863b. Photograph of the Solar Spectrum, showing its absorption lines, by George Rutherfurd, of New York.

Robert James Mann, M.D.

The entire blue part of the spectrum is divided into sections, which are mounted above each other. When these are placed together in their proper continuation, the spectrum is nearly 8 feet long.

1863c. Photographic reproduction of the Solar Spectrum in its natural colours. First proofs obtained by M. E. Becquerel in 1848. (This proof, enclosed in a box, must be protected from the light.) *M. E. Becquerel.*

1864. Photographs of the less refrangible parts of the sun's spectrum, from line E downwards, and photographs of a larger spectrograph, which, being in use, could not be spared for this exhibition. *Prof. H. W. Vogel, Berlin.*

The accompanying photographs of the solar spectrum had been taken on silver chloride and bromide, which had been made sensitive to the less refrangible rays through addition of light-absorbing media.

1864a. Sun spots photographed at Wilna with Dallmeyer's Heliograph. *The Observatory, Wilna.*

These photographs are made with the Dallmeyer heliograph, constructed for the Observatory of Wilna, on the designs of De la Rue. Six of these belong to the period of maximum of sun spots in September 1870, 12 other represent the largest sun spots observed during the years 1871–1875. Similar photographs are made at Wilna every bright day, under the direction of Colonel Smysloff, for promoting the study of the surface of the sun.

1865. Photographs of different parts of the Sun's spectrum. *Dr. H. C. Vogel and Dr. Osw. Lohse, Potsdam.*

1866. Photographs of the Sun. *Dr. H. C. Vogel and Dr. Osw. Lohse, Potsdam.*

1867. Photographs of Jupiter. *Dr. H. C. Vogel and Dr. Osw. Lohse, Potsdam.*

1868. Drawing of the Spectrum of the Sun between Fraunhofer's lines H_1 and H_2, made from a photograph. *Dr. H. C. Vogel and Dr. Osw. Lohse, Potsdam.*

1869. Specimens of Photographic Multiplication and **Reversion** of astronomical drawings of nebulæ and comets (Dr. Vogel's method).

Dr. H. C. Vogel and Dr. Osw. Lohse, Potsdam.

1870. Lunar and Solar Photographs.

Warren De La Rue, D.C.L., F.R.S.

The lunar photographs consist of:—

1st. A series of original lunar negatives obtained in the focus of an equatorial reflector of 10 feet focal length and 13 inches aperture. These should be examined with a lens in order to render the details visible. It will be noticed that the diameter of the lunar images varies in the several negatives, this arises from the moon being sometimes nearer to and sometimes more distant from the earth according to her position in her orbit.

2nd. A series of first enlargements 9-inches in diameter, collodion on glass transparent positives.

3rd. A series of paper positives 18 inches in diameter printed from negatives obtained by enlargement from the 9-inch positives.

4th. One paper positive 38 inches in diameter. The solar photograph is an original negative on a scale of 3 feet to the sun's diameter, obtained with the same reflector with the addition of a secondary magnifier placed at the eye end of the telescope.

1870a. Enlarged Solar Photographs, by Mr. Rutherfurd of New York. *J. Norman Lockyer, F.R.S.*

VIII.—CHRONOGRAPHS.

1871. Wheatstone's Magnetic Chronograph, for measuring very small intervals of time.

The Council of King's College, London.

1872. Groves's Chronograph, for astronomical calculations, for railway speed, and speed of machinery. *W. Groves.*

1873. Yvon Villarceau's Astronomical Chronograph.

Bréguet, Paris.

1874. Carrington's Astronomical Chronograph, made by Smith and Beck. *Dr. Stone.*

1875. Electro-magnetic Registering Apparatus.

M. Th. Edelmann, Physico-Mechanical Institute, Munich.

IX.—EDUCATIONAL.

1876. Sphere, moved by clockwork, of Just Burgh (1580).
Conservatoire des Arts et Métiers.

1877. Sphere, moved by clockwork, of Jean Reinhold (1588).
Conservatoire des Arts et Métiers.

1884. Planetarium or Orrery, designed by Ch. Huygens,
constructed by J. Van Ceulen, set in motion by clockwork.
Prof. Dr. P. L. Rijke, Leyden.

1887. Model devised by the Rev. **James Bradley,** Savilian
Professor of Astronomy, &c., and used by him for illustrating his
discovery of **Aberration.** *R. B. Clifton.*

For a description of this model, see Phil. Mag., Dec. 1846, vol. 29, p. 429.

1879. Apparatus for demonstrating the **Retrogression** of
the **Superior and Inferior Planets,** also the **Synodic Revolutions, the Transits of Venus and Mercury, &c.**
J. J. Oppel.

The long wire represents the line of vision, the small shield at one end
the apparent position of the planet, the fixed ball at the other end the earth,
and the movable ball the planet. The latter is fixed on the pivot of the
smaller or larger turn-table, according as it is wished to demonstrate the
retrogression of an inferior or superior planet. The twelve signs of the zodiac
are hung up on the walls of the lecture room; the handle must be turned
from left to right; the angular movement of the line of vision from left to
right demonstrates the retrogression.

1880. Cosmographical Apparatus, to explain various natural phenomena, made by M. Robert, of Paris, and purchased for
the South Kensington Museum in the Paris Exhibition of 1867.

1. The seasons.
2. The seasons.
3. Phases of the moon.
4. Eclipses.
5. Librations of the moon.
6. Real and apparent motion of the planets.
7. Fall of bodies.
8. Inequality of the seasons.
9. Precession of the equinoxes, physical.
10. Precession of the equinoxes, geometrical.
11. Precession of the equinoxes, mechanical.
12. Star to indicate a point in space.

1881. Nutoscope. Apparatus showing the laws of preces-
sion and nutation, and the conservation of the plane of rotation.
With diagrams, constructed by the aid of the apparatus.
Prof. Carl Wenzel Zenger, Prague.

1882. Orrery, lighted with gas, for the demonstration of eclipses, and, by the aid of a "sablier," tracing the real orbit of the moon.   *Ernest Recordon, Geneva.*

This apparatus shows:

1. By means of a jet of gas behind globes, representing the celestial bodies, a sufficient shadow is cast to give a clear idea of eclipses and the phases of the moon.
2. The orbits of Venus, the Earth, and Mars.
3. The difference in length of planetary years.
4. The diurnal rotation of the Earth.
5. The two classes of planets; the inferior represented by Venus, the superior by Mars.
6. The phases of the moon. Demonstration effected by means of a gas flame.
7. The real orbit of the moon. By means of a special contrivance, an epicycloidal line of fine sand is traced, which perfectly represents the lunar orbit.

1882b. Orrery.   *The Earl of Cork, K.P.*

1884b. Azimuthal Planisphere.
        Le Vicomte Duprat, Consul-General of Portugal.

1883. Selenographia, for showing all the effects of libration, rotation, and elongation on the surface of the moon.
                                John S. Marratt.

This instrument, the invention of Mr. John Russell, illustrates the various Lunar phenomena, the libration in latitude and polar obliquity, the libration in longitude, the mean state of libration, diurnal and monthly, the periodical and synodical revolutions, and how to determine the position of polar axis, &c.

1884a. Planisphere, with glass globe.   *A. Herbst, Berlin.*

1885. Model of the Solar System, made by Professor Kaiser for his popular lessons on astronomy. The orbits of the planets from Mercury to Jupiter are represented in their relative dimensions.
        H. G. Van de Sande Bakhuyzen, Director of the Observatory, Leyden.

1885a. Cosmographic Clock, reproducing all the astronomical phases of our globe, in relation to the sun.
                                M. Mouret, Paris.

1888. Planetarium, with clockwork. *Ernst Schotte, Berlin.*

1889. Tellurium and Lunarium, with clockwork.
                                Ernst Schotte, Berlin.

1890. Tellurium and Lunarium.
        F. Hornung, Langenbeutingen, Würtemberg.

1890a. Earth and Sun Instrument.
Horatio Allen, Homewood, South Orange, New Jersey, U.S.A.

This instrument presents clearly—
1st. The relations of the earth to the sun.
2nd. The relations of the surface of the earth in any parallel of latitude, in all positions of rotation, and at all places of the earth in its orbit, to lines of light and heat from the sun.

1891. Apparatus intended to elucidate the **Apparent Motions of Planets** seen from the earth.
H. G. Van de Sande Bakhuyzen, Director of the Observatory, Leyden.

The apparent motions of a superior planet are depicted on the inner surface of a cylinder. This apparatus was made by Professor Kaiser for his popular lessons on astronomy.

1892. Apparatus for demonstrating the **Path of the Moon** round the Sun, as an epicycloid, without cusp or loop.
Dr. Charles Oppel, Frankfort-on-the-Maine.

When the handle is turned the moon will mark, by means of a pencil, to be inserted in the socket under it, its serpentine path on a sheet of paper laid underneath.

1892a. Solarium, a mechanism for Kinematic demonstration of the real orbits in the solar or the terrestrial system.
Royal Academy of Industry, Berlin, Director, Prof. T. Reuleaux.

In ascertaining the polar orbs (centroids) for the rotation of the earth about the sun, an ellipse is obtained in one focus of which the sun is situated. The polar orb of the earth is a circle, and the latter travels on the ellipse. The curves actually described by the earth are roulettes of this system. The mechanism is arranged for Kinematic adjustment. so that on the one hand the roulettes of the earth in the solar system, and on the other those of the sun in the terrestrial system, can be demonstrated.

1893. Model of the Paths of the Earth and of Venus, with movable balls on a stand, for demonstrating the position of the nodes and apses, inclination of the orbit, the period of Venus, culminations, &c. *J. J. Oppel.*

1894. Armillary Sphere of brass, to take to pieces, with horizon and azimuth, meridian, equator, ecliptic, tropics, and polar circles, movable sun, &c. *J. J. Oppel.*

This sphere demonstrates many of the definitions of spherical astronomy: zenith, altitude, and azimuth culmination, circumpolar stars, right ascension and declination, longitude and latitude, the seasons, hour-angles, sunrise and sunset according to time and place, length of the day, &c., &c.

1895. Apparatus for demonstrating (*a*) **Foucault's Pendulum Experiment,** and (*b*) the relation between the **Period of Revolution of the Pendulum** and **Geographical Latitude.** *J. J. Oppel.*

a. The apparatus, placed on a common centrifugal machine and turned slowly from right to left, shows the maintenance of the plane of oscillation of the pendulum with respect to the spectator and its apparent revolution relatively to the apparatus, a revolution which the ball of the pendulum (painted half black half red) does not itself accomplish. *b.* With a movable tangent cone the instrument demonstrates, by means of some large diagrams, that the fact and the reason of the apparent angular velocity of revolution of the pendulum being proportional to the sine of the geographical latitude.

1896. Apparatus for demonstrating the alteration of the date in journeys round the world, from west to east. Property of His Highness Prince Pless, Fürstenstein.

The Breslau Committee.

This instrument dates from the first quarter of the 18th century.

1897. Sidereal Atwood's Machine, with a ball, which represents either the moon or a planet. *Chr. Trunk, Eisenach.*

The peculiarity of the apparatus and its object are explained in the description which accompanies the model.

By its means a sphere, placed on a free vibration axis, is caused to oscillate with changeable velocity round a focus in an elliptic orbit, as the satellites round the planets, and the planets round the sun.

The apparatus shows the libration of the satellites and of the planets, causing the precession of the equinoxes.

1898. Ring Sphere. *Dr. H. Löckermann, Hamburg.*

This armillary sphere, of which a more detailed explanation accompanies the instrument, is to be used for instruction in mathematical geography. It is to serve for object lessons, and makes therefore no pretence to scientific accuracy. The instrument demonstrates the apparent motion of the sun and moon, and of the more important constellations (49 constellations with 359 stars of from first to fifth magnitude) at any given place and at any given time.

1899. Projection Apparatus. *J. and A. Molteni, Paris.*

1900. Descrivani's Orrery, by M. Pierret.

Conservatoire des Arts et Métiers, Paris.

1901. Wall Maps (11) for teaching Cosmography :—
1. The Ptolemaic system.
2. Tycho Brahé's system.
3. The Copernican system.
4. Comparative sizes of the sun and earth.
5. Comparative sizes of the planets (with map of Mars).
6. The seasons.
7. The phases of the moon.
8. Eclipses.
9. Parallax.
10. Comets.
11. Nebulæ. *Ernest Recordon, Geneva.*

1902. Three Astronomical Diagrams and Two Rules, with scales, for the solution of problems in spherical trigonometry.

Michael Elbe, Ellwangen.

The graphic representations drawn on the maps are called astronomical webs, and the rules contain scales. By the assistance of a diagram and a-scale any problem in spherical trigonometry can be solved without calculation, a great advantage in navigation. It serves also on land for the determination of time and azimuth by means of one observation of a star.

In order to obtain the necessary accuracy in navigation, the drawing must be made as exact as possible by a machine, so that the accuracy of the solution, so far as it depends on the accuracy of the observing instrument; will be fully attained. Far greater precision will be arrived at by repetition, namely, by the easy reading of dozens of results which depend upon as many observations. Even. the most extensive table for nautical calculation cannot effect this; besides which the inverse problem, often so difficult of solution, becomes a pastime by means of this apparatus.

1903. Specimen of Transparent Astronomical Chart, southern evening sky in winter, Central Europe.

Prof. J. J. Oppel, Frankfort-on-Maine.

1904. Specimens of Astronomical Diagrams, for teaching. White figures on black ground. *J. J. Oppel.*

In both instruments the circular plate (white on one side and black on the other) represents the plane of illumination, at *a* at the time of the equinoxes. The arrangement demonstrates as a necessary effect of a secular revolution from left to right of the plane of inclination of the earth's axis,—A the increase of the longitude of the stars; B the difference of the sidereal and tropical year; C the change of the pole star. The long wire at *a* can be fixed either in the direction of the pole (for A) or of the earth's axis (for C). The plane of the ecliptic is supposed horizontal; the appliance *a* must be turned from left to right slowly on its pivot.

1906. (1.) **Diagram for Nautical Astronomy,** engraved on stone ; with a printed explanation. (2.) Diagram of Nautical Astronomy. Handbook of Practical Nautical Astronomy.

Prof. Prestel, Emden.

1907. Orrery, by Cole ; explanatory of eclipses.

Royal Society.

1907a. Working Model, showing the annual and diurnal revolutions of the earth, thus illustrating the changes of seasons and the successions of day and night, designed by W. Adcock. Made by J. H. Adcock. *Thomas Murby.*

1908. Celestial Globe, 34 cm. in diameter.

Dietrich Reimer, Berlin (Reimer and Hoefer).

1909. Celestial Globe of 80 cm. diameter, with complete equipment. *Dietrich Reimer, Berlin (Reimer and Hoefer).*

X.—ASTRONOMICAL, DRAWINGS.

Drawing made by William Temple of the **Nebula of Orion,** observed by him in the first months of the present year, by means of the great refractor of Amici.

The Royal Institute of " Studii Superiori." Florence.

1910. Unfinished **Chalk Drawings** of **Lunar Craters,** made with the reflector of 3 feet aperture, at Parsonstown, by Mr. Samuel Hunter, assistant in 1860 to 1864.   *Earl of Rosse, F.R.S*

1910a. Chart of the Moon, drawn by hand by Tobias Mayer.
Prof. Dr. Winnecke, Strassburg.

The highly interesting chart of the moon is the original drawing by Tobias Mayer, executed in the year 1750, which served for more than half a century as copy for all the maps of the moon used in nearly all the text-books. The autograph remarks on it, by the well-known Professor Lichtenberg, of Göttingen, show how and where the chart was preserved during the last century. It came into the possession of the exhibitor by a legacy of the late Privy Councillor Eisenlohr, of Carlsruhe.

1910b. Landscape of the Moon in relief, by Witte.
Prof. Dr. Winnecke, Strassburg,

This view of the moon was executed by the celebrated Lady Frau Hofrath Witte, in Hanover, from her own observations. After her death it was presented to the exhibitor by her daughter Frau Staatsrath von Mädler.

1910c. Series of Astronomical Engravings, from the Observatory of Harvard College.   *J. Norman Lockyer, F.R.S.*

MODELS, &c. OF ASTRONOMICAL INSTRUMENTS.

1911. Model of one of the three **Smaller Domes** for the new Imperial Observatory at Vienna, now in course of construction at Mr. Howard Grubb's Works, Rathmines, Dublin. Scale, 1 inch to a foot.   *Howard Grubb, F.R.A.S., Dublin.*

This dome is supplied with Mr. Grubb's improved shutter, by means of which (being perfectly balanced in all positions) the shutters of dome roofs are as easily managed as those of drum roofs.

This is accomplished by a set of counterpoises, equal in the aggregate to the whole weight of the shutter, which are lowered one after another into the place prepared for them. When the shutter is half open all the weights are deposited, the shutter being then balanced in itself. The chains then lap round a roller prepared for them, and as the shutter opens still further the weights are again raised up one by one as the shutter gets heavier and heavier towards the back.

If desired, this form of roof can also be made to open beyond the zenith by placing a pair of doors at the base of the shutter " chase," which open automatically, and allow the shutter to roll back.

1912a. 11 Photographs of the buildings of the Observatory and its principal instruments.   *The Pulkowa Observatory.*

1912b. 10 Photographs of several auxiliary instruments lately constructed by M. Herbst, at the mechanical workshop of the Observatory.   *The Pulkowa Observatory.*

1912c. Photographs of Mr. Newall's Observatory.
J. Norman Lockyer, F.R.S.

1914c. Two Photographs of the 25-inch Refractor constructed by Messrs. Cooke & Sons, of York, for Mr. Newall, of Gateshead-on-Tyne.   *J. Norman Lockyer, F.R.S.*

1912d. Photograph of Galileo's Tribune at Florence.
J. Norman Lockyer, F.R.S.

1912e. Photographs of the Old Astronomical Circles at Delhi. *Mrs. Norman Lockyer.*

1912f. Photographs of the Lamp in the Cathedral at Pisa (interesting in connexion with Galileo's observations).
Mrs. Norman Lockyer.

1914. Photographs of Astronomical Universal Instrument. *F. W. Breithaupt and Son, Cassel.*

Astronomical universal instrument, portable. The movable circles have each two micrometers for reading the seconds; the vertical circle is 33 cm., and the horizontal 50 cm. in diameter. The broken telescope has an aperture of 67 mm. and a focal length of 80 cm., and is illuminated through the axis. The instrument revolves on the vertical axis; the horizontal axis is balanced on one plate only, and can be inverted on spring rollers. One level rests on the horizontal axis, a second is attached to the carrier of the micrometer, and a third can be inverted on the same. The second vertical axis which serves as a counterpoise is graduated. The instrument itself was made in the year 1873 for the Japanese Government at Yokohama.

Photograph of Astronomical Universal Instrument (portable). *F. W. Breithaupt and Son, Cassel.*

This instrument is provided with two movable circles, of 25 cm. diameter, each having two micrometers, with a side telescope of 27 mm. aperture and opposing vertical circle, the carrier of the micrometer being in the middle. By this arrangement the upper part of the instrument is kept low; it has also the advantage that, without alteration of position, the telescope, the two circles, as well as the numerating circle, can be observed. This instrument was constructed in 1875 for the Royal Mining Academy at Schemnitz.

Photograph of Universal Instrument (portable).
F. W. Breithaupt and Son, Cassel.

The circles are 20 cm. in diameter, the vernier reads to 10 seconds, the telescope of 40 mm. aperture is at the side, the azimuth circle is movable and the vertical circle attached to the telescope. All the verniers are covered with glass, and the alhidada of the vertical circle has a separate level. The instrument was made for the Imperial Observatory at Strasburg.

1914a. Photographs of Chinese Astronomical Instruments, enlarged by the Autotype Company from the original photographs by J. Thomson, F.R.G.S. *Autotype Company.*

No. 1. Ancient armillary sphere in the court of the observatory, Pekin. This instrument was made under the direction of Ko-show-king (during the Yuen or Mongol dynasty, about the close of the 13th century), one of the most renowned astronomers in Chinese history, and at the time chief of the astronomical board. The instrument is solid bronze, of huge dimensions and exquisite workmanship. A substantial metal horizon crossed at right angles by a double ring for an azimuth circle forms the outer framework. The upper surface of the horizon is divided into 12 equal parts marked with cyclical characters, the names of the 12 hours into which the Chinese divide the day and night. These are paired with eight characters of the denary cycle and four of the famous eight diagrams of the Book of Changes. The inside of the ring bears the names of the 12 states into which China was in ancient time divided. An equatorial circle is fixed inside the frame, within which a sphere turns on two pivots at the poles of the azimuth. This is made up of an equatorial circle and double ring ecliptic, an equinoctial circle, and double ring solstitial colure. The equator is divided into 28 unequal portions marked by the names of a like number of constellations of unknown antiquity. The ecliptic is divided into 24 equal parts. All these circles are divided into 365¼ degrees, corresponding to the days of the year, and each degree is divided into 100 parts, as the centesimal division prevailed for everything less than degrees, till the arrival of Father Verbiest in the 17th century.

No. 2. Armillary sphere on the terrace of the observatory at Pekin, made under the direction of Father Verbiest ; see Thomson's "Illustrations of China " and its People."

No. 3. Celestial globe on the terrace of the observatory at Pekin ; see Thomson's "Illustrations of China and its People."

1914b. Model of Hipparchus' Astrolabe, showing how that astronomer observed longitudes, and was enabled to determine the precession of the equinoxes.   *J. Norman Lockyer, F.R.S.*

1914f. Photograph of the 9·62 inch Equatorial of the U.S. Naval Observatory, Washington, U.S.A.
Rear-Admiral C. H. Davis, U.S.N.

This instrument was made by Merz & Mähler and has been in use since 1845 in the observations of planets, comets, double stars, etc. The objective has been polished and refigured by Alvan Clark & Sons. The instrument differs in no important particular from those of the same makers at Berlin, Dorpat, etc. Its focal length is 14 feet 4·5 inches, it aperture is 9·62 English inches (9 French inches).

Photograph of the 26-inch Equatorial of the U.S. Naval Observatory, Washington, U.S.A.
Rear-Admiral C. H. Davis, U.S.N.

This instrument was constructed by Alvan Clark & Sons of Cambridge, U.S., and has a focal length of about 390 inches and an aperture of 26 inches, and is thus the largest refractor in the world. A full description of it may be found in the Washington Astronomical Observations for 1874, Appendix I. Its work has been, since 1873, the observation of the faint satellites of the outer planets, difficult double stars, and of nebulæ.

Photograph of the Transit Circle of the U.S. Naval Observatory, Washington, U.S.A.
Rear-Admiral C. H. Davis, U.S.N.

This instrument was mounted in 1866, and was made by Pistor & Martins of Berlin. Its focal length is 12 feet 0·1 inch and its aperture is 8.52 inches. A full description of it may be found in Washington Astronomical Observations for 1865, and also in the same publication for 1874, Appendix I.

Photograph of the Transit Instrument and Mural Circle of the U.S. Naval Observatory, Washington, U.S.A.
Rear-Admiral C. H. Davis, U.S.N.

These instruments have been in use since 1845. The Transit is by Ertel & Son, and has a focal length of 7 feet 0·4 inch, and an aperture of 5·33 (English) inches. The Mural Circle is by Troughton & Simms, the circle being 5 feet in diameter. This telescope has a focal length of 5 feet 3·8 inches and an aperture of 4.10 inches. It is upon the work of these two instruments that the positions given in the Washington Catalogue of 10,600 stars mainly depend.

Photograph of a Pastel Drawing of the Omega Nebula (G. C. 4403), made with the 26-inch Equatorial of the U.S. Naval Observatory, Washington, U.S.A.

Rear-Admiral C. H. Davis, U.S.N.

The drawing was made by Professor Holden and M. Trouvelot, and represents the appearance of this remarkable nebula in October 1875. It is fully described in the American Journal of Science for May 1875. There is a probability that changes are going on in the structure of the west end of this nebula, similar to those suspected in the nebula of Orion.

Photograph of a Pastel Drawing of the Ring Nebula in Lyra (G. C. 4447), made with the 26-inch Equatorial of the U.S. Naval Observatory, Washington, U.S.A.

Rear-Admiral C. H. Davis, U.S.N.

The drawing was made by Professor Holden, U.S.N. It is fully described in the Monthly Notices of the Royal Astronomical Society for November 1875. By close attention a small star can be detected within the ring which must have been much brighter when first seen by Von Hahn before 1800.

Photograph of a Pastel Drawing of the Planet Saturn, made with the 26-inch Equatorial of the U.S. Naval Observatory, Washington, U.S.A. *Rear-Admiral C. H. Davis, U.S.N.*

This drawing was made by M. Trouvelot, and is fully described in the Proceedings of the American Academy of Arts and Sciences for 1875.

Photograph of a Pastel Drawing of the Central part of the Nebula of Orion (G. C. 1179), made with the 26-inch Telescope of the U.S. Naval Observatory, Washington, U.S.A.

Rear-Admiral C. H. Davis, U.S.N.

This drawing was made from measures by Professor Holden, U.S.N., and M. Trouvelot. It is a preliminary chart of the Huyghenian region, which it is intended to complete. In the photograph the fainter portions of the nebula appear too faint, as compared with the drawing.

These photographs of instruments and drawings exhibited have been presented to the South Kensington Museum by Rear-Admiral C. H. Davis, U.S.N., on behalf of the United States Naval Observatory at Washington.

1915. Atlas Cœlestis Novus. Stellæ per mediam Europam solis oculis conspicuæ secundum veras lucis magnitudines e coelo ipso descriptæ ab Eduardo Heis D. Math. et Astron. Prof. P. O. in

Academia regia Monasteriensi. Coloniæ ad Rhenum 1872, impensis
M. Du Mont Schauberg.   Catalogus Stellarum.
  I. Two volumes bound.
  II. Thirteen plates for hanging on the wall.
                                            Prof. E. Heis, Münster.

The Atlas Cœlestis Novus, the result of observation extending over 27 years,
gives the appearance of the starry heaven as it is seen at the present day with
the naked eye.   It is more especially remarkable for containing, besides the
stars of the 1st to 6th magnitude, those also between the 6th and 7th magnitude, which the author himself can easily distinguish.   All the stars, without
exception, are compared with one another in respect of magnitude by the
naked eye, with the additional employment of other means of assistance;
thus, among others, has been used the "method of sequences" of Sir John
Herschel (*see* Results of Astronomical Observations at the Cape of Good
Hope).   The total number of stars observed by the author is 5,421, or 2,153
more than will be found in Argelander's Nova Uranometria.   As no single
star has been entered which has not been many times observed and compared,
future observers will be able to judge whether in the course of centuries the
sky has changed, whether any of the stars increase or diminish in brightness,
whether some have disappeared, or others come into view.

The author has paid particular attention to drawing the milky-way with the
greatest accuracy, and to show the brightness of the different parts in
5 degrees.   For this purpose, the drawings made by Sir John Herschel of the
milky-way in the southern sky were taken as models.   The figures of the
old constellations are copied from the classic figures on the ancient celestial
globe in the Royal Museum at Naples.   In the catalogue of the stars arranged
according to the 57 constellations, their right ascensions and declinations,
(Aug. 1855) are given; there are added also the numbers of Bayer and of
Flamsteed (according to Miss Caroline Herschel); and also the numbers in the
catalogue of the British Association for the Advancement of Science, and
other catalogues.

1916. Chronometrograph, for the determination of true time.
(Original drawing.)   *Prof. Dr. Prestel, Emden.*

1917. Pictorial Representation of the Solar System, for
the demonstration of the relative sizes of the sun and planets, also
of the relative distances of the planets from the sun, and of the
inclination of their orbits to the ecliptic.   (Original drawing.)
                                            Prof. Dr. Prestel, Emden.

**1918. Chart of the whole Celestial Sphere in epicycloidal
projection.**   *Dr. F. August, Berlin.*

This map gives a simultaneous view of the whole sidereal heaven.   Each
constellation preserves its proper form, for the representation is conformable,
that is, proportional in the smallest parts.   There is no want of conformity at
any point, not even at the poles, so that even there the meridians cut each
other at the correct angle.   By this means the spherical form is always
pictured to the eye; the arrangement of the map is easily imagined, by
supposing an elastic envelope to be stretched about a celestial sphere, then cut
open along a meridian and stretched on a frame.   The course of the milky-
way which follows one of the great circles of the heavens, and the parts comparatively free from stars which are at the poles of this great circle, are very

well represented by means of this map. The map contains the stars from the 1st to the 6th magnitude.

The mathematical considerations which are necessary for accurately understanding the construction of the map will be found in the accompanying treatise: Ueber eine conforme *Abbildung der Erde nach der epicycloidischen Projection*. (Extract from the *Zeitschrift für Erdkunde*, Vol. IX., Berlin, 1874, published by Dietrich Reimer.

1919. Treatise on a **Conformable Representation** of the **Earth** by **Epicycloidal Projection.** (Extract from the Zeitschrift für Erdkunde, vol. IX., Berlin, 1874.)

Dr. F. August, Berlin.

XI.—MISCELLANEOUS.

Photographs of the Tribune of Galileo.
The Royal Institute of " Studii Superiori," Florence.

1. A view of the whole Tribune of Galileo. Galileo's statue was sculptured by the Professor Aristodemo Costoli. In the lunette at the end is represented Galileo while presenting his telescope to the Venetian Senate. The four greater instruments that can be observed in the angles of the middle part of the tribune are, at the right hand the odometer of the Accademia del Cimento, and the lens of Benedetto Bregans, at the left hand a Florentine astrolabe and the quadrant of Rinaldini. The other parts of the tribune are explained in the special photographs.

2. The left side of the further end of the tribune.—The lunette represents Galileo observing the oscillations of the lamp in the Cathedral of Pisa. The busts are those of Benedetto Castelli and Bonaventura Cavalieri, disciples of Galileo. The shelf contains the two telescopes of Galileo, and the broken object-glass.

3. The right side of the further end of the tribune.—The lunette represents Galileo blind, dictating to his disciples Torricelli and Viviani the geometrical demonstration of the law of the fall of heavy bodies. The busts are those of Evangelista Torricelli and Vincenzo Viviani. The shelf contains the natural magnet armed by Galileo, his compass of proportion, the first design conceived by him of the application of the pendulum to clocks, and the first finger of his hand.

4. Left side of the middle part of the tribune.—The lunette represents Galileo who, helped by his disciples, repeats the famous experiments on the descent of heavy bodies on an inclined plane. The shelves contain various ancient instruments, partly of the Accademia del Cimento.

5. Right side of the middle part of the tribune.—The lunette represent a séance of the Accademia del Cimento in which the experiment is being made to prove whether the cold of ice can be reflected by a mirror, like the heat of burning coals and light. The shelves contain ancient instruments, partly of the Accademia del Cimento.

6 and 7. Represent two shelves with ancient instruments, partly belonging to the Accademia del Cimento.

1924f. Old View of the Interior of Flamsteed House, Greenwich, with Flamsteed engaged in taking an astronomical observation. *Gardner Collection.*

Exterior View of Flamsteed House, showing position of astronomical instruments. *Gardner Collection.*

Ancient Sundial with Planetary Arrangements erected by Charles II. in the gardens of Whitehall Palace.

Gardner Collection.

1919a. Apparatus consisting of Thermopiles and Concave Mirrors employed in measuring the Moon's radiant heat. *Earl of Rosse, F.R.S.*

The reflecting mirrors were made alternately to receive the moon's image formed by the speculum of the three-feet telescope at Parsonstown, and the heat was concentrated by each in succession on the face of the corresponding thermopile, of which the diameter is about half an inch. The thermopiles are inserted in reverse positions in the circuit of a Thomson's Reflecting Galvanometer.

1920. Vinot's Sideroscope.   *J. and A. Molteni, Paris.*

1921. Apparatus, constructed by Professor Kaiser, for determining the absolute value of personal errors in observations on the transit of stars.

H. G. Van de Sande Bakhuyzen, Director of the Observatory, Leyden.

The moment of the transit is registered by the action of a current. The construction of this instrument dates from 1858 ; the first observations were taken in 1859. (Dutch Records, Tome I., p. 193.)

1922. Observing Seat for Reflecting Telescopes, invented by the contributor. *E. B. Knobel.*

The observer sits as on horseback, and by simply raising himself off the seat, standing on the ground or on the movable footrests, as if in his stirrups, he can easily pull the seat up under him, and adjust it to the required height, without dismounting or moving from the eye-piece of the telescope. Releasing the ratchet wheel allows the seat to be lowered to any position.

1922b. Fittings for **Astronomical Telescopes.**

M. Lutz, Paris.

1855a. Perfect Diagonal Planes (2), $3\frac{5}{8}$ in. across the minor axis, for reflecting telescopes. *Adam Hilger.*

1855b. Right-Angle Prisms (9), for total reflection.

Adam Hilger.

1923. Cooke's Lamp for illuminating the micrometric spider webs of astronomical telescopes. *A. A. Pearson, Leeds.*

The lamp is inserted in the brass body of the instrument, where it is held by two projecting catches. The light, after passing through a condensing lens, is received by a rectangular prism placed at such an angle that the beam is totally reflected downwards into the window of the telescope, where its intensity and colour are modified by diaphragms. The lamp is suspended on a pivot, and also the framing and prism-box revolve from the bottom of the supporting pillar, so that it has a universal motion accommodating it to the position of the telescope. The weight of the end counterpoises the lamp, and the one at the side is the gravity poise. The top of the lamp is movable, and has attached internally a small tin chimney, which assists in promoting a draught and keeping it cool.

Perpetual Almanack.
The Royal Institute of " Studii Superiori," Florence.

1924. Calendarium perpetuum mobile.
Gust. Schubring, Erfurt.

1924c. Calendarium Perpetuum Mobile, eight Tables in glazed frame and in a stand. *Ch. A. Kesselmeyer, Manchester.*

A perpetual calendar, which gives the solution of any chronological problem during a period of from 10,000 years before to 100,000 years after Christ. The tables, which are still in course of construction, will contain the principles of a "Standard Calendar," as invented by the author, the object of which is to demonstrate the errors and inaccuracies both of the Julian and Gregorian calendars.

Tab. I. Adjustable universal calendar key.
Tab. II. Adjustable annual calendar.
Tab. III. Adjustable astronomical calendar of the northern zone.
Tab. IV. Table for finding the theoretical epacts.
Tab. V. Table for finding the epacts to be applied.
Tab. VI. to VIII. contain : explanations and examples of the Calend Perp. Mob.; adjustable universal calendar ; adjustable indicator of dates; adjustable cylindrical indicator of week days; adjustable perpetual pocket calendar ; annual pocket calendar for the year 1877 ; calend of the week days.

The principal object of this work is to solve all problems connected with chronological calculations, and ultimately to present an improved Calendar in accordance with present astronomical research. The first five tables are completed, the next three, now in the press, are to appear next year, and will then replace the three blank tables now in the frames. The first five tables treat of the Julian, Gregorian, and improved Gregorian Calendars, all calculations being extended over a period from 10,000 years B.C. to 100,000 A.D. The subsequent tables VI., VII., VIII. will treat of the *Normal Calendar of the Christian Era* (Reform of the Calendar), extending over a period from 2,000 B.C. to 3,000 A.D., containing various astronomical data.

TABLE I.—*The object of this table* is to find the Calendar Number corresponding to Easter Sunday (No. 1 standing for the 22nd of March, No. 35 for the 25th of April), the Dominical Letter, Epact, Golden Number, New and Full Moons, &c. for any given year of the Julian, Gregorian, and *improved* Gregorian Calendars from 10,000 years B.C. to 10,000 years A.D.

How to use the table.—For each century of the above-mentioned eras look for the Dominical Letter, the Golden Number, and the Number (Stellungszahl) giving the Cycle of 19 Epacts corresponding to the 19 Golden Numbers to be used during the century, in the catalogue situated under the movable frame, and place the same against these letters and figures, care being also taken to move the seven vertical strips of cardboard so that years either B.C. or A.D. are visible, as the case may require.

For the Julian Calendar always place the little black square ■ corresponding to the Golden Number 1 against Epact 8 or "Stellungszah. " (Number of Cycle of 19 Epacts) 1, and only in case the proper dates of the mean astronomical New and Full Moons are required use the indicated

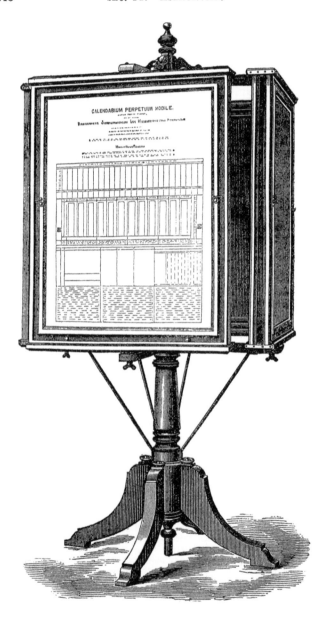

" Stellungszahl " or Number of Cycle of 19 Epacts. After the movable
parts have been *properly* placed, the Number of the Calendar of the year,
the New and Full Moons, &c. will be found on the same horizontal line, and
the Dominical Letter, &c. in the same vertical column. The Solar Cycle
can, however, only be ascertained by using the Julian Calendar.

TABLE II.—This table gives *the Calendar of a whole year* with all movable feasts, &c. as now generally in use on the Continent (especially Germany
and Austria) by Protestants and Roman Catholics, and also contains a short
sketch of the Calendar of the Greek Church, in which case the Julian
Calendar or old style is to be used. The English edition would of course
not be a mere translation of this calendar, as various data, such as the indication of the " terms," &c. (English and American) would have to be added,
and feasts not used in England or America be suppressed. The top part
contains a catalogue of Calendar Numbers for 2,000 years old style, and all
Calendar Numbers new style since 1582 up to Anno 2,000.

How to use the table.—Look for the Calendar Number of the year either
in the above-mentioned catalogue or, generally speaking, in Table No. I. for
any year between 10,000 B.C. and 100,000 A.D., and then place the movable
parts (left and right of the table) against the given Calendar Number, care
also being taken to place the months of January and February so that the
year may become a common or a leap year, as may be required.

TABLE III.—This table, besides giving the average position of the sun for
Greenwich and Leipsic for the second half of the present century, the Solar and
Lunar Eclipses for 2,000 years, and also the tide-tables for 400 places in Europe.
gives the age and position, &c. of the mean astronomical moon for every day in
the year corresponding to the " improved Epact," as devised by the author in
the subsequent tables. It must be borne in mind that the mean astronomical
New Moon precedes the New Moon of the Epact by one day, and the mean
astronomical Full Moon falls one day later than the Full Moon of the Epact,
which is always supposed to be the 14th day of the lunar month, New Moon
being the 1st day.

How to use the table.—Look for the " improved Epact " of the year either
in Table III. (for 2,000 years) or in Table I. by using the " Stellungszahl "
(Number of Cycle of 19 Epacts) in Table V. (I^c Gregorian Calendar,
I^D improved Gregorian Calendar, I^μ Julian Calendar), and then place the
movable parts against these Epacts, then the Lunar Calendar will be nearly
correct. The dates may vary one day (rarely two days), this arising from
differences of meridian, &c.

TABLES IV. and V.—These tables not only give the calculations of the
Epacts as devised by Lilius and Clavius, but also give an improved method
of the author's for finding them still more correctly. Supposing the length
of the lunar month as used by Lilius to have been correct, his *method* of
computing the Epact would still lead to a mistake of seven days in 100,000
years, always supposing that the length of the mean lunar month does
not vary. Lilius having taken a mean lunar month slightly different in
length from the one *now used* in chronological (not astronomical) research
(*i.e.* the lunation by Tobias Meyer of 29 days 12 hours 44 minutes, 2^s·8283),
a mistake of two days would ensue in 100,000 years. Had he used
Tobias Meyer's lunation his *method* would still have led him wrong by five
days. The double error of the length of lunation and of method compensate
each other up to two days error in 100,000 years. *Owing to the fact that the
average length of the lunar month is gradually decreasing for a lengthened
period of time*, and will ultimately lengthen again, to decrease later on (these
periods being of very different lengths and in direct connexion with the position of the planets and the variable lengths of the minor axis of their orbits,

which also act on the average length of the mean solar year of the earth), it can be proved that a *much greater mistake than two days will ensue for the now existing period of decrease of the average length of the lunar month.*

TABLE VI.—In this table the calculation according to Epacts will be advantageously replaced by a new calculation, in which lunar numbers will be used.

TABLES VII. and VIII.—In these tables greater simplicity and accuracy will be attained for our Calendar, as all necessary " corrections " will be adopted every secular year, as occasion may require, and not in accordance with a predetermined cycle as in the Gregorian Calendar. Easter can then either be maintained as a movable feast or become a fixed one, as it may be found desirable by modern astronomers and statesmen.

1924a. Calendar for Two Thousand Years.
M. Georges Sarasin, Geneva.

Lithographed sheet, framed and glazed, permitting the sight, by three openings, of portions of a second lithographed sheet which is capable of movement round a fixed centre. These lithographed sheets are divided into sectors of a circle radiating from a common centre, which is at the same time the centre of motion of the second. They are covered with figures and explanations. An inscription denotes briefly the method of use.

If, by the motion given to the central disc, the two figures which express the tens and units of a year, and the figure which constitutes—or the two figures which constitute—the hundreds (whether according to the Gregorian or Julian style), be brought into the same sector in such a way that the latter be to the left and the former to the right, the calendar of that year will be given on the lower portion of the sheet. The days of the week will correspond to the days of the month in the radial direction, and to the months in the circular direction, whichever of the two styles may have been chosen. There is no occasion either to give a new movement to the disc, or to take into consideration the dominical letter, except as serving to indicate the groups of years. Two of these *data* being given, the third may be found. When the three *data* are given, the years may be found, which, since the Christian era, have possessed them together.

The months of January and February are distinct according to whether it be a bissextile or ordinary year that is in question. In the former case, the tens and units figures, divisible by four, are separated by an empty space from the preceding in the table of years. A third designation of the two abovementioned months is also perfectly suitable to the two classes of years, if the date of the year immediately preceding be formed by the movement of the disc.

It may also be ascertained to what day of the ordinary week corresponds any date during the thirteen years of the Republican style which followed the year 1792, by taking for the hundreds portion the zero of the Julian style.

1924d. Reproduction of the Books on Astronomy, written by Dr. Alonso el Sabio, 13th century, from the original MS. at the Escorial.   *Academia de Ciencias, Madrid.*

"Libros del Saber de astronsmia del Rey Don Alonso el Sabio" 5 vol., gr. in fol., Madrid.

These volumes contain an extensive account of astronomical science in the 13th century, the plates reproduce among other details the constellations there known, and the astronomical instruments used at the time.

1924e. Reproduction in plaster of a fragment of the Zodiac (Aries). The original, in stone, is at the Archæological Museum at Madrid. *Archæological Museum, Madrid.*

This was found, with other objects of a very remarkable kind, at Yecla, in the province of Alicante (Spain). It has an inscription in old Greek characters relating to the subject.

1924f. Astronomical Symbol in the form of a Phœnix, probably of the beginning of the Christian Era, found at Yecla, Spain. *Archæological Museum, Madrid.*

The original in stone is at the Archæological Museum of Madrid. It has been illustrated and published in Sr. Rada y Delgado's study on his reception at the Academy of History of Madrid.

SECTION 12.—APPLIED MECHANICS.

SOUTH GALLERY.—GROUND FLOOR, ROOMS B. C.

I.—PROPERTIES OF MATERIALS.

217a. Machine for Testing the Strength and Ductility of Metals. *Sir Joseph Whitworth.*

The cylinder, pump, ram, and head are made of Whitworth's fluid compressed steel, so as to obtain the maximum of strength with the minimum of weight.

The machine is designed to test cylinders of one half a square inch in area, and it registers up to 100 tons per square inch.

217b. Cast Hoop of Whitworth's Fluid Compressed Steel. *Sir Joseph Whitworth.*

Strength as cast and compressed, 28 tons per square inch.
Strength when forged, 40 tons per square inch.
Ductility as cast and compressed, 18 per cent.
Ductility when forged, 30 per cent.
A hoop or ingot of the fluid metal while under pressure is shortened about $\frac{1}{8}$ of its length.
A portion of the hoop exhibited has been turned to show the nature of the metal.

428b. Pieces of Steel Cylinders, torn by traction in experiments made to ascertain the influence of mode of treatment on the mechanical properties of steel.

Imperial Technical Society, St. Petersburg.

Annexed is a report on the experiments. Out of a block of unforged soft cast steel were cut, in identical positions parallel to the axis, 16 bars which were treated under heating or forging in eight different manners, two samples in each way. The samples were then turned to the shape of cylinders for tensile proof, and elastic and permanent elongations for each of them determined for a series of increasing tensile forces, density, hardness (by indentation), and other elements. Results indicated in annexed diagrams and tables.

420. Cement Testing Apparatus, for ascertaining the tearing strain of Portland and other cements, in sections of $1\frac{1}{2}$ square inches. Originally designed for the Metropolitan Board of Works. **Press** for removing bricks from mould ; and **Moulds** for making test bricks. *Patrick Adie.*

420a. Michele's Patent Cement Testing Machine.
De Michele, Rochester.

The block to be tested is placed in the jaws prepared to receive it ; the handle is then turned, which raises the weighted lever by exerting a pull on its short end through the medium of the cement block. When the leverage is so increased as to exert a force too great for the cement to sustain, it breaks, and the lever falls, leaving the index-pointer at the spot to which it had been raised. The arc along which the pointer moves is graduated to show in number of pounds the tensile stress applied. A suitable arrangement, when the cement block breaks, prevents the lever from falling more than half an inch.

These machines are now in general use, nearly one hundred of them having been sent to different parts of the world. They are principally used by the leading royal and civil engineers in this country, and by a large number of contractors and cement manufacturers.

420b. Drawings of Machines and Apparatus for testing Materials.
Charles Jenny, Vienna.

Five sheets of drawings, representing :—
1. Machine for testing, by means of traction and pressure, the elasticity and rigidity of materials.
The machine of the Imperial Institute was constructed by C. Paff, Vienna.
2. Machine for testing materials. Werder's system constructed for testing elasticity and flexional rigidity. Executed by the Machine Factory Company, Kell & Co., Nuremberg.
3. Machine for testing materials. Werder's system. Constructed for testing the torsional elasticity and rigidity. Executed by the Machine Factory Company, Kless & Co., Nuremberg.
4. Machine and apparatus for ascertaining and determining the elasticity and rigidity of wire, leather straps, thin ropes, &c.
Executed in the former workshops of the Imperial and Royal Polytechnic Institute.
5. Optical apparatus for determining the modulus and the limits of elasticity by the application of the results obtained on tensile and compressive force, flexural elasticity, and rigidity.
The original apparatus of the Imperial Polytechnic Institute were constructed by G. Starke and Kammerer, mechanicians, Vienna.

426. Thurston's Testing Machine, invented by Professor Thurston, of Steven's Institute of Technology, Hoboken, U.S.A.
W. H. Bailey and Co., Manchester.

This machine is used for testing the limit of resistance, ductility, and homogeneousness of iron, brass, steel, and other materials of construction. By means of an ingenious but simple arrangement, a permanent diagram of the behaviour under varying conditions of the materials can readily be obtained.

423. Apparatus for determining the elasticity, tensile strength, and columnal strength of woods and timber.
Prof. Dr. Nördlinger, Hohenheim, Würtemberg.

423a. Apparatus used for experimenting on the Flexural and Torsional Rigidity of Solids, by Professor Everett, and described in the " Transactions of the Royal Society," 1866, p. 185.
Sir William Thomson.

424. Apparatus for determining the relative **Resistance to flexure,** and the **elasticity** of woods and timber.
Prof. Dr. Nördlinger, Hohenheim, Würtemburg.

427. Phroso-dynamic Apparatus for testing wires, by M. Alcun. *M. Digeon, Paris.*

1927. Cast-iron Test Bars. Specimens to illustrate the form and position of fractures when exposed to a breaking load.
W. J. Millar, C.E.

The bars were of 36″ span, 2″ deep, and 1″ broad.
The load was applied at centre of bars. Straight fractures occurred when bars broke at, or close to, centre of span ; but curved fractures when bars broke at points more or less removed from centre.

1927a. Metallometer for Testing Metals and Alloys by bending backwards and forwards a number of times through a certain angle. *Lewis Olrick.*

1957b. Drawings of Hydraulic Apparatus for the study of the extension, compression, and flexion of prismatic bars. Constructed by Professor Wischnegradski.
Mechanical Laboratory, Technological Institute, St. Petersburg.

This apparatus consists of a hydraulic cylinder, the piston of which bears a table supporting four iron columns, connected at the top by a strong cast-iron cross-head, that has at the centre a conical opening, and below a spherical recess within which pivots a hemispherical cast-iron piece, traversed by a powerful iron screw fixed by two nuts; to this screw is attached the upper end of the bar subjected to the experiment of extension ; the lower end of the bar is fastened to the large lever placed at the base of the apparatus and which pivots round an axle fixed in an immovable bearing. This lever is connected with the upper lever, by means of two iron braces suspended to one end of this lever ; at the other end is hung a scale ,pan for the weights used in calculating the tension effected. The ratio of the two arms of the lower lever is 5, and that of the two arms of the upper lever is 20. So that each lever being perfectly balanced the tension of the bar is exactly 100 times that of the weight on the scale board. By means of the screw described above, and by inverting the intermediate iron pieces, it is possible, with this apparatus, to experimentalise upon bars of any length up to 10 feet English.

No. 1 represents the apparatus arranged for experiments of extension.

No. 2 represents the apparatus arranged for experiments of flexion.

No. 3 represents the apparatus arranged for experiments of the compression of long bars.

The deformations in the bars are measured with a cathetometer constructed by Mr. Brauer, and the original section of the bar with an apparatus by the same engineer. The readings are given to the $\frac{1}{200}$ of a millimeter.

II.—SPECIAL COLLECTIONS.

COLLECTION OF THE ORIGINAL MODELS OF STEAM ENGINES AND OTHER MACHINES OF JAMES WATT. PRESENTED TO THE SOUTH KENSINGTON MUSEUM BY MESSRS. J. WATT AND CO.

1928a. Imperfect Model of method of converting reciprocating into rotatory motion by means of teeth or pins fixed to the connecting rod, which gear with the teeth in a wheel, and cause it to revolve. Some point of the connecting rod is guided by a pin, moving in a groove, so as to keep the teeth or pins always engaged in the teeth of the wheel.

This method of converting reciprocating into rotatory motion is included in Specification of Patent granted to James Watt, dated October 25th, 1781.

1928b. Two Fragments of a Model, consisting of wood rods with oval holes geared internally, and apparently belonging to one of the models selected from the Soho Works by the late Sir Francis Smith, as an illustration of one of the methods of converting reciprocating into circular motion.

1928c. Model of Grinding Mill, 6 pairs of stones in two sets of 3 pairs each, each set driven by a spur wheel with bevil gearing. The two fly wheels are connected and driven by pin and connecting rod.

1928d. Model of Grinding Mill, with six pairs of stones, in two sets of three pairs each. Each set driven from one spur wheel by bevelled gearing.

The two fly wheels are connected, and driven by one connecting rod, fitted with two sets of stepped sun and planet wheels.

1928e. Model of Rolling and Slitting Mill, driven by two connecting rods, on one beam, and fitted with sun and planet stepped gearing.

This improvement, consisting of new methods of applying the power of steam engines to drive mills for rolling and slitting iron and other metals, is included in Specification of Patent, granted to James Watt of Birmingham, and dated April 28th, 1784.

1928f. Model of Rolling Mill, driven by a connecting rod, fitted with stepped sun and planet motion, and with two fly wheels.

1928g. Model of two Tilt Hammers, at right angles to each other, one hammer actuated at the tail by cams, the other by lifting cams, driven by one connecting rod fitted with stepped sun and planet motion.

NOTE.—Part of the above model is missing, and the helve of one tilt hammer is broken.

1928h. Model of Wheel (probably for grinding). With sliding axle.

1928i. Fragment of Model (probably a pump bucket).

1928k. Models on a Stand, of four trussed beams, probably used experimentally for testing the strength of different methods of trussing.

1928l. Fragment of a Model of a Frame for a Machine.

1928m. Fragment of a Frame.

1928n. Model of a Horse Mill, with roller and trough, apparently designed for crushing material.

1928o. Model of a Train of Wheels.

1928p. Model of Beam and two connecting Rods with universal motion at their upper ends, and connected to transverse hinged links at their lower ends.

1928q. Model of Beam Pumping Engine, single acting and condensing, worked by tappet valve motion.

1928r. Model of double acting Beam Condensing Engine, conical valves worked by eccentric.

1928r. Model of inverted Cylinder direct-acting Pumping Engine, with tappet valve motion.

1928s. Sectional Model of Beam Engine, worked by eccentric and hollow valve.

1928t. Sectional Model of Engine, with shifting eccentric for altering valve.

1928u. Model of a Pair of Tilt Hammers, alongside each other. Two beams and connecting rods, with cranked pins at an angle to each other, and one of the wheels provided with a balance weight.

(NOTE.—Part of the above model missing.)

1928v. Fragment of a Model with part of Sun and planet motion.

1928w. Fragment of a Model with Sun and planet motion and weighted disc.

1928x. Fragment, an arch head.

1928y. Model of a Water Wheel.

1928z. Measuring Apparatus, with Micrometer Screw, for taking end measures.

1928aa. Model of Garnet's Patent Friction Rollers.

1928bb. Model used for Testing Pressure due to Vacuum.

1928cc. Model of Valve with Universal Joint.

1928dd. Brass Model in two Pieces.

1928ee. Model used in experiments on Governor.

1928ff. Experimental Model.

1928gg. Experimental Model.

1928hh. Experimental Model.

1928ii. Original Model of Cylinder with separate Condenser.

1828jj. Model of Surface Condenser.

III.—PRIME MOVERS.

a. STATIONARY ENGINES.

2019d. Papin's Steam Cylinder.
Royal Museum, Cassel (Director, Dr. Pinder).

This cast-iron cylinder was to have formed part of a large pumping engine, which, however, was never completed. The object was to supply a canal at the level of Hofgeismar with water, whereby the Landgraf Charles hoped to draw the traffic of the Weser to Cassel. An explosion which took place in Papin's laboratory when the Landgraf was contemplating a visit, led to the bold investigator withdrawing from the influence of his enemies. He came to England (1707), but did not succeed with his plans, and died in poverty. Papin's sketch of his contemplated pumping engine is exhibited with the cylinder. It was a peculiar combination of the Savery engine and the piston engine recommended by Papin for other purposes. In the closed boiler A (with safety-valve of Papin's design), steam was generated, which (on opening the cock C) could pass through pipe B to cylinder D. Here it pressed down the close fitting piston or float E which rested on water that had been supplied through the funnel I from a reservoir. The water was thus forced into the chamber F; its return was prevented by a valve at H; and the steam-cock C being now shut and the condensed steam allowed to escape from the upper part of D, water from the reservoir was admitted anew, and the process repeated. The water raised into F could be further directed through the tube G. Papin proposed to add to the effect by introducing red hot irons through the opening in the cover of the cylinder D. Of the two cylinders it is probably D that is exhibited.

F f

The cylinder went to the Government foundry at Cassel, and was used there for many years as a receptacle for the chips under a machine for boring pump barrels. When, in the year 1836, this establishment, at that time under the direction of the superintendent of mining works, Henschel, was burnt to the ground, the cylinder was saved, and bought by the manufacturing firm of Henschel and Son, founded by them after the fire. Here it was at first used for the same purposes as before; but afterwards, about the year 1837, being provided with the label "Papin's steam cylinder," it was placed opposite the main entrance to the factory. Behind it lay the stock of old iron, moulds, &c., which was destined to be melted down. For anyone who may have seen it, as it stood there for many a long year, the following passage in the work of the latest biographers of Papin must, to a certain extent, appear very strange. Speaking of the cylinder, they say:—

"Alors on le relégué dans un coin obscur, sur un tas de ferailles, destinées au fourneau. C'est là que, dans le mois d'Avril de l'année 1863, l'un de nous, voyageant en Allemagne, eut le bonheur de le retrouver."

How could the traveller, without having read the label, or received any information on the subject, have been able to perceive the original purpose of the cylinder, resting as it was in the midst of pieces of old iron? Nor was there ever the slightest intention of treating this valuable relic as an object fit only to be melted down. On the contrary, as the prosperity of the factory increased rapidly since 1866, and the arrival of so many new workmen seemed to threaten the danger that through their ignorance of the value of the cylinder it might suffer harm, the present proprietor, the Privy-Councillor of Commerce, Henschel, grandson of the late Superintendent of Mines, caused it to be removed to a more secure corner. Now the biographers of Papin continue :—

"À son retour il (l'un de nous) s'empressa d'instruire le Général Morin, directeur de notre Conservatoire des Arts et Métiers, de la destruction imminente de ce monument de ce travaux de Papin. Le général se hâta d'écrire au successeur de M. Henschel [his grandson] pour lui en proposer l'acquisition ou l'échange. Une négociation suivit; malheureusement les prétensions exagérées du détenteur et empêchèrent d'aboutir. L'œuvre de notre compatriote existe-t-elle encore ? Hélas ! nous ne saurions le dire."

This question, thrown out by *de la Saussaye*, who has in the meantime died, can be answered by the cylinder itself. The circumstance that it is in the possession of the museum at Cassel will show in their proper light the "prétensions exagérées." Herr Henschel did not wish to sell the cylinder, he wished to keep it for his native town. In fact, in 1869, he presented it to the Cassel Museum, to which institution the gift is especially valuable, inasmuch as it shows to what degree, even at the beginning of the last century, the art of iron foundry had reached in Hesse.

1943. Original Model of Newcomen's Steam Engine.
The Council of King's College, London.

1942a. Drawing of the Newcomen Engine, in the possession of the University of Glasgow.   *Thomas Ledstone.*

1944. Model of Captain Savery's Steam Engine. This form is a modification by Dr. Desaguliers, constructed about 1717. The first complete engine of this kind was made for the Czar of Russia (Peter the Great), for his garden at Petersburg. 1717 or 1718.   *The Council of King's College, London.*

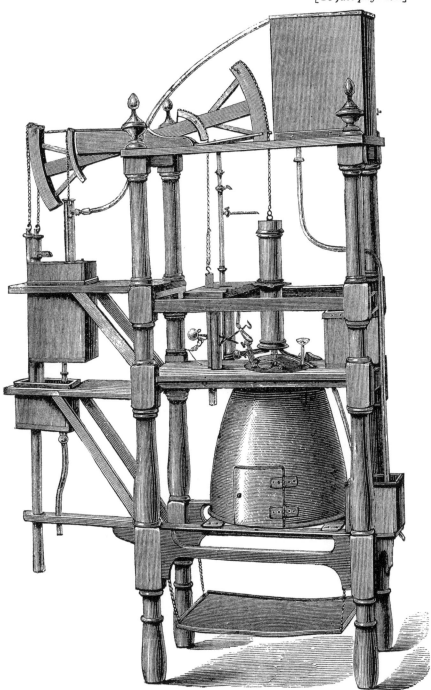

Original Model of Newcomen's Engine, described under No. 1943.

Newcomen Engine, repaired by Watt, the property of Glasgow University. *Sir William Thomson.*

Engraving by T. Barney (1719) of the steam engine near Dudley Castle, invented by Captain Savery and Newcomen, erected by the latter 1712. *Thos. Dow, Exeter.*

Engraving of Newcomen and Savery's Engine.
Bennet Woodcroft, F.R.S.

1932. Sectional Model of a Cabinet Steam Engine.
H.M. Commissioners of Patents.

This is a sectional model of a steam engine in the Patent Office Museum, and was made for the purpose of showing the following improvements in the steam engine made by James Watt. (The engine in the Patent Office Museum was the property of James Watt.) Improvements above referred to :—

a. Making the engine double acting.
b. Keeping the cylinder heated while the engine is at work by surrounding it with steam.
c. Using a separate condenser and air pump.
d. Parallel motion.
e. The governor.
f. The D slide valve.

1828kk. Model of a Steam Engine, made by James Watt, which since 1799 has been used to pump water from the mines of Almaden. This engine was the second set up in Spain. The first steam engine employed in Spain was placed at the arsenal of Ferrol, on Newcomen's design, improved by Watt and Boulton. *Royal School of Mines, Madrid.*

182811. Plan of the Steam Engine used to pump water from the mines at Almaden, drawn in 1830, by Vicento Romoro. *Royal School of Mines, Madrid.*

1828mm. Plan, Outline, and Sections of the Boilers of the Steam Engines used in the quicksilver mines at Almaden, drawn by Vicento Romoro, in 1830. *Royal School of Mines, Madrid.*

1930. Original **Model** of **Stirling's Air-engine.** Made by the inventor. *University of Edinburgh.*

1940. Working Model of **Stirling's Air Engine,** presented by the inventor, the Rev. Robert Stirling of Galston, to the Natural Philosophy Class of Glasgow University, and used constantly for lecture illustrations. *Sir William Thomson.*

1960. Working Model of Atmospheric Engine, with sun and planet motion. 111 *Glasgow Mechanics' Institution.*

1935. Wood Model of **Disc Engine.** (Taylor and Davies' Patent, 1836.)   *Bennet Woodcroft, F.R.S.*

1939. Models (2) of **Rotary Engines.**
                                        Bennet Woodcroft, F.R.S.

1931. Rotary Steam Engine. Designed and made by the Rev. Patrick Bell.   *H.M. Commissioners of Patents.*

2136. Stationary Direct-acting Steam Engine (model).
Royal Geological Institute and Mining Academy (Director, Prof. Hauchecorne), Berlin.

The cylinder and valve chest are opened so as to show the various parts. The eccentric is adjustable with reference to throw and lead; accordingly the valve rod and valve are changeable.

This model shows the general arrangement and essential details of a stationary direct-acting engine, and is arranged specially to demonstrate the relative motions and positions of the piston and the slide valve, and the mechanisms connected with them.

It shows :—

1. The dead points of the machine ;
2. The necessary relative positions of eccentric and crank ;
3. The way in which a steam engine is compelled to move in one direction.
4. The lead of the valve and the eccentric, and their influence upon the steam admission ;
5. The lap of the valve and its connexion with the angular advance of the eccentric and the expansion of the steam ;
6. The irregularities in steam distribution and in the transmission of motion to the fly-wheel, caused by the obliquity of the connecting-rod ; and
7. The effects upon the steam distribution of an eccentric rod of wrong length or an eccentric wrongly adjusted.

2137. Model of a **Direct - acting Cornish Pumping Engine,** with cataract.
Royal Geological Institute and Mining Academy (Director, Prof. Hauchecorne), Berlin.

This (also with open cylinder) has a cataract of simple construction, and a systematically arranged valve motion.

The model serves, in the first place, to illustrate the general nature of click-trains used as valve gear, and their application by means of a plug rod and tappets worked from a beam. It also shows in particular the mode of employing a condenser in a single-acting engine, where three valves (admission, exhaust, and equilibrium) are necessary, with their three separate weigh shafts and wipers, clicks, weights, and levers. The commencement of the expansion is shown very distinctly by the closing of the inlet valve. The pause at the end of the "indoor" stroke is effected by means of the cataract, which is filled with petroleum; the action of this mechanism can be very distinctly observed. With a slow motion of the cataract, it can also be easily noticed that the exhaust valve opens a little sooner than the steam valve, in order that a sufficient vacuum may exist on one side of the piston before the steam is admitted on the other.

The condenser itself is omitted in order to simplify the model and to make the complicated valve gear somewhat more easy to understand. A lever for the injection valve only is shown, to show that this valve must be opened before the engine can start.

For simplification, the cataract which determines the short pause at the end of the " outdoor " stroke is also omitted.

2141. Model of a Horizontal Steam Engine, with reversing gear (Gooch's link).

Royal Geological Institute and Mining Academy (Director, Prof. Hauchecorne), Berlin.

This model is intended to illustrate the action of the link motion generally, and especially that with adjustable block.

The link can be worked in two ways. Either its centre can be suspended and an eccentric rod connected with each of its ends, or its centre can be fixed, and one eccentric rod only used; an arrangement often employed, for example, in hoisting engines.

It shows very distinctly that with two eccentric rods with their eccentrics placed 180° apart, the centre of the link moves to and fro, and that this error can be almost entirely prevented by giving the eccentric a little advance.

2143. Horizontal Steam Engine, with reversing gear (Stephenson's link).

Royal Geological Institute and Mining Academy (Director, Prof. Hauchecorne), Berlin.

With the above model this shows the two chief systems of link reversing gear, their differences, and their comparative advantages and defects. The link has here a different form to that employed in the last case, partly simply for the sake of variety, and partly to show the influence of the position of the point of suspension upon the motions of the link.

Models of mechanisms :—

The eccentric is here also adjustable for variations of lead.

1941. Model of **Dawes' Compound Stationary Engine.**
In this the low-pressure cylinder is horizontal, and the high pressure is arranged over it, at an angle of 30° to the centre line of low pressure. The connecting rods couple to a single crank ; the air pumps and condensers are driven off the low pressure crosshead, and being two in number are arranged on each side of connecting rods. *Henry P. Holt, C.E.*

1937. Model of a **Caloric Engine** (unfinished).
Bennet Woodcroft, F.R.S.

1933. Drawing of " Head's Patent Prime-mover."
Jeremiah Head, M.Inst.C.E.

Being an inverted, direct-acting, non-condensing steam-engine, with steam-jacketed cylinder and covers, cylindrical slide valves, and variable expansion gear, controlled by a liquid-cataract parabolic governor, and balanced throughout for running at a high speed.

1944a. Working Model of latest improved horizontal high-pressure coupled winding engines, with winding drum as supplied for coal, copper, iron, salt, and other mines.
Messrs. Robert Daglish and Co.

2211. Steam Engine. Tangye's (Willan's Patent) four-inch three cylinder steam engine with horizontal multitubular boiler, feed pump for same, and all fittings complete.

Tangye Brothers and Holman.

This engine is of the simplest construction possible; it is self-contained, has neither eccentrics, separate slide valves, nor piston rod guides, and can be driven at a very high rate of speed without the slightest noise.

2212. Steam Engine. 6-horse power expansion portable steam engine, fitted with Head and Schemioth's patent straw burning apparatus and patent automatic governor expansion gear.

Ransomes, Sims, and Head.

By means of this patent invention all kinds of vegetable substances can now be used as fuel in a portable steam engine, such as straw, reeds, dry grass, cotton and maize stalks, brushwood, &c., and by removing the patent apparatus the engine can also be fired with wood or coal in the ordinary manner.

This engine is also fitted with a separate expansion slide valve, and Brown's patent automatic governor expansion gear, which consists of a link motion attached direct to the expansion valve, and under the control of the governor, by means of which the amount of steam admitted into the cylinder is varied instantaneously in exact proportion to the work to be performed by the engine ; an arrangement of the utmost importance in all cases where the load on the engine is suddenly increased or diminished, or where exact regularity of motion in the machine which is being driven is essential to success. The engine has also a simple and efficient arrangement for heating the feed water by means of the exhaust steam, and is provided with two safety valves, steam pressure guage, and all the most modern and complete fittings and accessories.

1954. Drawing, on a scale of ½-inch to 1 foot, of a patent horizontal high-pressure condensing steam engine; designed and made by the Reading Iron Works Co., Limited, Reading.

South Kensington Museum.

This drawing shows a side elevation of the engine and a through plan.

It represents an engine of 25 horse-power nominal; having variable expansion gear, fly wheel, governor, feed pump, and condenser.

A similar engine was employed to drive a part of the British machinery in motion at the Vienna Universal Exhibition for 1873.

1955. Photographs, two, of a compound horizontal-cylinder condensing steam engine. Constructed by the donors in 1873. 120 indicated horse-power. W. and J. Galloway and Sons, Engineers, Knott Mill Iron Works, Manchester.

South Kensington Museum.

One photographic view is of the cylinder end of the engine ; the other shows the fly-wheel, crank shaft, and governor motion, &c.

The high-pressure cylinder is 14 inches in diameter. The low pressure cylinder is 24 inches in diameter. The stroke of the piston is 2 feet 6 inches.

This engine was employed in driving a portion of the British machinery in motion at the Vienna Universal Exhibition of 1873.

1956. Photograph of Brotherhood's patent three-cylinder high-pressure steam engine, arranged as a stationary engine.

The engine was designed and patented by Mr. P. Brotherhood in 1872–73. Brotherhood and Hardingham, Engineers, London.

South Kensington Museum.

1956a. Model of **Brotherhood's** patent **Three-cylinder Hydraulic engine** arranged for turning a capstan, the pressure being supplied to the engine by a Brotherhood's patent three-cylinder pump. *Hydraulic Engineering Company, Limited.*

1967. Model of horizontal engine (novel girder pattern), with portion of cylinder removable to show the action of a variable automatic expansion valve gear (Rider's patent), controlled directly by the governor. The expansion valve works on the back of the lower valve by a separate eccentric in the ordinary manner, but owing to its triangular shape, and the form of the parts, the point of cut-off changes according to the angular motion of the valve round its spindle. This angular rotation is produced by the rise and fall of the governors through rack and quadrant. Any acceleration in speed thus affects the rise and fall of the governor balls, and accelerates or delays the time of steam admission.

Hayward Tyler & Co.

1981. Model of Robey and Co's. semi-portable mining and winding engine. Richardson's patent. Robey and Co., Lincoln.

South Kensington Museum.

1957. Working Model of a stationary steam engine
Royal Trade School, Halle (Director, Dr. Kohlmann).

2178ggl. Drawing of a Portable Engine, 1828.
Maudslay, Sons, and Field.

1945. Atmospheric Gas Engine. Otto Langen and Crossley's joint patents. Actuated by the vacuum resulting from the explosion of common coal gas and air. *Crossley Brothers.*

In this engine, which works by the vacuum resulting from the explosion of common coal gas and air, the piston is not, as is usual, connected with the shaft on both up and down stroke, but on down stroke only. It is thus at liberty to fly up freely from the force of the explosion, which takes place at the bottom only, and by driving the piston before it empties the cylinder of air through its open upper end. The return of the air on the down stroke yields the driving power, and turns the shaft by means of a friction clutch, to which the piston is geared by the rack. The vacuum beneath the piston in equal to about 11 lbs. per square inch for the greater part of the down stroke. The governor does not act, as is usual, by increasing or decreasing the power of each stroke, but by varying the number of strokes, each being of the same power. This is done without materially changing the speed of the shaft. Three or four explosions per minute are generally sufficient to turn the engine itself, and as a maximum of 30 to 35 may be made, there is a balance of, say, from 26 to 32 strokes or explosions per minute left to be applied to useful work under the regulation of the governor. As this engine can be started and stopped at a moment's notice, giving full power at once, and is free from the risks of a boiler explosion, it is peculiarly suited for use as a motor in a laboratory. The consumption of gas is seldom over 2s. 6d. worth per week for a 1-HP. engine. The engine as here exhibited contains many quite recent and very important improvements.

1968a. Model of a Steam Engine with **Glass Cylinders,** for demonstration, 1852.   *M. Eugène Bourdon, Paris.*

1946. Sectional Model of a **Steam Engine,** with expansion.
Paul Lochmann, Zeitz.

1948. Wall-diagrams illustrating the **Hot-air Engine.**
Prof. von Gizycki, Aix la Chapelle.

1949. Wall-diagrams illustrating the **Gas Engine.**
Prof. von Gizycki, Aix la Chapelle.

1950. Wall-diagram illustrating the **Steam Engine,** with continuous expansion.   *Prof. von Gizycki, Aix la Chapelle.*

These diagrams are used in Prof. von Gizycki's lectures on description and theory of machines.

1968. Bailey's Patent Quadruple Engine House Recorder registers on a diagram, which is removed, examined, and replaced every 24 hours, the varying pressure of the boiler and speed of the engine during that time. It consists of a steam pressure gauge, and a rotary speed indicator which registers on the diagram round the revolving drum, an eight-day timepiece which actuates the drum and indicates the time, and a thermometer, all complete in French polished mahogany case, with closet for the safe keeping of tools, scientific instruments, &c.
W. H. Bailey & Co.

1978. Holt's Automatic Cylinder Drain Valves.
The object of this is to let out condensed or priming water from steam-engine cylinders. The valves open automatically at each exhaust, or when the engine stands, and remain open until the admission of steam, when they close, and prevent waste of steam.
Henry P. Holt, C.E.

1979. Model of Dawes' Balanced Slide Valve.
The peculiar advantage of this consists in the mode of making an elastic joint between the relief frame and back of valve by means of a steel plate, secured to both in such a manner as to form practically one piece, thus avoiding leakage and the necessity of frequent attention.   *Henry P. Holt, C.E.*

1982. McCarter's Patent Condenser, applicable to steam engines, and other purposes where a vacuum is required by the condensation of steam, without an air-pump being applied, and drawing its own injection water.   *J. Wood.*

The condenser consists of two chambers, one above the other. The upper chamber (H) is for condensing the steam, the lower one (G), with the two

" Puffing Billy," described under No. 1934a.

"The Rocket," described under No. 1934.

tappet valves (C and D) opening into it, removes the condensed water from the upper chamber into the hot water cistern, whence it flows away.

The exhaust steam from engine enters at A, meets the injection water entering at B, and is condensed, thus forming a vacuum, the water falling to the bottom of chamber (H). To remove this water, a vacuum is alternately created and destroyed, six times per minute only, in the lower chamber (G), by alternately raising the steam or water tappet valve (the steam supplying the tappet valve being reduced by reducing valve to $2\frac{1}{2}$ lbs. pressure). On vacuum being created in lower chamber, the water collected in upper chamber is drawn down through india-rubber foot valve (E); and on vacuum being destroyed in lower chamber, the water falls out through the delivering valve (F) into waste water cistern.

b. Locomotives.

1936. Original Model of **Trevithick's Locomotive Engine.** (Trevithick's Patent, 1802.)
Bennet Woodcroft, F.R.S.

1934a. Puffing Billy, the oldest locomotive engine in existence, and the first which ran with smooth wheels on smooth rails, was constructed in 1813 by *Jonathan Foster,* under *William Hedley's* patent, for Christopher Blackett, Esq., the proprietor of the Wylam Collieries near Newcastle-upon-Tyne. This engine, after many trials and alterations, commenced regular working in 1813, and with tender and two trucks, a total load amounting to fifty tons ran, at an average rate of six miles an hour. This engine worked until the 6th June 1862, and was then purchased for the Patent Museum.
H.M. Commissioners of Patents.

1934. The **Locomotive Engine " Rocket,"** constructed, by Messrs. Stephenson & Co. in 1829, to compete with other engines on the Liverpool and Manchester Railway, where it gained the prize of 500*l.* The Liverpool and Manchester Railway was formally opened for passenger traffic on the 15th September 1830.
H.M. Commissioners of Patents.

Stephenson's Model of the " Rocket."
G. R. Stephenson.

2210d. Photographs of the first engine employed on a public railway, of the first and of the most modern railway coaches, and of the first two railway bills. *Alfred Marshall.*

2212g. Model of Central Rail Locomotive with its rail (1842). *Late Baron Séguier, Membre de l'Institut.*

2212f. Model of Railway with central rail, 1843.
Late Baron Séguier, Membre de l'Institut.

2212h. Model of Central Rail Locomotive, by Baron Séguier and Dumery, with part of the road, 1862.
Late Baron Séguier, Membre de l'Institut.

1952. Working Model, on a 1½ inch to 1 foot scale, of a four wheel locomotive engine. Built at Alexandria in 1862 for service of Egyptian Railway between Alexandria and Suez. Jeffrey Bey, C.E., Great George Street, Westminster.

South Kensington Museum.

The model represents an engine of the outside-cylinder "Stephenson" type, on four wheels, and is a tank engine of a peculiar form.

The water tank is hung beneath the boiler; the coal boxes are placed over the fire-box of the boiler.

To the model are attached the necessary accessories of a locomotive engine, viz., lifting screw jack with traverser, screw keys, fire bars, lights, stoking irons, &c., complete.

1953. Model, in wood and brass. Sectional working model of the cylinder, piston, slide-valve, eccentrics, link motion, and other parts of a locomotive engine. Jeffrey Bey, C.E., Great George Street, Westminster. *South Kensington Museum.*

This model also indicates the variable expansion and cut-off of steam in the engine cylinder.

1957a. Model of Goods Locomotive.
Museum of the Technological Institute, St. Petersburg.

1938. Model of Reversing Apparatus for **Locomotive Engines.** *Bennet Woodcroft, F.R.S.*

1939a. Model of Locomotive, of great adherent power, working by means of six clogs.
M. Adolphe Fortin Herrmann, Paris.

(This model belongs to the Conservatoire des Arts et Métiers.)

1939b. Frame containing photographs of the **first Engine employed on a public railway.** *A. Marshall.*

1934b. T. R. Crampton's Express Locomotive Engine. The peculiarities being a very low centre of gravity and large driving wheels with a minimum of overhanging weight, the whole of the moving machinery being on the outside of the engine. Designed in 1847. *Thomas Russell Crampton.*

Description of Locomotive Model.

The locomotive was designed by T. R. Crampton for high speed.

Previous to this engine being designed no express trains were run on the Continent. For this purpose it was selected in 1849 to commence an express service between Calais and Paris, since which period to the present time, 1876, the Northern of France express trains have been worked almost exclusively with this system of engine.

2212b. Dignity and Impudence (after Landseer).
F. W. Webb.

This photograph represents the largest and smallest locomotives employed by the London and North-western Railway Company. The following dimensions of them may be interesting : —

	Dignity.	Impudence.
Size of cylinders - - -	$17\frac{1}{2}$ in. × 24 in.	$4\frac{1}{4}$ in. × 6 in.
Diameter of driving wheels	8 ft. 6 in.	$15\frac{1}{4}$ inches.
Gauge - - - -	4 ft. $8\frac{1}{2}$ in.	18 inches.
Weight in working order -	$28\frac{1}{2}$ tons.	$2\frac{1}{2}$ tons.

2212h. Six Wheels coupled Mineral Engine. Size of cylinders, 17″ dia., 24″ stroke. Diameter of wheels, 4 ft. 3 in. Total weight in working order, 29 tons 11 cwt. *F. W. Webb.*

The barrel of the boiler and fire-box casing are of steel, and the axles and many of the working parts are also of steel.

2212i. Four Wheels coupled heavy Express Passenger Engines. Size of cylinders, 17″ diameter, 24″ stroke. Diameter of coupled wheels, 6 ft. 6 in. Total weight in working -order, 32 tons, 15 cwt. *F. W. Webb.*

The frames, barrel of boiler, and fire-box casing, are of steel, and the axles and many of the working parts are also of steel.

2212j. Four Wheels coupled Passenger Engine, for heavy gradients. Size of cylinders, 17″ × 24″ stroke. Diameter of coupled wheels, 5 ft. 6 in. Weight in working order, 31 tons, 8 cwt. *F. W. Webb.*

The barrel of the boiler and fire-box casing are of steel, and the axles and many of the working parts are also of steel.

1942. Model of Agricultural Locomotive Engine, fitted with patent side-plate brackets. *Aveling and Porter.*

This represents one of Aveling and Porter's road locomotive engines. The *single* cylinder is placed on the forward part of the boiler, and is surrounded by a jacket in direct communication with it ; the steam is taken into the cylinder from a dome connected with the jacket. Priming is by this means prevented, the use of steam-pipes either inside or outside the boiler is rendered needless, and a considerable economy in fuel is effected. The crank-shaft brackets are formed out of the side plates of the fire-box extended upwards and backwards in one piece, so as not only to carry the crank-shaft, but to provide bearings also for the counter-shaft and driving-axle, in the most convenient position. This arrangement produces a combination of much strength and lightness, reduces to a minimum the loss and annoyance from leakage at strained bolt holes, and unites all parts, peculiarly exposed to injury by jarring, with such firmness as to give almost absolute security against such injury on even very rough roads. The driving wheels are of wrought iron, and are fitted with compensating motion for turning sharp curves without disconnecting either wheel ; they carry about 85 per cent. of the weight of the engine. The engine is steered from the foot-plate. The boiler is made of best quality plates, and tested with cold water to 200 lbs. on the square inch ; the fire-box is of Lowmoor iron.

1947. Sectional Model of a Locomotive.

Paul Lochmann, Zeitz.

c. Turbines and Waterwheels.

1951. Turbine to act as prime mover for physical laboratories. Head of water necessary, 10–20 met. ; measure of water, 1 lit. per sec. ; effective power about 10 meterkilo.

Prof. Wüllner, Aix la Chapelle.

This turbine, with constant water pressure, the plan of which was made by Prof. Hermann, Aix la Chapelle, is exceedingly steady in its action, and thus is specially suited for apparatus that require a constant velocity of rotation. With the fall of 18 mètres available in air, and a water supply of about one litre per second, the effect of the machine is equivalent to one man's power.

1983. Fourneyron Turbine, $\frac{1}{5}$th scale, by M. Clair.

Conservatoire des Arts et Métiers, Paris.

1990. Working Model of Whitelaw and Stirrat's Patent Water-Mill Turbine. *Glasgow Mechanics' Institution.*

The water-mill acts on a principle similar to that of the well-known " Barker's mill," but the arms are bent, and otherwise shaped, so as to allow the water to run from the central opening out to the jet-pipes.

1991. Working Models of three sets of Waterwheels, viz., undershot wheel, overshot wheel, and breast wheel.

Glasgow Mechanics' Institution.

2151. Photograph of a waterwheel with paddles, floating by itself, and capable of being utilised on streams and navigable rivers. *Prof. Daniel Colladon, Geneva.*

A wheel on the above system has been at work for the last ten years on the Rhone near Geneva, with satisfactory results.

IV.—HYDRAULICS.

1994. The **First Hydraulic Press** ever made. Patented by Joseph Bramah, A.D. 1795, No. 2,045.

H.M. Commissioners of Patents.

409a. Professor James Thomson's V Gauge Notches, for measurement of water flowing in rivers or streams, shown together with the ordinary gauge notch of rectangular form.

Prof. James Thomson.

The **V** notch has been devised and brought forward chiefly for use in hydraulic engineering, as being more suitable than the rectangular notch for gauging the greatly varying flows of rivers and streams. If a rectangular notch is made wide enough to allow the water to pass in times of flood, the water flows in it, during long periods of dry weather, too shallow to be well suited for trustworthy measurement; but in the **V** notch the width of the flowing water varies proportionally with the depth. In the **V** notch the flow, while varying in quantity per unit of time, remains similar in its external configuration, and form of stream lines, at and near the notch, and it is only the magnitude of the configuration, and the velocity at homologous points, that vary when the quantity flowing varies. It results that the quantity flowing is proportional to the $\frac{5}{2}$ power of the vertical depth from the still water surface level down to the vertex of the notch. The apparatus exhibited shows two V notches, one *without floor*, and the other *with a level floor starting from the vertex of the notch*. The introduction of the floor allows of trustworthy gaugings being attained without the requirement of so deep a pool as is otherwise necessary. For further information reference may be made to British Association Report, Part I., Manchester Meeting, 1861.

1995. Weisbach's Apparatus, for illustrating experimentally the laws of **Hydraulics,** and for the determination of hydraulic co-efficients. *The Royal Indian Engineering College.*

Large Reservoir, with fittings for attaching mouth-pieces under different heads. With hook gauges for marking the change of water level. *The Royal Indian Engineering College.*

Smaller Reservoir, for attachment to larger reservoir ; used for experiments on large orifices and notches under constant head. *The Royal Indian Engineering College.*

Gauging Tank, fitted with **Hook Gauges.**
The Royal Indian Engineering College.

Reservoir and Long Channel, for illustrating the laws of flow in open channels. *The Royal Indian Engineering College.*

Series of Mouth-pieces and Orifices for use with the reservoirs. Comprising—
3 large rectangular orifices and notches.
6 thin edged orifices of different forms.
6 conoidal and conical orifices.
6 orifices with partially suppressed contraction.
5 elbows, bends, and sudden enlargements.
1 pair of plates for radial current.
The Royal Indian Engineering College.

409g. Darcy-Pitot Gauge, or current meter used in Darcy and Bazin's researches on the flow of water in pipes and channels. This gauge gives the velocity at a definite point of the stream and without a time observation. *Prof. W. C. Unwin.*

2064b. Professor James Thomson's Jet Pump, with intermittent reservoir for drainage of low lands or shallow lakes by water power. *Prof. James Thomson.*

The jet pump, while working, must always have its full supply of water; but the intermittent reservoir allows of its varying its work according to the wetness of the weather.

The jet pump with intermittent reservoir is a contrivance for the drainage of flat marshy lands, or shallow lakes, in cases in which there is no outlet available low enough to allow of the water flowing away by its own gravitation in drains of ordinary kinds, but in which there is water power available from rivers or streams descending from higher ground in proximity to the low wet ground. The jet pump is not necessarily arranged in conjunction with an intermittent reservoir, but may be used alone in many cases for raising water by the power of other water descending to the available outfall level from above. A general notion may be formed of the mode of action of the jet pump by conceiving the action as being somewhat like that in a locomotive steam-engine chimney, with the substitution of an impelling jet of water instead of the impelling jet of steam, together with the substitution of water drawn from the low land instead of smoke drawn from the fire. The part corresponding to the chimney must have a narrow throat at the place where the jet enters, and must thence widen very gradually towards its outlet end. It is usually convenient to have it inverted, as compared with the chimney, so that the jet shall shoot downwards; and the discharging end of the pipe corresponding to the chimney must be immersed in the outfall water, so as to prevent any admission of air during the pumping action. The water is drawn up from the low ground by a suction pipe, terminating in a chamber surrounding the jet nozzle. Now, since the jet pump, when wanted to be applied for drainage, must be made large enough to do the work of flood times, it would be much too large to work continuously in dry weather, and, therefore, it is made to work intermittently, by the arrangement, in conjunction with it, of an intermittently flowing reservoir. The reservoir is made to receive the continuous but variable supply of water for power coming from the higher ground, and to give it out intermittently to the jet pump. The intermittent action is brought about in a very simple way by means of two syphons. The whole arrangement here described was contrived many years ago by Professor Thomson, and brought, with entire success, into practical use on the lands of William Forster, Esq., of Ballynure, Clones. It is free for general use wherever applicable, not being the subject of a patent.

1996. Working Model of a Hydraulic Ram, arranged with glass air vessels, so as to show the action of a column of air and pulsations of delivery valve.   *K. W. Hedges & Co.*

2213. Apparatus for showing the **Motion** of **Fluids** through **long Tubes.**   *T. Hawksley.*

V.—FIRE ENGINES AND PUMPS.

2053. Pneumatic Pump, with taps.   Invented by 'sGrave-sande.   *Prof. Dr. P. L. Rijke, Leyden.*

The taps are at each stroke of the piston turned 90° by an appliance in the form of a cross fixed to the handle. (*See* 'sGravesande's " Physices Elementa Mathematica," ed. III. vol. II., p. 591.)

2020. " A new Water Engine for Quenching and Extinguishing Fires."   *H.M. Commissioners of Patents.*

This engine is made under patents No. 439, A.D. 1721, and No. 479, A.D. 1725, granted to Richard Newsham, pearl button maker, of London, and is one of the first engines in which two cylinders and an air vessel are combined and worked together so as to ensure the discharge of a continuous and uniform stream of water with great force. This invention of Newsham still exists in all fire-engines of the present day, with improvements in materials, workmanship, and the application of steam power.

2020a. Complete Working Model of the most Improved Form of London Brigade Steam Fire Engine, as now in daily use in the metropolis, constructed entirely by H. Nagy Effendi, Egyptian Government Pupil in the establishment of Shand, Mason, & Co. *Shand, Mason, & Co.*

2020b. Complete Working Model of the most Improved Form of London Brigade Manual Fire Engine, as in daily use in the metropolis. *Shand, Mason, & Co.*

1943a. Pulsometer (Hall's patent).
Hodgkin, Neuhaus, & Co., London.

Self-acting steam pump, a novel application of the general principle involved in Savery's engine, A.D. 1702. The result is produced by the pressure of the steam from the boiler upon the surface of the water in each chamber of the pump alternately, without the intervention of any steam piston or plunger, and the water is lifted into the chambers by a vacuum produced without injection or surface condensation. The action of the steam ball which governs the pulsations is purely automatic, and the moving parts, including four valves, are only five in number.

1967a. Pumping Machinery (being largely used for raising fluids, and the engine used as a prime mover.)
Hayward Tyler & Co.

The pumping machinery, Nos. 1967a and 1963, is distinguished for great simplicity of construction and durability of parts, Nos. 1964 and 1965, combined with the above, for obtaining a longer stroke, means of starting from the outside by a lever, and the obtaining of a "rest" at each end of the stroke to allow time for the pump valves to close easily.

1963. Model of a patent direct acting "Universal" steam pump, as used for pumping water from mines, or for other purposes where simplicity of construction and economy of space are matters of importance. *Hayward Tyler & Co.*

1964. Drawing showing a longitudinal and a cross section through the steam cylinder of "Universal" steam pump, showing steam piston with slide valve therein, and the arrangement of ports. *Hayward Tyler & Co.*

1965. Wood Model to show the action of a recent improvement in the mechanism of the "Universal" steam pump for high "lifts;" the slide valve being contained in a valve chest outside the cylinder, and allowing of the use of an ordinary steam piston, thus allowing a longer stroke without lengthening the cylinder. *Hayward Tyler & Co.*

1966. Drawing illustrative of the arrangement of slide valve, &c. for the "long stroke" "Universal" steam pump for high lifts. *Hayward Tyler & Co.*

1977a. Photographs (2) of Frank Pearn's Double and Quadruple acting Steam Ram Pumps for feeding steam boilers with water.   Frank Pearn & Co., Manchester.

South Kensington Museum.

1993a. Davey's Compound Differential Expansive Pumping Engine.   *Hathom & Co.*

The main slide valve spindle is attached to the centre of a differential lever, one end of which derives its motion from the engine, and reciprocates with the steam piston, whilst the other end is actuated by means of a subsidiary piston, caused to reciprocate in equal times by means of a double acting cataract governor. The movement of the cataract end of the lever is constant, and independent of the motion of the main piston of the engine, and has necessarily a lead in advance of that of the opposite end at the commencement of each stroke. With a constant load on the engine the lead will be constant and the cut off constant, but the slightest variation in the load causes a corresponding instant variation in the lead, and as a consequence the cut off takes place earlier or later, as the load is decreased or increased. The engine thus automatically varies the expansion to suit the varying conditions of resistance.

1993b. Davey's Hydraulic Pumping Engine. Worked by means of a natural or an artificial head of water.

Henry Davey.

In this engine there are no pistons, but the power is applied and the work done entirely with plungers. The power plungers are stationery, and are made to serve as pipes to convey the water from the valve box (to which they are fixed) to the inside of the pump plungers ; these latter forming the power cylinders and being connected to each other by side rods passing outside the valve box. In this way the forcing stroke of one pump plunger causes the suction stroke of the other, and *vice versâ*. In "dip" workings in collieries these engines are used to raise water to the main pumping engines, the motive water in this case being supplied from the rising main of the main engine. In hilly mining districts, too, water drawn from a high level is conducted by pipes into the mines and then used to raise water to the surface, thus avoiding the necessity for steam power. The valves of the engine are worked without any metallic connexions by means of water pressure through a small valve actuated by Davey's differential gear.

2062. Model of Force Pump.   *Prof. W. F. Barrett.*

2063. Model of Ordinary Water Pump.

Prof. W. F. Barrett.

1989. Model of Sand Pump.

The Council of King's College, London.

1989a. Air Water Pump, by Jagn.

R. Nippe, St. Petersburg.

2064. Model (to scale) **of Centrifugal Pump.**

Lawrence and Porter.

This pump has been patented by Messrs. Lawrence and Porter. The chief feature of the patent is the arrangement of making one side of the casing removable. The advantages of this system are as follows : By taking off the movable side, the disc or "fan" can readily be examined or removed in a few minutes without in any way disturbing or interfering with the suction or delivery pipes. This is found in practice to be a very great advantage.

Also by making the side removable the amount of both machine and hand work in fitting up the pumps is greatly reduced. The size and weight of any pump (to raise a given quantity of water) are considerably diminished, so that pumps made on this system are far more compact and portable than any centrifugal pumps of the ordinary construction.

One of Lawrence and Porter's pumps with discharge pipe six inches diameter, and weighing only 3¼ cwt., is capable of raising 900 gallons of water per minute, or 54,000 gallons of water per hour. Many of these pumps are now actually at work with highly satisfactory results, and they have received favourable notice by the scientific press.

In the model the side can be removed by simply pulling it gently in a horizontal direction, as the nuts are merely for show, and do not hold it on, and by loosening the small screw in the pulley the disc and spindle can be instantly withdrawn for examination. The patentees will be glad to give further information.

2064a. Appold's Original Centrifugal Pump and Four trial Discs.   *The Council of King's College, London.*

2066. Archimedean Screw, with glass screw to show the raising of the water.   *Elliott Brothers.*

VI.—RESERVOIRS OF ENERGY.

BOILERS, INJECTORS, PRESSURE GAUGES, ETC.

1969. Drawing, water-colour, on a ½-inch to 1 foot scale. A pair of double flue tubular Cornish boilers for high-pressure. Adamson's patent. D. Adamson & Co., Engineers, Hyde Junction, Manchester.   *South Kensington Museum.*

Two of these boilers were lent to H.M.'s Commissioners for the Vienna Universal Exhibition of 1873, for use in supplying steam to drive the British machinery exhibited in motion.

The drawing shows front or firing and elevation of boilers; longitudinal elevation with brick setting.

Longitudinal sectional elevation, showing arrangement of flues; blow-off, feed, and other pipes. Brick settings.

Two cross sections.—One through centre of boilers; the other through back end, showing brick setting, flues, &c.

1970. Sectional Model, in brass, showing the tubular arrangement, water spaces, and circulation of Richardson's patent vertical high-pressure steam boiler. Robey & Co., Limited, Engineers, Lincoln.   *South Kensington Museum.*

1971. Drawing of Richardson's patent vertical high-pressure tubular steam boiler. Made by Robey & Co., Limited, Engineers, Lincoln.   *South Kensington Museum.*

The drawing shows a sectional elevation indicating the water circulation and the direction of the fire and products of combustion. Also a sectional plan of the boiler.

40075.   G g

1972. Drawing, on a $\frac{3}{4}$-inch to 1-foot scale, of Howard's patent tubular safety land boiler; for high-pressure.   J. and F. Howard, Engineers, Bedford.   *South Kensington Museum.*

The drawing illustrates a front view of the boiler; longitudinal section, plan, and cross section.

On a scale of $\frac{1}{4}$ full size is shown the detail of the water tube connexions.

These boilers are made by the Barrow-in-Furness Shipbuilding Company.

1973. Drawing, on a $1\frac{1}{2}$-inch to 1-foot scale, of Messrs. A. Chaplin & Co.'s patent vertical tubular high-pressure steam boiler.   Alexander Chaplin & Co., Engineers, Glasgow.

South Kensington Museum.

The drawing shows vertical sections.   Two plans of the disposition of the upper and lower tubes.

1974. Drawing, sectional, on a 3-inch to 1-foot scale, of an improved vertical high-pressure steam boiler, having horizontal water tubes with "Nozzle" ends, to assist the water circulation. From the construction of the tubes with Nozzle ends this boiler is called the "Nozzle" boiler.   Reading Iron Works Company, Limited, Reading.   *South Kensington Museum.*

The drawing shows a sectional elevation of the boiler, and a sectional plan of the arrangement of the tubes; the circulation of the water, together with the direction of the fire and products of combustion.

1975. Steam Pump.   Horizontal direct-acting steam engine and pump for feeding steam boilers with water, or for pumping and draining purposes.   Cope and Maxwell's Patent.   Hayward Tyler and Co., 84, Whitecross Street, E.C.

The steam cylinder is 5 inches in diameter.   The stroke is 7 inches.

The pump plunger is 3 inches in diameter.

The valves are balls of india-rubber.

The pump will raise 2,000 gallons per hour to a height of 120 feet vertically.   *South Kensington Museum.*

1976. Three **Models,** in brass, showing in section the arrangement of Giffard's patent injector for feeding steam boilers with water.   Sharp, Stewart, and Co., Engineers, Manchester, and Victoria Street, S.W.

a. Giffard's patent injector, in section.

b. Giffard's injector, in section, with the patent adjustment of Messrs. Robinson and Gresham.

c. Giffard's injector, in section, with Seller's patent adjustment.

South Kensington Museum.

1977. Accessories.—Pressure Gauges, for **Engines, Boilers, &c.**   Schäffer and Budenberg, 23, Lower King Street, Manchester.

a. 5-inch Pearson's patent lubricator for steam cylinders, and other working parts of machinery.

b. Mercury vacuum gauge for condensing steam engines.

c. Thermometer for measuring high temperatures.

d. Bourdon's patent steam-pressure gauges, for high and low pressure boilers.

e. Bourdon's patent vacuum gauges.

f. Schäffer's patent steam-pressure gauges for high and low pressure boilers. Two of these gauges are in section showing interior arrangement.

g. Schäffer's patent vacuum gauges, for condensing steam engines, &c.

h. Schäffer's patent hydraulic-pressure gauges, with maximum indicators.

i. Blast furnace gauge, mercury; indicating 6 lbs. pressure.

k. 7-figure counter, in section, for counting steam engine revolutions and speeds of machinery. *South Kensington Museum.*

327a. Patent Water Gauge for steam boilers, independent of level or distance. *John Nicholas.*

This instrument is for indicating in an office or ship's cabin the quantity of water in the boiler.or other vessel to which it is attached. The boiler may be any distance from the office, and upon any relative level. The small tank represents a portion of a boiler to which the stand pipe is attached; in the centre of the stand pipe is a brass tube, open at the top into the steam space, and communicating at the bottom with the right-hand union. The left-hand union opens directly into the water space. The right-hand union is connected by a lead pipe with the top part of the gauge glass, the other by a similar tube with the bottom of the glass, forming a continuous tube, one end of which is open to the steam and the other to the water space. This system is now entirely filled with water, which will always have the same level inside the brass tube as in the boiler, and any movement in the boiler will cause a corresponding motion in the brass tube, such movement being continuous throughout the system. A small quantity of oil is placed in the gauge to show readily this movement, and the line of contact in the glass tube represents the position the of water in the boiler.

1977a. Indicator of Pressure, marking in points, thus avoiding the difficulty of propulsion.
Marcel Deprez, 16, Rue Cassine, Paris.

1977b. Model of Indicator, for steam-engine.
Marcel Deprez, 16, Rue Cassine, Paris.

1977c. Dianemometre, or rule, showing the graphic solution of indicator diagrams. *Marcel Deprez, 16, Rue Cassine, Paris.*

VII.—REGULATORS.

1997. Spherical Governor for Steam Engines. Patented by John Bourne. *H.M. Commissioners of Patents.*

1998. Governor for Steam Engine. *Gros, Paris.*

1998a. Gyrometric Governor for Steam Engines.
Messrs. Siemens Brothers.

It consists of an open cup of parabolic shape, fixed upon a vertical spindle, and caused to revolve within the closed chamber containing the liquid, the bottom of the cup being open and always immersed below the surface of the liquid. When the cup is made to revolve rapidly, the liquid contained in it rises round the sides of the cup and sinks in the centre, the surface of the liquid assuming the inverted parabolic form ; and on reaching the edge of the cup it overflows into the surrounding chamber, while at the same time a fresh supply of liquid is drawn into the cup through the opening in the bottom ; and the power absorbed in putting the overflowing liquid into motion offers a continuous resistance to the rotation of the cup. On a level with the edge of the cup, a series of fixed vanes are placed round the circumference of the external chamber, and a corresponding set of blades are also fixed round the outside of the cup just below the rim, so that the sheet of liquid overflowing from the edge of the revolving cup is thrown against the vanes, and by these is thrown back against the blades on the cup, whereby the overflowing liquid is made to offer an additional resistance to the rotation of the cup.

The internal radial arms uniting the shell of the cup to the centre boss serve to communicate the rotary motion to the liquid inside the cup, while the bottom of the external chamber is provided with a number of radial ribs, for the purpose of checking rotary motion in the liquid outside the cup.

So long as the cup is driven at a constant speed, the overflow is constant, and produces an absolutely constant resistance ; and, hence, if the cup be driven by a constant driving power, independent of the engine, its speed is as uniform as that of a chronometer, within a very small margin of variation, which is definitely fixed ; and it continues revolving at an unchanging speed, totally independent of the engine, and consequently affords the means of forming a governor for controlling the speed of the engine to a constantly uniform rate.

1998b. Chronometric Governor for steam, land, and marine engines, water wheels, turbines, &c.   *Dr. C. W. Siemens.*

1998c. Gyrometric Governor for a steam engine (size adapted for a 4″ steam pipe).   *W. D. Scott-Moncrieff.*

This is an apparatus in which the centrifugal force of a fluid is applied so as to regulate the aperture of a throttle valve through the movements of a loaded piston. It consists of two chambers containing fluid, and communicating with each other through a turbine wheel with four or more straight radial arms. The speed of this wheel depends upon that of the prime mover by which it is driven, and the pressure in the front chamber is dependent upon its velocity, and its variations open and shut the throttle valve.

In the pendulum governor the centrifugal force and the speed necessary to maintain it vary with the different planes of rotation due to different positions of the throttle valve. Such an apparatus can only be correct for one position, and errors of speed must occur for every change in the power of the prime mover or the work it is performing, unless these vary simultaneously and equally. In this governor the centrifugal force of the fluid remains in a fixed ratio to the speed of the engine, and the length of the connexion to the throttle valve varies with the altered length of the fluid column supporting the piston. By adjusting the load upon the piston any speed which the engine is capable of maintaining can be given and will remain constant for all variations of power and work. An index of the speed may be obtained by attaching a glass or Bourdon gauge to show the varying pressures.

1998d. Governors for **Turbines** and **Steam Engines** (J. G. Bodemer's), with apparatus for graphic representation.

Bock & Handrick, Dresden.

1999. Model of Holt's Injection Water Regulator.

This is an automatic valve for regulating the supply of injection water admitted into steam-engine condensers, according to the requirements of the vacuum. It is useful in engines having a very variable load, and where water is taken from water-works for condensation. *Henry P. Holt, C.E.*

1999a. Series of **Models** of **Governors,** &c. for **Steam Engines,** invented by Thomas Silver, of Philadelphia.

H.M. Commissioners of Patents.

a. Differential marine governor.
b. Method of adjusting pneumatic governors when in motion.
c. Marine governor.
d. Combined isochronal and centrifugal governor.
e. Model showing T. Silver's earliest attempts at combining centrifugal and isochronal principle in his marine governor.
f. Reversible link motion.
g. Method of equalising the tension of a spring when in action.
h. Marine governor.
i. Marine governor.

2000. Model showing the effect of hanging the **Arms** of a **Governor** from different points with respect to the axis of rotation. *Jeremiah Head, M.Inst.C.E.*

By turning the horizontal sheave upon the model with gradually increasing velocity, it will be seen that the cross-armed, or approximately parabolic governor, goes through its range with the least variation of speed. Next in efficiency is that wherein the arms are hung from the central axis, whilst the very common form wherein the arms are hung externally is the least efficient, or, in other words, permits the greatest variation in speed between fully opening and fully closing the throttle-valve.

Apparatus for feeding Steam Boilers.

M. Cleuet, Paris.

2001. Drawing, half the actual size, of a **Regulator** for a **25 Horse-power Boiler.** *M. Cleuet, Paris.*

This appliance is fixed to the inside of the furnace with an inclination of a few centimetres, in such a manner that the plane of the water level proper to be maintained passes through the upper tube at a point about half its length.

Connected with the boiler, on one hand, and with the feed pipe on the other, this appliance constitutes a kind of weight thermometer, the expansion and contraction of which depend upon the position of the water level in the boiler, and determine the flow by a discharge of the excess of water injected by the feed pump, which works uninterruptedly.

2001a. Two Proell's Patent Regulators for machinery, and three diagrams. *Dr. R. Proell, Görlitz.*

2010. Gyrograph.   *Prof. von Gizycki, Aix-la-Chapelle.*

This instrument serves in investigating the degree of inequality in the velocity of rotation of machine-shafts. The vertical deflections of the pencil from its lowest position are proportional to the increase of angular velocity of the shaft under examination. The instrument is driven by the latter by means of disc and cord.

VIII.--THE APPLICATION OF THE PRINCIPLES OF MECHANICS TO MACHINERY AS EMPLOYED IN THE ARTS.

2019b. Drawing of a connecting motion as applied to double screw boats.   *M. Chas. Bourdon, Paris.*

2019c. Model of Connecting Motion, joining the apparatus recommended in the Mulhouse manufactories.
M. Engel Dolfus, Paris.

This model belongs to the "Conservatoire des Arts et Métiers." Each apparatus of which it is composed has a special direction for its erection.

2021. Machine for **Winding Cotton** into **Balls.** Invented by Sir Marc Isambard Brunel in 1802.
H.M. Commissioners of Patents.

By the invention of this machine the use of cotton for sewing became universal, as before its invention linen thread or occasionally cotton, always in skeins, had been used.

2022. Model of **Machine** for **Carving Wood** and other **Materials.** Patented by Thomas Brown Jordan, A.D. 1845, February 17th, No. 10,523.   *H.M. Commissioners of Patents.*

2025. The first **Machine** constructed for **Printing** and **Numbering Railway Tickets.** Invented and patented by Thomas Edmondson.   *H.M. Commissioners of Patents.*

2031. Model of **Vauloue's Pile-driving Machine,** made by Jas. Ferguson, Esq., F.R.S. (the astronomer).
Bennet Woodcroft, F.R.S.

Vauloue's engine was used for driving the piles of old Westminster Bridge in 1739 and following years.

2113. Working Model of a steam pile-driving engine for submarine foundations, and other work. Sissons and White's patent. Sissons and White, Hull.   *South Kensington Museum.*

This model, on about ½-inch scale, is a complete working model of a steam pile-driver. The winch to raise the monkey by an endless chain is driven by frictional gearing by the engine, which represents a high-pressure inverted-cylinder direct-acting engine, having slide valve, eccentric, flywheel, and force pump for feeding the boiler with water. The boiler represents an upright tubular boiler for working at high pressure.

2026. Model of **Large Iron Shears.** Constructed by Messrs. Day & Co., of Southampton, for the Government Dockyards. *H.M. Commissioners of Patents.*

2027. Model of a **Hoist** or **Lift.** Patented by Thomas Silver in 1872. The principle of this invention is applicable to ascending gradients. *H.M. Commissioners of Patents.*

2027a. Model of Spouts used in Sunderland Docks for loading vessels with coal. *River Wear Commissioners.*

2027b. Model of Coal Drops used in Sunderland Docks for loading steam colliers with coal. *River Wear Commissioners.*

2027c. Model of **Apparatus** for **raising Heavy Spherical Bodies.** *Bennet Woodcroft, F.R.S.*

2028. Original Traversing Lifting Jack. Patented by George England in 1839. *H.M. Commissioners of Patents.*

2029. Four original Models showing methods of **converting Rectilinear** into **Circular Motion** as substitutes for the crank. Patented by James Watt in 1781.
H.M. Commissioners of Patents.

2030. Bar Lathe used by James Watt.
Bennet Woodcroft, F.R.S.

2030a. Circular Rest, for turning spheres nearly up to the full diameter of the lathe. The Rest being also adapted for holding all kinds of plain tools or the " drill," " universal," and " eccentric " " cutting frames," &c., with overhead motion. Invented by the exhibitor. *Tyssen Amhurst.*

2030b. Spherical Chuck (with six extra collars), for finishing spheres and chucking them for ornamental and other purposes. Invented by the exhibitor. *Tyssen Amhurst.*

2030c. Specimen in ivory turned by the above apparatus, consisting of a series of spheres detached one within the other, in Chinese fashion. *Tyssen Amhurst.*

2030d. Specimen of ivory turned with the above circular rest in conjunction with spiral apparatus, &c., by Mrs. Amhurst.
Tyssen Amhurst.

2032. Instrument for **dividing** and **ruling** the **Brass Meridian Rings** of **Globes,** made by Jas. Ferguson, Esq., F.R.S. (the astronomer). *Bennet Woodcroft, F.R.S.*

2037. Photograph and Model representing an Hydraulic Canal Lift at Anderton, in Cheshire, constructed by Messrs. Emmerson, Murgatroyd, & Co., of Stockport and Liverpool, for the Trustees of the River Weaver Navigation, from the designs and under the superintendence of Mr. Edwin Clark and Mr. Sidengham Duer.   *Sidengham Duer, B.Sc.*

This lift affords an easy and expeditious means of transferring laden barges between the Trent and Mersey Canal, and the River Weaver, instead of the tedious and costly process, previously in use, of transhipping goods from one set of barges to another. The canal is on the top of a bank, and the river is 50 feet 4 inches below it. By means of the lift two barges can be transferred from the river to the canal, and two others from the canal to the river, in eight minutes; whereas in a chain of locks, where the difference of level is the same, only half that work can be performed in an hour and a half. It is pre-eminently useful wherever water is scarce, as it only takes about one per cent. of the water from the upper level which would be used if a chain of locks had been employed. The photograph is from the work itself, while the model is only intended to show how one of the troughs, having taken a depth of 6 inches of water over its area from the upper level, descends to the river, and in doing so lifts the other or lighter one nearly to the level of the canal by means of a central vertical hydraulic ram under each of the troughs without the employment of any other power. The rest of the operation is performed by a small steam engine. It was opened for public traffic by the trustees of the Weaver in July last, and has been in constant and successful operation since that time. The whole apparatus and other works in connexion with it are fully described by the exhibitor, and its applicability for lifting large ships is discussed in the " Minutes of the Proceedings of the Institution of Civil Engineers."

2038. Somerville's Machine for charging and drawing gas retorts by steam power.   *John Somerville.*

It is constructed to run along the floor of retort house in front of retorts upon a line of rails or tramway, and consists of a platform on wheels, upon which is fixed a boiler and engine, which propels it and gives motion to the various parts. On the same platform is erected an upright frame, which serves as a support to the cradle or secondary platform carrying the scoop for charging or filling, and the rake for drawing the retorts ; on top of frame is a receptacle (over the scoop) which is supplied with coals from another receptacle below by means of an elevator or Archimedean screw, whereby the scoop is filled with coals. The rake is attached in a similar manner to the scoop, and is propelled and withdrawn in the same way.

2039. Model of a **Californian Stamping Mill.**
                    Royal Saxon Mining Academy, Freiberg.

Model of Stone Breaking Machine.
                    G. H. Goodman.

2041. Machine for **Engraving** duplicates of **Medallions, Sculpture, &c.** (J. Bates' Patent, 1823.)
                    Bennet Woodcroft, F.R.S.

The object of this machine is to copy or engrave on metal plates an exact representation of medals, sculpture, and other works of art executed in relief.

2044. Model of **Nasmyth's Direct-action Steam Hammer.** (Nasmyth's Patent, 1842.)
                    H.M. Commissioners of Patents.

2044a. Photograph of the original sketch of Nasmyth's Steam Hammer. *James Nasmyth.*

This rapid sketch made on the morning of November 24th, 1839, embodies the original ideas of the invention of the steam hammer; it contains the main features and details, which have proved so satisfactory in practice as to be retained to the present time.

2045. Drawing of a **50-ton Double-action Steam Hammer,** supplied to the Russian Government for their gun factory at St. Petersburg. *Thwaites and Carbutt.*

Diameter of cylinder, six feet six inches; length of stroke, twelve feet six inches; total height from ground line, fifty feet.

Allen's Steam Striker. *J. W. Thomas.*

2045a. Model of a Friction-Hammer.
John Tille, Prague.

This model is 72 cm. high, 42 cm. broad, and 21 cm. in length, and has been executed after the pattern of the friction-hammer constructed in the Royal Prussian Machine Workshops at Dirschau, where the exhibitor, in 1858 and 1859, was employed with the construction of the incline between Elbing and Osterrode.

With reference to this model it is to be observed that the small cog wheels fixed on the revolving shafts of the friction-rollers have only been attached for the purpose of putting the model in motion with a crank.

2045b. Model of a Spring or Elastic Hammer.
John Tille, Prague.

These favourite spring-hammers, if in quick motion, effect a very powerful stroke; if in slow motion, a moderate stroke, the regulation of which is obtained by a treadle and a draw-pole, by means of a tension roller supported by levers. If the movement is stopped, the tension roller acts as a brake-weight.

In recent times, spring-hammers of this form are manufactured by Messrs. Auth. Fetn, and Deliege, at Liége.

2045c. Model of a Punching Machine.
John Tille, Prague.

This model is constructed after Borsig's pattern, in which the balancing punch is moved forwards by a crank and wheels, by means of a movable press-bar.

2045d. Model of Parallel Shears. *John Tille, Prague.*

The movable cutting blade is fastened to a well regulated sliding piece, and moved by a crank and wheels by means of a draw-bar. This model is arranged at the same time to be worked by means of leather straps.

2045e. Model of Circular Shears, with cast ribbed frame.
John Tille, Prague.

2045f. Model of Circular Shears, with concave cast frame.
John Tille, Prague.

These two models, the circular cutting blades of which are fastened on spindles, and put in motion by means of wheels and cranks, illustrate chiefly the solidity and elegance of the concave casting as compared with the ribbed casting.

1980. Model of Dawes' and Holt's Hydraulic Shears. In order to avoid buckling or bending either half of plate cut in two, a strip of metal equal to the thickness of the plate is sheared out of the plate, and the gap thus formed is utilised to allow the plate to pass the tie connecting the upper and lower blades ; this machine can thus cut any size of plate.   *Henry P. Holt, C.E.*

2046. Circular Knitting Machine. (J. A. Tielen's Patent, 1842.)   *Bennet Woodcroft, F.R.S.*

2047. Models of **Chinese Agricultural Implements** (9), small mill, &c.   *Bennet Woodcroft, F.R.S.*

2048. Model for showing the **Curves** of **Screws.**
Bennet Woodcroft, F.R.S.

2050. Traction Dynamometer, used for ascertaining the draught of carts, waggons, and all agricultural implements that are drawn by horses. Also, for determining the resistance of roads and streets.   *Royal Agricultural Society of England.*

The instrument is harnessed just like a horse to the implement it is desired to test, and being itself drawn along by one or more horses it registers the total work done in passing over any given distance, the mean and extreme tractive force, the pressure on the back of the horse, and the lateral pressure in such implements as reaping and mowing machines.   First used at Bedford, 1874.   Designed and made for the society by its consulting engineers, Messrs. Eastons and Anderson.   *See* Journal of the Royal Agricultural Society of England, No. XX., part 2, page 678.

2051. Appold Friction Dynamometer, of 100 horse-power, used for measuring the work done by steam engines and other prime movers.   *Royal Agricultural Society of England.*

The prime mover to be tested is coupled to the main shaft of the dynamometer, the friction breaks of which are loaded in proportion to the power it is desired to develop.   The instrument registers the number of revolutions made in a given time, and this, together with the known weight on the breaks, furnishes the data for calculating the work done by the prime mover.   This powerful instrument was constructed by the consulting engineers of the society, Messrs. Eastons and Anderson, for the trial of steam ploughing and traction engines at Wolverhampton in 1871.

2053a. Pneumatic Intercepting Apparatus for sending and receiving message carriers in pneumatic despatch tubes of 3 inches diameter, to be worked on the circuit system.
Messrs. Siemens Brothers.

2052. Diplograph. Writing machine for the **Blind,** by which writing in relief and ordinary writing are performed at the same time.   *Ernest Recordon, Geneva.*

2054. Dr. Thursfield's Patent Writing Frame, for enabling the Blind, and persons with defective or failing sight, to

write ; and for facilitating writing in darkness or an unsteady light.
Elliott Brothers.

By means of Dr. Thursfield's apparatus anyone can write with equal facility in light and darkness. For those with defective or failing sight it is of the utmost value at all times, as when used in ordinary daylight no writing is visible, and the involuntary following of the writing by the eye is prevented. In cases of incipient cataract, and other ophthalmic diseases requiring rest, or after operations on the eye, it has been found a most valuable therapeutic agent. To literary men, travellers by railway or steamboat, and others compelled often to write in a bad or unsteady light, the invention will prove or great service. No ink or pencil is required, but the writing is equally legible and indelible. The mode of using the apparatus is very simple, and where any sight remains will at once be comprehended and found easy of execution.

2054a. Peters's Machine for Microscopic Writing, combining Ibbetson's Geometric Chuck.
Royal Microscopical Society, London.

With this machine any combination of bicycloid curves can be produced on a scale wonderfully minute. Many beautiful and complex designs of this kind have been engraved on glass with remarkable precision, in the space of a circle the fiftieth of an inch in diameter.

A disc the one-hundredth of an inch in diameter appears to the unaided eye as a mere point, yet that point, not larger than the full-stop of ordinary print, will contain five circles each, the three-hundredth of an inch in diameter, and in a circle of that size (that is, about the size of a transverse section of a hair of the human head) the Lord's Prayer is written and can be read. It has also been legibly written in the three hundred and fifty-six thousandth part of an inch. In this specimen the writing is so small, that, in similar characters, the Bible and Testament together (said to contain 3,566,480 letters) could be written twenty-two times in the space of one English square inch.

The name and address of Mr. "Matthew Marshall, Bank of England," has been written in the two-and-a-half millionth part of a square inch.

2054b. Facsimile Drawing and Writing Apparatus.
S. F. Pichler, London.

2055. Apparatus for filling Manometer Tubes with Mercury to any height.
Prof. Dr. R. A. Mees, Director of the Physical Laboratory of the University of Groningen.

The end of an india-rubber tube is screwed into an opening in the base of the manometer. A wooden vessel filled with mercury is elevated by means of a vertical iron rod to a height nearly corresponding with that desired in the manometer tubes. With the aid of a piece of wood, which is plunged into the wooden vessel, the exact adjustment of the mercury to the required height in the manometer can be effected.

2056. Diagrams illustrating the principles adopted in constructing Wood-planing Machines. These diagrams are used in lectures on Applied Mechanics at the I. and R. High School of Agriculture and Forestry, Vienna.
Dr. William Francis Exner, Vienna.

In the forestry section of the above school, series of lectures on the mechanical technology of wood-working are delivered annually. These are attended by foresters who wish to become qualified for the post of manager of wood-working establishments. For these lectures, diagrams of all sorts of tools and machinery for wood-working, are prepared. The examples shown form only the first part of the series.

2057. Model of Lever Plough, with 6 shares, for ploughing by steam. (Fowler's system, reduced to one-tenth.)

M. Digeon, Paris.

2059. First Type-distributing Machine, invented by Alexander Mackie. *John Maclauchlan.*

591. Small Wooden Model of Prof. Colladon's new Air and Gas Compressors (patented), used for the great St. Gothard Tunnel. *Prof. Daniel Colladon, Geneva.*

This system of compressors refrigerates air and gases simultaneously with the compression, and all the movable parts are kept cold, whatever may be the rapidity and force of compression. This system has been applied to several industrial purposes. It is the system of air compressors used exclusively in the boring of the great St.Gothard tunnel (length 14,920 mètres). At the extremity of this tunnel 30 of these compressors give 40 cubic mètres of air per minute, under the pressure of seven atmospheres, for the aëration of the tunnel, and the working of 60 boring machines and other apparatus, and eight other compressors supply air at 14 atmospheres for compressed air locomotives. *See* **2060.**

2060. Drawings of the air compressors on the Colladon system, employed in boring operations at the St. Gothard.

Prof. Daniel Colladon, Geneva.

2061. Models, Twenty-seven, for teaching machinery, executed by M. Schröder, of Darmstadt. *Prof. Pigot, Dublin.*

2065. Hook's Universal Joint, by which a shaft can be kept in rotation at any angle. *Elliott Brothers.*

2067. Steam Thermometer, for testing temperature of steam in the supply pipes. *Elliott Brothers.*

It consists of a thermometer the bulb of which dips into an iron cup containing mercury and oil. By this means the bulb is protected from breakage by pressure, and the thermometer can be removed at any time without letting off steam.

2068. Air Bell.

The Committee, Royal Museum, Peel Park, Salford.

An apparatus consisting of a metal tube 30 feet long, with a percussion pump at one end and a clapper at the other, which rings a bell whenever the piston is struck.—[Supposed to be the invention of the late Mr. Gavin M'Murdock, of Soho, Birmingham.] Made by the late Richard Roberts, C.E., of Manchester, 1840.

2069. Parallel Motion.

The Committee, Royal Museum, Peel Park, Salford.

Consisting of one wheel revolving within another of double its size, and carrying a pin whose centre, coinciding with the pitch line of the lesser wheel, traverses a straight line equal to the diameter of the larger.—[The invention of the late Mr. James White, of Manchester.] 1842.

2073. Apparatus for supposed **Perpetual Motion,** used by Dr. Thomas Young. *The Royal Institution of Great Britain.*

A wheel supposed to be capable of producing a perpetual motion, the descending balls acting at a greater distance from the centre, but being fewer in number than the ascending ones. "Lectures on Natural Philosophy," by Thomas Young, M.D., 1807.

2074. Model of Parallel Motion, consisting of three parts only. *William Hayden.*

This parallel motion depends on the properties of a right-angled triangle where the sides are as 3, 4, and 5. The longer radial arm being the less of the two sides containing the right angle, and the shorter radial arm half the longer of such sides, while the side of the triangle connecting the arms corresponds to the remaining half of such longer side. The distance between the fixed centres corresponds to the longest side of the triangle.

The line of motion is perpendicular to the straight line joining the fixed centres, and on the centre of the shorter radial arm; it is limited by the motion of the less radial arm through the less of the acute angles of the triangle first before referred to. The point giving the parallel motion is found by drawing a perpendicular from one of the acute angles of an isosceles triangle, where each of the equal sides is to the base as 5 to 8 (the other acute angle and the obtuse angle being connected to the two radial arms), to the perpendicular on the centre of the less radial arm.

2075. Parallel Motion; invented by the late Richard Roberts, C.E., Manchester ; the peculiarity consists in the fixed centres being in the plane of the parallelism.
The Committee, Royal Museum, Peel Park, Salford.

2076. McKay's patent Equilibrium **Drilling Tools,** with specimens of work. *Menzies & Blagburn.*

This apparatus is designed to permit circular holes of any size to be bored out with mathematical accuracy and absolute precision as to their relative positions.

This object is attained by maintaining the centre point, around which the cutters revolve, immovably fixed during the process of boring, while the cutters are advanced, and are held in equilibrium with the centre points by a hydraulic medium contained within the chambers of the tool, the pressure being conveyed by the action of the feed given to the boring machine.

This apparatus is exhibited for the purpose of showing how accurate work can be obtained with a minimum of skilled labour and cost.

Machine for drilling Watch Frames.
H.M. Commissioners of Patents.

2076a. Isometric Drawing of an Expanding Mandril, or tool for fixing the tubes in the plates of boilers by expanding the ends so that the tubes are firmly fixed in their places.
W. H. Prosser.

Invented and made by the late Richard Prosser of Birmingham in 1845.
Two of these expanders were sent to the United States in 1847 ; an improvement on them is now being imported from that country.

2077. Bates' Anaglyptograph. Machine for producing drawings or etchings in relief from models, coins, medals, &c.
George Hogarth Makins.

The machine consists of two portions :—The first a solidly framed oblong base, in which is fixed a long double screw, right-handed at one and left-handed at the other half, and of 20 threads per inch ; rotation of this gives opposite motion to two tables, upon the front one of which the object to be copied is fixed, and upon the other the plate to receive the etching or drawing. At the winch end of the screw is fixed a wheel cut with 100 teeth, and also a stop arrangement by which any number of the teeth can be taken, and thus equal partial turns given to the screw.

The second portion of the machine is a framed apparatus carrying the tracing and etching points. This is provided with wheels for travelling across the base, and a groove is formed in the latter for their direction. The tracing point is attached to a bar which in rising does so at an angle of 45°. In the centre of the bar is a piece which, being between friction wheels attached to the diamond or etching point frame, will, as the bar rises by passing over any elevation in the object, cause the diamond frame to move at right angles to the motion of the machine, and thus the line forming is curved or waved just in proportion to the height of the elevation passed over ; hence the appearance of relief given to the etching.

2078. Model of a **Steam-boiler** to show Herschel's boiler arrangement. *Bock and Handrick, Dresden.*

2079. Model of **Valve Motion,** to show the link motion of Stephenson, Gooch, and Allan. *Bock and Handrick, Dresden.*

Model of Reversing Gear. *Bock and Handrick, Dresden.*

2080. Drawing of a **Model** for **testing** the **Link Motions** of Stephenson, Gooch, and Allan, for locomotive manufactories. Greatly improved and quite new construction.
Bock and Handrick, Dresden.

2081. Model, for the demonstration of **Centroids.**
Bock and Handrick, Dresden.

2082. Model of a **higher pair** of **Elements.**
Bock and Handrick, Dresden.

2083. Vierkurbelkette. (Four crank-chains.)
Bock and Handrick, Dresden.

2083a. Slider Crank Chain.
Bock and Handrick, Dresden.

2083b. Mechanism of Slider Crank Chain.
Bock and Handrick, Dresden.

2084. Model of a **Hanger.** *Bock and Handrick, Dresden.*

2085. Model of a **Wall Bracket.**
Bock and Handrick, Dresden.

2086. Model of a **Plummer-block.**
Bock and Handrick, Dresden.

2087. Three Models of **connecting Rod-ends.**
Bock and Handrick, Dresden.

2088. Two Models of **rivetted Joints** for **Boiler Plates,** to take to pieces. *Bock and Handrick, Dresden.*

2089. Model of a **Double-beat Valve.**
Bock and Handrick, Dresden.

2090. Model of a **Throttle Valve.**
Bock and Handrick, Dresden.

2091. Model of a **Cup Valve.**
Bock and Handrick, Dresden.

2094. Drawing Instruments and Scales.
Bock and Handrick, Dresden.

2095. Tachometer with registering apparatus, after Dr. Pröll. *Bock and Handrick, Dresden.*

Wood Model of Clutch Gear. *Jeremiah Head.*

2095a. Printing Machinery. This model of an ordinary printing machine illustrates the inventions of Edward Cowper in 1818, and Augustus Applegath in 1823, as applied to the printing of books, newspapers, &c., which had up to that time been commonly printed at hand presses.
The distribution of the ink transversely as well as longitudinally removed the difficulties previously felt, and gave perfect distribution, and consequently good printing. *Edward Alfred Cowper.*

2096. Model of Cowper's Cylinder Printing Machine.
The Council of King's College, London.

2058. First Type-composing Machine, invented by Alexander Mackie.
John Maclauchlan, Dundee Free Library and Museum.

Mr. Mackie's Type-composing Machine, of which this is the germ, is used in offices of some of the London daily newspapers, and various books have been printed by its aid.

2058a. Type composing Machine. *John Walter, M.P.*

2058b. Type casting Machine, distribution being dispensed with. *John Walter, M.P.*

2097. Model of Revolving Screw, with Apparatus illustrative of the Inclined Plane. *The Council of King's College, London.*

2099. Working Model of a patent silk throwing machine. Thomas Dickens, Edgemoor House, Higher Broughton, Manchester.   *South Kensington Museum.*

The bobbin, flyer, and reel are driven by friction gear, which secures steady and certain action.

2100. Working Model of a strand-making machine for machine cotton rope ; with a 49 reel frame for cotton yarn. Henry Cotton.   *South Kensington Museum.*

2101. Model, on a ⅓rd size, of a pirn winding machine. For making weaver's bobbins from cotton hanks. Robert Hall, Hope Foundry, Bury.   *South Kensington Museum.*

2102. Model, on a ⅓rd size, of a drum winding machine. For making warper's bobbins from cotton hanks. Robert Hall, Hope Foundry, Bury.   *South Kensington Museum.*

2104. Model of power weaving loom by George White, Glasgow, 1830.   *South Kensington Museum.*

The shuttle in this loom is arranged with a peculiar even power movement, by which, for the propulsion of the shuttle, one uniform power is exerted, thereby enabling fine fabrics, such as cambrics, lawns, faconets, &c., to be manufactured.

2105. Model of a power weaving loom, showing arrangement for working a double shuttle-box, and other features. Designed about 1840.   *South Kensington Museum.*

2106. Model of power weaving loom, with Jacquard arrangement attached, for weaving or working figured stuffs or pattern stuffs.   *South Kensington Museum.*

This model shows the arrangement of the cards in the loom after they have been cut for the desired pattern to be worked. It further illustrates the general movement of the several parts of a Jacquard loom.

2106a. Model of **Power Loom** constructed to show the ordinary manner of working the Beam and Taylor's patent method of performing the same.   *H.M. Commissioners of Patents.*

2106b. Improved Jacquard Apparatus and **Fittings,** worked by pegs.   *Bennet Woodcroft, F.R.S.*

In this improved Jacquard apparatus, those warp threads which are not raised to form the shed or opening for the shuttle to pass through and deposit a weft thread are depressed to the same extent that the others are raised. (Woodcroft's Patent, 1838.)

2106c. Full-size **Section Tappet** for **Looms,** for weaving a variety of patterns.   *Bennet Woodcroft, F.R.S.*

2106d. Model of **Loom,** with improved tappet plates and Jacquard apparatus. (Woodcroft's Patent, 1838.)
        Bennet Woodcroft, F.R.S.

2106e. Mule for **Spinning Cotton** and other fibrous substances. (Jas. Smith's Patent, 1833.) *Bennet Woodcroft, F.R.S.*

2106f. Improved Jacquard Machine worked by paper cards. *Bennet Woodcroft, F.R.S.*

The improvement in the jacquard machine consists in its being so constructed and worked as to depress some of the warp threads as much as it elevates others, whereas in the machine invented by Jacquard, and called after him, none of the warp threads were depressed, the opening for the shuttle being made by elevating some of the warp threads only. (B. Woodcroft's Patent, 1838.)

2106g. Steam Power-loom. *Bennet Woodcroft, F.R.S.*

2107. Model of a hand loom, for weaving sacks, hop pockets, &c. This loom is designed to weave sacks or pockets without a seam either at the sides or end. Invented by T. Clulow.
South Kensington Museum.

2108. Model of a hand loom for weaving fishing nets. G. Roberts, inventor. *South Kensington Museum.*

2109. Model, on a ⅓rd scale, of a plain and fancy goods weaving loom, having 12-inch reed space. The model can be driven either by hand or power. Robert Hall, Bury, Lancashire.
South Kensington Museum.

2110. Model, on ⅓rd scale, of a plain and fancy goods weaving loom, having 12-inch reed space. The model is arranged to be driven by power. Savill and Woolstenhulme, machine makers, Oldham. *South Kensington Museum.*

2111. Series of Temples (20 in number), of various sizes. Used in power looms for stretching the woven cloth. The temples are self-acting, and are suitable for woollen, cotton, and other heavy or light fabrics. R. Hall, Hope Foundry, Bury, Lancashire. *South Kensington Museum.*

2112. Drawing of patent machinery for preparing chemically pulp from wood, straw, and fibrous material for the manufacture of paper of all kinds and qualities. Sinclair's system. J. McNicol, C.E., 97, Buchanan Street, Glasgow.
South Kensington Museum.

The drawing represents the following portions of the patent machinery by W. Sinclair, for paper manufacture.
On a ½-inch to 1 foot scale :—
Fig. 1. Longitudinal through section of Sinclair's patent high-pressure tubular steam boiler.
Fig. 2. Front or firing end elevation of Sinclair's high-pressure tubular steam boiler.
Fig. 3. Sectional elevation of wood-pulp boiler.
Fig. 4. End elevation of blow-off pulp receiving tank.

40075. H h

Fig. 5. General plan of apparatus.
Fig. 6. Front and side elevation of wood-chopping machine.
Fig. 7. On a scale of $\frac{1}{4}$ inch to 1 foot. The soda ash (used to dissolve the wood into pulp) recovery apparatus. A longitudinal and cross section of the apparatus.
Fig. 8. On a full size scale. The Sinclair's patent conical plug joint for high-pressure steam boiler tubes.
Fig. 9. On a full size scale. The section of a hollow conical boiler tube joint. Sinclair's system.

2114. Working Drawing, on a $\frac{1}{3}$ scale, or 4 inches to 1 foot. Sheave or pulley. For a hoisting crane.
Side and end elevations.
Two through sections. *South Kensington Museum.*

2115. Working Drawing, on a $\frac{1}{2}$ size scale, or 6 inches to 1 foot.
Spur wheel. Tooth wheel for gearing.
Side elevation, showing method of laying out pitch of tooth and tooth circle.
Section of arm of wheel.
Section through centre of wheel.
Plan of wheel. *South Kensington Museum.*

2116. Working Drawing, on a $\frac{1}{3}$ scale, or 4 inches to 1 foot.
Mitre wheel, gear wheel. Showing plan, elevation, and sections of wheel. *South Kensington Museum.*

2117. Working Drawing, on $\frac{1}{12}$ scale, or 1 inch to 1 foot.
Eight feet cast-iron fly-wheel.
Side elevation with part of rim in section.
End elevation. Three through sections.
South Kensington Museum.

The wheel is cast in two segments, and united by wrought-iron hoops, shrunk on the boss, and by dowels and cotters at the rim.

2118. Working Drawing, on a $\frac{1}{4}$ scale, or 3 inches to 1 foot.
Cast-iron engine crank.
Side elevation. Plan throughout.
Plan when turned $\frac{1}{4}$ of a revolution.
Section through firm and web.
Longitudinal section, and a section when turned $\frac{1}{4}$ of a revolution. *South Kensington Museum.*

2119. Working Drawing, on a $\frac{1}{3}$ scale, or 4 inches to 1 foot.
Connecting rod end, for a 25 horse-power steam engine.
Elevation. Plan. Two sections.
South Kensington Museum.

2120. Working Drawing, of the governor of a steam engine.
Front and side elevation.
Elevation and plan of slide.

Pendulum rod, showing ball in section, and method of attachment.

Elevation and plan of forked rod.

Section through slide.

The front and side elevations are on $\frac{1}{4}$ scale, or 3 inches to 1 foot.

The details are on $\frac{1}{2}$ scale, or 6 inches to 1 foot.

South Kensington Museum.

2121. Working Drawing, on a $\frac{1}{2}$ scale, or 6 inches to 1 foot. Pillar block, plummer block, or pedestal; for a $4\frac{1}{2}$-inch shaft or journal.

Side elevation.

End elevation. Plan. Sheet No. 1.

South Kensington Museum.

2122. Working Drawing, on a $\frac{1}{2}$ scale, or 6 inches to 1 foot. Pillar or plummer block for a $4\frac{1}{4}$-inch shaft or journal.

Various sections. Sheet 2.   *South Kensington Museum.*

2123. Working Drawing, on a $\frac{1}{4}$ scale, or 3 inches to 1 foot. Hanging bracket and pillar block, for a $3\frac{1}{2}$-inch shaft; attached to a 16-inch cast-iron girder.

Front and side elevations.   *South Kensington Museum.*

2124. Working Drawing, full size. Steam whistle.

Elevation. Plan. Sections.

The vertical section shows by arrows the passage of the steam through the whistle.   *South Kensington Museum.*

2125. Working Drawing, on a $\frac{1}{4}$ scale, or 3 inches to 1 foot. Movable head stock for a turning lathe.

Side elevation, plan, and section.

End elevation and section.   *South Kensington Museum.*

2126. Working Drawing, full size.

Water cock or tap.

Side elevation. End elevation.

Plan. Side and end elevation of plug.

Through section of tap.   *South Kensington Museum.*

2127. Working Drawing, on a $\frac{2}{3}$ scale, or 8 inches to 1 foot. Stop-cock or straightway cock, for steam or water.

End elevation. Side elevation. Plan. Two sections.

Fourteen working drawings (lithographs), parts of machinery and steam engines. Thomas Busbridge, Plumstead, S.E.

South Kensington Museum.

2178gg. Bobbin Winding Machine, by the late Sir Marc Isambard Brunel, 1830.   *Maudslay, Sons, and Field.*

H h 2

2827ff. Original Micrometer Screw Gauging Machine, by the late Henry Maudslay, 1805.   *Maudslay, Sons, & Field.*

Each division on the milled head of micrometer screw shows $\frac{1}{10000}$ of an inch.

2127a. Machine for Originating Screws, with micrometer adjustment applied to tangent screw which sets the guide at any angle suitable to required pitch (adjustable to $\frac{1}{100}$ degree). The cutter is carried in sliding rest. Specimens of screws originated in machine.   *Maudslay, Sons, and Field.*

2127aa. Machine for Originating Screws of any required pitch and diameter (by the late Mr. Henry Maudslay).
          Maudslay, Sons, and Field.

Inside the cylindrical hole (below the table) is a steel knife edge which is attached to the divided circle (above). This circle and knife edge can be adjusted by a micrometer tangent screw to the calculated angle of the thread. The knife edge presses on and acts as a guide to a cylindrical rod passed through the hole, and gives the required feed while the thread is cut by a small tool carried in the slide rest seen at the side of the machine. The bar on which the screw is to be cut is carried in the centres of a lathe, while the screw machine is allowed to travel freely along the bed of the lathe. The bar is removed from the machine so as to enable the knife edge inside to be seen. When the circle and knife edge are set at the required angle they can be clamped to the frame. Date 1800–1805.

2127b. Three Guide Screws originated in the above machine; each screw contains 50 threads to the inch. The largest screw is $1\frac{1}{2}$ in. diameter and has 3,144 threads in the entire length. The nut for this screw is shown by its side ; it is 12 inches long and has 600 threads.   *Maudslay, Sons, and Field.*

2127c. Hand Plane for Metal, by the late Mr. Henry Maudslay.   *Maudslay, Sons, and Field.*

2127d. Metal Plough for Moulding, by the late Henry Maudslay.   *Maudslay, Sons, and Field.*

2127f. Original Screw Cutting Lathe, made and used by the late Henry Maudslay. It has 28 change wheels, and sector frame for carrying an intermediate wheel for cutting left hand screws. Four guide screws are shown, 100 threads, 50 threads, 35 threads, 30 threads to the inch. Also a small cone chuck and hollow mandril, screw tools for slide rest, and the original working handle used by Mr. Maudslay, below the bed. Date 1800 to 1810.
          Maudslay, Sons, and Field.

2127g. Skeleton Stocks and Dies, four sets, about 1805 to 1810, made and used by the late Henry Maudslay. The extreme delicacy and finish of the work are worthy of notice.
          Maudslay, Sons, and Field.

2127h. Coining Machinery. A cutting-out press for blanks, with self-acting feed rolls to deliver fillet. Date about 1814.

Maudslay, Sons, and Field.

297a. Micrometer Gauge and Screw, unfinished; divisions adjusted to show $\frac{1}{10000}$th of an inch.

Maudslay, Sons, and Field.

297b. Micrometer (Screw) Gauging Machine, by the late Mr. Henry Maudslay, divided on the head of micrometer screw to show $\frac{1}{10000}$th of an inch. *Maudslay, Sons, and Field.*

2178go. Drawing of Machinery for Boring Cannon, for Rio Janeiro, 1813. *Maudslay, Sons, and Field.*

2135. Three Elliptical Guides.
Royal Geological Institute and Mining Academy (Director, Prof. Hauchecorne), Berlin.

2138. Elliptic Guide, with long radius bar.
Royal Geological Institute and Mining Academy (Director, Prof. Hauchecorne), Berlin.

2139. Approximate Elliptic Guide.
Royal Geological Institute and Mining Academy (Director, Prof. Hauchecorne), Berlin.

2142. Kinematic Pillar Vice, with triangular link-motion.
Royal Geological Institute and Mining Academy (Director, Prof. Hauchecorne), Berlin.

This stand is so arranged that any link of the kinematic chains can be fastened in it easily and securely, so that they can be shown in every possible position.

WHEEL GEARING.

2140. Spur Wheel Gearing.
Royal Geological Institute and Mining Academy (Director, Prof. Hauchecorne), Berlin.

2143a. Model of a Machine for Cutting the Teeth of Bevel Wheels. *F. Engel, Hamburg.*

In the usual cutting and planing machines for bevel wheels, each space between the teeth is produced by at least two operations; first one side of a tooth is shaped or planed, then the opposite side of the next tooth. The present machine cuts both profiles simultaneously with the same cutter. A lateral motion is communicated to the cutter, the axis of the tool oscillating about the cutting point of the conical surface, and working alternately the one and the other side of the tooth profile, while the feed of the tool is self-acting. The machine is more accurate than previous ones, and other kinds of wheel can be shaped with it.

2143b. Samples of Wheels which have been cut by the Machine. *F. Engel, Hamburg.*

2143c. Drawing of the Machine.   *F. Engel, Hamburg.*

2144. Bevel-Wheel Gearing.
Royal Geological Institute and Mining Academy (Director, Prof. Hauchecorne), Berlin.

The models exhibited are a small portion of the collection used for machine-instruction in the Königlichen Berg-Akademie (Royal School of Mines). They have been made specially for this purpose by Herr Maiss, in the workshops of the Academy, upon the designs of Prof. Hörmann. The models of separate mechanisms have been constructed upon the principles laid down in Prof. Reuleaux's "Theoretische Kinematik"; they are all arranged for "inversion," that is, so that either of their links may be fixed in the screw stand exhibited with them, and the various properties and applications of the inversions easily shown.

The models are all made with the special view of illustrating in the most complete, clear, and simple manner the principles of the machines and mechanisms which they represent. All details not required in this relation are omitted or made subordinate, the parts which it is important for the students to observe are polished, the other parts, in order not to attract attention unnecessarily, are made a dead black, so that all disturbing reflections of the light are prevented. Care has been taken also to arrange them so that all essential parts and their combinations may be visible at a glance in all parts of the class-room, so that the alterations which it is necessary to make during the lectures may occupy a minimum of time, and altogether that teachers may find their use in demonstrations very convenient. In the steam-engine models have been made, the relative dimensions and arrangements, as far as possible, to resemble those of actual practice.

2144a. Pair of Worm-Wheels, with parallel axes.
Royal Geological Institute and Mining Academy (Director, Prof. Hauchecorne), Berlin.

2144b. Worm-Wheel, with inclined teeth.
Royal Geological Institute and Mining Academy (Director, Prof. Hauchecorne), Berlin.

2144c. Annular Wheel and Pinion.
Royal Geological Institute and Mining Academy (Director, Prof. Hauchecorne), Berlin.

2144d. Hypocycloidal Gearing.
Royal Geological Institute and Mining Academy (Director, Prof. Hauchecorne), Berlin.

2144e. Worm-Wheel, with straight teeth.
Royal Geological Institute and Mining Academy (Director, Prof. Hauchecorne), Berlin.

2144f. Pair of Worm-Wheels, with axes at right angles.
Royal Geological Institute and Mining Academy (Director, Prof. Hauchecorne), Berlin.

2144g. Skew Mitres.
The Committee, Royal Museum, Peel Park, Salford.

A pair of toothed mitre wheels mounted on shafts that cross each other. The invention of the late R. Roberts, C.E., of Manchester, 1836.

2144h. Intermittent Wheels.
The Committee, Royal Museum, Peel Park, Salford.

The larger wheel performs six revolutions for one of the lesser. The invention of the late R. Roberts, C.E., of Manchester.

2145. Model of Cugnot's Steam Carriage, 1783.
Conservatoire des Arts et Métiers, Paris.

2146. Model of a centrifugal smoke purifier.
Dr. Otto Braun, Berlin.

This model shows the action of centrifugal force on bodies suspended in gases, by means of it sparks, soot, tar, ammonia, &c. can be removed from smoke.

2146a. Tyndall's Smoke Filter Respirator, suitable for firemen, fire-escape men, and for persons entering deletérious atmospheres. *James Sinclair.*

The filtration is by means of layers of dry cotton wool, cotton wool dipped in glycerine, and charcoal.

2146b. Tyndall's Smoke Filter Respirator, combined with elastic tube, suitable for mining operations, chemical works, breweries, &c. *James Sinclair.*

This respirator is dèsigned to enable the wearer to enter and breathe freely in mephitic gases by means of the tube, which can be used in lengths up to 90 feet, the wearer being enabled to communicate verbally with those outside and they with him.

1925. Chinese Steel Helmets (2).
Bennet Woodcroft, F.R.S.

1926. Chinese Cane Helmets (2).
Bennet Woodcroft, F.R.S.

IX.—SHIPPING, NAVAL ARCHITECTURE, AND MARINE ENGINEERING.

2147. Models made of hard paraffin for ascertaining the Resistance of Ships by measuring the resistance of their models.
W. Froude, F.R.S.

The models, from 6 to 16 feet in length, and from 120 lbs. to 700 lbs. displacement, are made of hard paraffin. The experimental apparatus employed in the treatment of the models includes appliances for designing, moulding, and casting them, shaping them by automatic machinery, moving them through the water at the required speeds, and automatically recording the leading phenomena of the trial, namely, the speed, the resistance, and the change of level induced by the speed at each end of the model.

The several processes are illustrated by the accompanying series of seven photographs and two specimens, which may be explained as follows :—

No. 1. The designer.

This consists of a pile of adjustable templates, the thicknesses of which represent the horizontal intervals between the successive water-lines of the intended models, shown on a reduced scale. One edge of each template is an elastic steel band held to a wooden base-piece by adjustable ordinates hinged to the band and sliding through mortices in the base-piece fitted with hinged metal clamps. One of these templates (No. 8) set up as for use is sent to aid this explanation.

The photograph shows them in combination; when thus placed they represent the intended model on a reduced scale, by a series of water-lines in steps, which, if either filled up solid and fair to the salient angle of each, or trimmed off fair to the re-entering angle, would constitute the finished form.

No. 2. The moulding box, the mould, and the core.

The former is a rectangular wooden box $16' \times 2' 9'' \times 1' 10''$. containing plastic clay.

In the clay the external form of the full sized model is moulded by help of a series of rough cross sections deduced from the small scale designer, and into the mould is fitted the core, which constitutes the figure of the inside of the model. The core is framed on a series of internal cross sections made good to a surface and rendered coherent, first by a series of laths nailed to them externally, and, secondly, by a skin of calico drawn tight over the lathed surface, and then coated with plaster-of-paris and clay. Between this "core" and the "mould" there is, of course, a space, equal to the intended thickness of the model, into which space the melted paraffin is run, and there allowed to remain until by cooling it has become solid enough to bear removal.

Nos. 3 and 4, the shaping machine.

This is what has sometimes been termed in technical phrase a "copying machine." The model, bottom upwards, and adjusted successively to a series of different levels, travels longitudinally between a pair of revolving cutters, which are caused by means of a hand lever to so recede and approach one another, as the model passes, as to cut upon the model the horizontal section or "water-line," correctly appropriate to the level at which the model is set. At the side of the machine, in full view of the operator, there is a vertical board, which carries either a drawing of the intended model, showing the series of water-lines to be cut, or one of the "designer" templates already described. In front of this board is a "tracer," and the board and the "tracer" severally imitate upon the appropriate scales (the former by longitudinal motion, the latter by vertical motion) the longitudinal motion of the model and the lateral motion of the cutters. Thus the drawing (or template) passing along beneath the tracer, is practically a small scale picture of the model travelling past the cutters, and if the tracer be made to follow the correct line on the drawing (or to follow the edge of the template) the revolving cutters will cut the correct water-line on the model.

The model is then finished by hand with spokeshaves and scrapers, an operation which takes a man about three hours.

No. 5. The hauling engine.

This is the instrument by which the required motion through the water at definite speed is given to the model. The dynamometric truck to which the model is attached, is connected by a wire rope with a winding drum, driven by a small stationary double-cylinder steam engine. The engine is regulated by an extremely sensitive governor, acting upon a delicate steam throttle valve, on what is known as the "differential" principle, in which the governor rotates at its own appropriate speed, independently of the engine, the steam valve being opened or closed according as the engine is lagging behind the governor or overtaking it.

By adjusting the centrifugal weights of the governor, with a right and left-handed screw, and by differently speeding the belt which connects it with the engine, any required speed may be assigned to the engine between the limits of about 150 and 350 revolutions per minute, and by further changing the gear wheels connecting the engine and winding drum, speeds varying from 60 to 1,200 feet per minute may be assigned to the dynamometric truck.

Nos. 6 and 7. The dynamometric truck with model under it.

The dynamometric truck runs on a straight and level railway about 200 feet in length, suspended over a waterway 36 feet wide and 10 feet deep roofed throughout. The model floating in the water is as it were "harnessed" to the truck, and travels with it. It is kept from diverging sideways by a knee-jointed frame or "guider" at each end, of such construction as to perfectly prevent the slightest sideways deviation of the model, but in no way to interfere with its rising or falling, or moving in a fore and aft direction with reference to the truck. The towing strain (i.e., the force necessary to make the model accompany the truck in its longitudinal progress) is taken during the experiment by a spiral spring, the extension of which, measuring the towing force, is indicated on a large scale (through the intervention of certain levers) by a pen which makes a line on a recording cylinder covered with a sheet of paper. The recording cylinder is driven by the truck wheels, and thus its circumferential travel indicates distance run; at the same time another pen, jerked at half second intervals by a clock, records time. Other pens actuated by strings led over pulleys, record the change of level of the ends of the model. Thus the diagrams made furnish an exact measure of the speed, and a continuous record of the resistances and of the change of level of the model throughout the experimental run at steady speed. While starting or stopping, the model is controlled by hand levers to prevent the dynamometric spring being overstrained.

No. 8. A "designer" template.

This consists of one of the pile of adjustable templates shown in photograph No. 1, and already described.

No. 9. A segment of a model.

This specimen segment of a model is partly in a finished condition and partly in the condition in which it is left by the shaping machine, Nos. 3 and 4. It thus shows the series of water-line cuts made by the machine, and a part of the original cast surface remaining between the cuts.

2147aa. Model of the solid of **"Least Resistance,"** by the late Andrew John Robertson, dated 1861.   *Michael Scott.*

2147ab. Model of the Steam Ship **" Sir John Lawrence,"** embodying to a considerable extent the form of least resistance, designed by Michael Scott, in conjunction with the late Andrew John Robertson. The performance of this ship was excellent.
Michael Scott.

2147ac. Three diagrams of a new type of **War Ship,** designed by Michael Scott in 1869, and published in 1870.
Michael Scott.

In this design the surface exposed to hostile fire is diminished by constructing the vessel with a central fort, armour plated all round, an armoured deck under water, and dividing the space above this armoured deck for a height of six feet into water-tight compartments, which would be filled with fuel or water when going into action. She is intended to carry both turret and broadside guns, and might be armed with a submarine gun.

2147ad. Three diagrams of new type of **War Ship,** designed by Michael Scott in 1870, and published in 1871.

Michael Scott.

In this design there is a central fort, armour plated all round; an armoured deck under water sloping downwards towards the bow, so as to prevent the vessel from being raked in a seaway, and strengthening the ramming stem. The ship is intended to carry sail, to be dismantled before going into action, her turrets to be placed abreast, and also to carry broadside guns.

Some of the most important features in these designs have been adopted in the most modern war ships.

2150. Parent Steam Engine, made for Patrick Miller, Esq., and used by him on the lake at Dalswinton, 1788.

Bennet Woodcroft, F.R.S.

For some years prior to 1787 Patrick Miller, Esq., of Dalswinton, Scotland, had been engaged in a series of experiments with double and triple vessels propelled by paddle-wheels, worked by manual labour. In the experimental trips of 1786 and 1787 he was assisted by Mr. James Taylor (the tutor to his younger sons), and at the suggestion of the latter it was determined to substitute steam power for manual labour. For this purpose, in the early part of 1788, Taylor introduced William Symington, an engineer at Wanlockhead Lead Mines, who had previously obtained letters patent (June 5, 1787, No. 1,610) for "his new invented steam engine on principles ' entirely new.'"

An arrangement was made with Symington to apply an engine, constructed according to his invention, to one of Mr. Miller's vessels, and consequently the engine which forms the subject of this notice was made, the castings being executed in brass by George Watt, founder, of Low Calton, Edinburgh, in 1788. At the beginning of October in that year the engine, mounted in a frame, was placed upon the deck of a double pleasure boat, 25 ft. long by 7 ft., and connected with two paddle-wheels, one forward and the other abaft the engine, in the space between the two hulls of the double boat. On the steam being put in action it propelled the vessel along Dalswinton Lake at the rate of 5 miles an hour.

2178bc. Original Whole Model of the Steam Boat "Comet." Built on the Clyde by J. Wood for Mr. Henry Bell, at Port Glasgow, 1811. Length 42 ft., breadth 11 ft., depth 5 ft. 6 ins.

John Reid and Co.

In August 1812, the steam passage boat " Comet," being the first steam vessel ever built in Europe, began to run between Glasgow, Greenock, and Helensburgh, with passengers only. She was advertised to leave the Broomielaw on Tuesdays, Thursdays, and Saturdays, at an hour suitable to the tide, and to return from Greenock on Mondays, Wednesdays, and Fridays. The fares were 4s. for the best cabin, and 3s. for the second, and no gratuities to the vessel's servants were allowed. The boat was driven by a condensing steam-engine of 4 horse-power. She had at first two sets of paddle-wheels on each side of the vessel (shown in the model). Her greatest speed was five miles per hour.

2178cb. Working Drawing, by John Wood, Port Glasgow, of the " Comet," built by him in 1811 for Henry Bell.
R. Napier.

The " Comet " was the first vessel propelled by steam that regularly traded in Europe.

This drawing was presented by John Wood to R. Napier.

The *engine* of this vessel was presented by R. Napier and Sons to the South Kensington Museum.

2178ct. Photograph of the " Comet " and " Iona " Steamers, 1811, 1874. *John Hamilton.*

From a painting by Wm. Clark-Greenock, to illustrate and keep on record the appearance of the first British steamer, and also to make a comparison between the past and present types of Clyde river steamers.

2148a. The **Original Engine** of **Henry Bell's Steamboat " Comet,"** which was the first steamboat in Europe advertised for the conveyance of passengers and goods.
H.M. Commissioners of Patents.

The engine was made and fitted on board the " Comet " in 1812.

2148b. Drawing of the " Elizabeth," built by Charles Baird in 1815, and run on the Neva.
George Baird, St. Petersburg.

She was constructed out of a barge, and the chimney was of brick. The floats of the paddle-wheels were kept in a vertical position by means of shafts and mitre wheels. Scale of drawing, $\frac{1}{4}$ inch to the foot.

2148c. Drawing of Paddle-wheel Steamer, built by Charles Baird in 1817, to carry passengers between Petersburg and Cronstadt, showing end view, longitudinal view, plan, and longitudinal section. *George Baird, St. Petersburg.*

A. Steam engine.

B. Boiler.

C. C. Crank shafts on either side, with fly-wheel.

D. D. Toothed wheels driving paddle-wheels.

E. E. Paddle-wheels with floats revolving 50 turns per minute, by which the vessel is propelled.

F. Funnel leading from furnace serving in place of a mast.

G. G. Fore and aft cabins.

H. H. Side decks, to protect the wheels from the blows of the waves.

I. I. Paddle-boxes.

K. L. Staircase and rudder.

Scale of drawing, $\frac{1}{4}''$ to 1 foot.

2148f. Picture of the " Great Western " steamship, the first steamer that traded regularly between England and America. Date 1838, *Maudslay, Sons, and Field, Engineers.*

2148g. Table of the first 50 Voyages of the "Great Western" steamship.

Maudslay, Sons, and Field, Engineers.

2175. Models (3) of Varying-pitch Screw Propellers.
Woodcroft's Patent, 1838.   *Bennet Woodcroft, F.R.S.*

2176. Varying-pitch Screw Propeller on shaft.
Bennet Woodcroft, F.R.S.

2177. Skeleton Model of part of a **Vessel** fitted with a
Screw Propeller. (Cummeron's Patent, 1828.)
Bennet Woodcroft, F.R.S.

2178. Model of **Vessel** fitted with **Screw Propeller,** by
Sir Francis Pettit Smith.   *Bennet Woodcroft, F.R.S.*

The propeller consists of two whole turns of a screw thread round its shaft,
and is placed in the dead wood or run of the vessel, but by a memorandum
of alteration the patentee limits himself to a screw of one turn or two half
turns.

2180e. Memoir, written by Don Fernandez Duro, upon the
models of metal plated ships made in the last century, preserved at
the Ministry of Marine at Madrid.   With plates.
Archæological Museum, Madrid.

2149a. Drawing of Steam Vessel, by the exhibitor.
The late Baron Séguier, Membre de l'Institut.

EXHIBITED BY THE SOUTH KENSINGTON MUSEUM.

2178fk. Model of the **After Body,** showing complete
framing, of the Screw Steamer **"Novelty."** Built 1839–40.
Mr. H. Wimshurst.   *South Kensington Museum.*

This vessel was the first fitted with direct-acting engines to drive the screw,
and having means also of shipping and unshipping the propeller.

2155. Model, on a $\frac{1}{10}$ scale, of the **Horizontal Condensing Screw Engines of H.M. Ships** "Nelson," built
1814, altered for the screw propeller 1860; "Conqueror," built
1833, altered for the screw propeller 1859; and "Tamar," built
1863.   The engines are of 500 horse-power, nominal.   Diameter
of cylinders, 71 inches; stroke, 3 feet.   Ravenhill, Easton, and
Co., Engineers, Ratcliff, London.   *South Kensington Museum.*

2160. Model of the **Engines of the Paddle-wheel
Steamer "Helen McGregor,"** of Liverpool.
Designed in 1843 by G. Forrester & Co., Engineers.
South Kensington Museum.

This engine has two inverted steam cylinders of very long stroke, driving
one crank on the paddle shaft.   It is a condensing low-pressure engine, it
occupies but little hull space, and is said to be still at work, 1873.

2161. Model of a **Paddle Marine Engine,** designed by J. Scott Russell, F.R.S. ; having three oscillating cylinders, all connected to one crank on the paddle shaft.   One of the cylinders is vertical, the other two are inclined inwards at about 45°.   J. Scott Russell, F.R.S.   *South Kensington Museum.*

2152. Model, on a scale of 1½ inches to 1 foot, of the **Horizontal Condensing Screw Engines of H.M. Turret Ship "Monarch,"** 8,164 tons.   Built, 1868.   1,100 horsepower, nominal.   Lent by Humphrys and Tennant, Engineers, Deptford.   *South Kensington Museum.*

The engines of H.M.S. "Monarch" are of 6,600 indicated horse-power, and were designed and built in 1868 by the lenders of the model. The engines make 60 revolutions per minute, and are on the direct-acting principle, with return connecting rods, and have surface condensers. There are four piston rods to each engine piston. The cylinders are 120 inches in diameter, having a stroke of 4 feet 6 inches.

The condensers are of wrought iron, their tops being of cast brass. They can be used either as surface or jet condensers. They contain 17,264 copper tubes, each 6 feet long, giving large condensing surface per nominal horse-power. The water is driven through the condensers passing outside the tubes, by a reciprocating pump, the inlet and outlet pipes being of ample diameter.

The cylinders of the engines are steam jacketed.

The crank shaft is 22 inches in diameter, and the propeller shaft 18 inches.

The starting, stopping, and reversing gear is placed on a central platform between the cylinders and condensers of the engines.

The crossheads to piston rods are forged solid; their guides have very large surfaces, and are easily adjusted.

The boilers are tubular, having brass tubes. They contain 21,000 square feet of heating surface, and about 770 square feet of grate surface.

The ship's propeller is a two-bladed Griffith screw, of gun-metal; 23 feet 6 inches diameter, with adjustable pitch from 23 feet 6 inches to 28 feet 6 inches, weighing 22 tons.

2153. Model, on a scale of 1 inch to 1 foot, of the **Horizontal Condensing Screw Engines of H.M. Turret Ship "Prince Albert,"** 2,529 tons. Built 1860. 500 horsepower, nominal. Humphrys and Tennant, Engineers, Deptford.
                                        South Kensington Museum.

The "Prince Albert" was built in 1864 by Messrs. Samuda. She is an armour-plated turret ship, 240 ft. in length, 48 ft. beam, 25 ft. 3 inches in depth. She is driven by engines, of which the model is a representation, and fitted with a four-bladed screw propeller.

2178fa. Whole Model of the Cunard iron paddle steamer **"Scotia."**   Built 1861.   Constructed for the British and North American (Cunard) Royal Mail Steam Packet Company, by R. Napier & Sons, Glasgow.   Length 366 ft., breadth 47 ft. 6 in.; tonnage, builder's measurement, 4,050; load displacement, 6,520 tons ; horse-power, 1,000 nominal; diameter of cylinders, 100 in.; length of stroke, 12 ft. ; diameter of paddle-wheels, 40 ft.; size of floats, 11 feet 6 in. by 2 ft.   *South Kensington Museum.*

2178fb. Whole Model of the Woodside ferry paddle steamboat "**Cheshire**," employed between Birkenhead and Liverpool. Licensed to carry 1,620 passengers. Draught of water 6 ft.— Designed by Mr. George Harrison, M. Inst. C.E.

South Kensington Museum.

2178fc. Whole Model of the screw steamship "**Faraday**." Built by Mitchell & Co., Newcastle, in 10 months, 1874, for Messrs. Siemens Brothers, for employment in carrying and laying electric telegraph cables for ocean telegraph lines.—Dr. C. W. Siemens, Queen Anne's Gate.   *South Kensington Museum.*

2178fd. Half Model of the iron sailing ship "**Victory**." Built 1863. Tons 1,198.—Designed and built by Laurence Hill & Co., Glasgow.   *South Kensington Museum.*

2178fe. Half Block Model of the Royal Mail screw steamship "**Boyne**." Built 1871. Length (keel) 358 ft. 6 in., breadth 40 ft. 5 in., depth 34 ft. 6 in.; tons o.m. 2,882.—W. Denny and Brothers, Dumbarton.   *South Kensington Museum.*

2178ff. Half Model of paddle steamer "**Glengyle**." Constructed for the navigation of the river Yangtzee. Tons 2,040; horse-power, 400 nominal.—W. Denny and Brothers, Dumbarton, N.B.   *South Kensington Museum.*

2178fg. Half Model of a corvette of the "**Alabama**" class. Proposed by Mr. George Turner, late master shipwright, Woolwich Dockyard.—Mr. George Turner.

South Kensington Museum.

2178fi. Whole Model of H.M. Indian relief steam troopship "**Jumna**." Built 1866, by Palmer's Shipbuilding Company. Length 365 feet, breadth 48 ft. 9 in., depth 42 ft.; tons gross, 4,174; horse-power, 700 nominal; speed, 14½ knots per hour; scale, $\frac{1}{48}$th real size.—Palmer's Shipbuilding Company, Limited, Newcastle-on-Tyne.   *South Kensington Museum.*

2178fj. Whole Model of the Mail Screw Steamer "**Montana**," Liverpool and New York line. Built 1873, by the Palmer's Shipbuilding Company, Limited. Length 412 ft., breadth 43½ ft., depth 42¾ ft.; tonnage gross, 4,320 horse-power, nominal, 900; speed, 15¼ knots per hour; scale, $\frac{1}{48}$th full size.— The Palmer Shipbuilding Company, Limited, Newcastle-on-Tyne.

South Kensington Museum.

2154a. Model, showing the complete **Framing** and **Construction** of the after body of the second "**Submerged Propeller**," or screw steamship "Novelty," built by Mr. H. Wimshurst, 1839–40.   *South Kensington Museum.*

This was the first vessel fitted with direct-acting engines to drive the screw propeller, and having means for shipping or unshipping the same.

2156. Model (working), on a 3 inch to 1 foot scale, of the **Condensing Vertical Screw Engines** of steamship " A. Lopez," Cadiz and Havannah Spanish Mail Service.

South Kensington Museum.

The engines are constructed on the hammer or inverted cylinder principle, and have condensers, air and feed pumps, variable expansion gear, &c. . The model is a complete working condensing engine of about 15 horse-power. It was made in 1866-7, and exhibited in motion at the Paris Universal Exhibition of 1867. W. Denny and Brothers, Engineers and Shipbuilders, Dumbarton.

The condensers of these engines are on Spencer's surface plan. They comprise a large central box, on the top edges of which rest the cylinders. The piston rods, two to each piston, work down by the sides of the condenser, and move in guides carried by the sides of the box. The pumps are worked off the cross heads, which are suitably prolonged for the purpose.

2158. Drawing, on a 1 inch to 1 foot scale, of the **Compound Inverted Cylinder Screw Engines** of the steamships " Edinburgh Castle " and " Windsor Castle," built and engined in 1872 by the donors of the drawing.—R. Napier and Sons, Engineers, Glasgow.   *South Kensington Museum.*

The engines are of 270 horse-power, nominal, having surface condensers, air and feed pumps, link motion for reversing, &c.
Diameter of high-pressure cylinder, 44 inches.
Diameter of low-pressure cylinder, 72 inches.
Stroke, 3 feet 6 inches.
The steamships belong to Messrs. Donald Currie and Co.'s Colonial Line of Mail Steamers, and run direct between London and the Cape of Good Hope.

2159. Model, on a scale of $1\frac{1}{2}$ inch to 1 foot, of the **Oscillating Cylinder Condensing Engines** of the Holyhead and Kingstown Royal Irish Mail **Paddle Steamer " Leinster."** 750 horse-power, nominal. Diameter of cylinders, 98 inches ; stroke, 6 feet 6 inches.   *South Kensington Museum.*

To the engines are attached, on the same scale, the feathering float paddle-wheels of the ship, which are 32 feet in diameter. The floats are 12 feet long by 4 feet 10 inches deep. Ravenhill, Easton, & Co., Engineers, Ratcliff, London.

The length of the " Leinster " is 350 feet over all. Beam, 35 feet; depth of hold, 21 feet. Tons, 2,000. The ship has 8 boilers, 4 before and 4 abaft the engines, having 40 furnaces fired in line with the keel. The draught of the ship is 8 feet 6 inches on an even keel, and her speed about 21 statute miles per hour. She was built by Messrs. Samuda Brothers, Poplar, in 1860.

2153d. Model of H.M. Steam Troopship " Orontes," on a $\frac{1}{4}$ scale, showing the arrangement of three canting bridges for lifeboats, with lifeboats built on Lamb and White's principle. Also 10 of Captain J. W. Hurst's patent life-rafts, lashed to ship's side.—J. White, Cowes.   *South Kensington Museum.*

2153e. Model of a Surf-boat, for service on West Coast of Africa. Built by Forrest and Son for the War Department, 1872–3. Scale 1 inch to 1 foot ; length 25 ft., breadth 5 ft., depth 2 ft. 3 in.—Forrest and Son, Limehouse.
South Kensington Museum.

2153f. Model of Pleasure Boat. Built for Lord Castlerosse in 1861 for the use of Her Majesty during her visit to the Lakes of Killarney.—Searle and Sons, Lambeth.
South Kensington Museum.

2153g. Model of the Lifeboat and its Transporting Carriage, on about a $\frac{1}{10}$ scale ; adopted by the Royal National Lifeboat Institution. Length 33 ft., breadth 8 ft., depth inside 3 ft. 4 in.   *South Kensington Museum.*

2178fm. Whole Model of the Australian lifeboat **" Lady Daly."** Built at Adelaide, from designs by Mr. W. Taylor, Government shipwright at that port, 1867. Length 43 ft. 1 in., breadth 9 ft., depth midships 4 ft. 1 in.; scale, $\frac{3}{4}$ inch to 1 ft. The model was presented to His Royal Highness the .Duke of Edinburgh, by the Marine Board of South Australia in 1868. —H.R.H. The Duke of Edinburgh. *South Kensington Museum.*

2178fl. Whole Model, rigged, of a **Thames Sailing Barge.** Built 1855. Length 70 ft., breadth 16 ft., depth 6 ft.—Searle and Sons, Lambeth.   *South Kensington Museum.*

2162. Drawing, coloured, of a **Patent Multi-flue High-pressure Marine Boiler ;** by Messrs. R. & W. Hawthorn, 1868. Scale, $\frac{1}{2}$ inch to 1 foot. R. & W. Hawthorn, Engineers, New-castle-on-Tyne.   *South Kensington Museum.*

The drawing shows a front view and cross section of the boiler, longitudinal and cross sections, and a sectional plan.

2163. Model of a set of **Patent High-pressure Marine Tubular Boilers ;** designed for the continuous use of fresh water. William Gray, Dawlish. *South Kensington Museum.*

These boilers possess the following arrangements :—They are fired from each end, the furnaces having special air-flues designed to assist in the combustion and the consumption of smoke. They are multi-tubular boilers, having steam superheaters, and over all large tanks to contain fresh water, which supply the boilers with heated feed water. The working pressure is 100 lb. per square inch.

2164. Drawing. Sections of Gray's Patent High-pressure Marine Steam Boilers, showing arrangement of the tubes, steam superheater, water heater, and flues. Scale, $\frac{1}{4}$ inch to 1 foot. W. Gray, Dawlish.
South Kensington Museum.

2165. Drawing, on a ¾ inch to 1 fôot scale, of an **Improved Marine Tubular Boiler;** for high pressure. Designed and patented by Messrs. J. and F. Howard. Made by the Barrow Shipbuilding Co., Barrow-in-Furness. J. and F. Howard, Engineers, Bedford. *South Kensington Museum.*

The drawing shows a front view of the boiler, side elevation, longitudinal section; and a section through tube plates showing method of coning joints, on a quarter full-size scale.

2166. Drawing of **Messenger's Patent High-pressure Vertical Water-tube Boiler;** for steam yachts and launches. Designed by Thomas Messenger about 1869. A. Verey and Co., Engineers, Dover. *South Kensington Museum.*

2167. Steam Donkey Engine and Pump, single-acting, for feeding steam boilers with water. Alexander Wilson and Co., Engineers, Vauxhall Works, Wandsworth Road, S.W.
South Kensington Museum.

Diameter of steam cylinder, 3½ inches; stroke, 4 inches. Diameter of pump, 1⅞ inches. Will pump 460 gallons per hour, and feed a 30 h.p. steam boiler.

2157. High-pressure Non-condensing Screw Engine of 3 horse-power, for screw steam launches. Constructed on the hammer or inverted cylinder principle. Diameter of cylinder, 5 inches. Stroke, 6 inches. A. Verey and Co., Engineers, Dover.
South Kensington Museum.

2153h. Pencil Drawing, by William Van der Velde, Amsterdam, 1663–1707. Hull of the ship of war "Sovereign of the Seas," 1687.—Mr. G. Smith. *South Kensington Museum.*

2178ehh. Model, Martin's patent self-canting anchors; adopted by the British Government and fitted to H.M. ships "Devastation," "Thunderer," and others.—C. Martin, King William Street, E.C. *South Kensington Museum.*

2180b. Model, in wood, on ¼-inch scale of the **"Lowe-Vansittart" screw-propeller blades,** as invented and fitted in 1869 by Mrs. Henrietta Vansittart, for trial on board H.M.S. "Druid," 350 horse-power. *South Kensington Museum.*

2148d. Model of **Screw Propeller,** driven by a pair of engines direct acting, having sun and planet motion, so as to give the propeller two revolutions for every revolution of the engine.
Glasgow Mechanics' Institution.

2148h. Model of the sailing ship **"Cairnsmore,"** length 223 ft., breadth 33 ft. 6 in., depth 20 ft. 6 in.; tonnage, O.M. 1211. Built by John Reid and Co., Port Glasgow, for Messrs. Nicholson and McGill, Liverpool, 1854. *John Reid and Co.*

This ship made her first passage from Clyde to Bombay in 65 days.

2149. Model of a **Direct-acting Marine Steam Engine.**
Patented by J. Miller, A.D. 1841, No. 9,107.

H. M. Commissioners of Patents.

2153a. Whole Model of an Iron Floating Dock, built
for the French Government by Messrs. Randolph, Elder, and Co.,
Glasgow, for Port Saigon, Cochin China.   *John Elder and Co.*

Dimensions of the dock:—Length 300 feet, breadth 94 feet, depth 42 feet.
Weight of dock, 2,800 tons.
The dock will lift a ship of 4,800 tons weight, drawing 27 feet of water.

2153b. Three Drawings of **Compound Marine Engines,**
constructed and fitted on board H.M.S. " Constance " in 1863, by
Messrs. Randolph, Elder, and Co., Glasgow.   *John Elder and Co.*

They were the first compound engines fitted in any of H.M. ships, and have
two 60 inch cylinders, four 78 inch cylinders, of a stroke of 3 feet 3 inches,
and are fitted with surface condensers.

No. 1, Plan, looking forward.
No. 2, Cross sectional elevation, looking forward.
No. 3, Cross sectional elevation, looking astern.

2153c. Model of Lifeboat Bridge and **Lifeboat,** as
fitted on board H.M.S. "Orontes" and "Tamar," showing the
method of carrying and launching ships' life-boats transversely.

John White.

This is a working model of Mr. J. White's plan, adopted by the Admiralty for H.M. ships of war and Indian relief troop ships, for carrying and
launching life-boats from ships' upper deck.

2154. Model of Dawes and Holt's Marine Engine.
Henry P. Holt, C.E.

This is of the vertical compound condensing type, without intermediate
receiver between high and low pressure cylinders, and very short passages,
reverse action of pistons, parallel motion, single crank and connecting rod,
one vertical and one horizontal air pump, the former forming a counterbalance
to connecting rod, and the latter is arranged to be used as a starting cylinder
when required.

2168b. Wave-Indicator. An instrument intended for the
direct tracing of the height of the waves, and for deducing their
form.   *Vice-Admiral Paris, Paris.*

2168c. Rolling-Indicator. An instrument indicating the
rolling of ships, not only at its extreme amplitude, but during all
its periods.   *Vice-Admiral Paris, Paris.*

These two instruments are due to the late Lieutenant Armand Paris, son of
the Admiral.

**2169. Set of (eight) Models of the Block-making
Machinery** invented by Sir Isambard Brunel, for the use of

Government, and set up in Portsmouth Dockyard by Messrs. Maudslay and Sons. in 1804, where it has remained in use to the present time.   *Royal Naval Museum, Greenwich.*

2169a. Model of Wrought-iron Ship's Block, by the exhibitor.   *The late Baron Séguier, Membre de l'Institut.*

2170. Plan for Feathering a Screw Propeller, proposed by Messrs. Maudslay, Sons, and Field, 1848.
Royal Naval Museum, Greenwich.

2171. Hirsch's Screw Propeller.
Royal Naval Museum, Greenwich.

2172. Four-bladed or Two-bladed Screw Propeller.
Royal Naval Museum, Greenwich.

2173. Working Model of a pair of **Inverted Cylinders Marine Engine** for driving a screw propeller. Cylinders $2\frac{3}{4}$ inches bore, stroke 4 inches, link motion and reversing gear complete, shut-off valve, &c. Full sized partly shaded drawings of same. Made by exhibitor during his apprenticeship.
Robert Lathbury.

2174. Model in mahogany of a **Three-bladed Griffiths' Patent Screw Propeller.** Made by exhibitor during his apprenticeship.   *Robert Lathbury.*

2178dn. Model of Small Steamer "Mab," the fastest on the Neva. Built in 1874, of brass, by George Baird, St. Petersburg.

Length	-	-	- 48 ft.
Breadth ·	-	-	- 6 ft. 6 in.
Depth at side	-	-	- 3 ., 6 ,,
Draught	-	-	- 1 ,, 7 ,,
,, over screw -	-	-	2 ,, 9 ,,
Speed	-	-	- 19 miles.
Diameter of high pressure cylinder	-	7 ins.
,, low ,,	-	11 ,,
Stroke	-	-	- 8 ,,
Working pressure	-	-	- 120 lbs.
Revolutions per minute	-	- 600
Nominal H.P.	-	- 8

George Baird, St. Petersburg.

2178b. Model of a Yarmouth Herring Fishing Lugger getting her nets in. Exact model and position of men when at their work.   *John Bracey.*

2178c. Model of a Yarmouth Trawling Smack. Without the trawl net.   *John Bracey.*

2178d. Model of a Bow of a first rate ship-of-war, showing the disposition of the frames and hawse pieces.   *Wm. Moody.*

2178e. Model of a Stern of a first rate ship-of-war, showing the disposition of the stern timbers and frames.
Wm. Moody.

2147l. Stem of a First-rate showing the disposition of the frame and stern timbers.   *Wm. Moody.*

2178h. Model of a Jury Rudder as fitted to the arctic schooner " Intrepid."   *Wm. Moody.*

2178i. Model of a Bermudian Sailing Boat, " Pearl."
Wm. Moody.

2178j. Model of the State Barge in which William III. landed at Greenwich, 1689.   *Wm. Moody.*

2178k. Model showing the method of framing a screw aperture of a line of battle ship, to enable the screw to be raised or lowered at pleasure.   *Wm. Moody.*

2178l. Model of a Midship Section, showing the method of framing a first rate ship-of-war.   *Wm. Moody.*

2178m. Model of the Bow of a first rate ship-of-war, " Prince of Wales."   *Wm. Moody.*

SERIES OF IRON MAIL PADDLE STEAMERS, DESIGNED AND BUILT BY THE LATE JOHN LAIRD, M.P., FROM 1840 TO 1860.

2178n. Model of H.M.S. " Dover," 1840. Length 113 ft., breadth 21 ft., depth 9 ft. 10½ ins.   228 tons.   90 horsepower.   *Laird Brothers.*

This was the first iron mail steamer, and was built for the Admiralty in 1840.   She carried the mails between Dover and Calais for many years, and afterwards did good service on the coast of Africa.

2178o. Model of " Lord Warden," 1847. Length 160 ft., breadth 24 ft., depth 10 ft. 9 ins.   446 tons.   160 horsepower.   *Laird Brothers.*

Built for the South-eastern Railway Company in 1847, and is still running as one of their despatch boats between Folkstone and Boulogne.

2178p. Model of " Cambria," 1848. Length 196 ft., breadth 27 ft., depth 14 ft. 6 ins.   716 tons.   370 horse-power.
Laird Brothers.

Built for the Chester and Holyhead Railway Company for their despatch service between Holyhead and Dublin, was lengthened in 1860, and is still running as a cattle boat.

2178q. Models of the "Ulster," "Munster," and "Connaught," 1860. Length 334 ft., breadth 35 ft., depth 19 ft. 2,039 tons. 750 horse-power. *Laird Brothers.*

Built for the City of Dublin Steam Packet Company, for the mail service between Holyhead and Kingston, in 1860.

The "Connaught" attained a speed of over 18 knots, or 21 statute miles, per hour on her official trial at Stokes Bay.

These three vessels, together with the "Leinster," built by Messrs. Samuda, still perform this service.

SERIES OF IRON VESSELS, DESIGNED AND BUILT BY THE LATE JOHN LAIRD, M.P., AND THE PRESENT FIRM, AS SURVEYING SHIPS, MEN-OF-WAR, &C.

2178r. Model of H.E.I.C. "Euphrates" and "Tigris," 1834. Length 105 ft., breadth 19 ft., depth 7 ft. 6 ins. 179 tons. 50 horse-power. *Laird Brothers.*

Built for the Hon. East India Company for General Chesney's expedition for the exploration of the River Euphrates. These vessels were built by Mr. Laird at Birkenhead, 1834, then taken to pieces and shipped to the coast of Syria, and after having been carried across the desert by camels, they were put together and launched on the banks of the Euphrates by artisans sent from Birkenhead for the purpose.

Three similar vessels of very light draught, the "Nimrod," "Nitocris," and "Assyria," of 153 tons, each carrying two 9-pr. pivot guns, were built for the navigation of the Tigris and Euphrates a few years later.

2178s. Model of H.E.I.C. "Nemesis," 1839. Length 169 ft., breadth 29 ft., depth 10 ft. 3 ins. 660 tons. 120 horse-power. *Laird Brothers.*

Built for the Hon. East India Company for service on the coast of India, and armed with two 32-pr. pivot guns.

This vessel, as well as the "Phlegethon," a similar but rather smaller vessel, though only drawing 5 feet of water, made the passage out to India round the Cape, a drop rudder and sliding keel, as shown on model, being fitted for that purpose.

Under the command of Captain (now Admiral Sir William) Hall, R.N., the "Nemesis" did distinguished service in the China wars, her light draught enabling her to perform service which no wooden vessel in the fleet was able to accomplish.

At the same time the "Medusa" and "Ariadne," of 432 tons, and each carrying two 24-pr. pivot guns, were built for the same service.

2178t. Model of the "Guadaloupe," 1842. Length 187 ft., breadth 30 ft., depth 16 ft. 788 tons. 180 horse-power. *Laird Brothers.*

The success of the above steamers (the first iron vessels armed with heavy guns) induced the agents of the Mexican Government to order the steam frigate "Guadaloupe," of 800 tons and 180 horse-power, armed with two 68 pounder pivot guns, one at each end, and four 24-pounder broadside guns. The satisfactory reports made upon the construction and trials of this vessel by the late Mr. Large and other officers induced the English Government to entrust Mr. Laird with the designing and building of the iron paddle-wheel frigate "Birkenhead," 1,400 tons, one of the first iron war ships in the English navy ; this ship was launched in 1845.

2178u. Models of the "Parana" and "Uruguay," 1873. Length 152 ft., breadth 25 ft., depth 12 ft. 6 ins. 455 tons. 80 horse-power. *Laird Brothers.*

Iron screw gunboats of modern type, built for the Government of the Argentine Confederation, armed with two 100-pr. Vavasseur pivot guns. Rigged as barques and fitted with Bevis' patent feathering screw propeller.

2178v. Models of the "Fu-Shêng" and "Chien-Shêng," 1875. Length 87 ft., breadth 26 ft., depth 8 ft. 8 ins. 256 tons. 40 horse-power. *Laird Brothers.*

Iron screw gunboats for coast and river defence, armed with one 18-ton Vavasseur gun, 450-pr., fitted with twin screws.
These vessels steamed out to China through the Suez Canal.

2178w. Models of the "El Plata" and "Los Andes," 1875. Length 180 ft., breadth 44 ft., depth 11 ft. 9 ins. 1,588 tons. 750 indicated horse-power. *Laird Brothers.*

Armour-plated twin screw turret vessels, built for the Government of the Argentine Confederation. Protected with armour 6 inches on the hull and 8 inches on the turret, and carry each two 12½-ton 300-pr. rifled guns.
They steam 9¼ knots on a load draught of water of 9 ft. 6 ins., and steamed from the Mersey to Buenos Ayres in about 50 days, including all stoppages.

SERIES OF IRON PADDLE-WHEEL VESSELS FOR COMMERCIAL PURPOSES, DESIGNED AND BUILT BY THE LATE JOHN LAIRD, M.P., AND PRESENT FIRM.

2178x. Model of the "John Randolph," 1834. Length 110 ft., breadth 22 ft., depth 7 ft. 6 ins. 249 tons...60 horse-power. *Laird Brothers.*

The first iron steamer ever seen on American waters, built at Birkenhead, taken to pieces, shipped at Liverpool, rivetted together on the Savannah river, where for many years after she did service as a tug boat.

2178y. Model of the "Garryowen," 1834. Length 130 ft., breadth 21 ft. 6 ins., depth 9 ft. 3 ins. 263 tons. 90 horse-power. *Laird Brothers.*

Paddle steamer built for the City of Dublin Steam Packet Company, for the navigation of the lower Shannon, and the largest iron vessel built at this time. After 30 years' service in Ireland the machinery was taken out of her and she was made into a sailing ship.
About the year 1841, at Mr. Laird's suggestion, this vessel was placed at the disposal of the Admiralty to enable them to institute a series of experiments on the variation of the compass in iron vessels; which were conducted by Capt. Johnson, R.N., and subsequently elaborated by the experiments carried out by Professor Airy on General Steam Navigation Company's steamer "Rainbow," built by Mr. Laird in 1837.

2178z. Model of the " Helen McGregor," 1843.
Length 180 ft., breadth 26 ft., depth 15 ft.   573 tons.   220 horse-
power.   *Laird Brothers.*
Built for carrying passengers and cattle between Hull and Antwerp; one
of the largest vessels of her class.

**2178aa. Model of the " Countess of Ellesmere,"
1852.** Length 160 ft., breadth 20 ft., depth 7 ft. 6 ins.   315
tons.   80 horse-power.   *Laird Brothers.*
Fast paddle passenger steamer, formerly running on the Mersey, afterwards
sold as a yacht to the Grand Duke Constantine of Russia.

2178ab. Model of the " Earl Spencer," 1874. Length
253 ft. 6 ins., breadth 29 ft., depth 14 ft. 9 ins.   1,067 tons.   350
horse-power.   *Laird Brothers.*
Showing present type of passenger and cattle steamers; built for the
London and North-western Railway Company for service between Holyhead
and Greenore.   Speed, 15 knots.

SERIES OF IRON SCREW STEAMERS, DESIGNED AND BUILT BY
THE LATE JOHN LAIRD, M.P., AND THE PRESENT FIRM,
FROM 1838.

2178ac. Model of the " Robert F. Stockton," 1838.
Length 63 ft. 5 ins., breadth 10 ft., depth 7 ft.   33 tons.   30 horse-
power.   *Laird Brothers.*
One of the first screw steamers ever built; fitted with Ericsson's screw
propeller.
The propeller was unshipped for the voyage made under canvas from
Liverpool to New York, where she was employed for many years as a tug
boat.

2178ad. Model of the " Forerunner," 1852. Length
161 ft. 6 ins., breadth 22 ft., depth 11 ft. 4½ ins.   381 tons.   50
horse-power.   *Laird Brothers.*
Built for MacGregor Laird, Esq., the founder of the African Royal Mail
Steam Navigation Company, of which she was the pioneer vessel.

2178ae. Model of the " Nubia," 1854. Length 292 ft.,
breadth 39 ft., depth 27 ft. 9 ins.   2,173 tons.   450 horse-power.
                                         Laird Brothers.
Type of screw mail and passenger steamer of her date. Built for the
Peninsular and Oriental Steam Navigation Company, and still carrying the
mails in their service.

2178af. Model of the " Africa," 1871. Length 285 ft.,
breadth 34 ft., depth 23 ft. 3½ ins.   1,627 tons.   200 horse-power.
                                         Laird Brothers.
One of the modern steamers of the African Royal Mail Company, for same
service as " Forerunner."

2178ag. Models of the "Corcovado," "Puno," and "Britannia," 1872. Length 375 ft., breadth 43 ft., depth 33 ft. 9 in. 3,434 tons. 600 horse-power.   *Laird Brothers.*

Type of screw mail and passenger steamer of present date.

Built for the mail and passenger service of the Pacific Steam Navigation Company.

The "Corcovado" made her first voyage from Liverpool to Callao, 11,000 knots, in 33½ days, equal to a mean speed of 13·54 knots.

SUNDRY MODELS AND PLANS.

2178ah. Model of Bow of Vessel, showing bow rudder, with guard on. Patent taken out by the late Mr. John Laird in 1843.   *Laird Brothers.*

Very generally adopted for all double-ended river steamers, and fitted in several paddle-wheel gun vessels.

2178ai. H. E. I. Co. "Napier," 1843. Length 160 ft., breadth 24 ft., depth 5 ft. 446 tons. 90 horse-power.
          Laird Brothers.

Built for the Hon. East India Company on a plan patented by Mr. Laird in 1843, with a spoon-shaped bow, and lifting dead wood and rudder.

This form of vessel combines speed, light draught of water, and good steering, with great carrying capacity, and was found to answer so well for the difficult navigation of the River Indus that a large number of river steamers were afterwards constructed by Mr. Laird for the Hon. East India Company on same system.

2178aj. Model prepared by the late William Laird, Esq., in 1836, to show application of the screw propeller to a frigate or troop ship of large size.   *Laird Brothers.*

2178ak. Picture of the "Rainbow," 1837. Length 185 ft., breadth 25 ft., depth 11 ft. 9 ins. 581 tons. 180 horse-power.   *Laird Brothers.*

Paddle steamer for passenger and cargo service, built for the General Steam Navigation Company of London, for service between London and Ramsgate. After being employed in London and Antwerp trade for some time, the "Rainbow," which was the fastest vessel of her day, ran for many years as a cargo steamer between Havre and London, and was in service till 1869.

2178al. Picture of H.M.S. "Soudan," "Albert," and "Wilberforce," 1840. Length 138 ft., breadth 27 ft., depth 8 ft. 8 ins. 459 tons. 70 horse-power.   *Laird Brothers.*

Built for H.M. Government for the exploration of the River Niger.

2178am. Picture of the Ferry Steamer "Nun," 1840. Length 105 ft., breadth 20 ft., depth 8 ft. 9 ins. 187 tons. 60 horse-power.   *Laird Brothers.*

The picture shows the "Nun" grounded on the stone pier at Birkenhead, her after end resting on the pier, and her bow on the bare rock below; the distance between the points of support being 81 ft. ; the whole weight of the machinery, 65 ¦tons, being in the middle of this unsupported space. She floated off the succeeding tide without having received the slightest damage. This incident, which occurred in 1842, went far to confirm the growing confidence in the strength of iron ships.

2178an. Model of the "Marajo," 1874. Length 221 ft., breadth 32 ft., depth 10 ft. 3 ins. 1,099 tons. 200 horse-power.
Laird Brothers.

Type of river steamer of modern construction, having large carrying capacity for passengers and cargo, on light draught of water, and great speed, and being fitted with compound oscillating engines.

2178ana. Model of H.M. Ironclad Turret Ship "Captain."
Built at Birkenhead, 1869.
Horse power, 900 nominal.
Engines by Laird Brothers.
Horse power, indicated 5,400.
Length, 335 feet. Breadth, 52 ft. 3 in.
Tons 4,272. *Laird Brothers.*

This vessel was ship rigged and had tripod iron masts and lower yards of iron; she carried two armour plated turrets containing two 600-pr. 25 ton guns each, one bow gun, 100-pr. 6½ tons, one stern gun, 100-pr. 6½ tons. The "Captain" was lost in a gale off the coast of Spain in the night of 6th September 1870.

2178bf. Drawings of the Sailing Ships "Falmouth," 1752; "Friendship," 1790; "Royal William," 1726. (Three drawings.) *R. and H. Green.*

2178bg. Half-block Model of SS. "Sultan." Built 1870. *R. and H. Green.*

2178ao. Half-block Model of S.S. "Viceroy." Built 1871. Length 320 ft., breadth 37 ft. 6 ins., depth, keel to upper beams, 20 ft. 3 ins., depth, keel to main beams, 32 ft. 3 ins. Tons B.M. 2,225. Tons net register 1,851. *R. and H. Green.*

2178ap. Half-block Model of Ironclad Ram "Arapiles." Built for the Spanish Government, 1863. Length 279 ft. 4 ins., breadth 54 ft., depth 32 ft. 5 ins. Tons burthen 3,547. Screw engines, horse-power 800 nominal. 34 guns.
R. and H. Green.

2178aq. Half-block Model of H.M. Ironclad Batteries "Glatton" and "Trusty." Built 1855. Length 172 ft. 6 ins., breadth 45 ft. 1 in., depth 14 ft. 9½ ins. Tons 1,546. Engines horse-power 150 nominal. *R. and H. Green.*

2178ar. Whole Model of East Indiaman "Falmouth." Built 1752. *R. and H. Green.*

2178as. Whole Model of Merchant Sailing Ship "Seringapatan." Built 1837.   *B. and H. Green.*

2178at. Half-block Model of Paddle Tug "Rienzi." Built 1875. Length 114 ft. 6 ins., breadth 20 ft., depth 11 ft. Tons B.M. 218$\frac{3}{94}$. Horse-power 70 nominal.
R. and H. Green.

2178au. Half-block Model of Sailing Ship "Agamemnon." Built 1855. Length 244 ft., breadth 36 ft. 6 ins., depth 16 ft. 6 ins. Tons 1,536$\frac{67}{94}$.   *R. and H. Green.*

2178av. Half-block Model of Sailing Ship "Nile." Built 1850. Length 173 ft. 8 ins., breadth 36 ft. 6 ins., depth 16 ft. Tons 1,038$\frac{29}{94}$.   *R. and H. Green.*

2178aw. Half-block Model of Sailing Ship "Carnatic." Built 1833. Length 130 ft. 6 ins., breadth 32 ft. 6 ins., depth 14 ft. 7 ins. Tons 595.   *R. and H. Green.*

2178ax. Half-block Model of Sailing Ship "Melbourne." Built 1874. Length 260 ft., breadth 40 ft., depth 24 ft. Tons register 1,965, tons O.M. 2,008.   *R. and H. Green.*

2178ay. Half-block Model of Sailing Ship "Carlisle Castle." Length 220 ft., breadth 38 ft., depth 22 ft. 10 in. Tons register 1,458, tons O.M. 1,51$\frac{1}{4}$.   *R. and H. Green.*

2178aya. Sectional Drawing and Plan of the Ship "Falmouth," 1752.   *R. and H. Green.*

2178ayb. Sectional Drawing, fore and aft view of the ship **"Royal William,"** 1726.   *R. and H. Green.*

2178ayc. Profile, Plan, and Body of the Ship "Friendship."
Laid down 1789.
Launched 1790.
Length, 102 ft. 7$\frac{1}{2}$ inches.
Breadth, 26 feet 6 inches.
Depth, 12 feet 2 inches.
Tons, 307·2294.   *R. and H. Green.*

2178az. Drawing, Profile of Captain Dicey's Channel Steamer "Castalia."
The Thames Iron Works and Shipbuilding Company.

2178ba. Drawing, Main Deck Plan of Captain Dicey's Channel Steamer "Castalia."
The Thames Iron Works and Shipbuilding Company.

2178bb. Drawing, Section of Captain Dicey's Channel Steamer "Castalia."
The Thames Iron Works and Shipbuilding Company.

2178bz. Glass Case containing Models of Ships built by the Company.
The Thames Iron Works and Shipbuilding Company.

CONTENTS.

H.M.S. "Serapis" (half model), English Government, troop ship.
"Victoria" (half model), Spanish Government, ironclad.
"König Wilhelm" (half model), Imperial German Government, iron clad.
"Mesoudiyè" (half model), Sultan of Turkey, ironclad.
"Pervenetz" (half model), Czar of Russia, ironclad.
H.M.S. "Waterwitch" (half model), English Government, ironclad.
"Vasco da Gama (half model), Portuguese Government, ironclad.
"Absalon" (half model), Danish Government, ironclad.
"Avni Illah" (whole model), Sultan of Turkey, ironclad.
"King George" „ Hellenic Government, ironclad.
"Castalia" „ iron steamer for channel service.
Glass case section of H.M.S. "Warrior," English Government.

2178ca. 13 Mounted Drawings, midship sections of various vessels built at the Thames Iron Works, and showing the advance in the thickness of armour plating from 1855 to 1876.
The Thames Iron Works and Shipbuilding Company.

2178caa. Photographs of Steam Ships built by the Thames Iron Works, of which models are in show case.
Twin ship "Castalia." Bow and side view.
Ship of war "Vasco da Gama." Portuguese ironclad.
"Messoudiyé." Turkish ironclad.
Thames Iron Works and Shipbuilding Company, Limited.

2178cab. Model of the Paddle Steamer, "Lady of the Lake," Isle of Wight, Southampton and Portsmouth boat.
Nominal horse power, 60.
Engines by J. Stewart, London.
Thames Iron Works and Shipbuilding Company, Limited.

2178cac. Model of the Paddle Dispatch Vessel, "Ijedden." Built for H.I.M. the Sultan of Turkey.
Nominal horse-power, 300.
Length, 250 feet. Breadth, 29 feet 6 inches.
Depth, 18 feet. Tons, 1,075.
Thames Iron Works and Shipbuilding Company, Limited.

2178bh. Half-block Model of Ship "Japanese."
T. Royden.

2178bi. Two whole Models, illustrating framings in wood of old ships. *T. Royden.*

2178bj. Model of the Iron SS. "Nina," built by the exhibitors. *Richardson, Duck, and Co.*

2178bk. Model of the Iron SS. "El Rahmanieh." Built for His Highness the Viceroy of Egypt by the exhibitors. *Richardson, Duck, and Co.*

2178bl. Whole Model in Bone of a Three-decker Line of Battle Ship, made by French prisoners in England during the Peninsular War, 1812. *Vaughan Pendred.*

2178bm. Model showing Interior of a Wooden Merchant Ship. *Lloyd's Register of British and Foreign Shipping.*

2178bn. Model showing the Framing of the Fore Body of a Wooden Merchant Ship. *Lloyd's Register of British and Foreign Shipping.*

2178bo. Model showing the Framing of the after Body of a Wooden Merchant Ship. *Lloyd's Register of British and Foreign Shipping.*

2178bp. Model of a Ship with radiating Frames. *Lloyd's Register of British and Foreign Shipping.*

2178bq. Model of a Ship Sheathed with Diagonal Doubling Plank. *Lloyd's Register of British and Foreign Shipping.*

2178br. Model of an Old Ship showing Framing of Deck-Beams, &c., also lower masts and bowsprit in place. *Lloyd's Register of British and Foreign Shipping.*

2178bs. Model of Thomas Bilbe's System of constructing Composite Vessels. *Lloyd's Register of British and Foreign Shipping.*

2178bt. Sectional Model of an Iron Vessel. *Lloyd's Register of British and Foreign Shipping.*

2178bu. Drawing of a Composite Vessel, showing the iron framework, and the mode of fastening the wood bottom. *Lloyd's Register of British and Foreign Shipping.*

2178bv. 20 Drawings, illustrating Lloyd's Rules for Shipbuilding on the Composite System. *Lloyd's Register of British and Foreign Shipping.*

2178bw. Model of H.M. Gun Boat "Staunch." Built by C. Mitchell and Co., Newcastle-on-Tyne. *C. Mitchell and Co.*

The "Staunch" was built in 1868, as the type of a new class, which has since been largely introduced with the British and other services. Length of vessel 75 feet, beam 25, depth 8 feet, armament one 12½ ton gun. Engines are of 40 horse-power nominal with twin screws.

2178bx. Model of Armour Clad Frigate " Prince Pojarski." Built for the Imperial Russian Government by C. Mitchell and Co., Newcastle-on-Tyne. *C. Mitchell and Co.*

The dimensions of vessel are, length 280 feet, beam 49 feet, depth 31 feet, armament eight 300-pr. contained in central battery. Machinery, 600 horse-power nominal, and speed of vessel about 13 knots per hour.

2178by. Model of Paddle Steamer " Baron Osy." Built by C. Mitchell and Co., Newcastle-on-Tyne, in 1875, for the Société Anversoise de Bateaux à Vapeur, to provide an improved passenger service between London and Antwerp. Dimensions are, length 240 feet, beam 30 feet, engines 250 horse-power, and speed 13 knots per hour. *C. Mitchell and Co.*

2178cc. Model of the Paddle Steam Yacht " Menai," supplied by R. Napier in 1830 to Thomas Assheton Smith. *R. Napier.*

The hull, 132 feet long, 20 ft. 6 ins. broad, 12 ft. 8 ins. depth of hold, and 146$\frac{11}{94}$ tons register, was built by John Wood, Port Glasgow, from Mr. Smith's design, the vessel's bottom having a hollow or channel on each side of centre keel.

The engines were made and fitted by R. Napier. Cylinders 42¾ ins. diam. × 4 feet stroke, 120 nominal horse-power.

2178cd. Album containing Drawings of the Engines of the Paddle Steamer " British Queen." Constructed in 1838 for the British and American Steam Navigation Company, London, by R. Napier, Glasgow. Also drawings of engines and boilers of other vessels of the Cunard line for Transatlantic service. *R. Napier.*

The hull, 275 ft. long, 39 ft. 8 ins. broad, 27 ft. 6 ins. deep (moulded), was built by Messrs. Curling and Young, London.

Cylinders, 77¾ ins. diameter × 7 feet stroke, 420 nominal horse-power.

These engines (the largest marine engines made at the time they were constructed) were fitted with Hall's surface condenser.

2178fh. Photograph of H.M. Indian Relief Troopship " Malabar." Built and engined, 1867, by R. Napier and Sons. Length 360 ft., breadth 49 ft., depth, 22 ft. 4 in. ; tonnage, 4,173 ; horse-power, 700. *R. Napier and Sons, Glasgow.*

2178ce. Model of the Paddle Steamer " Faid Zafar." Length 225 ft. 6 ins., breadth 20 ft., depth 8 ft. 7 ins. 360 tons gross. *David and William Henderson and Co.*

Built and engined by the late firm of Tod & McGregor in 1864, for His Highness Ismail Pacha, Viceroy of Egypt, being completely composed of steel for light draught of water; schooner rigged.

Engines oscillating, cylinders 44 ins. diameter, stroke 3 ft., horse-power 120 nominal, speed 20 miles per hour.

2178cf. Model of the Paddle Steamer "Mukhbir Suroor." Length 220 ft., breadth 20 ft., depth 9 ft.

David and William Henderson and Co.

Built of iron and engined by the late firm of Tod & McGregor in 1864, for His Highness Ismail Pacha, Viceroy of Egypt, as tender to the "Faid Zafar," schooner rigged.

Engines oscillating, cylinders 40 ins. diameter, length of stroke 3 ft., horse-power 100 nominal.

2178cg. Model of the Screw Steamer " City of Manchester." Length 262 ft., breadth 36 ft., depth 25 ft. 2,096 tons gross.

David and William Henderson and Co.

Built and engined by the late firm of Tod & McGregor in 1851, for the Inman Company, to trade between Liverpool and New York ; barque-rigged, with jigger, about the first vessel ever fitted with *iron masts.*

Engines beam-geared, cylinders 71 ins. diameter, length of stroke 5 feet, horse-power 370 nominal.

2178ch. Model of the Screw Steamer "City of Richmond." Length 440 ft., breadth 43 ft. 6 ins., depth 34 ft. 4,606 tons gross.

David and William Henderson and Co.

Built and engined by the late firm of Tod & McGregor in 1873, for the Inman Line to trade between Liverpool and New York. Ship rigged.

Engines, direct acting compound inverted, cylinders 76 ins. and 120 ins. diameters, length of stroke 5 ft. ; propeller 22 ft. diameter ; horse-power 670 nominal ; speed at trial trip, 17 knots per hour ; 10 boilers, 30 furnaces.

2178ci. Model of the Screw Steamer " Princess Royal." Dimensions, length 200 ft., breadth 28 ft., depth 15 ft. 527 tons gross.

David and William Henderson and Co.

Built by the late firm of Tod & McGregor in 1861, for the Glasgow and Liverpool trade, schooner rigged.

Engines, direct acting inverted, compounded in 1873 by having 2 high pressure cylinders 24 inch diameter placed on top of the 45 inch cylinders— length of stroke 2 ft. 6 ins., horse-power 170 nominal, propeller 12 feet diameter, speed 12 knots per hour.

2178cj. Model of the Screw Steamer " Lady Nyessa." Dimensions, length 150 ft., breadth 14 ft., depth 6 ft.

David and William Henderson and Co.

Built by the late firm of Tod & McGregor in 1861, for the late Dr. Livingstone, for exploration in Central Africa. To facilitate transportation of the vessel from the Zambesi to Lake Nyessa, she was constructed in 12 segments in her length, each segment being in two pieces which were bolted together through angle iron frames. Dr. Livingstone on his return home from Africa,

on his first exploration, sailed in her from the Zambesi to Calcutta. The
" Lady Nyessa" was fitted with two propellers, one abaft the other on the same
shaft.

Engines high pressure, cylinders 14 in. diameter, length of stroke 12 inches,
horse-power 25 nominal.

2178ck. Model of the Screw Steamer "Bengal."
Dimensions, length 297 ft., breadth 39 ft., depth 27 ft. 6 ins. 2,235
B.M. tons.   *David and William Henderson and Co.*

Built by the late firm of Tod & McGregor in 1852, for the Peninsular and
Oriental Steam Navigation Co., lately trading in the China Seas.

Engines beam geared, cylinders 80 inches diameter, stroke 5 feet, horse-
power 470 nominal.

2178cl. Model of the Paddle Steamer "Trafalgar."
Dimensions, length 190 ft., breadth 28 ft. 6 ins., depth 16 ft.
4 ins. 689 tons gross. *David and William Henderson and Co.*

Built and engined by the late firm of Tod & McGregor in 1847, for Messrs.
Wm. Watson & Co., Dublin, for their Liverpool and Dublin trade, schooner
rigged.

Engines oscillating, cylinders 69 inches diameter, length of stroke 5 feet
6 inches, horse-power 360 nominal.

**2178cm. Model of the Paddle Steamer "Princess
Royal."**   *David and William Henderson and Co.*

Built and engined by the late firm of Tod & McGregor in 1841, for the
Glasgow and Liverpool trade ; wrecked in 1856, and a new paddle steam
vessel built in 1857 to replace her.

Engines, steeple, cylinders 72 inches diameter, length of stroke 6 feet 6
inches, horse-power 400 nominal.

**2178cn. Model of the Paddle Steamer "Countess of
Galloway."** Length 165 ft., breadth 24 ft., depth 14 ft. 3 ins.
                              David and William Henderson and Co.

Built and engined by the late firm of Tod & McGregor Two vessels were
built from this model and of the same name, one in 1835 and one in 1843,
fitted up as passenger and cattle-boats plying between Liverpool and Wig-
town ; the latter vessel is still on her station, and is considered one of the best
cattle-boats afloat. Owned by the Galloway Steam Navigation Company,
rigged 3-masted schooner.

Engines vertical, direct-acting, cylinders 54 inches diameter, length of
stroke 5 feet, horse-power 220 nominal, paddle wheel 22 feet diameter.

**2178co. Model of the Twin Paddle Steamer "Alli-
ance."** Dimensions 140 ft. by 9 ft., breadth of each hull, depth,
8 ft., distance between hulls, 12 ft., total breadth, 30 ft.
                              David and William Henderson and Co.

Propelled by one paddle wheel, 18 feet diameter, placed between hulls,
27 ft. 6 ins. abaft centre of vessel. The paddle wheel was afterwards taken
out of the centre and the two halves rivetted together ; since that alteration
she made several successful passages as a blockade runner during the late

American war. The "Alliance" was built by the late firm of Tod & McGregor in 1856, for Clyde river traffic.

Cylinders 34 inches diameter, length of stroke 3 feet, horse-power 70 nominal.

2178coa. Drawing, Rigging Plan of four-masted Screw Steamer, "City of Manchester," 1876. Scale, ⅛-inch to 1 foot.   *D. & W. Henderson & Co.*

The vessel was fitted with iron masts.

2178cob. Drawing, Profile Plan, and Sail Draft of Screw Steamer "No. 113." 1861.

Length, 115 feet. Breadth, 14 feet.

Depth, 6 feet. Tons, 111.

Scale, ¼-inch to 1 foot.   *D. & W. Henderson & Co.*

2178coc. Drawing, Rigging Plan of two-masted Screw Steamer "Princess Royal." 1871. Scale, ⅛-inch to 1 foot.
D. & W. Henderson & Co.

This steamer was lengthened 20 feet, and the rigging to suit.

2178cod. Drawing, Midship Section of Screw Steamer "Princess Royal." 1871. Showing position of boilers and decks. Scale, ½ inch to 1 foot.
D. & W. Henderson & Co.

2178coe. Drawing, Rigging Plan of three-masted Screw Steamer "City of Richmond." 1873. Scale, ⅛-inch to 1 foot.   *D. & W. Henderson & Co.*

2178cof. Drawing, Profile Plan and Lines of Paddle Steamer "Alliance." Twin ship. 1876.

Length, 140 feet; breadth of each turn, 9 feet; breadth between hulls, 12 feet; depth of hulls, 8 feet.
D. & W. Henderson & Co.

This ship driven by a single wheel aft between the hulls.

2178cog. Drawing, Rigging Plan of Steel Paddle Yacht No. 135, "Faid Zafar." 1865. Scale, ⅛-inch to 1 foot.   *D. & W. Henderson & Co.*

2178coh. Drawing, Profiles of Stern and Bow, and Plan of the Paddle Steamer "Countess of Galloway." 1843.

Length, 150 feet; breadth, 24 feet; depth, 14 feet 7 inches. Scale ¼ inch to 1 foot.   *D. & W. Henderson & Co.*

2178cp. Model of the Merchant Vessel "Brotherly Love." Built in the year 1764, and still running. The mark on the side is from damage in a collision with a steamer some years ago.   *James Young.*

This is the ship in which Captain Cook first sailed from the port of Whitby.

2178cq. Model of the Merchant Vessel " Antelope."
Built in the year 1757.   *James Young.*

2178cr. Model of a Shields Pilot's Coble.
                                            James Young.

2178cs. Model of the Mast of the Man-of-war " Nelson," in seven pieces, without the pieces which form the top.
                                            James Young.

DIMENSIONS, WEIGHT, AND EXPENSE OF THE "NELSON'S" MAIN
MAST.

				£	s.	d.
No. 1. Expense of seven trees	-	-	-	860	16	6
2. Hoops, bolts, and nails	-	-	-	61	11	6
3. Smith work	-	-	-	16	4	0
4. Mast makers	-	-	-	55	0	7¾
				993	12	7¾

			ft.	ins.
Length	-	-	127	2¼
Greatest diameter	-	-	0	41
Smallest diameter	-	-	0	30⅗

		tons	cwt.	qr.	lbs.
Weight	-	26	0	3	24

2178cu. Half Model of an Iron Screw Steamer,
showing about 23 ft. of midship body, with a Price's patent self-trimming hatchway fitted.   *William Denton.*

The adoption of this new style of hatch opening in the vessels already fitted with Price's patent has effected very large saving in the cost of loading, as trimming is entirely dispensed with. The pamphlet containing testimonials explains the opinions of the owners, who have adopted the invention. The lithograph shows profile of vessel fitted with Price's hatchways, also detailed sketches of manner of fitting half-beams, hatch sides, &c.

2178cv. Half Model of Iron Merchant Vessel " Donnybrook." Length 260 ft., breadth 40 ft., depth 24 ft. Designed in 1876 by Messrs. Austin and Hunter. Tonnage 1,700 tons, water draught 21 ft.   *Austin and Hunter.*

2178cw. Half Model of Iron Cargo Steamer " Decapolis." Length 260 ft., breadth 32 ft., depth 23½ ft., tonnage 1,000 tons, draught of water 20. Built 1876 by Messrs. Austin and Hunter.   *Austin and Hunter.*

2178cx. Model of Merchant Ship " Cherubim."
Length 156 ft., keel 136 ft., breadth 28 ft., depth 18 ft.
                                            Thomas Wrightson.

2178cy. Model of Passenger Steamer " North Star."
Length 316 ft., depth 23 ft., breadth 36 ft., tonnage 2,500 tons.
                                            Thomas Wrightson.

2178cz. Model of the "Crest of the Wave," a noted China clipper, built by the late William Pile, Sunderland, and owned by John Hay, Esq., and others, Sunderland. Built in 1853 of wood. Length 180 feet, breadth 32 feet, depth 20 feet. Classed 13 years A1. in Lloyd's registry ; 776 tons register.

R. H. Hay.

2178da. Model of the wood clipper vessel " Aurora Borealis," built by the late William Pile, Sunderland, and owned by John Hay, Esq., Sunderland, in 1850.   *R. H. Hay.*

2178db. Model of the wood clipper vessel " Herald of Light," built by the late William Pile, Sunderland, in 1850, and owned by Henry Ellis, Esq., London.   *R. H. Hay.*

2178de. Model of the " Dover Castle " (wood). Built for Richard Green, Esq., Blackwall, in 1858. Length 185 ft., depth 22 ft., breadth 34 ft., tonnage 1,002 nett.   *John Haswell.*

This vessel was draughted and model made by Master George Haswell, son of the builder, when 15 years old.

2178df. Model of the " Min " and " Sumida," iron sister ships. Built for the Japanese Government by R. Thompson, junior, in 1874. Length 268 ft., breadth 32 ft. 1 in., depth 21 ft. 4 ins.

Gross tonnage	-	-	- 1,411
Net "	-	-	- 896
H.P. - "	-	-	- 200

Robert Thompson, junior.

2178dg. Model of the " Lady Louisa," iron composite. Built for Wilson and Co., London, by R. Thompson, junior, in 1865. Length 152 ft., breadth 28 ft. 3 ins., depth 17 ft. 9 ins. Net tonnage 542.   *Robert Thompson, junior.*

2178dh. Model of the " Life Brigade," iron. Built for E. Shotton and Co., North Shields, by R. Thompson, junior, in 1873. Length 249 ft. 5 ins., breadth 32 ft. 4 ins., depth 17 ft. 15 ins.

Gross tonnage	-	-	- 1,512
Net "	-	-	- 977
H.P. - "	-	-	- 150

Robert Thompson, junior.

2178di. Model of the " Graham " (wood). Built for Edmund Graham, Newcastle, by Robert Thompson and Sons, designed by Robert Thompson, junior, in 1850. Length B.P. 132 ft., breadth 30 ft. 4 ins., depth of hold 20 ft. 10 ins. Net tonnage 668.   *Robert Thompson, junior.*

2178dj. Model of the " Opal." Composite corvette, by William Doxford and Sons. Length 220 ft., breadth 40 ft., depth 21 feet. *Wm. Doxford and Sons.*

2178dk. Model of the Screw Steamer "Adalia." Built for E. T. Gourley, Esq., M.P., by William Doxford and Sons. Length 231 ft. 6 ins., breadth 32 ft., depth 24 ft. 8 ins. *Wm. Doxford and Sons.*

2178dl. Model of Composite Gunboats " Cygnet," "Express," and "Contest." By William Doxford and Sons. Length 125 ft., breadth 23 ft. 6 ins., depth 12 ft., horse-power 360. *Wm. Doxford and Sons.*

2178dm. Model of Passenger Screw Steamer " Gloria " built for Olano Larrinaga & Co., Liverpool, by William Doxford and Sons. Length 322 ft., breadth 38 ft. 3 ins., depth 30–55. *Wm. Doxford and Sons.*

2178dp. Set of Original Block Models from which the following vessels were built in wood :—

" Victory," barque, built in 1847, length 113 ft., breadth 22 ft. 8 ins., depth 13 ft. 6 ins. Classed 7 years A1, tonnage 239 tons.

" St. George," Snow collier, built in 1847, length 78 ft., breadth 23 ft., depth 12 ft. 3 ins. Classed 8 years A1, tonnage 153 tons.

" Providence," schooner, fruit trade, built in 1850, length 69 ft., breadth 17 ft., depth 10 ft. Classed 7 years A1, tonnage 83 tons.

" Bury St. Edmunds," full-rigged ship, built in 1853, length 155 ft., breadth 31 ft., depth 21 ft. Classed 13 years A1, tonnage 701 tons. *Peter Forster.*

2178em. Model. Midship section of a composite vessel experimentally built and classed at Lloyd's for 11 years. *Nelson Dock Company, Limited.*

2178en. Model. Half midship section of a composite vessel experimentally built and classed at Lloyd's for 15 years. *Nelson Dock Company, Limited.*

2178eo. Model. No. 3. Half model of an old paddle steamer. *Nelson Dock Company, Limited.*

2178ep. Model. No. 4. After body and stern framing of an old East Indiaman. *Nelson Dock Company, Limited.*

2178eq. Half-block Model of a Yarmouth Trawling Smack, drawings of " Lord Peto," and other Yarmouth vessels. *I. H. Fellowes.*

2178er. The " Mary Taylor," New York pilot boat, built by George Steers, an Englishman in New York, about 1850. He afterwards built the celebrated " America " yacht. He told the exhibitor in New York, in September 1854, that he considered the "Mary Taylor" the faster vessel of the two, particularly in rough water.   *Henry Liggins.*

2178do. Half Block Model of the Popovka "Admiral Popoff."   *Russian Embassy.*

This ship is the second of the series of circular ironclads in course of construction by Russia.

The principal dimensions are the following :—

Extreme diameter, 121 ft.   Mean draught of water, 13 ft.   Height of upper deck at side from load water line amidships, 1 ft. 6 ins.   Height of barbette turret from load water line, 13 ft. 3 in.   Displacement in tons, 3,550.

This vessel carries two 41-ton guns, which can be trained completely round inside the turret, their carriages being arranged so as to admit the guns to be lowered down for loading and aiming behind the protection of the thick turret armour, and to be brought up only for firing.

Besides these heavy guns there are, on each side of the superstructure containing cabins and quarters for crew, two smaller guns throwing projectiles of sufficient power to penetrate an unarmoured enemy.

The vessel and turret are protected by means of armour made in two solid layers, having a total thickness of 1 ft. 6 in. (including equivalent thickness for the hollow iron girders behind armour).   Side armour of the vessel extends from the edge of the vessel's upper deck down to 4 ft. 6 in. below load water line amidships.   The upper deck is protected by deck armour 3 in. thick, extending from the circumference of the turret to the circumference of the vessel.

The propelling power is developed by eight compound vertical engines, each of 80 horse-power (nominal); four of these work each an independent screw of 10 ft. 6 in. in diameter; and four remaining engines arranged and worked in pairs propel the middle screws on each side of the vessel.   These two screws are made larger and their shafting lowered down so as to increase the speed of the vessel when steaming in deep water.

This ship, though launched, has not been yet completed for sea, and accordingly no results of her trial with regard to speed have yet been obtained, but it is expected that she will attain from 9 to 10 knots, which for coast defence purposes is considered quite sufficient.

The vessel is built of iron, with double bottom, and is sheathed with wood and copper.

The chief principal objects of the design of the Popovka are :—1. The extremely large armour and gun-carrying power attained upon a very limited draught of water.   This condition renders such ships extremely valuable for coast defence. 2. Extraordinary steadiness of gun platform, ensuring the efficiency of working the guns at sea, when guns of an ordinary ship must remain silent. 3. Complete protection of hull and battery by uniform thickness of armour throughout, which does not exist in any other ironclad. 4. Great cheapness of construction as compared with other vessels. 5. Number of compartments around the central vital parts of the vessel which are specially effective for the purpose of saving the vessel in case of torpedo attack.

2178dp. Half model of Russian Circular Ship of War

mounted on glass support, showing the position of her large and small guns.   *The Institute of Naval Architects.*

2178dq. Model of the Composite Clipper Sailing Ship " Torrens " (Elder line of packets to Adelaide). Built by James Laing, Sunderland, in 1875. Length 222 ft., breadth 38 ft., depth 21 ft. 6 ins. 1,334 tons, gross. *James Laing.*

2178dr. Model of the Peninsular and Oriental Steam Navigation Company's Vessels " Khiva " and " Kashgar." Built by James Laing, Sunderland, in 1874. Length 370 ft., breadth 36 ft. 8 ins., depth 27 ft. 4 ins. 2,600 tons gross. *James Laing.*

2178ds. Model of the Hamburg South American Steam Ship Company's S.S. " Buenos Aires." Built by James Laing, Sunderland. Length 340 ft., breadth 36 ft. 4 ins., depth 25 ft. 6 ins. 2,400 tons gross. *James Laing.*

2178dt. Model of Sailing Ship " Vimiera." Built in 1850 by James Laing, Sunderland, for the late Duncan Dunbar. Length 165 ft., breadth 33 ft., depth 23 ft. 1,037 tons. *James Laing.*

The side of this model can be opened to show the construction of the ship.

2178du. Model of the Peninsular and Oriental Steam Navigation Company's S.S. " Poonah," as altered and refitted by James Laing, Sunderland. Length 414 ft., breadth 41 ft. 6 ins., depth 27 ft. 4 ins. 3,118 tons gross. *James Laing.*

2178ek. Half Models of the Ship " Ellen Rodger," China clipper, belonging to Alexander Rodger, Esq., Glasgow; launched June 1858; length 153 ft., breadth 29 ft. 3 in., depth 19 ft. 6 in.; tonnage, $584\frac{83}{100}$; wooden ship. **Ship " Taeping,"** China clipper, belonging to Alexander Rodger, Esq., Glasgow ; launched December 1863 ; length 185 ft., breadth 31 ft., depth 20 ft. ; tonnage, $764\frac{22}{100}$; composite ship. **Ship " Ariel,"** China clipper, belonging to Messrs. Shaw, Maxton, and Co., London ; launched June 1865 ; length 195 ft., breadth 33 ft. 9 in., depth 21 ft. ; tonnage, $852\frac{87}{100}$; composite ship. **Ship " Zahloo,"** belonging to Alexander Rodger, Esq., Glasgow; launched July 1867 ; length 190 ft., breadth 33 ft., depth 20 ft. ; tonnage, $799\frac{21}{100}$; composite ship. *Robert Steele and Co., Greenock.*

2178el. Half Models and Drawings of the Paddle Steamer " Hibernia," belonging to the Cunard Company, Glasgow ; launched September 1842 ; length 219 ft. 8 in., breadth 35 ft. 10 in., depth 24 ft. 2 in.; tonnage, gross $1,421\frac{2790}{3500}$, net $791\frac{800}{3500}$; wooden ship. **Screw Steamer " Polynesian,"**

belonging to the Montreal Ocean Company, Glasgow; launched February 1872; length 400 ft., breadth 40 ft., depth 35 ft. 9 in.; tonnage, gross $4,282\frac{58}{100}$, net $2,023\frac{21}{100}$; iron ship.
Robert Steele and Co., Greenock.

2178dv. Model of Screw Steamer " Lestris." Built by Joseph L. Thompson and Sons, Sunderland. Length 227 ft., breadth 30 ft., depth of hold 17 ft. *Joseph L. Thompson and Sons.*

2178dw. Model of Sailing Ship " Baroda." Built by Harland and Wolff, Belfast.

	ft.	in.
Length, extreme - - -	251	0
,, between perpendiculars -	231	5
Breadth - - - -	36	5
Depth - - - -	23	5
	tons.	
Tonnage builders, measurement -	1,299	
Gross register - - -	1,364	

Launched 1864. *P. Phorson.*

2178dx. Model of a Vessel built about 1844 by James Leithead, Sunderland. *P. Phorson.*

2178dy. Model of the wooden Ship " Chowringhee." Built by the late Wm. Pile, of Sunderland, for the late John Hay, Esq., in 1857. Length 156 ft. 3 ins., breadth 31 ft. 6 ins., depth 21 ft. 6 ins. *Sunderland Corporation.*

2178dz. Plans and Elevations of Steamships " Marquess de Nunez " and " Wear." Built by the late Mr. Wm. Pile, Sunderland. *Skinner, Sunderland.*

2178ea. Model of Screw Collier " David Burn." Built in 1873 by Messrs. Wm. Doxford and Sons, Sunderland.
Alfred Simey and Co

This vessel was sunk off the Tyne by collision on her first trial trip. The visitors (chiefly ladies) were saved by the captain of the vessel which struck her, keeping his engines going at easy speed, and thus filling the hole made by the collision.

2159h. Drawing of the " London Engineer," 70 horse-power, a vessel fitted with a central paddle wheel, and air-tight paddle case, in the year 1818, intended to run to Margate. The scheme did not succeed, owing to the difficulty of keeping a sufficient supply of air in the paddle case to keep down the level of the water. *Maudslay, Sons, and Field.*

2178ga. Model of Mr. Joseph Maudslay's **Original Oscillating Cylinder Engines.** Date 1827. (The arrangement for working the air pump is not that originally fitted.)
Maudslay, Sons, and Field.

2178fn. Drawing showing the first iron vessels built in London, designed for the navigation of the River Ganges, with oscillating engines of 60 horse-power collective; date 1832.

Maudslay, Sons, and Field.

2178fo. Drawing showing another set of same, with beam engines; date 1838.   *Maudslay, Sons, and Field.*

2178gm. Drawing of 30 horse-power oscillating **Engine** fitted in the **" Sylph,"** Woolwich packet, 1836.

Maudslay, Sons, and Field.

2178gb. Model of Patent Annular Cylinder Engines, by Maudslay, Sons, and Field, as fitted in first improved Channel Packet " Princess Alice " (Dover to Calais), 1843.

Maudslay, Sons, and Field.

2178gc. Model of Double Piston-rod Steeple Engines, fitted in the Hon. East India Company's vessels on the River Indus. 1840.   *Maudslay, Sons, and Field.*

2178gd. Sectional Model of Side by Side Cylinder Horizontal Compound Engines. John Milner, C.E., 1853. J. Warriner.   *Maudslay, Sons, and Field.*

2178gh. Drawing. First direct-acting screw engines fitted by Maudslay, Sons, and Field, in 1846 (H.M.S. " Ajax " and " Edinburgh," 450 horse-power nominal collective).

Maudslay, Sons, and Field.

2178gi. Drawing, showing arrangement of machinery in vessels.   *Maudslay, Sons, and Field.*

2178gk. Drawing, showing **Engines** of H.M.S. **" Retribution,"** 1842 (similar to those of the " Terrible," 800 horse-power nominal collective); patent 4-cylinder engines.

Maudslay, Sons, and Field.

2178gl. Drawing of Beam Engines of 220 horse-power collective of H.M. ships **"Rhadamañthus," " Phoenix," " Salamander," " Medea,"** date 1832.

Maudslay, Sons, and Field.

2159d. Model of Beam Engines, similar to those of the " Great Western " steamship. This model was made by the late Henry Maudslay.   *Maudslay, Sons, and Field.*

2159e. Model of Feathering Screw, by the late Joseph Maudslay. Date 1848.   *Maudslay, Sons, and Field.*

2159f. Model of Paddle Wheel of " Great Western " steamship, designed by the late Joshua Field, F.R.S.

Maudslay, Sons, and Field.

2159g. **Two Drawings** showing the **four cylinder com-pound Engines** designed and fitted by Maudslay, Sons, and Field in the steaemrs of the " White Star " line, and in those of the " Compagnie Générale Transatlantique."

Maudslay, Sons, and Field.

2159a. Model of Direct-acting Screw Engines, date 1850, made for the Danish Government by Maudslay, Sons, and Field. *Maudslay, Sons, and Field.*

2159b. Model of Direct-acting Screw Engines, as fitted in 1851 by Maudslay, Sons, and Field in the mail steamers from London to Calcutta, and from London to the Cape of Good Hope. *Maudslay, Sons, and Field.*

2159c. Model of Horizontal, Double Piston Rod, Direct-acting Screw Engines, as fitted in H.M.S. " Agin-court," 1,350 horse-power nominal, and in H.M. ships " Prince Consort," " Caledonia " and " Ocean," each 1,000 horse-power nominal. *Maudslay, Sons, and Field.*

2178a. Photograph of the " Germanic " and " Bri-tannic " mail steamers, belonging to the " White Star " line from Liverpool to New York. An abstract log of one of the voyages of the " Germanic " is attached to the frame. Length 470 ft., breadth 45 ft.; tons register, 5,004; 760 horse-power nominal, four cylinder compound engines by Maudslay, Sons, and Field. Ships built by Harland and Wolff, of Belfast.

Maudslay, Sons, and Field.

2148ea. Model, showing $\frac{1}{10}$ of the length of the **Floating Dock,** built for the Spanish Government for the Royal Dock-yards at Carthagena and Ferrol, 1856–7.
Dimensions :

Total length	-	-	350 feet.
Extreme breadth		-	105 feet.
Depth inside	-	-	37 feet 6 ins.
Displacement	-	-	13,000 tons.
Total weight	-	-	5,000 tons.

Weight of ship lifted, 6,500 tons.
Scale of model, 1 inch to 1 foot.

J. & G. Rennie.

2148eb. Photograph of the Iron Floating Dock at Bermuda, built by J. Elder and Co., Glasgow; carrying H.M.S. "Royal Alfred," 6,720 tons displacement.

Campbell, Johnstone, & Co.

The dock is capable of sustaining H.M.S. " Fury," of 10,400 tons dis-placement.

2148ec. Photographs (2) **of Iron Floating Docks for Bermuda,** constructed by Campbell's Shipbuilding and Floating Dock Company, 1868.

Length, including cutwaters, 381 feet; width, outside, 128 feet 1 inch; height, outside, 74 feet 5 inches; length, inside, 330 feet; width, inside, 83 feet 1 inch; height, inside, 54 feet 5 inches.

Campbell's Shipbuilding and Floating Dock Company.

2178fq. Model of the inverted cylinder compound engines, P. and O. Co.'s ship " **Pera** " on the compound system of 2,000 horse-power, 1872.   *J. and G. Rennie.*

2178fr. Model of the first screw-steamer in the British Navy, " **Mermaid,**" afterwards named the " **Dwarf.**" Built in 1840. Purchased from J. and G. Rennie by the British Admiralty, according to Sir George Cockburn's advice, on the condition she should steam 12 miles per hour (7th March 1842); tried 15th May 1843. Mean speed of 6 runs—12·142 miles.   *J. and G. Rennie.*

2178fs. Model of H.M. Gun Boats " Arrow " and " Bonetta." Built by J. and G. Rennie, 1871. Length 85 ft., breadth 26 ft., depth 8 ft. 10 in.; tons, 244. To carry one 18 ton gun.   *J. and G. Rennie.*

2178ft. Model of the Iron Paddle-wheel Steamer " Queen." Built and fitted with engines by J. and G. Rennie, 1842. Length 160 ft., breadth 17 ft., depth 9 ft. Speed between 17 and 18 miles per hour, which exceeded that of any vessel on the Thames.   *J. and G. Rennie.*

2178fu. Model of Indian Famine Relief Steamers. Built by J. and G. Rennie, 1874. Date of order, 24th February. Date of launch, 30th March. Tried under steam, 4th April. Length 90 ft., beam 14 ft., depth 5 ft. 6 in.; indicated horse-power, 100. Built complete with engines in 35 working days.

J. and G. Rennie.

2178fw. Model of Twin-Screw Gun Boats. Built for the East Indian Government by J. and G. Rennie, 1857. Length, 70 ft., breadth 11 ft., draught 2 ft. 6 in.; nominal horse-power, 20; indicated horse-power, 76; speed, 9¼ miles. Armament, one long brass 12-pr., 18 cwt.   *J. and G. Rennie.*

2178fx. Model of Brazilian Twin-Screw Ironclad Gun Boats " Colombo " and " Cabral." Built by J. and G. Rennie, 1866. Length 160 ft., breadth 34 ft., depth 17 ft.; tons, 858 B.M.; nominal horse-power, 240; speed, 10 knots. Armament, 4 68-prs.   *J. and G. Rennie.*

2178fy. Model of Twin-Screw Gun Boats. Built for the Spanish Government by J. and G. Rennie, 1859. Length 90 ft., breadth 14 ft., draught 2 ft. 6 in.; horse-power, 30.

J. and G. Rennie.

2178gp. Model of the **Engines** of H.M. ships **"Boadicea"** and **"Bacchante,"** on the compound system, of 5,250 horse-power indicated, 1875 and 1876.   *J. and G. Rennie.*

2178gq. Model of Horizontal Marine Engines, with injection condensers. Made by J. and G. Rennie, 1860.

J. and G. Rennie.

2178gr. Model of Reversed Horizontal Marine Screw Engines. Built 1860.   *J. and G. Rennie.*

2178gs. Drawing, design for **60 horse-power Low Pressure Condensing Disc Engines for Screw Steamship.**
The drawing shows front, side, and back elevation and plan.
Section of H.M.S. "Cruiser," fitted with the disc engines, 1853.

J. and G. Rennie.

2178gu. Compound Surface-Condensing Marine Engines.   *T. Richardson and Sons, Hartlepool.*
Three drawings of marine engines of the most modern construction.
One drawing represents engines of the largest class fitted with steam reversing gear.
Another drawing represents engines of a more moderate size, and gives the names of the vessels into which they have been fitted.
A third drawing represents marine engines of the smaller class.
The three drawings all represent compound surface-condensing engines.

2178ei. Whole Model of the German ironclad ships of war **"Kaiser"** and **"Deutschland."** Built 1874 for the Imperial German navy. 7,500 tons (displacement); 8,000 indicated horse-power; engines by Penn and Sons, Greenwich.

Samuda Brothers.

2178ej. Whole Model of the **"Mahrousse,"** state paddle steam yacht. Built for His Highness the Khedive of Egypt tons burthen, 3,150; horse-power, 800 nominal.   *Samuda Brothers.*

2168. Helm Indicator for the prevention of Collisions at Sea.   *John James Nickoll.*
This is a lantern which is hoisted under the mast-head light of a steamer, and connected with the helm by rods or chains, and acts automatically with the rudder like a tell-tale, showing a green light when she is starboarding, and a red light when she is porting her helm, thereby showing the course a vessel is *about* to steer, and *before* she answers her helm or deviates from her course.

2168a. Pair of Ship's Side Lights with lenses.
John James Nickoll.

Guaranteed to show the light at two miles, that being the distance required by the regulations for preventing collisions at sea. The lenses are so constructed as to permit the smallest amount of absorption of the rays of light of the colour, and the burner requires no chimney.

2178ec. Transparency showing the effects of the helm indicating signals for preventing collisions at sea.
John James Nickoll.

2178eca. Set of Ships' Lights ; Port, Starboard, and Masthead Light, arranged to indicate by night the position of ships' helms, to avoid collisions. *Nickoll & Crewe.*

2178ed. Pair of Coloured Lenses for railway and other signals, calculated to show the signals upwards of two miles.
John James Nickoll.

The green signals can be seen at a much shorter distance than the red ; this green lens will show a distance of *three miles.*

2178ee. Model of an Anchor invented by Sir Edward Belcher when a midshipman, 1815.
Admiral Sir Edward Belcher, K.C.B.

2178ef. Model of a method of forming an improved Anchor when the shank has been broken, proposed by the exhibitor in 1830. The method is by means of 3 pieces of ballast iron. *Admiral Sir Edward Belcher, K.C.B.*

2178eg. Model in Silver. Patent "Stockless" anchor.
Wasteneys Smith, C.E.

2178eh. Model in Brass. Patent "Stockless" anchor.
Wasteneys Smith, C.E.

These anchors are said to possess—

1. Great holding power, with less weight, besides being diminished in the weight of the stock.

2. Extraordinary strength, proved at Lloyd's test.

3. It is always canted, no matter how it falls, and requires no stock to keep it canted.

4. Being always canted when on the ground, and by the assistance of the horns or toggles, it takes hold as soon as any strain is put upon the cable.

5. Spare, and wider, and different shaped arms for various grounds may be carried on board.

6. It will not foul or get fouled, and when holding there is nothing above ground, nor is there any stock to cause accidents.

7. It trips with great ease, because there is no stock to lift, and the crown end has so large a surface that good purchase is obtained for weighing.

8. It is easily fished, and can be stowed in-board on deck, thus clear of the bow, and avoiding risk of damage in case of collision of any description.

9. A ship may speedily be brought up by it, and ride with very short cable ; the steadying power being at the crown end, it is of no object if the shank is raised off the ground, which stocked anchors will not allow.

10. In shallow water no damage can occur to a ship's bottom, as no part of the anchor projects above ground.

11. It is at least one-third shorter than ordinary anchors, therefore soon clear of the water, and more convenient to manage.

12. It can readily be disconnected, thus convenient to stow and easy to transport in case of need, its heaviest part being less than one-third its total weight.

13. It is made without welding, thus of great soundness.

14. It is worked with only one davit (being hoisted and let go by the fish shackle), therefore considering the saving of first cost and future maintenance of one davit and blocks, &c., it is by far the cheapest anchor to use, besides greatest safety and simplicity in working.

15. It is not dangerous when at "wash," for in the event of a collision, the arms would simply be flattened to the ship's side instead of being driven in.

16. Should the anchor be difficult or dangerous to weigh, from having got fast in rock or wreckage, a runner with messenger attached may be slipped down the cable to the crown end, and by this means the anchor can be drawn out freely, as there are no barbs or palms to retard it.

17. Being of such greater strength and holding power, and requiring less cable than other anchors, shorter and stronger cables may be carried, thus increasing the safety of the ship without additional weight or cost.

2178ha. Tube connections (Roots' Patent) for Water Tube Boilers, for use on land. The Patent Steam Boiler Company, Birmingham.   *J. F. Flannery.*

2178hb. Full sized representations of the Tube connections (Roots' Patent) of the Water Tube Boilers, fitted in the screw steamer "Malta," of Liverpool, 1869. The Patent Steam Boiler Company, Birmingham.   *J. F. Flannery.*

2178hc. Full sized representations of the Tube connections (Roots' Patent) of the Water Tube Boilers, fitted in the Mersey ferry steamer "Birkenhead," 1872. The Patent Steam Boiler Company, Birmingham.   *J. F. Flannery.*

2178hd. Drawings (7) representing Sections of Water Tube Boilers, fitted by the Patent Steam Boiler Company, Birmingham.   *J. F. Flannery.*

a. Ordinary water tube boiler on Roots' system.
b. Sections and elevations (2) of the water tube boilers in the Mersey ferry steamer "Birkenhead."
c. Water tube boilers fitted in the screw steamer "Montana."
d. Elevation and section of the water tube boilers fitted in the screw steamer "Propantis."
e. Sections of the water tube boilers of the steam flat "Gertrude."
f. Drawings of Wigsell's patent water tube marine steam boilers.

2178he. Models of a Water Tube Boiler and improved Steam Engine.   *T. Moy.*

2178hf. Drawings of the High Pressure Tubular Boilers of the paddle-wheel steam yacht "Galatea." Front and end elevations. Section through firing end. Cross section. Plan.

Nominal horse-power, 200. Scale of drawing, 1 inch to 1 foot. Heating surface of two boilers of ship, 308 square feet; fire-grate surface, 186 square feet. Water space, 1,392 cubic feet. Steam space, 732 cubic feet. *William Willis.*

These drawings represent reproductions of the originals by Willis' Aniline Process. Executed by Vincent Brooks & Co.

2178hg. Drawing of Compound Marine Screw Engines. Side elevation and end elevation. *J. Dickinson.*

2178hh. Drawing of Dickinson and Mace's Patent Marine and Land Boilers. *J. Dickinson.*

Two mid-sections ; front and longitudinal elevation. Sectional plans.

2178hi. Drawing representing proposals by Messrs. E. J. Reed, C.B., and W. Wimshurst, for a method of lowering and raising the screw propeller at ship's stern to suit her variable draft of water, and to avoid "racing" of the engines. *J. F. Flannery.*

2178hj. Model and Drawings of improved Water Tube Marine Steam Boilers. *E. Wigsell.*

2178hk. Willan's Patent Three Cylinder High Speed Engine for steam launches. *Hunter & English.*

2178hl. Steam Launch Screw Propeller for the three cylinder engine. *Hunter & English.*

The three cylinders are placed side by side. The three pistons work on to three cranks on the shaft set at angles of about 120°. Each cylinder is single-acting the steam acting on the top only of each piston.

The connecting rods are jointed at one end to the pistons, at the other end to the cranks on the shaft.

Each piston acts as a steam valve, by which steam is supplied to one of the other pistons. On the return stroke of the pistons towards the top of the cylinder, the lower part in the cylinder is uncovered to allow the steam to pass back out of the cylinder into the casing surrounding the crank shaft.

2178hm. Small size High Pressure Quick Speed Three Cylinder Horizontal Engine for screw launches, P. Brotherhood's patent. *Brotherhood & Hardingham.*

2178hn. Models of Hirsch's Screw Propellers as fitted to H.M.S. "Minotaur," and SS. "Pereine," French Atlantic service. *H. Hirsch.*

2178ho. Four block Models of the "Hirsch" Ship (Patent) illustrating the "Hirsch" principles of shipbuilding.

Hermann Hirsch.

The SS. "Paun-ting" was built on Mr. Hirsch's principle, by Messrs. John Elder & Co. for Paul S. Forbes, Esq., New York, in 1873.

Dimensions of SS. " Paun-ting."

Length between perpendiculars, 210 feet.
Breadth, moulded, 33 feet.
Depth, moulded, 22 feet 9 inches.
Draught on trial, 12' 8" forward, 13' 9" aft, 13 feet 2½ inches.
Displacement, 1,520 tons.
Mid-area, 392·2 feet.
Immersed surface, 12,344 feet.
Area of immersed midship section, 5,239 feet.
Displacement per inch, 12·4 tons.
Co-efficient of displacement, ·568.
Weight of hull and machinery, 790 tons.
Weights on board, 730 tons.
Maximum of Ind. H.P. 800.

Result.

Speed 11 knots.
Displacement co-efficient of performance, 220.
Mid-area co-efficient of performance, 652.

The Hirsch Ship (Patent).

CALCULATIONS and PRINCIPAL DIMENSIONS of SHIPS as per Models.

—	Mail Steamer.	Cargo. Steamer.
	ft. ins.	ft. in.
Length between perpendiculars -	500 0	450 0
„ of keel for tonnage - -	470 0	414 0
Breadth extreme and for tonnage - -	50 0	60 0
Depth moulded to gunwale - - -	37 0	38 0
Tonnage B. O. M. - - - -	6,250	7,927$\frac{62}{94}$
	ft. ins.	ft. in.
Draught of water { forward - - -	21 0	20 0
{ aft - - -	21 0	20 0
Displacement in tons - - - -	8304·8	8221·22
Area of midship section in sq. feet -	948	1091·8
„ loadwater line „ „ -	18304·166	19,602·4
Tons per inch immersion - - -	43·58	46·67
Estimated weight of hull and machinery -	4,300	4,410
„ „ cargo and coals -	4004·8	3,811
Indicated H.P. - - - -	5,200	5,200
Estimated speed in knots - - -	17	16
Co-efficient of fineness - - -	·550	·527
Displacement coefficient of performance -	387	321
Mid-section „ „ „ -	895	859·4
Metacentre above centre of buoyancy -	10·024	16·154.
„ „ water line - -	1·520	8·151
Centre of buoyancy below water line - -	8·504	8·003

2178hp. Photograph of the Twin Hull Passenger Steamer "Prinzessin Charlotte," built near Spandau, 1816.

G. P. Rubie.

This was a double hull or twin steamer. She had an engine of 14 horsepower, built by J. B. Humphreys, engineer, which drove one paddle-wheel between the hulls. Dimensions were, length, 130 feet 5 inches; width, 19 feet 4 inches; tons B.M. 236. The steamer was built by John Rubie, naval architect, and began to run on the Elbe, November 1816. The Photograph represents the steamer lying at "die Zelten" near Berlin.

2178hq. Drawing, showing longitudinal section, upper and lower deck plans of the SS. "Gloria" and "Victoria," built by W. Doxford & Sons, Sunderland.

Length, 320 feet.
Breadth, extreme, 38 feet.
Hold, depth to top of floors, 31 feet.
Scale, $\frac{1}{3}$ inch to 1 foot.

W. Doxford & Sons.

2178hr. Drawing, showing elevation and sail draft, longitudinal section and upper deck plan of the iron sailing ship "Delscey," built by W. Doxford & Sons. *W. Doxford & Sons.*

2178hs. Drawing, showing longitudinal section and upper deck plan of Her Majesty's gunboats "Cygnet," "Express" and "Contest," built by W. Doxford & Sons.

Scale, $\frac{1}{4}$ inch to 1 foot. *W. Doxford & Sons.*

2178ht. Drawing, showing longitudinal section and upper deck plan of SS. "Autocrat," built by W. Doxford & Sons.

Scale, $\frac{1}{8}$ inch to 1 foot. *W. Doxford & Sons.*

2178hu. Model of the Paddle Steamer "Aloungpyah," built for the navigation of river Irrawaddy in Burmah.

Length, 245 feet.
Breadth, 26 feet.
Depth, 8 feet.
Engines, compound diagonal, 150 horse-power.

R. Duncan & Co.

The model represents one of a fleet of steamers of the same type belonging to the "Irrawaddy Flotilla Company," which are used for the general traffic of the river Irrawaddy from Rangoon, in British Burmah, to Bhamo, in Upper Burmah, near the frontier of China, the whole distance by river being nearly 1,000 miles.

Each steamer carries light goods and passengers, and takes in tow two iron barges each 190 feet long and carrying from 300 to 500 tons weight of general cargo according to the season. In the dry season the river is very shallow in many parts, and in the wet season the current is very strong; the vessels have therefore to be all of very light draft and great power.

The " Aloungpyah " is 245 feet in length, 26 feet in breadth, and 8 feet in depth of hull. Above the main deck, raised upon iron stanchions as shown, there is a light passenger deck and small cabin, and above that again, on a continuation of the same stanchions, a light galvanized iron roof, watertight, under which a canvas awning is stretched to intercept the heat, and curtains along the sides for the same purpose. With the exception of the upper passenger deck and small cabins shown, the whole structure is of iron, so arranged as to give the greatest strength with the least material.

The engines are compound, diagonal, of 150 nominal horse-power, with two cylinders 31 and 54 inches diameter by 54 inches stroke.

Two boilers, cylindrical, tubular, horizontal, carrying a working pressure of 70 lbs. per square inch.

The coal bunkers on both sides of the engine and boiler space have capacity for 100 tons of coal, which in practice is found sufficient for the voyage from Rangoon to Mandalay and back, a distance of 1,400 miles, with two barges in tow lashed one on each side, and loaded as stated.

2178hv. Model of the Composite Sailing Ships " James Nichol Fleming " and " Otago," showing framework of composite ships of 1000 tons register.   Built 1869.

R. Duncan & Co.

The model is intended to show the arrangement of the framework of a composite sailing ship of 1,000 tons register, and of the highest class at Lloyd's. The keel, stem, sternpost, rudder, and planking are of wood.

The frames, floors, beams, keelsons, stringers, and ties are of iron. The vertical black lines show the frames. The longitudinal and diagonal plates coloured blue show the outside ties on the frames, and the longitudinal black lines show the seams or edges of the planks.

The planking is put on outside of the frames and plate ties, and is bolted to the frames and plates by brass bolts and nuts, the seams are caulked with oakum, and then the whole wooden skin is sheathed with copper or yellow metal. The iron framework is intended to ensure the rigidity and strength of an iron ship, and the coppered bottom to ensure freedom from sea grass and barnacles, to which iron ships are subject on long sea voyages.

Whole Model, full-rigged, of the Sailing Ship " Lanoma."   Built at Sunderland, 1876, for Thos. B. Walker, Esq., London.   Length 185 feet, breadth 31 feet, depth 18 feet; tons, 665 register.   *Austin and Hunter.*

The ship is barque rigged, and is on her first voyage.

Whole Model of the Four-Masted Sailing Ship " Shakespeare." Built on the composite system, iron framing, with exterior wood planking, by Messrs. Short Brothers, for W. Adamson, Esq., 1876. The ship is intended for the Australian trade.   *Short Brothers.*

This four-masted sailing ship is designed for the utmost speed. From a voyage made already by a sister ship to Australia, it is expected that the " Shakespeare " will average 55 days on the passage. She is already at sea on the first voyage.

The officers and crew are housed on the upper deck, leaving the hull entirely free for cargo carrying only.

The ship's dimensions are—

Length	-	-	-	278 feet.
Breadth	-	-	-	40 feet 2 inches.
Depth	-	-	-	24 feet.
Tons	-	-	gross	1,814
,,	-	-	nett	1,757
She is classed at Lloyd's			-	100 A 1.

Sectional Models (three in number), showing various methods carried out by Messrs. Short Brothers in building ships on the composite principle. *Short Brothers.*

Model No. 1 illustrates a method of diagonal bracing both for the framing and deck beams of composite vessels.

Model No. 2 illustrates a trial of filling in between the ship's frame with vertical iron plating.

Model No. 3 illustrates the iron framing of a composite ship with improved stringers for tying the framing, and carrying the beams.

2178hw. Photograph from a painting of the paddle steamer **" Iona,"** 1874, compared with the "Comet," the first steamer on the Clyde. *The Institute of Naval Architects.*

2178hx. Model of the " Shakespeare," a four-masted sailing ship, and three sections illustrating composite construction. *Messrs. Short Bros.*

2178xy. Set of Annual Reports and Proceedings of the Institution of Naval Architects, 1860 to 1875. *Institution of Naval Architects.*

2179. Model of J. Ericsson's Screw Propeller Engines, applied to the American and Swedish Monitors, patented in America in 1858. *H.M. Commissioners of Patents.*

2180. Models of Screw Propellers. *The Council of King's College, London.*

2167a. Peraon, for automatically fixing the course of a ship, and for use with the bathometer. *Dr. F. Schöpfleuthner, Vienna.*

This apparatus replaces the compass and tiller of a ship, as when it has once been arranged for a particular course the ship is compelled accurately to follow that course, at the same time that it enables any tacking or turning to be executed with great ease and precision. This peculiarly constructed compass is fitted with hand-wheel and vernier, and by means of it the whole ship can be directed as required. The declination-needle, of which the arc of oscillation is very small, turns between the two movable electrodes of an electrical battery, which on coming into contact with the magnetic needle control the movement either backwards or forwards of the admission valve of a steam or hydraulic motor. This acts on the steering-gear, and when that is turned to its extreme point (which only occurs in turning the ship) its cuts off the admission, so as to avoid breakage.

2167b. Bathometer, for submarine levelling with graphic registration on board the surveying vessel.

Dr. F. Schöpfleuthner, Vienna.

This instrument is used in submarine levelling, and for the construction of charts of the water-covered portions of the globe. When in use it is attached to the surveying ship by a cable running through an insulated copper tube. It is towed after the ship, and is connected with a registering apparatus on board in such a way that the portion of the sea bottom passed over is represented in profile. The principle on which it acts is that of the equilibrium between a closed volume of air, and the pressure of the column of water resting on it, and the registration is effected by means of an electric telegraph. When it is employed the course of the ship is fixed by the Peraon described above; in order that the instrument may not be damaged against the rocks on the ground it is provided with a separate horizontal rudder.

2167c. Apparatus for guarding against the **Stranding** of **Ships.**   *Dr. F. Schöpfleuthner, Vienna.*

This apparatus is placed on the bow of the ship, or on a steam launch which moves several hundred metres in advance of the ship; it serves for warning against rocks, sand banks, or steep coasts, and generally is used when in unknown waters. The determination of the sounding is effected by means of a lead which is towed at about a depth of 30 metres under water, and which is connected with a free swinging pendulum in such a way that, as an alteration of the angle between the two, a needle makes contact on the respective side. A tell-tale shows the depth, and by means of a cable connected with the ship sets a signal apparatus in action. The latter is only used when the steam launch is sent out in advance.

2178be. Russian and Manilla Hemp Ropes, used in shipping, and galvanized iron wire ropes for ships' rigging, &c.

Thomas & William Smith.

Tarred Russian hemp ropes for ships' rigging, running gear, lanyard ropes, towlines, warps, bolt rope for binding sails, and hauling ropes.

White manilla hemp rope, for towlines, warps, running gear, point line, &c., &c.

Houseline and marline for serving rigging, signal halyards, log lines, lead lines.

Galvanized wire rope, for ships rigging, stays, &c., with specimens of the different modes of fitting shrouds and stays with deadeyes, thimbles, and sockets.

Galvanized hank to run on the stays, different sizes of seizing strand used in fitting wire rigging, lanyard, and shackle thimbles.

Copper wire rope lightning conductor, fitted with a thimble to pass over the copper rod when driven into masthead.

2178bea. Specimens of Iron Wire Rope, for ships' rigging.   *Newall & Co.*

2178beb. Case of Samples of Canvas Sail Cloth, for naval and commercial purposes.   *Baxter Brothers.*

2180a. Model of a Propeller for Ships.

S. F. Pichler, London.

2180b. Model of Bevis's Patent Feathering Screw Propeller.   *Laird Brothers.*

Mr. R. R. Bevis, managing engineer to the firm of Messrs. Laird Brothers, of Birkenhead, in 1868 patented an arrangement for altering the pitch or feathering the blades of a screw propeller in a fore and aft direction, which claims to be a great advantage for screw steamers, making them faster and more handy when under sail alone, and when under steam and sail allowing of adjusting the pitch to obtain the best result. A screw propeller of the ordinary kind, whether fixed or revolving, is a heavy drag against speed and handiness for sailing, and "lifting" it is a laborious operation, and requires a large hole or well through the ship's counter to admit of so doing.

The arrangement of this new screw propeller is free from many of the objections which have been made to feathering screws previously tried. The gear for feathering the blades is well protected, the levers and other gear that move the blades being enclosed within the boss of the screw propeller, and attached to a rod passing through the centre of the shaft, which is worked in the screw shaft tunnel. This system is admirably adapted for ships of war or sailing ships with auxiliary power or yachts, where it is as important to have a good result under sail alone as under steam. The operation of altering the pitch, or of feathering the blades to any angle, is done in a few minutes, without in any way putting the engines into a position that they may not be used in an emergency.

This feathering propeller has now been successfully tried in 15 vessels, and is at present being fitted to four vessels of H.M. Navy, one of them of 2,100 ind. horse-power.

2180c. First Helmet made for Diving Purposes, date A.D. 1829.   *Siebe and Gorman.*

2180d. Patent Helmet for Diving, fitted with segmental neck ring and safety locking arrangement, inflating valve for bringing diver to the surface. Fitted with speaking apparatus to enable the diver to communicate with his attendant. Used on board H.M. Ships of the Royal Navy.   *Siebe and Gorman.*

2180e. Diving Spectacles, fitted with air-lenses, having a focus in water of 1½ inch, for subaqueous vision.
    R. E. Dudgeon, M.D.

The air-lenses are prepared from two sections of a thin glass globe 1½ inch in diameter, fixed in a spectacle frame with their convexities towards one another, and their concavities outwards. Under water these spectacles have a focus of 1½ inch, and supply the refractive power lost by the eye when submersed in water, owing to the aqueous humour which forms the anterior lens of the eye having the same index of refraction as water. Exhibited in the International Exhibition of 1871.

2180f. Boat moved by A. H. Garrod's Undulating Propeller.   *H. C. Ahrbecker.*

This model is moved by an undulating propeller, of which the component staves (which carry the undulating membrane) successively receive their transverse motion from the rotation of a screw with which they are in lateral contact.

Invented by A. H. Garrod, in imitation of the undulating propelling fins of the pipe fish and sea horses.

X.—LIGHTHOUSES AND FOG SIGNALS.

2181. Model of a **Lighthouse** built upon the **Bishop Rock,** 7 miles from land, forming part of the outermost reef S.W. of the Scilly Islands.   *Trinity House, London.*

The tower is of Cornish granite (Carnsew), and is surmounted by a lantern of gun-metal, containing lenticular apparatus to exhibit a fixed light of the first order, whose focal plane is 110 feet above high-water spring tides. The structure measures from base to vane 147 feet. It was built from designs by the late James Walker, M.I.C.E., under the superintendence of Nicholas Douglass, for the Corporation of Trinity House, London, at a cost of 36,500*l.*, and occupied six years in construction; it was completed in 1853. A sectional drawing shows the method of bracing by vertical and radiating wrought-iron ties, lately adopted for strengthening the structure.

2181a. Model of the Old Lighthouse on the Smalls Rocks, about 17 miles off the coast of Pembrokeshire. Erected 1776; replaced in 1861 by a stone lighthouse. The model is made out of one of the oak pillars of the original lighthouse. —Executors of the late Captain Pickering Clarke, R.N.
South Kensington Museum.

2181b. Model of the Lighthouse on the Needles Rocks, Isle of Wight. Built 1857–1858. Light shown 1 January 1859.—Trinity House, London.   *South Kensington Museum.*

2181c. Model of the Gunfleet Lighthouse. Iron. Built on Mitchell's patent screw piles, of which a model is also shown. The piles screw 40 ft. into the sand, and have screws 4 ft. in diameter. James Walker, engineer.—The Trinity House.
South Kensington Museum.

2181d. Series of Models, illustrating mark buoys, used by the Trinity House Corporation round the British coast.—The Trinity House.   *South Kensington Museum.*

2181e. Harbour Light. Chance's patent dioptric lens of the fourth order, for fixed light.—Chance Brothers and Company, Glass Works, Birmingham.   *South Kensington Museum.*

2181f. Babbage's occulting Light Apparatus. Designed by the late Charles Babbage, F.R.S., for distinguishing lighthouses from one another by numbers.
It is adapted in the Exhibition to a dioptric light exhibited by Messrs. Chance.   *Major-General Babbage.*

The system of occulting lights for distinguishing lighthouses from one another by numbers was invented by the late Charles Babbage about the year 1850. It was published in "The Exposition of 1851," and was favourably reported on by the Lighthouse Board of the United States in January 1852. It was subsequently published in the "Times" of Wednesday, July 11th,

1855, and elsewhere. It is capable of being introduced into existing light-houses without disturbing the existing lenses, &c., and can be adapted to any single lighthouse without interfering with the adjacent ones. The number here shown is 587, thus :

The light is visible during a long interval; five occultations are made :

The light is then visible during a short interval ; eight occultations are made :

The light is then visible during a short interval; seven occultations are made :

The light is then visible again during the long interval as before, and the same series is repeated continually from sunset to sunrise : thus the seaman by counting the occultations becomes at once aware of the number of the lighthouse, and knows from his chart his exact position.

The lighthouses along the coast would, of course, not be numbered consecutively ; in fact, no two adjacent lighthouses could have numbers containing the same digit : for example, the lighthouses on either side of No. 587 might be 839, 614, 587, 293, 418, so that a mistake could not occur in identifying any one of them from the others.

2182. Model of a **Lighthouse** now building upon the **Little Basses Rock,** part of a reef about 7 miles S.S.E. of the coast of Ceylon. *Trinity House, London.*

The tower is of Scotch granite (Dalbeattie), each stone of which was dressed, fitted, and marked in this country, freighted to Galle, and thence carried to the rock and fixed in its place. The light is intended to be of the first order, dioptric, on the group-flashing principle, showing two flashes in quick succession every minute, at an elevation of 110 feet above high-water spring tides.

The rock is awash at low water, and is exposed to heavy seas during both the N.E. and S.W. monsoons, and while the latter prevails is inaccessible for work. The drawings show the methods of landing stone in a seaway by steam-power.

This lighthouse, as well as its fellow on the Great Basses just completed, is building from designs by James N. Douglass, M. Inst. C.E., under the superintendence of William Douglass, M. Inst. C.E., for the Corporation of Trinity House, London, acting on behalf of the Home and Colonial authorities.

Its cost is estimated at 73,000l., and completion is anticipated within five years from date of commencement.

2182a. Lantern and Apparatus intended for the Little Basses Lighthouse, Ceylon. *Trinity House, London.*

The lantern is of the cylindrical type adopted by the Trinity House, its form gives maximum strength, and secures greater optical accuracy than the earlier methods of flat glazing. The gun-metal framing is inclined about 30° from the perpendicular, and is helically curved throughout, thus reducing to a minimum the obstruction offered to the light sent forth from the lenses.

The optical apparatus, constructed upon the group-flashing principle, designed by J. Hopkinson, B.A., D.Sc., at the glass works of Messrs. Chance Brothers, is the first dioptric instrument of its kind adopted by the British lighthouse authorities. It is 12-sided, and makes a completed revolution in six minutes, so that the panels being arranged in pairs, a double-flash meets the eye of the observer once a minute.

The lantern and apparatus prepared for this structure are exhibited in working order in the grounds outside the Museum.

2183. Drawing of a **Light Vessel** with deck plans showing internal arrangements and disposition of Syren **Fog Signal** machinery.   *Trinity House, London.*

The hull is designed after that of the vessel now at South Sand Head (Goodwin), built last year, of about 212 tons, and fitted with a syren fog-signal, giving one blast every two minutes, by means of compressed air, at a pressure of 30 lbs. to the square inch, the apparatus being driven by a caloric engine, which also works the windlass.

The illuminating apparatus represents that in use at the Royal Sovereign Shoal, off the coast of Sussex ; it is catoptric, and is upon the " group-flashing " principle, giving three flashes in quick succession every minute. The crew space is for seven men, including the officer in charge. The hollow iron mast affords access to the lantern, and allows of the lamps being trimmed in all weathers without danger of extinction.

2184. Two **Syrens,** each a portion of the present **First Class Fog Signal,** and a diagram showing the method by which they are put in action.   *Trinity House, London.*

The disc syren, with the trumpet by which its sound is directed, is shown in the diagram. It is composed of a fixed disc, forming one end of the chamber into which steam or compressed air is forced, and a movable disc rotating rapidly by separate mechanism outside it. Both are perforated by 12 radial slits exactly corresponding each to each, and the rotation of the moving disc, close to and upon a common axis with its fixed associate, permits the compressed air or steam to escape when the slits coincide, and shuts it off when they do not. The vibrations thus produced being repeated in the instrument described, at the rate of more than 400 per second, emit a sound of very great intensity, which is directed by the trumpet towards any desired point.

The cylindrical syren is a later form of the instrument, in which the chamber for compressed air surrounds a fixed cylinder having 24 slits, within which another cylinder coincidently perforated rotates, and the vibrations pass through the open end of the inner cylinder to the trumpet.

Syrens sounded by steam have for some time been used for fog-signalling on the coasts of America, and have lately been adopted, with the substitution of compressed air for steam, in Great Britain as a result of experiments made by the Trinity House with the assistance of Professor Tyndall during the winter of 1873.

2185. Fog Signal Apparatus. Designed by Dr. G. Amadi, of Trieste.

The Imperial and Royal Maritime Government at Trieste.

By this apparatus deep tones, like those of an organ, are produced by metallic tongues, driven by steam, and sent through a trumpet in a given direction. These, from experiments, have extended as far as 16 nautical, or nearly four German miles.

In working this apparatus, of which already three are in use, at Trieste, Salvore, and Grado, the sounds are made self-producing at certain intervals by means of a steam-engine.

2185a. Holmes' Shipwreck Distress Signal Flare and Life Buoy Rescue Lights.   *N. J. Holmes.*

These have the remarkable property of bursting into flame when placed in contact with water, and when once ignited are absolutely inextinguishable by either wind or water. They emit a most powerful white light, as brilliant as the magnesium light, and continue to burn over 30 minutes. The shipwreck distress signal flare is visible on a dark night with a clear atmosphere at a sufficient elevation for over ten nautical miles, and burns with greater brilliancy the more seas sweep over it.

The light is a chemical light, and produced by the action of the water upon phosphuret of calcium, giving off phosphuret of hydrogen, which, combining with the oxygen in the atmosphere, spontaneously ignites. These distress signals are free from danger, are not affected by heat, friction, or percussion, and contain no explosive compound whatever.

2185b. Holmes' Mechanical Compound Reed Fog Horn. *N. J. Holmes.*

These mechanical fog alarms are constructed upon the most approved acoustical principles, and emit a most powerful sound. The "aurora" fog horn can be heard over three nautical miles, and the note produced is the 8 foot C of the musical scale. The tone is produced by the vibrations of two metal tongues, placed together in absolute contact, and closing the same reed, by which means (the split tongue) a powerful vibration is set up with a minimum pressure of air. The air bellows consist of two metal cylinders, one working inside the other; and the compressed air upon the return of the cylinder is driven through the reed into an inner trumpet-shaped tube contained within and a part of the external cylinder.

2185c. Drawing of the Steam Syren or Fog Horn, erected on the South Foreland, coast of Kent. *Prof. W. F. Barrett.*

2186. Parabolic Reflector of 21-inch aperture. *Trinity House, London.*

Composed of copper coated with pure silver in the proportion of 3 ozs. (troy) pure silver to 1 lb. (avoirdupois) copper. Its focal distance is 3 inches.

Improvements in the construction of light vessels' lanterns have permitted the introduction of this large sized reflector into that service, in which a 12-inch aperture had hitherto been the limit of size.

2187. Centrifugal Governor. A mechanical arrangement used for controlling the movements of the clockwork machinery, by means of which a light is made to revolve on board a light ship. *Trinity House, London.*

Before this contrivance was designed by Mr. Slight of the Trinity Workshops, there was always a tendency to irregularity in the periods of duration of the light and the interval of darkness, but the centrifugal governor ensures the working of the revolving machinery with very great accuracy.

2188. Improved Six-Concentric-wick Lamp, for burning vegetable or mineral oil. *Trinity House, London.*

The burner hitherto in use for dioptric apparatus of the first order has carried four concentric wicks, and in burning has been maintained at full power in all weathers. In the improved six-wick burner, designed by J. N. Douglass, M. Inst. C.E., only the three outer wicks are to be used in ordinary weather, and in thick weather the three inner wicks (at other times cut off by a concentric reflector) are also brought into action. The full light-producing power of the six-wick lamp is equal to 722 standard sperm candles, attained by a consumption of 1 gallon of oil in 1 hour 50 minutes; its half-power equals 342 candles, with 1 gallon consumed in 2 hours 45 minutes. By a simple arrangement regulating the level of oil in the burner, and the position of the air-deflectors, the lamp can be made to burn any description of oil at pleasure.

2189. Panel of **Cata-Dioptric Apparatus, One** of a set of **Polyzonal Lenses** manufactured in 1836 by Messrs. Cookson and Sons, of Newcastle, for the Trinity House of London, and by them fixed in the Start Point Lighthouse, Devonshire.

Trinity House, London.

The first lenticular apparatus used in an English lighthouse, with a central lamp upon Fresnel's principle.

2190. Plano-Convex Lens, used at Portland Lighthouse in the year 1789. *Trinity House, London.*

It is 22 inches in diameter, and was placed in front of an argand burner and reflector. It is believed to be one of the lenses first used in combination with an oil lamp and reflector for lighthouse illumination.

2191. Facet Reflector. Specimen of a reflector and lamp used first at Liverpool about the year 1763, and afterwards at Lowestoft and other lighthouses. *Trinity House, London.*

It is made of wood with facets of silvered glass, is nearly paraboloidal in form, and is the earliest kind of reflector known to have been used in aid of an oil lamp in lighthouse service.

2192. Lamp and **Reflector,** as used in English floating lights about the year 1809. The curve of the reflector is spherical.

Trinity House, London.

2193. Dhu Heartach Lighthouse, situated on a rock on the West Coast of Scotland, exposed to the force of the Atlantic Ocean, and 14 miles from Iona, the nearest land. Designed and executed by D. & T. Stevenson, Engineers to the Board. Commenced 1867, finished 1872.

The Commissioners of Northern Lighthouses.

2195. Model of a **First-class fixed Dioptric Light;** scale, one-fifth of full size.

The Commissioners of Northern Lighthouses.

This apparatus consists of a central lenticular band, and an upper and lower set of reflecting prisms. The cylindrical belt with diagonal joints and the upper and lower reflecting prisms were substituted by Mr. Alan Stevenson, in 1836, for the segmental belt, and upper and lower silvered mirrors of Fresnel's first-class apparatus.

2196. Model of Fresnel Revolving Apparatus, as made for Skerryvore Lighthouse in 1843 ; scale, one-fifth of full size.
The Commissioners of Northern Lighthouses.

The light is received and collected into eight horizontal beams by the principal lenses—the light which would escape above is collected into eight inclined beams by small lenses, and reflected to the horizon by inclined mirrors. The lower part of the light is sent equally to all parts of the horizon by prismatic rings of glass, which act as mirrors. The rings at Skerryvore are the first that were made of the largest or first order size, and were undertaken by M. Soleil, on the proposal of Mr. Alan Stevenson.

2197. Model of Stevenson's Holophotal Revolving Apparatus ; scale, one-fifth of full size.
The Commissioners of Northern Lighthouses.

The central part of this apparatus consists of eight of Fresnel's lenses. The light which passes above and below these lenses is collected into eight horizontal beams by reflecting prisms. These reflecting prisms were substituted for the inclined lenses and mirrors of Fresnel's first-class revolving apparatus by Mr. Thomas Stevenson, and were first used by him at Singapore, in 1849, on a small scale, and adopted on a large scale at North Ronaldshay, in Orkney, in 1851.

2198. Stevenson's Dioptric Holophote.
The Commissioners of Northern Lighthouses.

This apparatus collects all the light of the lamp into one beam of parallel rays solely by means of glass. The apparatus constituting the front half of the instrument bends the light that falls upon it into a beam of parallel rays, while the prisms which constitute the back half are so formed as to prevent any light from passing through, and to cause every ray to return back to the flame, and to be finally transmitted through the front half, so as to increase the intensity of the emergent beam. A large red ball is fixed on a rod, so as to be in focus, to illustrate the action of the instrument. To an observer, the front half of the apparatus will appear full of red light, but in the back half no red is to be seen, though the lower part of the rod which carries the ball, not being in focus, is distinctly visible.

2199. Stevenson's Fixed Azimuthal Condensing Light.
As used at the leading lights of the River Tay.
The Commissioners of Northern Lighthouses.

It is remarkable from its employing every kind of dioptric apparatus. The whole of the light coming from the flame is spread equally over a horizontal arc of 45° by means of the following instruments ; viz., Fresnel's fixed-light apparatus and annular lens, and Mr. T. Stevenson's condensing prisms, holophote, right-angled conoidal prisms, and dioptric spherical mirror, with Mr. J. T. Chance's setting.

2200. Model of Stevenson's Apparent Light.
The Commissioners of Northern Lighthouses.

A beam of light, projected on the apparatus in the lantern on the beacon from a lighthouse on the shore, is reflected or refracted in such a manner as to indicate the position of the beacon at night. It was first used at Stornoway, in Scotland, in 1852.

2201. Model of the **Lamlash Lighthouse Apparatus,** showing the new twin prisms for preventing the loss of light by absorption, and for saving room in the lighthouse, lately described by Mr. Thomas Stevenson in "Nature," which are now for the first time being constructed, and the new back prisms first introduced at Lochindaal Lighthouse, in Islay in 1869.

The Commissioners of Northern Lighthouses.

2201a. Revolving Light, with Fresnel Apparatus, condensing lenses, and electric lamp, 1876.

Sautter, Lemonnier, & Co., Paris.

It is proposed to use this light on Cape Grisnez, near Calais.

2201b. Catadioptric Light, with "projector" apparatus and electric lamp, 1876.   *Sautter, Lemonnier, & Co., Paris.*

This apparatus is designed to throw or "project" a flash of light seaward at certain intervals of time, or at will. It is fitted with mechanism for the purpose of varying the altitude and azimuth of the beam of light projected towards the sea, so as to ensure the utmost effect in all states of the atmosphere.

2202. Lighthouse Apparatus.

1. Lens with échelons. 1st essay. Fresnel, inventor, 1819.
2. Polygonal lens, of the first order. Do.   do.   1820.
3. Annular lens, of the first order. Do.   do.   1821.
4. Apparatus for fixed lights.   Do.   do.   1824.
5. Apparatus with catadioptric rings. Do.   do.   1825.
6. Lens with catadioptric rings, constructed in   . 1825.
7. Model of apparatus with catadioptric rings. Fresnel, 1825.
8. Burner with 4 wicks, constructed after experiments made by Arago and Fresnel in   .   .   . 1820.
9. Burner with 2 wicks, with external covering, by Henry-Lepaute   .   .   .   .   . 1845.
10. Burner with 4 wicks, storied, for mineral oils.
11. Large annular lens of Barbier and Tenestre   . 1867.
12. Lenticular panel for flashing lights of Henry-Lepaute 1876.
13. Apparatus for revolving electric lights, of Sautter, Lemonnier, & Co.   .   .   .   . 1876.

Lighthouse Service of France, Paris.

Echeloned Lenses.

No. 1. First essay of echeloned lens, polygonal form. Invented by A. Fresnel, and constructed under his direction in 1819.

No. 2. First echeloned lens, polygonal form, for flashing lights of the first class. Invented by A. Fresnel, and constructed in 1820.

No. 3. First echeloned lens, annular form, for flashing lights of the first class. Invented by A. Fresnel, and proceeding from the lenticular apparatus fixed on the tower of Cordouan in 1821.

When Fresnel conceived the idea of substituting in lighthouses large glass lenses for metallic reflectors, he thought of composing these lenses of several pieces, and of calculating the curves of these different pieces so as to rectify their spherical divergence. He demonstrated his plan before the Lighthouse Committee in August 1819, three months only after his appointment on the Committee, and on the 19th of October following he was granted the sum of 500 fr. for constructing a trial lens. He consulted the optician Soleil, who seconded him with much good will, but who could only put at his disposal the limited appliances then in use. Glass was at this time worked still by hand, and shaped only into plane or spherical forms. Fresnel admitted that the lens should be flat on one side ; that the different gradients, instead of forming circular rings, should be defined by polygons and divided into a certain number of pieces, each of which should receive on its echeloned side a spherical surface properly calculated. Another difficulty arose from the glass factories being unable to supply in sufficient size pieces of crown glass free from bubbles and striæ ; but M. Soleil discovered the way of re-smelting glass without altering its transparency.

He first constructed a trial lens of 35 centimetres diameter (the one exhibited under No. 1). It was given by Soleil to the Academy of Sciences, and deposited at the " Conservatoire des Arts et Métiers." It is composed of 21 pieces, cemented together, and fixed upon a pane serving as a support.

Emboldened by this first success, Fresnel proposed to the Lighthouse Committee, at their sitting of 31st December 1820, to order the construction of a lenticular revolving light apparatus for the Cordouan lighthouse. The principal part of this apparatus was to include eight square lenses of 76 centimetres, forming together an octagonal prism inscribed within a cylinder of 2^m diameter. This proposal was adopted, and M. Soleil undertook the construction of these eight polygonal echeloned lenses. (One of them is exhibited under No. 2.) It is to be seen that it was composed of 100 pieces of glass, cemented together, and that the flat pane, which in the trial lens serves as a support, has been done away with. One of these new lenses was first tried in public on 13th April 1821. It was placed on the top of the Observatory buildings, together with two large reflectors, one by Lenoir, the other by Bordier-Marcet. The Lighthouse Committee, of which Mr. Becquey, director-general of the " Ponts et Chaussées," was chairman, went to the summit of Montmartre to judge of the effect. The result confirmed the inventor's previsions, and every one allowed the superiority of the lens over the reflectors. Meanwhile, Fresnel had already thought of improving these first essays. He had invented a machine for constructing circular rings, and M. Soleil was instructed to make eight large lenses constructed on annular principles. He soon finished some of them, and in September 1821 the Lighthouse Committee determined to try their effect at long distances. Fresnel had fitted up on the top of the " Arc de l'Etoile " a revolving apparatus upon which were fixed two of these annular lenses, four polygonal lenses previously constructed, and four semi-polygonal lenses. At the focus, a four-wicked lamp was burning. The committee then went to Chatenay, a village situated N.N.E. of Paris, $24\frac{1}{2}$ kilometres distance from the " Arc de l'Etoile." The experiment took place during the night of 7/8 September 1821, and the results were adjudged as very satisfactory. The eight annular lenses that had just been constructed form part of the first flashing-light apparatus of the 1st class that Fresnel himself fixed on the watch-tower of Cordouan, and which has lighted the entrance to the Gironde for more than 30 years. (The lens exhibited under No. 3 comes from this apparatus.)

If an idea is to be formed of the progress made in the science of lenticular lighthouses from its origin to later times, the three lenses before mentioned should be compared with the lenses of modern construction exhibited under Nos. 11, 12, and 13.

Fixed Light Apparatus.

No. 4. First fixed light apparatus of 0·50ᵐ diameter, invented by A. Fresnel, and constructed in 1824.

Fresnel, after his appointment to the Lighthouse Committee of 1819, first gave his attention to flashing lights ; meanwhile, he had thought about obtaining fixed lights, and in the first design of the lenticular lighthouse that he submitted to the committee on 31st October 1820, he indicated, as a solution, the use of cylindrical lenses ; but the Lighthouse Committee had thrown aside the fixed light system as possessing less power than revolving lights, and as being liable to be mistaken for the incidental lights of the coast. The committee altered this decision later, and Fresnel then invented the system of fixed light apparatus (0.50ᵐ diameter), exhibited under No. 3. In this apparatus the lenticular drum, which should be cylindrical, so as to give an uniform subdivision of light, shows a polygonal form of 16 fascets, because no lathe was then known for making cylindrical pieces. The upper part is made up of two lenticular zones in the shape of a 16-panel cupola, every element of which is coupled with a plane mirror. The lenses unite in parallel fasces the rays emitted by the light, and the mirrors reflect them in the direction of the horizon.

A similar system, but having one lenticular zone only, is fitted at the lower part. The lamp has two wicks, and stands upon a plate raised or lowered by a jack between three leaders. With the polygonal form that had to be adopted, there were 16 directly receiving more light than the intermediate parts, but Fresnel, while constructing the instrument, discovered the means of greatly lessening this inequality, by alternating the shining directions of the lenticular cylinders with those of the two other parts. This first trial of a fixed light apparatus was demonstrated by Fresnel before the Academy of Sciences of Paris, at their sittings of 3rd May 1824. It was then inaugurated on the 1st February 1825 in the port of Dunkirk.

Catadioptric Rings.

No. 5. First apparatus containing catadioptric rings, as well for fixed light as for flashing lights, invented by A. Fresnel, for lighting the St. Martin Canal, and constructed in 1825.

No. 6. Annular lens, composed of dioptric and catadioptric elements, similar to those of apparatus No. 5, and constructed at the same date.

No. 7. Models in wood of an apparatus similar to No. 5, but on a larger scale. Researches of A. Fresnel in 1825.

The last invention of A. Fresnel, that of the catadioptric rings, was promoted by a request for information addressed to him by the Prefect of the Seine in 1825. It was a question of applying to the lighting the quays of the St. Martin's Canal more powerful lamps than those used commonly in the city of Paris. This problem, to which Fresnel's attention was called, was the same as that of the port-lights apparatus, of which he had postponed the study because the sidereal reflectors of Bordier-Marcet were sufficient to supply the wants of the service.

The principal part of these small apparatus, that is, the lenticular cylinder, offered no theoretical difficulty. It was to proceed out of an echeloned section turning around the vertical axis ; the only thing was to construct it in circular shape, because the polygonal shape would have been impossible for rings of 20 to 25 centimetres diameter. The question was not so easily solved as regards the accessory parts intended to utilise the luminous rays passing outside the cylinder, because the reflectors used in the other classes must now be reduced to too small dimensions. It was then that Fresnel thought of the phenomenon known in optics under the name of "total reflection," and proposed to substitute for the common reflectors glass rings, within which the luminous rays should be reflected without appreciable loss.

Fresnel's first conception for these circular rings was to place the fascets through which the luminous rays pass perpendicularly to these rays, so as not to alter their direction; the reflecting surface would then have preserved the shape of the mirrors to be replaced, but hence resulted inconvenience, and a too great weight of glass. Fresnel found out that a direction inclined to the rays could be given to the fascets, and thus these inclinations be combined, as well as the shape of the reflecting surface, so as to cause the rays to emerge horizontally. The transverse section of the rings then became triangular, instead of showing four sides, and the dimensions were reduced.

The apparatus exhibited is that to which Fresnel first applied this invention. Its diameter is reduced to $0 \cdot 20^m$; the cylinder is generated by an echeloned section composed of three elements and fills up a half circumference. The rays passing above this cylinder are gathered by four total reflection rings, and this is effected by turning around the vertical point of the focus the section of the catadioptric triangles just spoken of. Thus is obtained a fixed light apparatus, lighting up half the horizon. The lamps of the St. Martin's Canal having to be erected at 70 metres distance, it became necessary to give them a greater lateral than frontal intenseness. Fresnel succeeded in this by placing on each side a half annular dioptric lens, generated by the rotation of the section of the cylinder around an horizontal axis, parallel with the longitudinal direction of the quay, but he had moreover the happy idea of making the section of catadioptric triangles to revolve around this axis so as to form an annular lens, collecting around the focus an angle of great amplitude, and comprising at the same time dioptric and catadioptric rings.

The manufacture of these different circular rings offered serious difficulty, and Fresnel was obliged to set up a factory.

A first apparatus was completed in 1826, and submitted to the Lighthouse Committee towards the end of December. Four of these new lamps were finished at the beginning of 1827, but they could not be tried until after the inventor's death.

This study shows how Fresnel came to invent not only the section of catadioptric rings, and the use of these rings in fixed light apparatus, but also to apply them to annular lenses for flashing lights or for fixed lights. By uniting the pieces of dioptric elements and of catadioptric rings manufactured in Fresnel's time for the apparatus of the St. Martin's Canal, the annular lens exhibited under No. 6 was formed, and it may be considered as the type of all the annular lenses used in the lighthouses of different order.

The model in wood, No. 7, represents an apparatus similar to the preceding, but having a diameter of $0 \cdot 25^m$ instead of $0 \cdot 20^m$. It is a study of Fresnel's which he did not carry out.

Lamp Burners.

No. 8. One of the first burners, with four concentric wicks, constructed after experiments made by Arago and Fresnel in 1819-20.

No. 9. Burner, with two wicks and outer casing for directing the draught, constructed by Henry-Lepaute in 1845.

No. 10. Burner, with five wicks, of graduated shape, for mineral oil, with the last improvements adopted in the lighthouses of France, 1876.

When Fresnel undertook the improvement of lighthouses, he had to solve not only the problem of construction of the lenses, but also that of lamps with several wicks. The chemist Guyton-Morvau had already studied the question. In a paper read by him at the Institute in 1797, he stated that he had constructed, 10 years before, a lamp on the Argand principle, with three concentric circular wicks, each having an inner and an outer draught. It gave great intensity, but the solderings of the burner were destroyed by the heat. About 1800, the watchmaker Carcel invented the lamp that bears his name, and in which the oil at the bottom is forced up by a pump towards the burner above which it overflows. This invention subsequently led to the solution of the problem of lamps with many wicks. Consequently, when Arago and Fresnel began, in 1819, their experiments with lamps, they forced up the burner oil in superabundance so as to replenish it, and thus avoid the inconvenience met with by Guyton-Morvau. The first trial took place in September 1819 with two-wicked and three-wicked burners, constructed after Fresnel's designs. After several deliberations, chiefly respecting the width to be adopted for the draught between the wicks, they succeeded in constructing a four-wicked burner that gave good results. It was tried 12th May 1820, in presence of the Lighthouse Committee. The burner exhibited under No. 8 is one of those that were constructed according to this first type.

The two-wick burner, No. 9, was constructed by Henry-Lepaute in 1845, for the lighthouse of Schevening in Holland. It has an outer cylinder for dividing the draught generated between the glass and the burner, and throwing back a part of it upon the light. It is the first application of this cylinder which exists in all modern burners.

The five-wick burner, No. 10, is a model of those now constructed for using mineral oil in the French lighthouses. It contains an appendage through which the oil must pass before reaching the upper part of the burner. This said piece, of which the arrangements were invented by M. Dénéchaux, acting engineer in ordinary at the lighthouse depôt, is intended to secure a continuous level, and comprises three tubes, juxtaposed, and open on the upper part at the proper height; the central tube springs from the small reservoir which forms the base of the burner, and in which the oil is forced by the machinery of the lamp; this oil, having no other exit, rises in the tube, and, arriving at the top, flows into the second tube, which carries it into the annular spaces containing the wicks; these it fills while keeping the same level as in the lateral appendage. As the quantity of oil forced up by the lamp is greater than the consumption, the excess comes down into the large reservoir of the lamp by flowing into the third tube over a fall rather higher than that cleared by the oil in reaching the burner. A horizontal disk of 20 millimetres diameter rises, at the height of 21 millimetres, above the central draught tube, and an outer cylinder divides in two the draught created between the burner and the glass. It is upon this outer cylinder that the glass-holder stands. In this burner the empty spaces between the wicks, intended for air passages, are $5\frac{1}{2}$ millimetres wide, while the spaces that contain the wicks are only $4\frac{1}{2}$ mill. In the burners constructed up to the present time, both widths are of 5 mill.; this new arrangement seems to give better results. The burner has, besides, on its upper part, a graduated shape, so that each wick is placed about 2 millimetres below the one which precedes it towards the centre. This arrangement, as yet adopted only for the Pilier lighthouse, exhibited under No. 12, has been found necessary since the burners, in each order of lighthouses, have had one burner added to them, and therefore are wider. Its object is to lower the edge of the burner, in reference to the centre of the light, so as to reduce as much as possible the portion of light obscured by this edge in the lower part of the lenses. (*See* description of apparatus, No. 12.)

Modern Apparatus.

No. 11. Great annular lens, of the first order, $1\cdot10^{m}$ in diameter, Messrs. Barbier and Fenestre, constructors, 1867. This lens was constructed by Messrs.

Barbier and Fenestre as a specimen of high class workmanship. Each ring is one single piece; the joints which divide the rings are inclined according to the direction of the refracted ray. The lens is mounted on a pedestal, and revolves around any horizontal axis.

No. 12. Lenticular panel, dioptric and catadioptric, for flashing lights of the second class, planned by the head engineer, Allard, and constructed by M. Henry-Lepaute, 1876.

This panel forms part of an apparatus intended for the Pilier lighthouse, situated at the mouth of the Loire, and of which the tower has just been rebuilt. The character given to it in 1829 has been preserved; it is a fixed light varied by flashes every four minutes   To produce this character a fixed light apparatus has been adopted, of which two sectors of $\frac{1}{8}$th horizon, opposed to one another, are replaced by perfect annular lenses; it revolves at the rate of one turn in eight minutes. In order that the two kinds of lenses may be adjusted upon the edges, and have a common pinion-jack, the focal distance, which is 0.700^m for the fixed lenses, has been reduced to 0.647^m for the annular lenses. The focal lamp has five concentric wicks, instead of four, as usual in lamps of the second class, because the light, being coloured red in certain directions, it was thought necessary to increase its intensity.

This panel shows several novel arrangements, some of which are now applied for the first time.

1st. In the central or dioptric parts of the section, the joints that divide the elements, and therefore the lower sides of these elements, instead of being horizontal, are inclined according to the direction of the refracted ray. This system has several advantages : it does away with a triangular part of glass which is useless, and thus lessens the weight of the apparatus; it reduces in a large proportion the loss of light caused by horizontal joints; it makes less acute, and consequently less fragile, the outer angles of the elements, and, besides, it diminishes their projection, thus enabling the dioptric lens to have a greater height.

2nd. The central lens (or dioptric) comprehends a vertical angle of 76 degrees, whereas, in the old sections, this angle was of about 60 degrees only ; its elevation is thus increased from 0.85^m to 1.10^m. This advantage is thus obtained, that the luminous rays meet the last dioptric element at the same angle as the first catadioptric ring, and suffer no more loss of reflection upon the one than upon the other.

3rd. The section commonly used in apparatus of the second class had been calculated for a three-wick lamp burner of 0.074^m diameter. With a five-wick burner of 0.110^m diameter, the inferior elements of the dioptric lens and the lower catadioptric rings, constructed after this old section, emit rays that are no longer in the proper direction, because the portion of light which the base of the burner leaves visible becomes perceptibly nearer to the lens than in the case of a three-wick burner. To lessen this defect, a graduated shape was given to the burner, by placing each wick 0.002^m below the one preceding it on the side of the centre. This arrangement reduces neither the regularity nor the intensity of the light, and the part of that light, visible from each of the lower lenticular elements, becomes somewhat increased. Moreover, these lower elements have been calculated by determining for each of them a particular focus taken on the brightest line of the apparent part of the light, instead of on the axis itself of the lamp. Similar arrangements might be advantageously adopted in many cases.

4th. The central lens and the lower rings are included in the same frame, the upper rings are set in a second frame, separated from the first by a metal cross-bar. In the annular lenses, this cross-bar takes the shape of the arc of a circle having, like the rings, its centre on the optical axis; the result is that the rings can remain intact, instead of having to be cut, as was the case until now.

5th. The lamp, placed at the focus of the lens exhibited, shows special arrangements, due to M. Dénéchaux.

Thus, the skin pockets or valvulæ, and the leathern valves, which are sometimes the cause of disorder, are replaced by ordinary pistons and metal valves. This system has produced good results in experiments made at the depôt, but it has not yet received practical sanction.

The lamp with five wicks for burning mineral oil has an intensity of 36 carcel burners, the fixed light apparatus produces an intensity of 640 burners, and the annular lenses of more than 5,000 burners.

No. 13. Apparatus for electric revolving light, constructed by Messrs. Sautter, Lemonnier, & Co., 1876.

This instrument is intended to produce, by electric light, a light revolving at intervals of 30 seconds. It includes a fixed light apparatus $0 \cdot 50^m$ diameter, lighting three-fourths of the horizon, around which revolves, in eight minutes, a tambour of $0 \cdot 62^m$ diameter, and composed of 16 vertical, lenticular elements.

In the section of the fixed light apparatus the central dioptric part subtends a vertical angle of 76 degrees, which is greater than in the old sections. This arrangement is adopted in order that the luminous ray may meet the last dioptric element at the same angle as the first catadioptric ring, and should suffer no more loss by reflection upon the one than upon the other. The apparatus having to be fixed on an elevated point, the section of the several parts, except that of the two lowest catadioptric rings, has been calculated so as to throw the focal line of the emergent rays, 30 minutes below the horizontal line; in the calculation of the two lowest rings, this angle is increased by three degrees for the last but one, and by five degrees for the last, so that the lighthouse may remain visible at a short distance, that is; by a navigator placed below the divergent cone emitted by the rest of the apparatus.

The sixteen vertical lenses are contiguous, and are each composed of a single element, about $0 \cdot 12^m$ wide, the curve of which has been calculated so as to give with the electric light an horizontal divergence of three degrees seven minutes. The duration of a flash is, accordingly, of about five seconds, and the interval between the end of a flash and the beginning of the following one is 25 seconds.

The maximum intensity of the flash rises to about 60,000 carcel burners, assuming at the focus an electric light of 200 burners power.

The electric light is produced in this apparatus, as in the lighthouses established on the coasts of France, by means of a Serrin regulator and an electrical machine of the Compagnie l'Alliance.

Experiments have been made with the Serrin regulator at the lighthouse depôt since the year 1860. A model on a large scale has been constructed especially for the lighthouse service, and has always given good results. The regulator exhibited is a counterpart of this model.

The electro-magnetic machine has been, as is well known, designed by MM. Nollet and Joseph Van Malderen, in accordance with the same principle as the scientific apparatus of Pixii and Clarke. It produces alternate currents, and, as it was in the first instance destined for the decomposition of water or for electro metallurgy, it was provided with a commutator for bringing the currents into one constant direction. When the question was raised of applying it to the production of light, M. Van Malderen, who had then become the mechanical engineer to the Compagnie l'Alliance, conceived the happy idea of suppressing the commutator, which is difficult to maintain, and has the effect of more or less weakening the current. The luminous intensity was found to be appreciably augmented, and the fact was soon acknowledged that alternate currents are, *cæteris paribus*, more favourable regulators than those in a constant direction. The machines of the Compagnie l'Alliance had originally six discs; these were reduced to four when the improvements introduced into the coils and the magnets permitted of a greater intensity being obtained with these smaller machines than with the former. In the case of lighthouses, where there cannot be too great intensity, the number of six discs has been preserved.

The central depôt in Paris has retained, since 1860, the first specimen constructed by M. Van Malderen of this machine, with the currents not brought into one constant direction. It has six discs, and carries 56 magnets; it is 1·63 metre high, and 1·43 metre in diameter; it gives less light than the present machines, but it works very well still, and serves for the experiments that are made at the depôt.

This first machine of the Compagnie l'Alliance may be regarded as the starting point of all the attempts which have since been made of economically transforming power into electricity, and consequently into light. On that account it is no more than right, although the machine is not included in the Exhibition, to make mention of it in the catalogue.

2203. The Original Model of the Eddystone Lighthouse.

The Eddystone Rocks, so named from the great variety of sets of tides and currents which surround them, are situated about 14 miles S.S.W. of the port of Plymouth, the sea being fully 30 fathoms in depth. A lighthouse was constructed on these rocks by Winstanley in 1696, and destroyed by a storm in 1703. A second was built by Rudyerd in 1709, and was totally consumed by fire in 1755. The present lighthouse was commenced in 1756, and completed in 1759, by Smeaton, F.R.S., civil engineer. This original model, made by Smeaton, was sent by royal command for the inspection of His Majesty George III. and the Royal Family, and has since then remained in the possession of Mr. Smeaton's family. *Mrs. Croft Brooke.*

2203a. Model of the Lighthouse on La Corbière Rock, Jersey. *Sir John Coode.*

The first lighthouse constructed in concrete; tower erected in 1874. Sir John Coode, engineer. Modelled by Mr. Joseph Thomas. Scale, $\frac{1}{128}$ of natural size.

2203b. Parabolic Reflector, rendered holophotal, according to Mr. Thomas Stevenson's design, by being fitted with a lens and reflecting prisms, and a portion of a spherical mirror, so as to parallelise all the light of the lamp. Introduced in 1849.
The Commissioners of Northern Lighthouses.

When the apparatus is to be cleaned the lamp is lowered out of the reflector on a sliding carriage, as arranged by the late Mr. Robert Stevenson, in 1814. The object of the sliding carriage is to insure the return of the burner to the proper focus.

40075. M m

2204. Model of First-class Fixed Dioptric Light. This apparatus consists of a central lenticular band, and an upper and lower set of reflecting prisms. The cylindrical belt, with diagonal joints, and the upper and lower reflecting prisms, were substituted by Mr. Alan Stevenson, in 1836, for the segmental belt and upper and lower silvered mirrors of Fresnel's first-class apparatus.
(One-fifth of full size.)

The Commissioners of Northern Lighthouses.

2206. Model of Mr. Thomas Stevenson's Marine Dynamometer, for ascertaining the force of waves during storms. The greatest force recorded in the German Ocean was $3\frac{1}{2}$ tons per square foot. *Messrs. D. and T. Stevenson.*

2207. Drawing of Storm Curve for Waves, showing the genesis of waves seawards from the windward shore, illustrative of Mr. Thomas Stevenson's formula $h = 1\cdot5\sqrt{\overline{D}}$, where $h =$ height of wave in feet, and $D = $ length of fetch in miles.
Messrs. D. and T. Stevenson.

2209. Historical Series of published engravings, showing the improvements in lighthouse illumination between 1787 and 1876, by the Engineers of the Northern Lighthouse Board.
The Commissioners of Northern Lighthouses.

2209a. Ship's Light, arranged to show a port, starboard, or anchor light.—J. S. Starnes, Broad Street, Ratcliff, E.
South Kensington Museum.

2209b. Ship's Lights, port, starboard, and mast head light.
—Stevens & Sons, Southwark Bridge Road.
South Kensington Museum.

XI.—MISCELLANEOUS.

Blenkinsop's Rail and two Chairs.

F. J. Bramwell, C.E., F.R.S.

This rail was designed and patented by Blenkinsop in 1811 to meet a difficulty propounded by Trevethick previously; as to the adhesion of the wheels of a locomotive engine to the rails.

Trevethick proposed that the periphery of the driving wheels of a locomotive should be made rough by the projection of bolts or cross-grooves, so that the adhesion of the wheels to the rail should be certain.

Blenkinsop, to meet Trevethick's notion, caused his rail to be made with studs or racks along one side of it, into which a toothed-wheel on his locomotive was to work. This toothed-wheel was fixed to the driving wheels of the engine, and merely served to gear into the rack on the side of the rail.

Blenkinsop's engines were carried out by Matthew Murray, of Leeds (died 1826). They had two cylinders instead of one, as in Trevethick's engine, and began to run from the Middleton collieries to Leeds, about three miles and a half, in August 1812.

The length of the piece of the Blenkinsop rail is 3 feet 5 inches. The face width $1\frac{3}{4}$ inches. Extreme depth $4\frac{1}{2}$ inches. From centre to centre of the hollow half-hoop studs on the side of the rail is 6 inches.

The rail is made on the form known as fish-bellied.

Diagrams for Lecture Rooms. Steam and other machinery.
F. J. Bramwell, C.E., F.R.S.

Amongst these diagrams for lecture purposes painted in colours on black glazed linen, the following representations of interesting subjects are to be found.

Drawing of Jonathan Hull's steamboat, 1737.
Stephenson's Darlington "No. 1" locomotive engine, 1825.
Stephenson's locomotive engine "Rocket," 1829.
Cornish pumping engine with single cylinder.
Cornish pumping engine with high and low pressure cylinders. Woolf's system, 1804.
Horizontal double cylinder high and low pressure "compound" engine.
Agricultural portable engine.
Modern traction engine.
Thompson's road locomotive engine, 1870–1871.
Hancock's steam carriage, 1826.
Modern London Brigade steam fire engine.
Oscillating cylinder marine paddle engines, 1827–1876.
Modern locomotive engine on six wheels, four wheels coupled, "Cias."
Longitudinal through section of the same locomotive, "Cias," showing fire box, boiler, and boiler tubes, cylinders, connecting rods, eccentrics, &c.

2210b. Model of Tumbler Lock and Key.
The Council of King's College, London.

2210c. Model of Ancient Egyptian Lock and Key.
The Council of King's College, London.

2210d. Model of Mangle Motion.
The Council of King's College, London.

2212a. Two views of Ramsbottom's Pick-up Troughs
Whitmore.
F. W. Webb, Locomotive Department, London and North-western Railway, Crewe.

These troughs are laid down to supply the tenders of the locomotives with water whilst running; a dip pipe on the tender is lowered into the water, which thus runs into the tank, by this means saving time during the journey.

2212c. Train used at the **Break Trials,** near Newark, June 9th to 15th, 1875 (2 photos.). *F. W. Webb.*

The train represented was one of the ordinary express passenger trains, fitted with continuous break, sent by the London and North-western Railway Company, to take part in the break trials at Newark in June 1875, before the Royal Commissioners.

2212d. Section of Tabular Results showing the Wear of Steel Rails. *F. W. Webb.*

2212e. Lovell's Patent Apparatus for recording the bad joints on railways and tramways. *Thomas Lowell, M.Inst.C.E*

This invention is for the purpose of showing the faulty joints in the permanent way, thus providing a means, hitherto unprovided for, for showing in the case of broken springs, or the oscillation of a train, on which side the fault rests, whether in weak springs or bad joints, and consists of a clockwork movement, arranged to draw an endless paper over suitable drums, in combination with a pencil. The indicator is attached between the spring and footplate of an engine or other vehicle, the pencil making a longitudinal line when no oscillation, and a vertical line when oscillation, takes place. The length of such vertical line is various, according to the amount of such oscillation.

The instrument exhibited is capable of taking a diagram over 120 miles of line, when attached to an engine travelling 60 miles per hour, without change of rollers, the rate of travel of the paper being 4 inches per minute; this can be reduced or regulated by adjustable flies for a lesser speed of engine.

2212k. Fog Signal. *Edward Alfred Cowper.*

This simple little instrument illustrates the application of the science of acoustics to a very useful object, namely the communication of information from a person on a railway to the driver of a passing train, in a dense fog or on a dark night; it is the only instrument that accomplishes the object, and has been the means of saving many thousands of lives. Invented by the exhibitor in 1841.

The principle consists in producing a very different sound from any that is constantly recurring in a railway train, and a sudden explosion or detonation is found to be the best for the purpose; it is caused by the explosion of a small quantity of gunpowder in a small tin box, by the firing of a match inside which is crushed by the wheel of the passing train.

Telegraph Steam Signal. *W. H. Bailey.*

2214. Model of London and its Environs. Scale 12 inches to a mile. *John Fowler, Westminster.*

Made for the purpose of showing before Committees of the Houses of Parliament the railways existing or in course of construction during the year 1864; also the proposed system of the inner circle railway, comprising the Metropolitan Railway, the Metropolitan and St. John's Wood Railway, and the Metropolitan District Railway.

2215. Improved Method of Reversing Rolling-Mills. *Jeremiah Head, M.Inst.C.E.*

A separate piece is introduced between the clutch and each clutch wheel, and connected with the latter only by elastic arms. The shock which ordinarily takes place when the clutch is thrown into gear is thus prevented.

2215a. Photograph of a Coal-testing Station.

2215b. Description of experiments on Coal relating to steam producing power; determination of the proper breadth for the spaces between the fire-bars, and treatment of the coal; ashes; weight per hectolitre; specific gravity and texture of the coal.

Berggewerkschaftskasse Bochum (Bergrath Heintzmann).

The coal experiment station is a central establishment for experiment on Westphalian coal, designed for giving exact scientific instruction as to the best uses of particular kinds of coal, and for ascertaining their comparative practical values.

2215c. Frisbie's Patent Feeder and Grate for feeding fuel up from underneath the fire into all descriptions of furnaces, fuel boxes, and fire grates, for saving fuel, securing an intense heat, and consuming the gas and smoke. *J. M. Holmes.*

This invention is designed for feeding fuel up from beneath the fire into furnaces and fire grates. No fuel is thrown upon the top of the fire, but into a charger which rocks forward to the front of the furnace under the grate bars. By turning a crank geared to a central shaft the charger is carried back under a central opening in the grate, when a piston rises in the charger and pushes the fuel up underneath the burning mass above. When the piston rises level with the top surface of the charger, it is retained in place by a catch until a reverse motion of the crank brings the charger again to the front, when the catch releases the piston and the charger is again ready for filling, the previous charge of fuel being sustained by a movable apron. The grate is constructed to revolve, so that any melted coal or slag can be brought to the door and removed quickly.

By this method of firing the ignition of the fresh coal is downwards, and all the volatilised coal, combustible gas, and carbonaceous matter pass from below through the live coals above, and break at once into flame, securing the greatest intensity of heat, and consuming the gas and smoke, thus effecting a great saving of fuel.

2216. Model of a Blast Furnace Boiler, upon elastic supports. *Jeremiah Head, M.Inst.C.E.*

Long plain cylinder boilers, when hung as ordinarily upon rigid supports, are found to rise clear of the latter at their ends when in use, and in the middle when out of use. This leads to seam rips, and frequently explosion. By turning the small hand wheel right or left, it will be seen that, by the method exhibited, the boiler is able to modify its form without straining, and consequently without danger.

2216b. Model of an Apparatus for Exchanging Despatches on Railways, without Stoppage of the Trains. *M. Cacheleux, Paris.*

Delineation of the original Steam Omnibus used for a short period on the city and other roads for passenger traffic. *Gardner Collection.*

2216c. Model of the late John Grantham's Patent Steam Tramway Car.

Also a detailed drawing of the above. *Mrs. John Grantham, Croydon.*

2010a. Hand Bill, dated 1st June 1804, announcing the completion of the first section of the Surrey Iron Railway.

Trovey Blackmore.

This railway, which is believed to have been the first railway established in England for public traffic, was made pursuant of an Act of Parliament passed in the year 1801, by which the company was empowered to raise a capital of 30,000*l.* in shares of 100*l.* The railway was first opened from Wandsworth to Croydon, and was subsequently extended to Merstham. It passed through the valley of the Wandle, a district abounding in mills of various kinds, and conveyed the manufactures and produce of the district to the Thames at Wandsworth, where a terminus was built with large store-houses and a wet dock for the convenience of lading and unlading goods carried by water. This terminus is still in existence, though now appropriated to other uses. Large quantities of coals were conveyed inland by this railway. The rails were laid upon stone blocks or sleepers, and the cars were drawn by horses and mules. The railway was in operation till about the year 1851, soon after which time the company was wound up and the plant and property sold.

2216d. Model Railway and Carriage, invented and made by Richard Roberts, C.E., Manchester, in 1824, to illustrate the nature of centrifugal force, in his lectures at the Manchester Mechanics' Institution. A practical model.

The Committee, Royal Museum, Peel Park, Salford.

2216d. Drawing of Steam Carriage, by the late Baron Séguier, with " sun and planet wheel motion," 1846.

Late Baron Séguier, Membre de l'Institut.

2216e. Incline Carriage-way, designed for conveying passengers up and down steep hills, inclines, or sea-side cliffs, and worked by balance, hydraulic, or other machinery. *R. Hunt.*

The incline carriage-way, designed for the purpose of conveying passengers or material up and down steep inclines by means of carriages adapted to the angle of inclination and worked by means of hydraulic or other machinery, is particularly applicable to sea-side cliffs, where the sea provides an easily available motive power.

2216f. Underground Hauling Machine.

Murray and Co.

2216g. Model of London and North-Western Railway Company's Travelling Post Office and Apparatus, for picking up and setting down mail bags without stopping.

H.M. Postmaster-General.

Model of Steam Hoist used for raising the bags, &c. in the General Post Office. *H.M. Postmaster-General.*

Letter Stamping Table and Machine.

H.M. Postmaster-General.

2216f. Twelve Designs of Engineering Works executed by students, Royal College of Science, Ireland. *Prof. Pigot.*

2216e. Asbestos Packed Cocks and Valves.
John Dewrance & Co.

This entirely new principle of making cocks and valves with a bearing of asbestos, which is indestructible by heat and most of the acids, instead of the usual metal surfaces, which have to be ground together to make them tight, has stood the test of one twelvemonth's trials with perfect success.

Model of 1½ in. cast-iron cock all nickel plated.
Drawing, framed, of 3-in. cock.
Model of 3-in. cock, cast iron, flanged.
Do.   2-in.   do.   do.
Do.   1¼-in.   do.   do.
Do.   gun-metal shell and two cups.
Do.   gun-metal 1-in. cock.
Do.   ¾-in. water gauge.
Do.   ¾-in.   do.   three-way.
Do.   nickel plated cast-iron plug.

2216h. Captain MacEvoy's Boat Torpedo, copper, full size. *London Ordnance Company.*

A percussion torpedo to be used by a steam launch or other boat. The torpedo is fixed at the end of a long pole which is driven before the boat on torpedo service. When the torpedo comes into contact with the object of attack, it is fired by the concussion through a percussion fuze arrangement.

This torpedo is also fitted with a separate and distinct apparatus for firing at will from the boat by means of electricity ; charge of powder, 50 lbs.

Captain MacEvoy's Floating Torpedo, copper, full size, for harbour defence. *London Ordnance Company.*

To be fired by electric wire communication with the shore, or by contact. The torpedo when charged carries 60 lbs. of powder.

Captain MacEvoy's Current or Tidal Torpedo, copper, full size. *London Ordnance Company.*

The torpedo floats into a harbour amongst the shipping with the tide, and when arrested by any object, the vane is set in motion by the current's flow. In a few seconds the vane releases a hammer which falls by a spring with much force upon a percussion cap. The cap fires a detonating charge which ignites the powder or dynamite or other explosive with which the torpedo is charged. The charge of powder is 50 lbs.

Set of Torpedos for Naval use.
London Ordnance Company.

"Harvey" torpedo complete with sustaining buoys, towing lines, trigger apparatus for firing, and contact levers.

The torpedo when charged carries 100 lbs. of powder. In towing, it diverges several points to port or starboard, and when in contact with the vessel attacked, explodes itself mechanically, or electrically.

Sustaining Buoys of Cork with lines or gear complete for the " Whitehead " Torpedo.

London Ordnance Company.

Original Drawing of the Semaphore at the Admiralty, Whitehall, with code of signals and their method of arrangement.   *Gardner Collection.*

Exterior View of the Admiralty with Semaphore in action.   *Gardner Collection.*

XII.—BRIDGE CONSTRUCTION.

2008. Diagrams for Bridge Construction :—

a. Resistances of materials, tension, and compression.
b. Girder on two supports.
c. Graphic treatment of arches for bridges.
d. Construction of iron bridges.

Prof. Heinzerling, Aix-la-Chapelle.

2009. Photographs for the study of Construction :—

a. Bauwaage (curves of construction), a symmetrical parabola.
b. Bauwaage (curves of construction), circle.
c. Bauwaage (curves of construction), segment of a circle.
d. Bauwaage (curves of construction), ellipse.
e. Bauwaage (curves of construction), aclinoid.
f. Bauwaage (curves of construction), cubic parabola—
as lines of equilibrium.   *Prof. Heinzerling, Aix-la-Chapelle.*

2092. Model of the roof of the new Polytechnic School at Dresden.   *Bock and Handrick, Dresden.*

2093. Three Models of railway bridges at Niederwarthe, Corvey, and Zserks.   *Bock and Handrick, Dresden.*

2093a. Cartoons (3), showing the **Terminus,** in London, of the **Midland Railway,** called St. Pancras Station.

W. H. Barlow, F.R.S.

1. General plan of station.
2. General section of station roof and flooring, with an enlarged section of one rib, showing mode of connexion with purlins. Elevation of part of one rib, showing the method of carrying the sash bars and ventilators.
3. The timber staging used by the contractors in erecting the roof.

The clear span of station roof is 240 feet; the distance between side walls, 245 ft. 6 inches; area of ground covered by the roof, 169,400 square feet; number of columns in basement supporting the flooring, 720.

The contractors for the iron work of the station and its erection were the Butterly Iron Company, Staffordshire.

The station was opened in the beginning of 1869.

2093d. Detail Drawing of Iron Lattice Girder Railway Bridges on the Pacasmayo, Guadalupe, and Magdalena Railway. 1872.   Scales, 1 inch to 1 foot, and 3 inches to 1 foot.

William Willis.

The drawing represents reproductions of the original by Willis' aniline process, executed by Vincent Brooks and Co.

2098. Model of Bowstring Bridge, thrown over the canal at Wormwood Scrubs, the tunnel of the Great Western Railway passing under the canal.   *Council of King's College, London.*

2216a. Model of a portion of the **Pontcysyllte Aqueduct,** which carries the Shropshire Union (late the Ellesmere) Canal across the river Dee, and the Vale of Llangollen.   *G. R. Jebb.*

This model was made under the superintendence and direction of Telford before the aqueduct was built. The aqueduct consists of 19 arches, each having a span of 45 feet; the total length is 1,007 feet, and the height from the river Dee to the surface of the water in the canal is 127 feet. The foundation stone was laid on the 25th July 1795, and the work was finished in the year 1803.

This model having fallen into partial decay was a short time ago restored and repainted.

XIII. — COLLECTION OF MODELS AND DIAGRAMS ILLUSTRATIVE OF THE PRINCIPLES OF MECHANICS, ROYAL SCHOOL OF MINES.

2217. Mr. Shelley's Educational Diagrams on the Steam Engine.

2218. Educational Diagrams for illustration of **Applied Mechanics.**

Reciprocating motion in bullet-making machine.

Cams for bullet-making machine.

Machine for shaping plugs for bullets.

Reversing motion in rifling machine (2).

A parallel motion applied for rolling under pressure.

Pulleys connected by belts.

Angle of repose for various substances.

Screw press.

Dutch crane.

Differential motion.

Train of wheelwork in a crane.
A crane.
The Whitworth measuring machine.
A lathe.
Headstock of lathe.
Train for screw-cutting lathe.
Drilling machine.
Planing machine (2).
Slide rest for face lathe.
Blanchard's lathe.
Other shaping machines (3).
Arrangements for advancing boring bar.
The Cordelier.
Clock train.
Chronometer.
Regulator of watch.
Going fuzee of watch.
Rack lever escapement.
Detached lever escapement.
Chronometer escapement.
Gravity escapement.
Compensation pendulum.
Siemens' steam jet and the pneumatic despatch tube.
Principle of Giffard's injector.
Mode of supplying water to trains while running.
Air-pump of the Allen steam engine.
Husband's atmospheric stamp.
Steam hammers.
Hydraulic ram.
Cast-iron girder beams (2).
Beam of steam engine, right and wrong construction.
Girder bridges (5).
Locomotive (2).

Cylindrical boiler (2).
Marine boiler.
Forms of rail and wheel tire for locomotive (2).
Indicator diagram and slide valve of locomotive.
Hornblower's double cylinder engine.
Double cylinder engine, with piston valves.
Watt's disc valve and connections.
Fairbairn's equilibrium valve.
High-pressure cylinder, piston, and valves.
Mode of receiving thrust of screw propeller.
Gauge, with corrugated plate, and india-rubber diaphragm.
Pendulum governor (2).
Pendulum governor applied to water-wheel.
Siemens' chronometric governor.
The forge bellows.
Blowing fan, with gauge.
Guibal's ventilating fan.
Centrifugal pump.
Turbine.
Water-wheel.
Hydraulic press.
The accumulator.
Pump worked by water pressure.
Hydraulic cranes (5).
Water pressure engine (as for crane).
Press for squirting metals.

2219. Mr. Anderson's Educational Diagrams on Mechanics.

Levers (12).
Wheel and axle (18).
Pulley (8).
Inclined plane (8).
Wedge (8).
Screw (10).

Dynamometers (3).
Toggle joint.
Virtual velocities.
Hydraulic press.
Elastic cords.
Pumps (2).

2220. Model to show the Conversion of Circular into Reciprocating Motion, in one direction only.

2221. Model to show the Conversion of Circular into Reciprocating Motion, in two perpendicular directions.

2222. Model to show Inequality of Motion, when a crank and connecting rod are used; also that the linear motion may be doubled when the connecting rod is equal in length to the crank.

2223. Diagram Model, with parts to illustrate the movements when the connecting rod is equal to the crank. Straight line motion.

2224. Three Models of an Eccentric Circle.

2225. Model to show Conversion of Circular into Reciprocating Motion, with intervals of rest.

2226. The Swash Plate.

2227. Crown Wheel Escapement.

2228. Simple form of Escapement.

2229. Recoil Escapement for Clocks.

2230. Dead-beat Escapement.

2231. Various forms of Escapement.

2232. Wheel, with one tooth and locking ring.

2233. Geneva Stop.

2234. Model of Recoil Escapement, for throwing an image on a screen by the electric lamp.

2235. Model of Dead-beat Escapement, for throwing an image on a screen by the electric lamp.

2236. Striking Movement of a Clock (locking plate).

2237. Striking Movement with Repeater.

2238. Model to illustrate Compensation Balance Wheel of Chronometer.

2239. Model of Wheel and Axle.

2240. Pulley Block.

2241. Disc and Roller Motion.

2242. Two forms of Screw Thread.

2243. Two Models to show the principle of Counting Machines.

2244. Counting Machine.

2245. Model to show Reversing Motion of the Table in Whitworth's Planing Machine.

2246. Model of Double Rack and Pin Wheel.

2247. Model of Cams in combination, taken from a machine for shaping plugs for bullets.

2248. Cam to Write the letters R.I. (Cowper's Model).

2249. Worm Barrel.

2250. Reversing Motion, with two and three spur wheels (quick return).

2251. Reversing Motion with Clutch.

2252. Model of Train of Wheels.

2253. Specimen of Spur Wheel.

2254. Two Models of Slit-bar Motion.

2255. Application of Slit Bar Motion in Whitworth's shaping machine.

2256. Two Models of Ratchet Wheels.

2257. Masked Ratchet Wheel.

2258. Silent Ratchet Wheel.

2259. Model to show Advance of a Ratchet Wheel through part of the Space of a Tooth.

2260. Ratchet Wheel, for traversing the cutter in a planing machine. Clement's Click.

2261. Machine for Printing Numbers.

2262. Drawing of a Lathe. (Whitworth.)

2263. Drawing of a Planing Machine. (Whitworth.)

2264. Apparatus to show that Friction is independent of Velocity.

2265. Apparatus to illustrate Friction Grips, and for experiments on the laws of friction.

2266. Model to show when a Drawer jams.

2267. Model to explain Weston's Friction Clutch.

2268. Silent-feed Motion (Worssam).

2269. Apparatus for illustrating the Angle of Repose between **Two Surfaces.**

2270. Plummer Block, showing Cylindrical Bearing.

2271. Form of Conical Bearing.

2272. Model of Friction Wheels.

2273. Model of Single Pulley Block.

2274. Two Models of Unbalanced Wheels.

2275. Model of an Arch.

2276. Model of Jointed Sword (Cowper).

2277. Model of Continuous Action Pump.

2278. Model of Pendulum, arranged for beating Time.

2279. Model for Tracing the Curves in the Slit Bar Motion.

2280. Two Models of Eccentric Rosette Motion.

2281. Various Models to show the production of Measuring Bars for End measure.

2282. Model to illustrate the conversion of Linear into End measure.

2283. Model to show Principle of Roberval's Balance.

2284. Balance on Roberval's Construction.

2285. Weighing Machine.

2286. Frame for experimenting with Epicyclic Trains, and with other trains, as in a screw-cutting lathe.

2287. Model of Combined Joints.

2288. Two forms of Ball and Socket Joint.

2289. Model of Hooke's Joint.

2290. Two Models of a Double Hooke's Joint, with adjustments.

2291. Mode of Connecting Parallel Axes.

2292. Parallel Axes, connected by a fork and grooved disc.

2293. Motion between Inclined Axes.

2294. Parallel Axes with Cranks and Links.

2295. Pair of Cranks, connected by a link, with adjustments.

2296. Mode of connecting Parallel Axes.

2297. Earliest Contrivance for feathering Floats in Paddle Wheels.

2298. Model of Link Motion in a Wool Combing Machine.

2299. Model of a Ventilating Fan.

2300. Model of Knuckle Joint.

2301. Knuckle Joint, with crank and connecting rod.

2302. Model of Stanhope Levers.

2303. Model of Lever Shears.

2304. Circular into Reciprocating Motion, with four beats for each rotation.

2305. Circular into Reciprocating Motion, with alternate intervals of rest.

2306. Model of Screw Surface.

2307. Model to show Difference of Inclination in Screw Threads of the same pitch, traced on cylinders of different diameters.

2308. Model of Endless Screw and Worm Wheel.

2309. Rolling Contact between Hyperboloids, with generating line mounted separately.

2310. Model of Skew Bevels.

2311. Two Axes, connected by a cord and grooved pulleys.

2312. Chain and Pulley.

2313. Pantograph, for drawing similar curves.

2314. Model of Watt's Parallel Motion.

2315. Model to illustrate the principle of Watt's Parallel Motion.

2316. Outline Model of Beam Engine, with Watt's parallel motion.

2317. Model of Roberts' Parallel Motion.

2318. Model of Peaucellier's Straight-line Motion.

2319. Sectional Model of a Trunk Engine.

2320. Model showing Expansion Valve.

2321. Model of Watt's Pendulum Governor.

2322. Model of the Cataract as applied to single-acting engines.

2323. Model of Silver's Marine Governor

2324. Cornish Crown Valve.

2325. Hawthorn's Safety Valve.

2326. Richard's Indicator.

2327. Diagram Model for setting the Slide Valve of a Horizontal Engine.

2328. Mode of reversing Marine Engine by a single eccentric.

2329. Model of Stephenson's Link Motion.

2330. Model of Steam Pump with Valve inside the Piston.

2331. Sectional diagram Model of the Pump.

2332. Giffard's Injector.

2333. Model of Slide Valve and Ports.

2334. Right and left handed Screw.

2335. Models to illustrate Ferguson's Paradox.

2336. Model to illustrate Silver's Marine Governor.

2337. Three Bevel Wheels forming Epicyclic Train.

2338. Model to explain the Differential Dynamometer.

2339. Differential Motion.

2340. A Geometrical Pen, showing application of an epicyclic train to the drawing of curves.

2341. Model to illustrate the principle of the Geometrical Pen.

2342. Model to illustrate Rope-making.

2343. The Cordelier.

2344. Model of expanding and contracting Crank.

2345. Model of Differential Pulley (Saxton's Patent).

2346. Lazy Tongs.

2347. Model for generating a Cycloid.

2348. Model of Headstock of a Lathe.

2349. Model of Collier's Planing Machine.

2350. Model of a Drilling Machine.

2351. Two Models illustrating Whitworth's Drilling Machine.

2352. Model of Eccentric Chuck.

2353. Model to illustrate Houldsworth's Differential Motion.

2354. Model of Screw Propeller.

2355. Model of a pair of Locomotive Wheels.

2356. Model of Hindley's Screw.

2357. Rolling contact between Ellipses.

2358. Elliptical Wheels, for obtaining a quick return in shaping machine.

2359. Variable motion from Wheels set eccentrically.

2360. Model of Roemer's Wheels.

2361. Method of altering or stopping the reciprocation of an Arm without stopping the prime mover.

2362. Model for comparing Bourdon's Gauge with Mercurial Pressure Gauge.

2363. Model to show principle of Bourdon's Gauge.

2364. Model of Wave Line Cam.

2365. Model of Heart Cam.

2366. Cam with Pulleys and Band.

2367. Model to illustrate Slit Bar Motion.

2368. Model of Spur Wheels.

2369. Model of Perrault's Wheel and Axle.

2370. Model of Blowing Fan.

2371. Model of Centrifugal Pump.

2372. Model of Screw.

2373. Model of Pump with revolving Belt.

2374. Model to explain the Hydrostatic Press.

2375. Whirling Table for Suspended Objects.

2376. Models of Pyramid, Cone, Cylinder, Ring, Chain, and weighted piece, for experiments on rotation.

2377. Whirling Table.

2378. Model to show effect of Rotation on a cork and a bullet in tubes of water.

2379. Flask arranged for experiments on Rotation.

2380. Model to explain Siemens' Water Governor.

2381. Model to explain Ramsbottom's Velocimeter.

2382. Various Models showing effects due to Rotation.

2383. Apparatus for showing the effect of the atmosphere in causing a spherical bullet to deviate laterally.

2384. Balls swinging in a Cycloid.

2385. Models of Cone and Paraboloid.

2386. Two forms of Gyroscope.

2387. Model of Pendulum, with adjustments.

2388. Model of Accumulator, with Press.

2389. Sectional Model of Locomotive.

2390. Belt on Pulleys at right angles.

2391. Action of Convex Pulley on a Belt.

2392. Models to show the principle of Lattice Girder Beams.

2393. Whitworth's Bench Measuring Machine, to measure intervals differing by one ten thousandth of an inch, and for constructing difference gauges.

2394. Apparatus for obtaining a Rectangular Measuring Bar with plane ends at right angles to its axis.

2395. Hexagonal Surface Plates.

Milton Keynes UK
Ingram Content Group UK Ltd.
UKHW032320161024
449665UK00001B/20